Supercomputer Algorithms for Reactivity, Dynamics and Kinetics of Small Molecules

NATO ASI Series

Advanced Science Institutes Series

A Series presenting the results of activities sponsored by the NATO Science Committee, which aims at the dissemination of advanced scientific and technological knowledge, with a view to strengthening links between scientific communities.

The Series is published by an international board of publishers in conjunction with the NATO Scientific Affairs Division

A Life Sciences **B Physics**	Plenum Publishing Corporation London and New York
C Mathematical and Physical Sciences **D Behavioural and Social Sciences** **E Applied Sciences**	Kluwer Academic Publishers Dordrecht, Boston and London
F Computer and Systems Sciences **G Ecological Sciences** **H Cell Biology**	Springer-Verlag Berlin, Heidelberg, New York, London, Paris and Tokyo

Series C: Mathematical and Physical Sciences - Vol. 277

Supercomputer Algorithms for Reactivity, Dynamics and Kinetics of Small Molecules

edited by

Antonio Laganà
Department of Chemistry,
University of Perugia,
Perugia, Italy

Kluwer Academic Publishers

Dordrecht / Boston / London

Published in cooperation with NATO Scientific Affairs Division

Proceedings of the NATO Advanced Research Workshop on
Supercomputer Algorithms for Reactivity, Dynamics and Kinetics of Small Molecules
Colombella di Perugia, Italy
30 August – 3 September 1988

Library of Congress Cataloging in Publication Data

NATO Advanced Research Workshop on "Supercomputer Algorithms for Reactivity, Dynamics, and Kinetics of Small Molecules" (1988 : Colombella, Italy)
Supercomputer algorithms for reactivity, dynamics, and kinetics of small molecules : proceedings of the NATO Advanced Research Workshop on "Supercomputer Algorithms for Reactivity, Dynamics, and Kinetics of Small Molecules," held in Colombella di Perugia, Italy, 30 August-3 September 1988 / edited by Antonio Laganà.
p. cm. -- (NATO ASI series. Series C, Mathematical and physical sciences ; vol. 277)
Includes index.
ISBN-13: 978-94-010-6915-1
1. Reactivity (Chemistry)--Data processing--Congresses.
2. Chemical reaction, Rate of--Data processing--Congresses.
3. Molecular dynamics--Data processing--Congresses.
4. Supercomputers--Congresses. I. Laganà, Antonio. II. Title.
III. Series: NATO ASI series. Series C, Mathematical and physical sciences ; no. 277.
QD505.5.N37 1988
541.3'94'0285--dc19 89-2457

ISBN-13: 978-94-010-6915-1 e-ISBN-13:978-94-009-0945-8
DOI: 10.1007/978-94-009-0945-8

Published by Kluwer Academic Publishers,
P.O. Box 17, 3300 AA Dordrecht, The Netherlands.

Kluwer Academic Publishers incorporates the publishing programmes of
D. Reidel, Martinus Nijhoff, Dr W. Junk and MTP Press.

Sold and distributed in the U.S.A. and Canada
by Kluwer Academic Publishers,
101 Philip Drive, Norwell, MA 02061, U.S.A.

In all other countries, sold and distributed
by Kluwer Academic Publishers Group,
P.O. Box 322, 3300 AH Dordrecht, The Netherlands.

Printed on acid free paper

Contents

Preface

The need for accurate computational procedures to evaluate detailed properties of gas phase chemical reactions is evident when one considers the wealth of information provided by laser, molecular beam and fast flow experiments. By stressing ordinary scalar computers to their limiting performance quantum chemistry codes can already provide sufficiently accurate estimates of the stability of several small molecules and of the reactivity of a few elementary processes. However, the accurate characterization of a reactive process, even for small systems, is so demanding in terms of computer resources to make the use of supercomputers having vector and parallel features unavoidable.

Sometimes to take full advantage from these features all is needed is a restructure of those parts of the computer code which perform vector and matrix manipulations and a parallel execution of its independent tasks. More often, a deeper restructure has to be carried out. This may involve the problem of choosing a suitable computational strategy or the more radical alternative of changing the theoretical treatment. There are cases, in fact, where theoretical approaches found to be inefficient on a scalar computer exhibit their full computational strength on a supercomputer.

The discussion at the NATO workshop "*Supercomputer Algorithms for Reactivity, Dynamics and Kinetics of small Molecules*" held by the end of August 1988 at the Villa Colombella, (Colombella di Perugia, Italy) has focussed upon these aspects. This book collects the papers of both invited and contributed lectures. The first part of the book deals with supercomputer strategies for the calculation of the electronic structure of small molecules and the investigation of potential energy features characterizing a reactive process. In the second part theoretical methods developed for the exact calculation of the dynamics of reactive atom diatom systems are described. Finally, in the last section, quantum reduced dimensionality as well as classical three dimensional (including semiclassical corrections) approaches are discussed for extension to more complex systems. Applications of artificial intelligence techniques are also presented. In all papers particular attention has been given to storage management and speed up problems related to the use of vector and parallel features.

The book shows how intense has been in recent years the work for designing parallel and vector algorithms. Accurate electronic structure of reactive systems as well as exact and high level approximate three-dimensional calculations of the reactive dynamics, efficient directive and declaratory software for modeling complex systems. In turn, new and more complex problems have been posed by these advances. Some of them are concerned with the definition of the computer architecture better suited for chemical calculations. Others are concerned with balancing within the application vector and parallel structures.

The workshop has been generously funded by the Scientific Affairs Division of NATO.

Antonio Laganà
Department of Chemistry
University of Perugia, Italy

RECENT ADVANCES IN ELECTRONIC STRUCTURE THEORY AND THEIR INFLUENCE ON THE ACCURACY OF *AB INITIO* POTENTIAL ENERGY SURFACES

Charles W. Bauschlicher, Jr., Stephen R. Langhoff
NASA Ames Research Center
Moffett Field, CA 94035
and
Peter R. Taylor
ELORET Institute
Sunnyvale, CA 94087

ABSTRACT. Recent advances in electronic structure theory and the availability of high speed vector processors have substantially increased the accuracy of *ab initio* potential energy surfaces. The recently developed atomic natural orbital approach for basis set contraction has reduced both the basis set incompleteness and superposition errors in molecular calculations. Furthermore, full CI calculations can often be used to calibrate a CASSCF/MRCI approach that quantitatively accounts for the valence correlation energy. These computational advances also provide a vehicle for systematically improving the calculations and for estimating the residual error in the calculations. Calculations on selected diatomic and triatomic systems will be used to illustrate the accuracy that currently can be achieved for molecular systems. In particular, the $F+H_2 \rightarrow HF+H$ potential energy hypersurface is used to illustrate the impact of these computational advances on the calculation of potential energy surfaces.

I. INTRODUCTION

The theoretical determination of purely *ab initio* reaction rates is becoming an important area of computational chemistry research. At NASA Ames Research Center there is considerable interest in determining rates for chemical reactions occuring at high temperatures and in exotic environments. These conditions will be encountered in the re-entry bow shock wave of aero-assisted orbital transfer vehicles (AOTV) [1] or inside the combustion chamber of the hydrogen-fuelled hypersonic craft National Aero-space Plane (NASP) [2]. It is important to have such rate data at the design stage in order to estimate what heating effects will be encountered during re-entry and the combustion efficiency that can be expected under hypersonic conditions. However, it is clearly very difficult (sometimes it is not even possible) to study such environments in the laboratory, and as a result theoretical determinations can provide data that is simply not obtainable by other means.

At present, there is a variety of methodologies [3] for carrying out calculations of reaction cross sections, rate constants and product state distributions. In general, these dynamical methods, either classical or quantum mechanical, are based on knowledge of the potential energy surface (PES), and as a result, the accuracy of the kinetic predictions ultimately depends on the PES itself. Futher, those dynamical

A. Laganà (ed.), Supercomputer Algorithms for Reactivity, Dynamics and Kinetics of Small Molecules, 1–21.

methods which rely on a global representation of the PES are conditioned not only by the accuracy of the computed energy points, but also by the techniques used to represents these points with a functional form.

In the present work we shall discuss recent advances in quantum chemical methodology that have improved the reliability of *ab initio* electronic structure calculations. These include full configuration interaction (FCI) calculations [4-13], which have given new insight into the errors associated with the common approximations for treating electron correlation, and atomic natural orbital (ANO) basis sets [14-15], which have reduced the error in the one-particle basis sets by allowing large primitive sets to be contracted with little loss in accuracy. In cases for which it would be unreasonably expensive to apply these techniques over the whole PES, we demonstrate that it should be possible to study the global surface by adjusting a PES based on a lower (and less expensive) level of theory using very accurate calculations performed at the critical points of the surface. It is hoped that surfaces generated in this manner will be sufficiently accurate that comparison with experiment will provide insight into the limitations of the dynamical studies rather than reflect the limitations of the PES itself. This, of course, presupposes that adequate methods for fitting the computed energy points are available [16]; this aspect of the problem is discussed briefly below, but is generally beyond the scope of the present work.

In Section II we give an overview of current theoretical methods. It is not our aim to provide detailed descriptions of methods and algorithms, but rather to discuss the techniques used in broad terms for reference in later discussions. In Sections III and IV we discuss FCI calibration calculations and ANO basis sets, respectively. In Section V, the accuracy of current methods is illustrated by comparing with selected diatomic and triatomic systems where accurate experimental spectroscopic constants are available for comparison. We consider the $F+H_2 \rightarrow FH+H$ reaction in Section VI, and Section VII contains our conclusions.

II. QUANTUM CHEMICAL METHODOLOGY

The determination of a PES to be used in computing reaction rates involves solving the non-relativistic time-independent Schrödinger equation for fixed nuclear positions in the Born-Oppenheimer approximation. A review of the general methodology of computational chemistry is given in Ref. 17. The first step in solving the Schrödinger equation is to select a one-particle basis set. This is generally a set of Gaussian-type orbitals (GTOs), grouped into fixed linear combinations called contracted functions. While this type of one-particle basis is universally referred to as an atomic orbital basis, it must be borne in mind that the description it provides of the individual atoms is often far from perfect. This can lead to problems in describing atom-atom interactions or binding, since deficiencies in the "atomic basis" for one atom can be compensated for by using part of the basis on another center, resulting in a completely spurious energy lowering referred to as superposition error (SE) [18]. The effects of superposition error on a computed PES are discussed in more detail below.

Once the one-particle basis has been chosen, a method for solving for the electronic motion must be selected. In principle, the correlation, or n-particle, problem can be solved exactly in a given one-particle basis set by a full configuration interaction (FCI) calculation, which includes all arrangements of the n electrons in the given one-particle basis, consistent with Fermi statistics and the desired spin and spatial symmetry. However, the length of the FCI expansion increases factorially with the number of orbitals and electrons: this generally necessitates the use of methods in which the n-particle expansion is truncated. The expansion is most commonly truncated by restricting the excitation level to single and double excitations (SDCI) from either a self-consistent field (SCF) or a multiconfigurational SCF (MCSCF) reference wave function. Such an SCF or MCSCF description is the best zeroth-order treatment of the system, and then it is single and double excitations relative to this treatment that formally enter in the next order of perturbation theory. Although much chemistry is well described using an SCF wave function as the reference, an MCSCF zeroth-order wave function is generally required for computing a PES where chemical bonds are formed and broken in order to account for near-degeneracy and multireference effects.

While the SCF description of a system is uniquely defined, the use of an MCSCF wave function introduces an additional degree of freedom, namely, the choice of configurations. Several schemes for choosing configurations have been devised, most of which are based on restricting the multiconfigurational treatment to the bonds or lone-pairs of interest, such as the generalized valence bond (GVB) model and its variants [19]. One simple and widely used scheme is the complete-active-space-SCF (CASSCF) approach [20], in which only a choice of active orbitals and electrons is required: the CASSCF configuration space is then a full CI in this active space. Choosing an active orbital space is usually simpler (and much less error-prone) than constructing a configuration list explicitly. Once the MCSCF configuration space has been defined, it is then necessary to decide on the configuration list for the CI expansion. Several possible routes to such a list exist and it is useful here to distinguish between them.

For the case of a closed-shell single-reference single and double excitation CI (SDCI) calculation, the configuration list is defined unambiguously by considering all configuration state functions (CSFs) in which no more than two electrons occupy virtual orbitals — this is exactly equivalent to including all singly- and doubly-excited CSFs relative to the reference configuration. One approach to multireference SDCI (MRCI) is simply to include all CSFs with no more than two electrons in the virtual orbitals, and if all electrons correlated in the CI calculation are active in the CASSCF calculation this so-called second-order CI (SOCI) corresponds again to all single and double excitations out of the CASSCF configurations as references. However, if there are some electrons correlated in the CI but not active in the CASSCF, these two prescriptions do not lead to the same CI spaces. Specifically, if we take the molecule N_2 as an example, treated with the $2p$-derived orbitals and electrons active in the CASSCF, but with all ten valence electrons correlated in the CI, the SOCI wave function would contain some CSFs with, say, two electrons in virtual orbitals and eight electrons in the active orbitals: these are quadruple excitations from the CASSCF configuration space and would *not* be included in a calculation defined as "all single and double excitations from the CASSCF configurations". While the distinction between these two approaches can be important in terms of computational expense (where SOCI can be considerably more expensive), it is seldom of quantitative significance in practice, and as described above no problem arises when only the CASSCF active electrons are correlated in a subsequent CI.

The construction of the CI configuration list frequently involves problems in addition to those described above. For example, it will often be the case that a CASSCF configuration space of several thousand CSFs, in conjunction with a large one-particle basis, will generate several million or several tens of millions of CSFs in the SOCI (or CASSCF reference MRCI), making such a CI calculation impossible. The most common approach is then to truncate the list of reference configurations, usually by selecting those configurations with a coefficient in the CASSCF calculation above some threshold. An additional, common, simplification is to eliminate for the CI configuration list any CSFs that have vanishing Hamiltonian matrix elements with all the reference CSFs [21]. This restriction to the so-called "first-order interacting space" can be justified formally by perturbation theory and in practice it seems to result in negligible effects on computed energies and properties. These various approaches to the construction of MRCI configuration lists are discussed in more detail, with numerical examples, below.

Even with truncation of the list of reference CSFs, MRCI calculations can be very costly. Several alternative schemes have therefore been suggested. One approach is to combine together ("contract") CSFs in the MRCI. The "internally contracted" CI method [22], in which the reference CSFs are combined with fixed coefficients to give a single effective reference function has proven to be quite reliable, and with recent improvements [23] is expected to be more commonly used. The "externally contracted" CI (CCI) method of Siegbahn [24], in which the singly- and doubly-excited CSFs are combined using coefficients derived from perturbation theory, has been extensively used and in general is in good agreement with the MRCI results. As we discuss below, a CCI surface, calibrated by very accurate MRCI calculations, may be one of the most cost-effective methods of computing a PES.

Finally, it should be pointed out that the perturbation-theoretic methods [25] (MP2, etc.), used with such success for molecules near their equilibrium geometries, are much less satisfactory when used to compute a PES. Obviously, where several reference configurations are required these single-reference treatments cannot be expected to perform well, and it does not seem possible to overcome such problems by the use of unrestricted Hartree-Fock (UHF) methods to define a single reference CSF: the UHF PES itself will often display discontinuities from spontaneous symmetry breaking, and this inevitably compromises the subsequent perturbation theory treatment. Recent efforts [26] to devise projected UHF-based schemes may overcome these problems, but this is simply another approach to generating a multireference wave function.

Correlation treatments will, of course, approach the FCI result as higher levels of excitations are included. Although explicitly including such higher-excited configurations in the wave function usually leads to a prohibitively long CI expansion, methods of estimating the effects of some of these higher excitations have been developed. For single-configuration-based wave functions these include the Davidson correction [27], the coupled-pair functional (CPF) method [28] (or its modified form, MCPF [29]), and coupled cluster (CC) methods [30]. For the multireference case an extension of the Davidson correction [31] is commonly used. Recent additions to this list of so-called "size-consistent" methods include the quadratic CI (QCI) method [32], which is an approximate (single-reference) CC treatment and the averaged CPF (ACPF) method [33], which is a simple multireference extension of CPF. Finally, the scaled external correlation (SEC) method [34] has been proposed to account for the remaining errors in both the n- and one-particle expansions.

Estimates of the higher excitations or other approximations can be useful in computing more accurate dissociation energies and barrier heights, especially in systems where more than about eight electrons are correlated. However, these approximate techniques are not bounded variationally like the MRCI, and if the quality of the approximation varies across the PES it may introduce significant "noise" into the computed energies [35]. This should be borne in mind when computed energies are used in fitting a PES, since obtaining a satisfactory fit may require a high degree of precision in the computed energies. It should be noted in this context that techniques [36] in which selection is performed on *all* the CSFs in the CI expansion, rather than just the reference CSFs (especially those methods in which perturbation theory is used to estimate the contribution of the neglected CSFs) may not always be capable of achieving the necessary precision.

From the above discussion it is clear that what is essentially a double basis set expansion technique for obtaining a wave function may suffer from two interrelated sources of error: the incompleteness of the one- and n-particle basis sets. FCI calculations, which solve the n-particle problem exactly for a given one-particle basis set, are therefore extremely useful in that they separate the contributions from these two sources of error. By designing a truncated CI treatment that reproduces the FCI results in a moderate sized basis set (thereby eliminating any error in the n-particle treatment), accurate results can be obtained by performing the same CI treatment in a very large basis (which minimizes the one-particle basis errors). This general approach has proved very effective in several cases in which accurate experimental data are available for comparison. Atomic natural orbital (ANO) basis sets are particularly useful for studying the saturation of the one-particle basis, as their definition provides a systematic procedure for improving basis set completeness. We next consider in more detail how the FCI method and the ANO basis sets have led to more accurate calculations.

III. FCI BENCHMARK CALCULATIONS

While this work is directed towards evaluating how recent computational advances have affected the reliability of a computed PES, it is not possible to accomplish this goal by a direct comparison with experimental rate constants. Instead, as the differential correlation effects on a PES are similar to those between different electronic states or between equilibrium and dissociated geometries for a given system, we will use accurate experimental data for diatomic and triatomic systems to evaluate our methods. In addition to comparing with experiment, we also use the FCI results as an absolute standard of comparison for a given one-particle basis set.

Early comparisons of this sort were based on work by Handy and coworkers [37], who developed an efficient direct FCI approach in terms of determinants. Using a double-zeta (DZ) basis set, they considered stretching the O-H bond lengths in the H_2O molecule to 1.5 and 2.0 times their equilibrium values. The FCI results showed that even the restricted Hartree-Fock (RHF) based fourth-order many-body perturbation theory (MBPT) approach [38], which includes the effect of single, double, triple and quadruple excitations, did not accurately describe the stretching of the bond; the error increased from 0.6 kcal/mole at r_e to 10.3 kcal/mole with the bonds stretched to twice their equilibrium values. Although the MBPT method is rigorously size-consistent and contains the effects of higher than double excitations, it does not describe the bond breaking process well because the RHF reference becomes a poor zeroth-order description of the system as the bond is stretched. Size-consistent methods that include double excitations iteratively — infinite-order methods such as the coupled cluster (CC) approach— do better. However, only methods that account for the multireference character in the wave function as the bonds are broken, such as the CASSCF/MRCI method, provide an accurate description at all bond lengths, or, correspondingly, at all geometries on the PES [39].

The ability to perform FCI calculations was significantly advanced by Sieg-

Table I. $^1A_1 - ^3B_1$ splitting in CH_2 (kcal/mole) using a DZP basis set and correlating six electrons.

Method	Splitting	Error
SCF[a]	26.14	14.17
SCF[a]/SDCI	14.63	2.66
SCF/SDCI+Q	12.35	0.38
CPF	12.42	0.45
TCSCF[b]/MRCI	12.20	0.23
TCSCF/MRCI+Q	12.03	0.06
CASSCF[c]/SOCI	11.97	0.00
CASSCF/SOCI+Q	11.79	−0.18
Full CI	11.97	—

[a] The SCF occupations are $1a_1^2 2a_1^2 3a_1^2 1b_2^2$ and $1a_1^2 2a_1^2 1b_2^2 3a_1^1 1b_1^1$.
[b] SCF treatment for $^3B_1$ state, two-configuration MCSCF treatment for $^1A_1$ state (SCF configuration and $3a_1^2 \rightarrow 1b_1^2$ excitation).
[c] Active space comprises the C $2s$ $2p$ and H $1s$ orbitals .

bahn [40], who formulated the FCI as a series of matrix multiplications to utilize the vector capabilities of current supercomputers, and by Knowles and Handy [41] who effectively eliminated the input/output requirements by formulating the problem in terms of determinants. Using this new FCI approach, a series of benchmark calculations has been performed [4-13], expanding the scope of the earlier studies [37] to include several additional aspects of the effects of electron correlation.

In Table I we compare the FCI $^1A_1 - ^3B_1$ separation in CH_2 with various truncated CI results [4]. Since the $^1A_1$ and $^3B_1$ states are derived nominally from the ^3P and ^5S states of carbon, respectively, the different bonding mechanisms result in a substantial correlation contribution to the separation: the SCF separation is over 14 kcal/mole too large. The error of 2.7 kcal/mole at the SDCI level is still relatively large; an error of this magnitude in a barrier height, for example, could result in significant errors in a computed reaction rate. The inclusion of the contribution of unlinked higher excitations through either the Davidson correction (+Q) or the CPF method reduces the error substantially. The origin of the error in the SCF/SDCI treatment is the second important configuration, arising from the double excitation $3a_1^2 \rightarrow 1b_1^2$, in the $^1A_1$ state. If the orbitals for the $^1A_1$ state are optimized in a two-configuration SCF (TCSCF) calculation, and correlation is included by performing an MRCI calculation based on both these reference configurations, the error is about half that of the SDCI+Q or CPF treatments. The error is reduced to only 0.06 kcal/mole if the multireference analog of the +Q correction is added. After the $3a_1^2 \rightarrow 1b_1^2$ excitation, the next most important correlation effect is that associated with the C-H bonds. If this correlation effect is accounted for in both the CASSCF zeroth-order reference and a subsequent SOCI calculation, essentially perfect agreement between the SOCI and FCI is observed. That is, a well-defined CASSCF/MRCI treatment accounts for all of the differential correlation effects. It is interesting to note that adding the multireference +Q correction to the SOCI energy results in an overcounting of the effect of higher excitations and

Table II. N_2 $^1\Sigma_g^+$ spectroscopic constants

Method	r_e (a_0)	DZP basis 6 electrons correlated ω_e (cm^{-1})	D_e (eV)
SDCI	2.102	2436	8.298
SDCI+Q	2.115	2373	8.613
SDTCI	2.107	2411	8.462
SDTQCI	2.121	2343	8.732
SDQCI	2.116	2361	8.586
CPF	2.112	2382	8.526
MCPF	2.114	2370	8.556
SOCI	2.123	2334	8.743
SOCI+Q	2.123	2333	8.766
Full CI	2.123	2333	8.750

the separation becomes smaller than the FCI result.

We next consider the spectroscopic constants [5] for the ground state of N_2, which are summarized in Table II. The SDCI calculation yields a bond length that is in good agreement with the FCI, but the error in D_e is 0.45 eV, even when size-consistency problems are minimized by using the $^7\Sigma_u^+$ state of N_2 to represent two ground state N(^4S) atoms at infinite separation. Although the addition of quadruple excitations, either variationally (SDQCI) or by the +Q, CPF or MCPF approximations, further reduces the error in D_e, it remains too large for chemical accuracy (1 kcal/mole). If both triple and quadruple excitations are included, the spectroscopic constants are all in good agreement with the FCI. However, this level of treatment is prohibitively expensive in a large one-particle basis set, and even this wave function does not dissociate correctly to ground state atoms, as this requires six-fold excitations relative the SCF configuration at r_e. The spectroscopic constants computed from an SOCI treatment based on a CASSCF wave function are in excellent agreement with the FCI. Furthermore, this treatment agrees with the FCI for all r values. The addition of the +Q correction does not affect r_e or ω_e, but it makes D_e too large compared with the FCI.

Dipole-induced dipole or dipole-quadrupole interactions can give rise to weakly bound complexes of considerable importance in dynamical studies. It is therefore important that properties such as the polarizability can be accurately determined. A FCI study [6] of the polarizability of F^- is summarized in Table III. As in the previous examples, the SDCI treatment is not sufficiently accurate for very high quality calculations. The inclusion of an estimate of higher excitations improves the results; in this case CPF is superior to the +Q correction. In the multireference case two different approaches were used. In the first, the CASSCF included the $2p$ electrons and $2p$ and $3p$ orbitals in the active space, and all CASSCF CSFs were used as references for the CI, in which the $2s$ and $2p$ electrons were correlated. Results obtained in this way are denoted MRCI in Table III. A more elaborate CASSCF calculation, with the $2s$ and $2p$ electrons and the $2s$, $3s$, $2p$ and $3p$ orbitals in the active space, was also performed: the use of all these CASSCF CSFs as references gives an SOCI expansion and the results are denoted SOCI in

Table III. Polarizability of F^- (a.u.)

Method	DZP + diffuse *spd* basis 8 electrons correlated, $\alpha = d^2E/dF^2$ α
SDCI	13.965
SDCI+Q	15.540
CPF	16.050
MRCI	16.134
MRCI+Q	16.346
SOCI	16.034
SOCI+Q	16.303
Full CI	16.295

Table III. The MRCI and SOCI results are not in as good agreement with the FCI as the MRCI+Q or SOCI+Q results. The +Q correction does not overshoot as it did for N_2 and CH_2, in part because of the larger number of electrons correlated here. As noted above, when only six electrons are correlated, the MRCI accounts for such a high percentage of the correlation that the +Q correction overestimates the missing correlation. For more than six electrons, or for cases where the zeroth-order wave function used is less satisfactory than was the CASSCF for N_2 and CH_2, the +Q correction may become a better approximation. This is especially true where quantities involving large differential correlation effects, such as electron affinities, are sought [7]. Thus the +Q correction substantially improves the agreement with FCI for the electron affinity of fluorine [8], even if large CASSCF active spaces and SOCI wave functions are employed.

It is almost always the case that only the valence electrons are correlated in quantum chemical studies, although for very accurate results core-valence and core-core correlation effects may be required. A recent detailed study [9] has shown that it is very difficult to find a truncated CI treatment that satisfactorily reproduces FCI correlation energies including core correlation. If the main interest is in a differential effect of core correlation (such as the contribution to an energy separation or to a bond length) it may be possible to neglect the core-core correlation, which is often almost independent of geometry, and to compute only the core-valence correlation. This is most easily accomplished by eliminating from the CI expansion any CSFs corresponding to a double (or higher) excitation from the core orbitals. Such an approach seems to recover most of the core correlation contribution to spectroscopic constants. While the effects of core correlation are often small (a few tenths kcal/mole in the CH_2 singlet/triplet separation, for example [9]), they may be important if accuracy better than about one kcal/mole is desired. In view of the expense incurred by including the core electrons in the CI treatment, and the additional problem that correlating more electrons increases the need for size consistency, it may be preferable to handle core-valence correlation effects by an effective potential approach like that developed by Meyer and co-workers [42].

Most scattering formalisms are developed in a diabatic representation, whereas a theoretical PES is computed in the adiabatic representation. Hence when curve crossings (or more complicated phenomena for polyatomic systems) occur both potentials must be accurately represented in the crossing region, and nonadiabatic coupling matrix elements (NACMEs) will be required to define the unitary transformation between the diabatic and adiabatic representations. Until recently, NACMEs were computed either using finite difference methods [43] or via approximations to avoid computing matrix elements between non-orthogonal wave functions [44]. However, Lengsfield, Saxe, and Yarkony [45] have recently developed an efficient method of evaluating NACMEs based on state-averaged MCSCF wave functions and analytic derivative methods. This should provide NACMEs of the same overall accuracy as that obtained for the adiabatic potentials.

In curve crossings where the molecular orbitals for the two states are similar, such as interactions between valence states derived from different asymptotic limits, the CASSCF/MRCI approach would be expected to describe both potentials accurately irrespective of which state is used for the orbital optimization. However, when the character of the two states is very different, such as valence/Rydberg mixing [10] or interaction between states derived from ionic and covalent limits [11], it is more difficult to achieve equivalent accuracy for the lowest adiabatic state on either side of the crossing point. This is commonly the case for charge-exchange reactions, such as $N^+ + N_2 \rightarrow N + N_2^+$, or chemi-ionization processes such as $M + X \rightarrow M^+ + X^-$, where the optimal molecular orbitals for the ionic and neutral solutions differ greatly. To gain additional insight into the computational requirements for describing the potentials in the region of curve crossings, we have studied [11] the $Li + F \rightarrow Li^+ + F^-$ chemi-ionization process using the FCI approach. In LiF the lowest adiabatic state at short r values, namely the ionic $X^1\Sigma^+$ state, dissociates adiabatically to neutral ground state atoms. There is an avoided crossing at the point where the energy difference between the F electron affinity (EA) and the Li ionization potential exactly balances the $1/r$ electrostatic stabilization. Since the CASSCF description of F^- is poor [8], the CASSCF estimate for the bond distance at the crossing point is unrealistically small. When orbitals from the ground-state CASSCF wave function are used to construct an MRCI wave function, this problem with the CASSCF description of the crossing point will compromise the accuracy of the MRCI description. This problem is not easily resolved by expanding the CASSCF active space, as very large active spaces are required to obtain a good description of atomic electron affinities. However, by performing instead a state-averaged CASSCF calculation, in which the orbitals are optimized for the average of the two lowest $^1\Sigma^+$ states in LiF (the ionic and neutral states), the orbital bias is eliminated and the MRCI treatment is in excellent agreement with the FCI. It is also important to note that this averaging does not significantly degrade the description of the system near r_e. Thus state averaging appears to be an excellent method of achieving equal accuracy for two potential curves in a curve crossing region, and should also perform well for polyatomic systems. The utility of state averaging as a means of obtaining a good compromise orthogonal set of molecular orbitals for use in an MRCI wave function has also been found to be an excellent route to computing accurate electronic transition moments [12]. It is thus seen that there is much in common between methods that account accurately for differential correlation effects on a PES and those that yield accurate spectroscopic constants and molecular properties.

A number of important conclusions can be drawn from these FCI benchmark studies. First, even in a complete one-particle basis set, it is unlikely that an accurate PES can be generated using single-reference-based treatments such as SCF/SDCI. If all regions of a PES are well described by an SCF reference, SDCI+Q, CPF or CC methods should yield acceptable results, but this is a very uncommon situation in practice. Second, the CASSCF/MRCI method is capable not only of achieving high accuracy for an individual PES, but also of achieving equivalent accuracy for several PES derived from different asymptotic limits. Some questions remain as to the utility of including the multireference +Q correction: it probably should not be included when six or fewer electrons are correlated and the CASSCF active space contains all the important correlation effects, since a large percentage of the higher excitations are already accounted for in the MRCI and any correction then overshoots the FCI. For more than six electrons correlated, the true answer may be closer to the +Q corrected value, although even here for systems that are very well described at the CASSCF level the +Q correction to MRCI may result in some overshoot. For relatively large numbers of electrons correlated (say, more than 12) or cases where there are large differential correlation effects, such as those encountered in the computation of electron affinities, the +Q may *underestimate* the importance of higher excitations. Thus, it is generally most important to include the +Q correction when there is a significant change in the character of the system that is not well described at the CASSCF level. Such a situation arises, for example, for the reaction $F+H_2 \rightarrow HF+H$, where F in the HF product has some F^- character. This is discussed further below when this reaction is considered in detail.

IV. ANO BASIS SETS

In the previous section we showed that the CASSCF/MRCI approach yields results in excellent agreement with FCI, that is, near the n-particle limit. We may therefore expect excellent agreement with experiment when the CASSCF/MRCI approach is used in conjunction with extended one-particle basis sets. It has become clear [46] that, until recently, the basis set requirements for achieving the one-particle limit at the correlated level were commonly underestimated, both in the number of functions required to saturate the space for each angular momentum quantum number and in the maximum angular momentum required. For the segmented basis sets that are widely used in quantum chemistry, improving the basis set normally involves replacing a smaller primitive basis set with a larger one. It is then seldom possible to guarantee that the smaller basis spans a subspace of the larger set, and it is thus difficult to establish how results obtained with different basis sets relate to convergence of the one-particle space. Ideally, the possibility of differences in primitive basis sets would be eliminated by using a single (nearly complete) primitive set, contracted in different ways such that the smaller contracted sets are subsets of the larger one. Such an approach requires a general contraction scheme, such as the one proposed by Raffenetti [47] for contracting valence orbitals at the SCF level. However, a contraction based on atomic SCF orbitals is not necessarily suitable for handling the correlation problem, and provides no means to contract polarization functions, large primitive sets of which are required for accurate results. Calculations on molecular systems have shown [48] that natural orbitals (NO) provide an efficient method of truncating the orbital space in correlated treatments. Almlöf and Taylor [14] have proposed a NO procedure for contracting atomic basis sets suitable for use in correlated molecular calculations: this atomic natural orbitals (ANO) approach is an efficient method for contracting large primitive valence and polarization basis sets. It has the advantage that the natural orbital occupation numbers provide a criterion for systematically expanding the basis set.

These ideas are illustrated for N atom and N_2 in Table IV. As the contraction of the ($13s$ $8p$ $6d$) primitive set is expanded from [$4s$ $3p$ $2d$] to [$5s$ $4p$ $3d$] to [$6s$ $5p$ $4d$],

Table IV. N/N_2 extended basis total energies (E_H) and "dissociation energies" (eV).

N atom

Basis set	$E_{SCF} + 54.$	ϵ_{corr}
(13*s* 8*p* 6*d*)	−0.400790	−0.111493
[6*s* 5*p* 4*d*]	−0.400779	−0.111321
[5*s* 4*p* 3*d*]	−0.400769	−0.110925
[4*s* 3*p* 2*d*]	−0.400725	−0.109066
(13*s* 8*p* 6*d* 4*f*)	−0.400790	−0.121385
[5*s* 4*p* 3*d* 2*f*]	−0.400769	−0.120499
[4*s* 3*p* 2*d* 1*f*]	−0.400725	−0.117584
[5*s* 4*p* 3*d* 2*f*] (2*g*)[a]	−0.400769	−0.122472
[5*s* 4*p* 3*d* 2*f* 1*g*]	−0.400769	−0.122138

N_2 molecule[b]

Basis set	$E_{SCF} + 108.$	ϵ_{corr}	D_e(SCF)	D_e(SDCI)
(13*s* 8*p* 6*d*)	−0.986307	−0.338118	5.03	8.16
[6*s* 5*p* 4*d*]	−0.985913	−0.337304	5.02	8.14
[5*s* 4*p* 3*d*]	−0.984833	−0.335395	4.99	8.08
[4*s* 3*p* 2*d*]	−0.983483	−0.329330	4.95	7.98
(13*s* 8*p* 6*d* 4*f*)	−0.989318	−0.365735	5.11	8.45
[5*s* 4*p* 3*d* 2*f*]	−0.988031	−0.362548	5.07	8.38
[4*s* 3*p* 2*d* 1*f*]	−0.986230	−0.353283	5.03	8.24
[5*s* 4*p* 3*d* 2*f*](2*g*)[a]	−0.988458	−0.370808	5.09	8.51
[5*s* 4*p* 3*d* 2*f* 1*g*]	−0.988322	−0.369270	5.08	8.48

[a] 2 uncontracted *g* sets.

[b] r(N–N) = 2.1 a_0, 10 electrons correlated.

the correlation energy systematically converges to that of the uncontracted results. The same is true for the (4*f*) and (2*g*) polarization sets. When these ANO sets are applied to N_2, the same systematic convergence of D_e is observed. In addition, by contracting the basis set for the atom, the superposition errors at the correlated level are minimized.

In order to treat atomic states with different character equally, e.g. F and F^-, the ANOs can be averaged to yield a compromise set. This is analogous to the state averaging used to define compromise orbitals suitable for describing molecular states of different character discussed above. A [5*s* 4*p* 3*d* 2*f* 1*g*] contraction based on the average of F and F^- has an SDCI level EA that agrees with the uncontracted (13*s* 9*p* 6*d* 4*f* 2*g*) basis set result to within 0.01 eV [49]. This can be compared with a 0.1 eV error for the same size basis set that is contracted for F alone, but with the outermost (the most diffuse) *s* and *p* primitive functions uncontracted.

Table V. $^1A_1 - {}^3B_1$ splitting in CH_2 (kcal/mole)

SOCI: six electrons correlated	
Basis	Separation
[3*s* 2*p* 1*d*/2*s* 1*p*]	11.33
[4*s* 3*p* 2*d* 1*f*/3*s* 2*p* 1*d*]	9.66
[5*s* 4*p* 3*d* 2*f* 1*g*/4*s* 3*p* 2*d*]	9.13
Expt (T_0)	9.02 (±0.01)
Expt+Theory[a] (T_e)	9.28 (±0.1)

[a] Ref. 53.

The results are better if the contraction is based on F^- alone, but they are still not as good as those obtained by using the average ANOs. ANO basis sets averaged for different states should thus supply a more uniform description in cases in which there is charge transfer or ionic/covalent mixing. It should be noted, however, that it may still be necessary to uncontract the most diffuse primitive functions and/or add extra diffuse functions to describe properties such as the dipole moments and polarizabilities [15,50] that are sensitive to the outer regions of the charge density.

V. CALIBRATION CALCULATIONS

The FCI benchmarks calculations discussed in Section III show that a CASSCF/MRCI treatment is capable of accurately reproducing the FCI results, at least when six electrons or fewer are correlated. Further, the ANO basis sets discussed in Section IV show that it is now possible to contract nearly complete primitive sets to manageable size with only a small loss in accuracy. Therefore, a six electron CASSCF/MRCI treatment performed in a large ANO basis set is expected to reproduce accurately the FCI result in a complete one-particle basis set, and hence should accurately reproduce experiment. By including the +Q correction, we believe chemical accuracy should be achievable for eight electron systems. In this section we illustrate several calculations that have achieved unprecedented accuracy by combining FCI benchmarks and ANO basis sets.

As discussed above, FCI calculations for CH_2 show that the SOCI treatment accurately accounts for the differential correlation contribution to the CH_2 $^1A_1 - {}^3B_1$ separation. In Table V, this level of treatment is performed using increasingly accurate ANO basis sets [51]. It is interesting to note that although the [4*s* 3*p* 2*d* 1*f*/3*s* 2*p* 1*d*] basis contains fewer contracted functions than the large segmented basis sets previously applied to this problem [52], it produces a superior result for the separation. The largest ANO basis set used gives a separation in good agreement with, but smaller than, the T_e value deduced from a combination of theory and experiment [53]. From the convergence of the result with expansion of the ANO basis set, it is estimated that the valence limit is about 9.05±0.1 kcal/mole. The remaining discrepancy with experiment is probably due to core-valence correlation effects. While FCI calculations have shown that a high level of correlation

Table VI. N_2 $^1\Sigma_g^+$ spectroscopic constants

	[5s 4p 3d 2f 1g] basis		
Method	r_e (Å)	ω_e (cm^{-1})	D_e (eV)
SOCI(6)	1.096	2382	10.015
SOCI(6)+Q	1.096	2382	10.042
MRCI(10)	1.101	2343	9.723
MRCI(10)+Q	1.102	2336	9.745
Expt[a]	1.098	2359	9.905

[a] Ref. 58.

treatment is required for an accurate estimate of the CV contribution to the separation, somewhat simpler theoretical calculations [9] place an upper bound on this quantity of a 0.35 kcal/mole increase in the separation. Therefore, it is clearly possible to achieve an accuracy of better than one kcal/mole in the singlet-triplet separation in methylene.

An analogous study [51] for SiH_2 indicates that the singlet-triplet splitting can also be accurately computed for this second-row molecule. However, it now becomes necessary to include the dominant relativistic contributions, namely the mass-velocity and Darwin terms [54], via first-order perturbation theory [55] or by using an effective core potential, if chemical accuracy is to be achieved. Once relativistic effects have been accounted for, an accuracy of about 0.2 kcal/mole is obtained. This incidentally establishes the ionization potential of the $^1A_1$ state of SiH_2 as 9.15 eV, the higher of two recent experimental values [56]. The ability to treat second-row systems accurately can have some advantages when comparing with experiment: for example, molecules containing Cl can be isotopically substituted, while the corresponding F species cannot. This will often mean that more information is available for comparison for some second-row systems.

The FCI study of the $X^1\Sigma_g^+$ state of N_2 showed that the SOCI treatment correlating the six $2p$ electrons, SOCI(6), accounts for essentially all of the correlation effects on the spectroscopic constants. However, this treatment [57] in a large ANO basis set produces a D_e value that is larger than experiment [58] (see Table VI). Since this basis set has virtually no CI superposition error, this conclusively shows that $2s$ correlation reduces the D_e value. The inclusion of $2s$ correlation, i.e. correlating ten electrons, results in a D_e that is 0.16 eV smaller than experiment. The decrease in D_e when the $2s$ electrons are correlated can be explained in terms of an important atomic correlation effect that has no analog in the molecular system, namely the $2s \rightarrow 3d$ excitation with a recoupling of the $2p$ electrons. At the MRCI(10) level (this is based on the CASSCF reference space from the six-electron calculation) the error in D_e is about 4 kcal/mole, or larger than the 1 kcal/mole desired for chemical accuracy. We may thus expect that computations on reactions involving multiply-bonded systems will have relative errors of several kcal/mole, in spite of the recent improvements in methodology. On the other hand, it is still important to apply the most accurate techniques to such problems, since certain

Table VII. FCI calibration of the classical barrier height of $F+H_2 \rightarrow HF+H^a$.

A. At the FCI saddle point

	barrier	exothermicity
FCI	4.50	28.84
MRCI(300)	5.18	28.57
MRCI(300)+Q	4.43	29.12
MRCI(322)(0.05)	5.00	29.12
MRCI(322)(0.05)+Q	4.32	29.21
MRCI(322)(0.025)	4.73	29.17
MRCI(322)(0.025)+Q	4.51	28.80
MRCI(322)(0.01)	4.71	29.19
MRCI(322)(0.01)+Q	4.54	28.84
MRCI(522)(0.025)	4.55	29.41
MRCI(522)(0.025)+Q	4.32	29.31

B. At the optimized saddle-point geometry[b]

	r(F-H)	r(H-H)	barrier	exothermicity
FCI	2.761	1.467	4.50	28.84
CPF	2.801	1.467	4.40	26.47
MRCI(300)	2.740	1.476	5.16	28.57
MRCI(300)+Q	2.795	1.467	4.42	29.12
MRCI(322)(0.025)	2.761	1.474	4.70	29.17
MRCI(322)(0.025)+Q	2.755	1.475	4.49	28.80

[a] Energies in kcal/mole and bond lengths in a_0. All calculations are done using the [4s3p1d/2s1p] basis set and correlating seven electrons. The barrier is referenced to $F\ldots H_2(50a_0)$, and the exothermicity is computed using $HF\ldots H(50a_0)$.
[b] Geometry optimizations were done using a biquadratic fit to a grid of nine points.

aspects, such as understanding the effect of $2s$ correlation in reducing the D_e of the $X^1\Sigma_g^+$ state of N_2, are a by-product of accurate, calibrated calculations. We hope that similar insights will occur in the study of dynamics that are based on a PES using these recent advances in electronic structure theory.

VI. THE $F+H_2 \rightarrow HF+H$ REACTION

In the previous section we showed that CASSCF/MRCI treatments of electron correlation in large ANO basis sets give spectroscopic constants that are in excellent agreement with experiment. We now discuss the application of these methods to computing the barrier height and exothermicity of the $F+H_2 \rightarrow HF+H$ reaction.

In Table VII, several different MRCI treatments are compared to the FCI barrier height and exothermicity [13,59]; in all of these treatments only the seven F $2p$ and H $1s$ valence electrons are correlated. The smallest MRCI treatment has the F $2p\sigma$ and H $1s$ orbitals active in the CASSCF and MRCI (denoted MRCI(300), since there are three active orbitals of a_1 symmetry). This calculation yields a barrier height that is 0.68 kcal/mole higher than the FCI. The inclusion of the +Q correction improves the barrier height, but it is now slightly too small. The sign of the error in the exothermicity also changes with the addition of the +Q correction. Such problems with the MRCI(300) treatment are not unexpected since the HF wave function is known to contain significant H^+F^- character. For accurate results it is therefore necessary to improve the description of the electron affinity (EA) of F by expanding the active space to (322) to include $2p \rightarrow 2p'$ correlation. With such an active space a very large number of CSFs would arise in a CASSCF reference MRCI (or an SOCI) wave function, so it becomes necessary to select reference CSFs according to their CASSCF coefficients as described in Section II: this expanded active space yields MRCI(322)+Q results that are in excellent agreement with the FCI, provided that the threshold for including CASSCF CSFs as references is no larger than 0.025. As noted above, further expansion of the active space improves the results, but the +Q correction may now overshoot the FCI value.

The MRCI(300)+Q and MRCI(322)(0.025)+Q saddle-point geometries are both in excellent agreement with the FCI value. It is interesting to note that although the CPF method is quite accurate for the barrier height and saddle-point geometry, it is significantly poorer for the exothermicity.

In order to compute an accurate barrier height, the basis set is expanded from the [4*s* 3*p* 1*d*/2*s* 1*p*] set used for the FCI calibration to a [5*s* 5*p* 3*d* 2*f* 1*g*/4*s* 3*p* 2*d*] ANO set. In this large basis set, the spectroscopic constants for H_2 are in almost perfect agreement with experiment. The MRCI(222)+Q treatment of HF, which is analogous to the MRCI(322) treatment of $F+H_2$, yields an excellent r_e, but a D_e which is 1.22 kcal/mole (0.05 eV) too small. The CI superposition error for F in the H_2 ghost basis set is 0.15 kcal/mole; this is even smaller than that obtained using the large Slater-type basis set from Ref. 60. An accuracy of better than 1 kcal/mole for the barrier height can therefore be expected.

The theoretical results for the classical saddle-point and barrier height are summarized in Table VIII. Based on the FCI calibration, the MRCI(322)(2*p*)+Q calculations, in which seven electrons (i.e. excluding F 2*s*) are correlated, are expected to reproduce the result of an FCI calculation in a nearly complete one-particle basis set. Since F 2*s* correlation decreases the barrier, this MRCI(322)(2*p*)+Q barrier, when corrected for the CI superposition error (SE), represents an absolute upper bound of 2.52 kcal/mole for the barrier.

The inclusion of F 2*s* correlation decreases the barrier height, and increases the magnitude of the +Q correction. Unfortunately, it is not possible to calibrate this level of treatment using the FCI approach. However, the nine electron +Q correction must be at least as large as the seven electron +Q correction, which *is* calibrated against the FCI. Therefore, a conservative upper bound of 2.26 kcal/mole is obtained using the MRCI(322)(2*p*)+Q treatment (corrected for SE). However, experience for nine electron systems, especially in calculations of electron affinities, has shown that the +Q correction for nine electrons is probably somewhat too small, making the actual MRCI(322)+Q value of 1.86 kcal/mole (corrected for SE) a more realistic estimate. Finally, the best estimate should also include an estimate of basis set incompleteness and account for the underestimation of the effects of higher excitations by the +Q correction. To accomplish this, we first omit the SE correction, assuming instead that the basis set incompleteness is 0.1 kcal/mole, and we further assume that the true +Q correction is 120% of that computed; this yields our best empirical estimate of 1.35 kcal/mole for the classical barrier height. Thus

Table VIII. Theoretical studies of the classical saddle-point geometry and barrier for the $F+H_2$ reaction.

Basis[a]	Level of treatment	saddle-point		barrier[b]	exothermicity[b]
A	MRCI(322)(2p)[c]	2.899	1.455	2.99	33.96
A	MRCI(322)(2p)+Q[c]	2.910	1.456	2.42	33.42
A	MRCI(322)	2.914	1.451	2.63	31.61
A	MRCI(322)+Q	2.950	1.450	1.66	30.47
A	CCI(322)	...[d]	...[d]	2.79	31.8
A	CCI(322)+Q	...[d]	...[d]	2.02	30.7
A+H(*f*)[e]	CCI(322)	...[d]	...[d]	2.73	
A+H(*f*)	CCI(322) +Q	...[d]	...[d]	1.95	
A−F(*g*)	CCI(322)	2.879	1.447	2.89	
A−F(*g*)	CCI(322)+Q	2.909	1.445	2.14	
	Expt.				31.73

[a] This letter "A" denotes the [5s5p3d2f1g/4s3p2d] basis described in the text.
[b] The barrier is referenced to $F\ldots H_2(50a_0)$, and the exothermicity is computed using $HF\ldots H(50a_0)$.
[c] These are seven-electron treatments (i.e. 2*s* correlation is excluded).
[d] The MRCI(300)+Q saddle point geometry is used, r(F-H)=2.921 a_0 and r(H-H)=1.450 a_0.
[e] Denotes that a function of this angular momentum type has been added.

based solely on estimates from *ab initio* calculation the barrier height should be between 1.35 and 1.86 kcal/mole.

While the MRCI(322)+Q calculations in the ANO basis set are more reliable than any previous results, considerable computer time would be required to compute a global surface at this level. The barrier height was therefore investigated using the contracted CI (CCI) approach. In the same ANO basis set, the CCI+Q barrier is 0.4 kcal/mole higher than the corresponding MRCI+Q value. Further extension of the basis set was also investigated at the CCI level: *f* polarization functions on H were found to lower the barrier by only 0.07 kcal/mole, while eliminating the *g* function on F increased the barrier by 0.12 kcal/mole. These observations are consistent with the contention that the basis set is nearly complete. The CCI calculation in this basis set is sufficiently inexpensive that much larger regions of the PES can be investigated. Of course, given the differences between the MRCI and CCI barrier heights some account would have to be taken of the errors in CCI treatment; this might involve adjusting the parameters in the fitted potential based on the MRCI(322)+Q calculation or on information deduced from experiment.

While the MRCI(322)+Q-corrected CCI+Q PES should be accurate, direct comparison with experiment is difficult. To facilitate comparison we have employed canonical variational transition state theory [61] at the classical and adiabatic barrier using the CCI+Q potential for both $F+H_2$ and $F+D_2$. These calculations account for the zero-point energy and include a tunneling correction. The results of these calculations are summarized in Table IX. As expected, the zero-point and tunneling corrections are different for H_2 and D_2. At the classical saddle-point,

Table IX. Zero-point and tunneling effects on the barrier height of the F + H_2 and F + D_2 reactions.

F + H_2 surface

	Classical Barrier		Adiabatic Barrier	
	CCI	CCI + Q	CCI	CCI + Q
r_{HF}, a_0	2.879	2.909	3.070	3.155
r_{HH}, a_0	1.447	1.445	1.425	1.421
Barrier, kcal/mole	2.888	2.143	2.639	1.860
Sym. stretch, cm^{-1}	3706	3768	4074	4178
Bend, cm^{-1}	68.5	45.9	68.5	45.9
Asym. stretch[a], cm^{-1}	692i	605i	530i	371i
Zero-point correction[b], kcal/mole	−0.602	−0.643	−0.076	−0.057
E barrier + zero point, kcal/mole	2.286	1.500	2.563	1.803
Tunneling correction, kcal/mole	−0.54	−0.47	−0.42	−0.29
Threshold, kcal/mole	1.75	1.03	2.14	1.51

F + D_2 surface

	Classical Barrier		Adiabatic Barrier	
	CCI	CCI + Q	CCI	CCI + Q
r_{HF}, a_0	2.879	2.909	3.010	3.075
r_{HH}, a_0	1.447	1.445	1.430	1.427
Barrier, kcal/mole	2.888	2.143	2.761	1.997
Sym. stretch, cm^{-1}	2623	2667	2811	2876
Bend, cm^{-1}	37.7	19.1	37.7	19.1
Asym. stretch[a], cm^{-1}	512i	448i	428i	334i
Zero-point correction[b], kcal/mole	−0.488	−0.532	−0.220	−0.233
E barrier + zero point, kcal/mole	2.400	1.611	2.541	1.764
Tunneling correction, kcal/mole	−0.40	−0.35	−0.34	−0.26
Threshold, kcal/mole	2.00	1.26	2.20	1.50

[a]From the normal mode analysis at the classical barrier, and computed from the curvature along the Eckart potential at the adiabatic barrier.
[b]For $H_2(D_2)$ we used ω_e=4401(3116) cm^{-1}, respectively, from Ref. 58.

the barrier heights for H_2 and D_2 differ by 0.2 kcal/mole, whereas at the adiabatic saddle point the barriers are the same. The observation [62] of nearly identical thresholds for H_2 and D_2 also provides strong support for using the adiabatic barrier. In order to bring the computed threshold into agreement with experiment [62], we must lower the CCI+Q classical barrier by 0.7-0.8 kcal/mole. This produces a barrier height of 1.3-1.4 kcal/mole, or after accounting for the errors associated with these approximations, a barrier height of 1.0-1.5 kcal/mole. This is in good agreement with the estimate made directly from the MRCI calculations, and also with that deduced in recent calculations by Truhlar and co-workers [63], although it disagrees with the value inferred by Schaefer [64] from most previous calculations.

VII. DISCUSSION AND CONCLUSIONS

We have shown that recent developments in electronic structure calculations have given new insight into the solution of the n-particle problem and ANO basis sets have reduced the error in the one-particle basis sets. This leads to more accurate calculations then previously possible. For systems with only one heavy atom, results with an error of less than one kcal/mole are now possible. For systems with multiple bonds, the error is still a few kcal/mole.

While current calculations are capable of high accuracy, they are still computationally intensive. Therefore, it is usually not possible to fully characterize a global PES by computing a closely-spaced grid of points, and the maximum information must be extracted from the available points by a fitting procedure. At present, most dynamical methods use only a fit to the total energies; such fits to a global surface can be very difficult even when sufficient precision exists in the computed energies. One suggestion for improving the fitting procedure (without a large increase in the number of computed points) is to compute the energy derivatives, as well as the total energy, at each point on the surface. Given that analytic derivative techniques have proven far more cost-effective than the use of finite differences for locating stationary points [65], it seems likely that this could represent a major improvement in the definition of the required surfaces [66].

As illustrated by the $F+H_2$ reaction, even the best calculations may require some scaling to reproduce experimental barrier heights, exothermicities, or reaction rates. The global surfaces obtained at a lower level of theory can also be adjusted by comparing with more accurate calibration calculations carried out at critical points on the surface. For example, the CCI+Q potential for the $F+H_2$ reaction could be modified by comparison with more accurate MRCI calculations.

The SEC method [34] has been proposed as a way to correct for both errors in the n- and one-particle basis sets. However, its application requires a one-particle basis set with approximately the same error for the reactant and product channel. This is not easy to arrange when the basis sets are defined by a single prescription, as is the case for ANO basis sets (as seen above for $F+H_2$), and is only applicable when the heats of reaction for reactants and products are known. The SEC method may be useful in estimating residual errors in less accurate calculations, especially in combination with the +Q correction so that the estimated contributions of n-particle and one-particle space incompleteness can be analyzed [67]. When ANO basis sets are used it seems preferable to use successively larger contracted sets and to obtain an estimate of the basis set limit from these results, as in the CH_2 calculations described above.

The recent improvements in electronic structure calculations now make it feasible to develop of a complete PES containing one heavy atom competitive with those deduced from experimental results. Accuracies of 1 kcal/mole are often achievable for the critical points on the surface. The availability of more accurate PES should also facilitate the evaluation of dynamical methods through comparison with experiment.

References

1. D. M. Cooper, R. L. Jaffe, and J. O. Arnold, J. Spacecraft and Rockets **22**,

60 (1985).
2. Pioneering the Space Frontier, The Report of the National Commission on Space, (Bantum Books 1986); also see G. Y. Anderson, AIAA Paper No. 87-2074, June 29-July 2, 1987.
3. See for example the volume of Chem. Rev. (and references there in) devoted to Chemical Dynamics; Chem. Rev. **87**, 1-288 (1987).
4. C. W. Bauschlicher, and P. R. Taylor, J. Chem. Phys. **85**, 6510 (1986).
5. C. W. Bauschlicher and S. R. Langhoff, J. Chem. Phys. **86**, 5595 (1987).
6. C. W. Bauschlicher, and P. R. Taylor, Theor. Chim. Acta **71**, 263 (1987).
7. C. W. Bauschlicher, S. R. Langhoff, H. Partridge, and P. R. Taylor, J. Chem. Phys. **85**, 3407 (1986).
8. C. W. Bauschlicher and P. R. Taylor, J. Chem. Phys. **85**, 2779 (1986).
9. C. W. Bauschlicher, S. R. Langhoff, and P. R. Taylor, J. Chem. Phys. **88**, 2540 (1988).
10. "Full configuration-interaction benchmark calculations for AlH", C. W. Bauschlicher, and S. R. Langhoff, J. Chem. Phys. (in press).
11. "Full configuration-interaction study of the ionic-neutral curve crossing in LiF", C. W. Bauschlicher and S. R. Langhoff, J. Chem. Phys. (in press).
12. C. W. Bauschlicher, and S. R. Langhoff, J. Chem. Phys. **87**, 4665 (1987).
13. C. W. Bauschlicher, and P. R. Taylor, J. Chem. Phys. **86**, 858 (1987).
14. J. Almlöf and P. R. Taylor, J. Chem. Phys. **86**, 4070 (1987).
15. J. Almlöf, T. U. Helgaker and P. R. Taylor, J. Phys. Chem. **92**, 3029 (1988).
16. J. N. Murrell, S. Carter, S. C. Farantos, P. Huxley, A. J. C. Varandas in "Molecular Potential Energy Surfaces" (Wiley, New York 1984).
17. "Advances in Chemical Physics: Ab initio methods in quantum chemistry" Vol 67 and 69 (Wiley, New York, 1987)
18. S. F. Boys and F. Bernardi, Mol. Phys. **19**, 553 (1970).
19. F. W. Bobrowicz and W. A. Goddard in "Methods of Electronic Structure Theory", H. F. Schaefer, ed., pp. 79, (Plenum Press, New York, 1977).
20. P. E. M. Siegbahn, A. Heiberg, B. O. Roos, B. Levy, Phys. Scr., **21**, 323 (1980); B. O. Roos, P. R. Taylor, P. E. M. Siegbahn, Chem. Phys. **48**, 157 (1980); B. O. Roos, Int. J. Quantum Chem. **S14**, 175 (1980): P. E. M. Siegbahn, J. Almlöf, A. Heiberg, B. O. Roos, J. Chem. Phys. **74**, 2384 (1981).
21. A. Bunge, J. Chem. Phys. **53**, 20 (1970); A. D. McLean and B. Liu, ibid. **58**, 1066 (1973); C. F. Bender and H. F. Schaefer, ibid. **55**, 7498 (1971).
22. H-J. Werner, E-A. Reinsch, J. Chem. Phys. **76**, 3144 (1982).
23. H-J. Werner and P. J. Knowles, J. Chem. Phys. (submitted).
24. P. E. M. Siegbahn, Int. J. Quantum Chem. **23**, 1869 (1983).
25. W. J. Hehre, L. Radom, P. von R. Schleyer and J. A. Pople "Ab initio molecular orbital theory", (Wiley, New York 1986).
26. See for example H. B. Schlegel and C. Sosa, Chem. Phys. Lett. **145**, 329 (1988).
27. S. R. Langhoff and E. R. Davidson, Int. J. Quantum Chem. **8**, 61 (1974).
28. R. Ahlrichs, P. Scharf and K. Jankowski, Chem. Phys. **98**, 381 (1985).
29. D. P. Chong and S. R. Langhoff, J. Chem. Phys. **84**, 5606 (1986).
30. R. J. Bartlett, C. E. Dykstra and J. Paldus, in "Advanced Theories and Computational Approaches to the Electronic Structure of Molecules", C. E. Dykstra, ed. p. 127 (D. Reidel, Dordrecht 1984).
31. M. R. A. Blomberg and P. E. M. Siegbahn, J. Chem. Phys. **78**, 5682 (1983).

32. J. A. Pople, M. Head-Gordon and K. Raghavachari, J. Chem. Phys. **87**, 5968 (1987).

33. R. J. Gdantiz and R. Ahlrichs, Chem. Phys. Lett. **143**, 413 (1988).

34. F. B. Brown and D. G. Truhlar, Chem. Phys. Lett. **117**, 307 (1985).

35. See for example R. J. Bartlett, I. Shavitt and G. D. Purvis, J. Chem. Phys. **71**, 281 (1979), where the addition of the +Q correction was shown to significantly increase the standard deviation for the fitted potential.

36. R. J. Buenker and S. D. Peyerimhoff, Theor. Chim. Acta **35**, 33 (1974).

37. P. Saxe, H. F. Schaefer, and N. C. Handy, Chem. Phys. Lett., **79**, 202 (1981); R. J. Harrison and N. C. Handy, *ibid.* **96**, 386 (1983).

38. R. J. Bartlett, H. Sekino, and G. D. Purvis, Chem. Phys. Lett., **98**, 66 (1983); W. D. Laidig and R. J. Bartlett, *ibid.* **104**, 424 (1984).

39. F. B. Brown, I. Shavitt, and R. Shepard, Chem. Phys. Lett., **105**, 363 (1984).

40. P. E. M. Siegbahn, Chem. Phys. Lett., **109**, 417 (1984).

41. P. J. Knowles and N. C. Handy, Chem. Phys. Lett., **111**, 315 (1984).

42. W. Müller, J. Flesch and W. Meyer, J. Chem. Phys. **80**, 3297 (1984).

43. M. Desouter-Lecomte, J. C. Leclerc, and J. C. Lorquet, J. Chem. Phys. **66**, 4006 (1977); C. Galloy and J. C. Lorquet, *ibid.* 4672 (1977); B. C. Garrett, M. J. Redmon, D. G. Truhlar, and C. F. Melius, *ibid.* **74**, 412 (1981).

44. R. Grice and D. R. Herschbach, Mol. Phys. **27**, 159 (1974); L. R. Kahn, P. J. Hay, and I. Shavitt, J. Chem. Phys. **61**, 3530 (1974).

45. B. H. Lengsfield, P. Saxe, and D. R. Yarkony, J. Chem. Phys. **81**, 4549 (1984); P. Saxe, B. H. Lengsfield, and D. R. Yarkony, Chem. Phys. Lett. **113**, 159 (1985).

46. K. Jankowski, R. Becherer, P. Scharf, H. Schiffer and R. Ahlrichs J. Chem. Phys. **82** 1413 (1985).

47. R. C. Raffenetti, J. Chem. Phys. **58**, 4452 (1973).

48. I. Shavitt, B. J. Rosenberg, and S. Palalikit, Int. J. Quantum Chem. **S10**, 33 (1976).

49. S. R. Langhoff, C. W. Bauschlicher, and P. R. Taylor, J. Chem. Phys. **88**, 5715 (1988).

50. S. R. Langhoff and C. W. Bauschlicher and P. R. Taylor, J. Chem. Phys. **86**, 6992 (1987).

51. C. W. Bauschlicher, S. R. Langhoff, and P. R. Taylor, J. Chem. Phys. **87**, 387 (1987).

52. H.-J. Werner and E.-A. Reinsch, in "Proceedings of the 5th European Seminar on computational methods in quantum chemistry", eds P. Th. van Duijnen and W. C. Nieuwpoort (Max-Planck-Institut für Astrophysik, Munich 1981).

53. A. D. McLean, P. R. Bunker, R. M. Escribano, and P. Jensen, J. Chem. Phys. **87**, 2166 (1987).

54. R. D. Cowan, and D. C. Griffin, J. Opt. Soc. Am. **66**, 1010 (1976).

55. R. L. Martin, J. Phys. Chem. **87**, 750 (1983).

56. J. Berkowitz, J. P. Greene, H. Cho and B. Ruscic, J. Chem. Phys. **86**, 1235 (1987): also see K. Balasubramanian and A. D. McLean, *ibid.* **85**, 5117 (1986).

57. S. R. Langhoff, C. W. Bauschlicher, and P. R. Taylor, Chem. Phys. Lett. **135**, 543 (1987).

58. K. P. Huber and G. Herzberg, in "Constants of Diatomic Molecules" (Van Nostrand Reinhold, New York, 1979).

59. C. W. Bauschlicher, S. P. Walch, S. R. Langhoff, P. R. Taylor, and R. L. Jaffe, J. Chem. Phys. **88**, 1743 (1988).

60. M. J. Frisch, B. Liu, J. S. Binkley, H. F. Schaefer and W. H. Miller, Chem. Phys. Letters **114**, 1 (1985).

61. B. C. Garrett, D. G. Truhlar, R. S. Grev and A. W. Magnuson, J. Phys. Chem. **84**, 1730 (1980).

62. D. M. Neumark, A. M. Wodtke, G. N. Robinson, C. C. Hayden and Y. T. Lee, J. Chem. Phys. **82**, 3045 (1985) and D. M. Neumark, A. M. Wodtke, G. N. Robinson, C. C. Hayden, K. Shobatake, R. K. Sparks, T. P. Schafer and Y. T. Lee, *ibid.* **82**, 3067 (1985).

63. D. W. Schwenke, R. Steckler, F. B. Brown and D. G. Truhlar, J. Chem. Phys. **84**, 5706 (1986) and references therein.

64. H. F. Schaefer, J. Phys. Chem. **89**, 5336 (1985).

65. P. Pulay in "Applications of Electronic Structure Theory" H. F. Schaefer, ed., pp. 153 (Plenum Press, New York, 1977).

66. B. C. Garrett, M. J. Redmon, R. Steckler, D. G. Truhlar K. K. Baldridge, D. Bartol, M. W. Schmidt, and M. S. Gordon, J. Phys. Chem. **92**, 1476 (1988).

67. D. W. Schwenke, R. Steckler, F. B. Brown and D. G. Truhlar, J. Chem. Phys. **86**, 2443 (1987).

MODERN ELECTRONIC STRUCTURE CALCULATIONS: THE ACCURATE PREDICTION OF SPECTROSCOPIC BAND ORIGINS

NICHOLAS C. HANDY
University Chemical Laboratory
Lensfield Road
Cambridge
CB2 1EW
UK

ABSTRACT. Derivative theory has been one of major advances in quantum chemistry in recent years, enabling the quantum chemist to make valuable predictions in the area of microwave and infrared spectroscopy. Here it is argued that a high accuracy model for predictive quantum chemistry is MP2 with sufficiently large basis sets (TZ2p+ f), for which it is possible today to calculate analytic second derivatives. Harmonic frequencies are often accurate to 1% in this model. If analytic SCF third derivatives together with finite difference fourth derivatives are also available, it is then possible to extend the model to the prediction of band origins. Small scaling procedures are suggested which yield an accuracy of the order of $15cm^{-1}$. Supporting calculations are presented, including an incomplete study of H_2O_2.

1. INTRODUCTION

In this lecture, we shall concentrate on the high accuracy determination of the shape of the potential energy surface in the region of the equilibrium geometry of a molecule. The purpose will be to show that it is now possible to devise a quantum chemistry model which yields high accuracy spectroscopic properties which have only a small degree of uncertainty. The idea of such a model is not new; the recent book by Hehre, Radom, Schleyer and Pople [1] demonstrates the success of a model based on the Self Consistent Field (SCF) and Møller-Plesset (MP) theory, all linked through the GAUSSIAN set of quantum chemistry codes.

Our model is based upon the successful development of algorithms for the following calculations:

(i) Møller-Plesset second order, for closed shell systems (MP2). Second order perturbation theory calculations have been available for many years, and it is well recognised that this is the most straightforward and simple method for the inclusion of dynamical correlation effects. It has also theadvantage that it is a size-consistent method, now recognised to be a much more important criteria than the variational criteria of configuration interaction calculations. MP2 only needs the integrals (ia| jb), where i,j are occupied orbitals and a, b are virtual orbitals, in the molecular orbital basis.

In 1979, Pople et al[2] gave the first implementation of an analytic MP2 gradient algorithm, and although not efficient, led the way to the optimisation of molecular geometries with a correlated method. Simandiras et al[3], in 1987, demonstrated in detail the efficient implementation of the MP2 gradient algorithm which encompassed the following

(a) its evaluation through a formula

$$\frac{dE}{dX} = \Sigma\, \Gamma_{\alpha\beta\gamma\delta}\,(\alpha\beta|\gamma\delta)^X + \Sigma\, \gamma_{\alpha\beta}\,(\alpha|h|\beta)^X + \Sigma\, W_{\alpha\beta}\, S^X_{\alpha\beta}, \qquad (1)$$

where $\underline{\Gamma}$, $\underline{\gamma}$ are $\underline{W}$ are predetermined a.o. basis quantities. Thus no derivative integrals have to be transformed.

A. Laganà (ed.), Supercomputer Algorithms for Reactivity, Dynamics and Kinetics of Small Molecules, 23–36.

(b) It is not necessary to solve the Coupled Hartree Fock equations [4] for the orbital perturbations U^x_{ia}, instead, following Handy and Schaefer[5], it is only necessary to solve one set of simultaneous equations, for what is now often called the response vector Z_{ia},

$$\underline{H}\,\underline{Z} = \underline{L} \qquad (2)$$

where $\underline{L}$ is a lagrangian vector, and $\underline{H}$ is a hessian matrix. The effects of $\underline{Z}$ are absorbed into $\underline{W}$ in eqn(1).

In 1985, at a meeting sponsored by NATO on Geometrical Derivatives[6], Pople, Bartlett and Handy discussed the implementation of analytic second derivatives for the MP2 method. Handy et al[7] then implemented an efficient algorithm for this calculation, which has been operational for two years. It is not an easy code; it demands the storage on disc of the first derivative m.o. integrals $(ia|jb)^x$, as well as the full determination of the first order wavefunction parameters U^x_{ia}. It does not require the solution of the second order CPHF equations, these being eliminated by The Handy-Schaefer[5] idea.

These MP2 codes form a central feature of the Cambridge Analytic Derivatives Package[8] (CADPAC). On a CRAY-XMP, it is often found that the MP2 second derivative algorithm runs at 80 Mflops, and with our disc limitation of 4 Gbytes, in effect we are limited to calculations with less than 200 basis functions. However, this enables substantial calculations to be performed on small molecules, as we shall demonstrate.

(ii) Considerable experience[3,9] has been gained with these MP2 algorithms, and in the next section it will be demonstrated that near MP2-basis set limit geometries and harmonic frequencies are obtained with triple-zeta plus double polarisation (plus f functions in some circumstances) basis sets, denoted TZ2p(+f), for first row atoms. The accuracy of these predictions is remarkably high, for reasons we do not fully understand.

(iii) Gaw et al[10] have successfully implemented, at the Self Consistent Field level, an analytic third derivative code. This has been in existence for four years, and many SCF DZP basis quality calculations have been performed. To obtain the anharmonic constants x_{rs}, it is also necessary to have available fourth derivatives, and today we calculate these from central differences of the analytic third derivatives.

(iv) There is a paucity of reliable experimental information for the anharmonic constants x_{rs}, and other spectroscopic constants derived from higher (than second) order force constants, but such information as is available[11] suggests that the SCF DZP anharmonic constants are accurate to approximately 10%, for reasons again which we do not fully understand, but are related to the fact that the SCF method is capable of representing the repulsive part of potential energy curves[39].

In the next section we discuss the evidence based on calculations using the above algorithms. In particular, we shall come forward with a model for the accurate prediction of band origins for fundamentals for first-row atom containing molecules. We shall also discuss how the model may be improved, and shall observe that an order of magnitude more effort is required if greater accuracy is to be achieved.

In section 3 an application of the model to H_2O_2 is reported, and a brief discussion is presented in section 4.

2. THE ACCURACY OF THE MP2 CALCULATIONS, AND THE HIGHER DERIVATIVE SCF CALCULATIONS

A philosophy behind the model we are suggesting is that it is an alternative to use a quantum chemistry method with as near the basis set limit as possible for that method, instead of using smaller basis sets on more advanced methods. Thus the successful model of Pople and his co-workers[1] which may be described as "MP2, MP3 and MP4 with 6-31G* basis set" should be compared with our approach of "MP2 with a near-complete basis set". In favour of the latter approach is that the (only!) uncertainty concerns the omission of MP3, MP4... effects, whereas the first approach certainly introduces the important effects of single and triple replacements at the MP4 level; on the other hand there are substantial basis set deficiencies with small sets at the MP2 level.

We have published [3,11,12] many of our calculations using the MP2 method with large basis sets, and so it is inappropriate to repeat these. We only summarise some data in Table 1, which is indicative of the accuracy we find for geometrical parameters. The following observations seem clear, from these calculations with TZ2p(+f) basis sets.
(a) XH bonds are predicted to be a little short, up to 0.003Å (X = O, N, C)
(b) Angles appear to have a remarkably high accuracy
(c) Multiple bonds (CO, CN, CC) are too long, up to 0.01Å for CN in HCN.
We note that the effect of these large basis sets is to reduce bond lengths, when compared to those calculated with a 6-31G* basis set. Thus the multiple bond lengths are much improved, and the single bonds change from being too long to being slightly too short. We must of course add that the uncertainty in "experimental" bond lengths can easily be as much as 0.002Å.

Once the reliability of geometrical parameters is established, then the next property of interest are harmonic frequencies, available immediately from the analytic MP2 second derivative program. Table 2 summarises our calculations with the TZ2p(+f) basis sets, for the same molecules, and the following conclusions may tentatively be made.

(i) XH (X= C, N, O) stretching frequencies are all too large, by a factor varying between 0.5% and 2.5%. This correlates with the fact that the XH bonds are slighly too short. On the limited evidence from Table 2, it is suggested that a suitable scaling of the TZ2p MP2 XH frequencies will substantially aid theoretical prediction, and that this scaling ought to be such that ab initio XH frequencies are reduced by the order of 1 ~ 2%. These are of course the largest frequencies, with consequently the largest error, and thus this is the important scaling to make.

(ii) It is important to include f functions in the basis if certain molecules are studied, because of their substantial effect on certain bending vibrations, the most drastic one being the π_g vibration of C_2H_2. In this list the important molecules for this effect are C_2H_2 and HCN.

(iii) From the previous discussion, there are some multiple bonds, most notably the CN triple bond, which MP2 predicts to be too long. Of course this has an effect on the associated stretching frequency, as evidenced by the CN stretching ω being 85cm^{-1}too low with the TZ2p+f basis set. In this situation other methods must be used to calibrate the accuracy of MP2.

(iv) All other vibrations are in error by an average of 15 cm^{-1} or less, with either the TZ2p or TZ2p+f basis sets, with this approximate procedure, there seems to be no argument for further refinement of these values at this level.

The conclusion of this discussion is that if the MP2 calculation is carried out with TZ2p (or TZ2p+f if appropriate), and a small scaling of the XH frequencies is carried out, then with the exception of the multiple bond stretching vibrations, the absolute accuracy of the predicted harmonic frequencies will be of the order of 20 cm^{-1} or better for a large number of such frequencies.

The second aspect of our predictive procedure is the accuracy of the higher derivatives calculated at the SCF level. This is a very much more difficult problem because there is very little firm evidence on which to judge the accuracy. The most important anharmonic constants are x_{rs}; the expansion formula for the vibrational levels of an asymmetric top polyatomic molecule are given by

$$G(v) = \Sigma\omega_r(v_r + \frac{1}{2}) + \Sigma x_{rs}(v_r + \frac{1}{2})(v_s + \frac{1}{2}) + \ldots \quad (3)$$

To determine x_{rs} from ab initio data, all cartesian third and fourth derivatives are required. The experimentalist however often starts from an internal coordinate force field, and keeps only those terms (cubic and quartic) which he considers most important and which therefore only involve a few parameters; he then transforms to normal coordinate force constants ϕ_{rst}, ϕ_{rstu}, from which with a knowledge of only ϕ_{rrss}, ϕ_{rrs} he can then determine x_{rs}. Because of the difficulty of assigning sufficient excited vibration states, other information (from the rotational lines) are used to determine some of the required force constants; the whole process is therefore based on a lack of evidence, and therefore there is a great paucity of reliable information for values of x_{rs}. For H_2O, there is firm evidence. For NH_3 and CH_4 there is some evidence, and for these molecules we have discussed[11] in detail the comparison between our computed values at the SCF DZP level, and the experimental values.

Our conclusions may be summarised from an examination of table 3, where we report the anharmonic corrections ($v_r - \omega_r$) for these molecules which is given by

$$v_r - \omega_r = x_{rr}(1+g_r) + 1/2\, \Sigma\, x_{rs}\, g_s \quad (4)$$

(g_r is the degeneracy of the rth mode)

A glance at this table suggests that there is nothing that should obviously be done to the calculated values by scaling or other procedures. For H_2O, the values for $v_r - \omega_r$ are each within 9cm $^{-1}$, and these are the most reliable comparisons; for CH_4 the worst discrepancy is 23 cm^{-1} (and the associated value for ω_4 is open to some question); for NH_3 the worst discrepancy is 46 cm^{-1}, about which we say more later.

Therefore our model is clear and we can test it for the molecules for which we have calculations. To calculate the positions of the band origins, we add the SCF-DZP values of ($v_r - \omega_r$) to the MP2 values of ω_r, as calculated and suitably scaled. The results of this calculation are given in Table 4, using both TZ2p and the TZ2p +f basis sets. To obtain the results in Table 4, the OH (MP2) harmonic frequencies have been scaled by 1% and the NH and CH (MP2) harmonic frequencies by 2%. The reasons behind this scaling are (i) the OH bond length is better predicted than the NH and CH bonds, and (ii) on a simple empirical basis a 2% scaling of the NH and CH harmonic frequencies gives a much superior set of results than the 1% scaling. Whether the 2% scaling is subsuming some error in the calculated anharmonic corrections is uncertain, but that could be the case. However, it is interesting to note that, from table 3, the anharmonic calculated correction for v_1 for NH_3 is in error by +46 cm^{-1}, for v_3 by -18 cm^{-1}, whereas from table 2 the error in the calculated harmonic frequency ω_1 is 14 cm^{-1}, and for ω_3, 93 cm^{-1}. Our calculations suggest that the current experimental value, 3506 cm^{-1}[24], for ω_1 is too high by as much as 40 cm^{-1}, because the uniform scaling of our MP2 values, when added to the SCF anharmonic correction, gives good agreement for both v_1 and v_3 (error -10 and -3 cm^{-1} respectively). If we ignore the HCN results in Table 4, the average absolute error of the predicted TZ2p +f based results for v_r is 13 cm^{-1}, for the 29 frequencies. It is reasonable to exclude HCN because of the incorrect treatment of the triple bond by the MP2 approach. The worst agreement is the 33 cm^{-1} for v_2 for C_2H_4, which corresponds to the double bond stretch (this result might have been anticipated).

Thus based on the results of Table 4, this ab initio model seems to offer a genuine prospect of a reliable prediction for band origins to an accuracy of approximately 15 cm^{-1}.

The idea of adding harmonic frequencies calculated by a correlated method to anharmonic contributions from the SCF method is not new. Indeed, Clabo et al[14] did this using the CISD method for the harmonic frequencies. However, the principal purpose of their paper was an examination of the anharmonic constants, rather than the accurate prediction of band origins. Also of course the idea of scaling is not new. The schools of Pulay and Boggs[15] have been very successful with scaling procedures of force constants which are usually based on SCF calculations. But the scaling factor involved is usually as much as 10%, rather than the much smaller 2% involved here, and which is only applied to XH frequencies. There has to be a much greater reliability with smaller scale factors, and therefore we commend this procedure which is based on MP2 with large basis sets.

There are two final comments to make in this analysis. Firstly, it it most important that further calculations are performed to gain a greater confidence, with further refinements, on a wider variety of molecules, for which the experimental fundamentals are known. Secondly, the value of such a scheme is its ability to predict, and to this end we must know when the model will fail.

In our case, the model fails when MP2 is a poor predictor i.e. when the higher order terms in the series are important. There are several aspects to this discussion:

(i) We have examined predictions using the series up to MP3 with large basis sets[13]. In many cases, we have found that MP3 predictions are inferior to MP2, and we conclude that the additional effort required for MP3 is not valuable. This may be unsatisfactory; all that it means is that MP2 is a good "Pauling" point for many calculations.

(ii) A better examination of the reliability of the MP2 prediction is through the perturbed wavefunction series. Thus, to obtain MP2 (and MP3), only ψ_1 is required, and this involves Double Replacement determinants. ψ_2 involves single, double, triple and some quadruple replacements. Recent discussion on spin contamination[16] and the importance of single replacements in Coupled-Cluster calculations[17] strongly suggest that an examination of the singles contribution to ψ_2 is important. It will be expensive to examine the contributions of the triple replacements to ψ_2, but we shall certainly be examining the various contributions to ψ_2 (through the sum of the squares of the coefficients), and comparing them with the similar quantity for ψ_1. It is important to note that the fact that $|\psi_1|$ is large does not necessarily mean that the MP2 predictions are poor, because as the number of electron pairs increases, the $|\psi_1|$ necessarily increases. However, it is often the case that a large $|\psi_1|$ indicates that one is studying a multi-reference problem for which MP2 is inappropriate.

(iii) If MP2 is a poor predictor, then an order of magnitude more work will have to be done, because there is today no other analytic second derivative program which works successfully with large basis sets. There are some MCSCF codes[18], but these are not appropriate for the inclusion of dynamic correlation, which is important for the prediction of accurate force constants. The only alternative is to use more advanced quantum chemistry codes, for which the gradient is available, and then use finite differences of these.

As an example of the use of this MP2 + third + fourth derivative SCF combination, we now report some investigations on H_2O_2, which form the first part of an attempt to construct an accurate potential surface for this molecule.

3. A COMPARISON OF THE AB INITIO PREDICTION FOR THE FUNDAMENTALS OF H_2O_2, WITH EXPERIMENTAL VALUES

The basis set used for this investigation was the Dunning [19] 5s 4p contraction of the Huzinaga[10] 10s 6p basis on Oxygen, with two sets of d functions with gaussian exponents 1.35 and 0.45. For the H atoms, the basis was the Dunning[19] 3s contraction of Huzinaga's[19] 6s primitive set, with a scale factor of 1.2, and two sets of p functions with exponents 1.5 and 0.5. This basis set is denoted TZ2p. The

TZ2p+f basis set was obtained by the addition of a set of f basis functions with gaussian exponents 0.9, on oxygen.

Further technical details are that the SCF fourth derivatives were determined by central differences of the analytic SCF third derivatives, using cartesian displacements of $\pm$ 0.001 a_0. These calculations, all carried out at the SCF optimised geometry, were performed with the codes developed by J.F. Gaw[10]. The determination of the spectroscopic constants was possible with the program SPECTRO[20], specially designed to derive this information from a quartic force field. The MP2 analytic derivative calculations were performed with the Cambridge Analytic Derivatives Package (CADPAC), this edition being provided by R.D. Amos and J.E. Rice[8].

Table 5 reports the optimised geometrical parameters for H_2O_2, determined at the SCF (DZP) and MP2 (TZ2p) and MP2 (TZ2p+ f) levels. There has been much experimental uncertainty for the equilibrium structure of H_2O_2, primarily because there are only three rotational constants with which to fix the four structural parameters. The other difficulty is that the torsional vibration is large amptitude and cannot be considered through perturbation theory. The latest evidence appears to be that due to Koput[21], who suggests that ϕ_e = 111.7 $\pm$ 0.19. The range of values of bond lengths considered by Koput is large. Our evidence is that (i) r_{OH} may be longer than previously thought, perhaps in the region 0.968 to 0.970Å, which contrasts with the recommended value of 0.965Å, (ii) r_{OO} may be shorter than previously thought. The MP2 calculations show a significant f basis function effect, so we cautiously suggest a value for r_{OO} a little longer than 1.454Å, (iii) Our value for θ at 99.5^o may be very close to the equilibrium value. (iv) Our calculations suggest a value for ϕ_e in the region of 114°.

Our calculations for the harmonic frequencies ω_r and fundamentals v_r for five vibrations of H_2O_2 are given in Table 6 (the torsional vibration is omitted). The model suggested in the last section is used to calculate predicted values for the fundamental frequencies, and these are compared with the observed values. It is seen that the agreement, based on the TZ2p basis set values for ω, is quite remarkable!

These tables report a small part of a detailed study in progress on H_2O_2. The principle difficulty lies in the fact that a quartic cartesian displacement force-field cannot represent the tunneling motion in the torsional coordinate, and thus it is necessary to have a mixed representation for the potential involving both small and large amplitude motion coordinates. It is interesting to note that the MP2 value for the trans-barrier to internal rotation is 384 cm^{-1}, within 1 cm^{-1} of the value of Koput[21], assuming all other geometrical parameters remain fixed. If they are allowed to relax, the barrier drops to 343 cm^{-1}.

Once such a potential has been determined, our plan is to examine its accuracy by calculating in a precise way all the low-lying J = 0 vibrational energy levels. This may be achieved using our tetra-atomic vibrational variational program[22], which has been tested previously with calculations on an acetylene surface. Some adjustment of some of the force constants may be necessary to obtain good agreement between theory and experiment, but in this way we hope to determine an accurate representation of the potential energy surface in the region of equilibrium. Details of this research will be published elsewhere.

4. CONCLUSION

In this review, the accuracy of a quantum chemistry model for the determination of spectroscopic molecular properties has been examined. We have concentrated on the determination of band origins for fundamental vibrations. Our model takes advantage of the high accuracy predictions of the MP2 approach, when large basis sets are used, together with the fact that analytic second derivatives may be determined at this level. We also use the SCF analytic third derivative program (and through it, fourth derivatives) to calculate anharmonic corrections to the harmonic frequencies. Evidence for the accuracy is presented from calculations on several molecular systems. The model is used as part of a calculation for the spectroscopic properties of H_2O_2. Without a supercomputer, calculations such as these would be impossible. The MP2 analytic second derivative code (fully vectorised by E.D. Simandiras and R.D. Amos) runs at 80Mflops on a CRAY-XMP. These programs form part of a specially vectorised quantum chemistry package, CADPAC, developed by R.D. Amos and J.E. Rice. The analytic SCF third derivative code was written by J.F. Gaw. The specific calculations on H_2O_2 were performed by A. Willetts. The variational vibrational programs were developed over the years by S. Carter and the author.

ACKNOWLEDGEMENT

This paper uses many results obtained by members of the quantum chemistry group in Cambridge in recent years, namely R.D. Amos, J.E. Rice, E.D. Simandiras, J.F. Gaw, K. Somasundram, T.J. Lee, A. Willetts, to all of whom the author is grateful.

TABLE 1 "Basis set limit" geometrical parameters determined at the SCF and MP2 level, and compared with available experimental data.

Molecule	Param	SCF	MP2	Expt [a]
H_2O	r_{OH}/Å	0.940	0.958	0.958
	θ/°	106.3	104.5	104.5
NH_3	r_{NH}/Å	0.998	1.009	1.012
	θ/°	108.1	107.2	106.7
CH_4	r_{CH}/Å	1.082	1.083	1.086
H_2CO	r_{CO}/Å	1.178	1.211	1.203
	r_{CH}/Å	1.092	1.097	1.099
	θ/°	116.1	116.5	116.5
HCN	r_{CN}/Å	1.124	1.164	1.153
	r_{CH}/Å	1.057	1.064	1.065
C_2H_2	r_{CC}/Å	1.179	1.207	1.203
	r_{CH}/Å	1.054	1.060	1.062
C_2H_4	r_{CC}/Å		1.328	1.334
	r_{CH}/Å		1.079	1.081
	θ/°		117.2	117.3

(a) Experimental geometries from refs. 23-29.

TABLE 2 Errors (%, cm^{-1}) in MP2 harmonic frequencies, when compared with 'experimental' harmonic frequencies

Molecule	mode	TZ2p (%, cm^{-1})	TZ2p+f (%, cm^{-1})	ω_{exp}^{a}	
H_2O	ω1	0.7 (+27)	0.2 (+7)	3832	OH str
	ω2	-0.5 (-8)	-1.2 (-20)	1649	bend
	ω3	1.1 (+43)	0.6 (+24)	3942	OH str
NH_3	ω1	0.6 (+21)	0.4 (+14)	3506	NH str
	ω2	2.5 (+26)	1.3 (+13)	1022	s-def
	ω3	2.7 (+95)	2.6 (+93)	3577	NH str
	ω4	0.1 (+1)	-0.9 (-16)	1691	d-def
CH_4	ω1	2.0 (+62)	1.8 (+56)	3025	CH str
	ω2	1.4 (+22)	0.7 (+11)	1583	d-def
	ω3	2.1 (+66)	1.9 (+61)	3157	CH str
	ω4	1.1 (+15)	-0.4 (-6)	1368	d-def
C_2H_2	ω1	0.9 (+31)	1.2 (+45)	3497	CH str
	ω2	-1.3 (-27)	-1.1 (-22)	2011	CC str
	ω3	0.7 (+23)	1.1 (+38)	3415	CH str
	ω4	-16.2 (-101)	0.2 (+1)	624	πg bend
	ω3	-3.2 (-24)	2.1 (+16)	747	πu bend
HCN	ω1	0.6 (+19)	0.9 (+31)	3442	CH str
	ω2	-4.5 (-96)	-4.0 (-86)	2129	CN str
	ω3	-8.1 (-59)	1.5 (+11)	727	bend
H_2CO	ω1	0.3 (+8)	0.2 (+7)	2978, 2944	CH str
	ω2	-1.0 (-17)	-0.7 (-13)	1778, 1764	CO str
	ω3	1.9 (+29)	1.5 (+23)	1529, 1563	CH_2 sc.
	ω4	1.0 (+12)	1.4 (+17)	1191, 1191	CH_2 wag
	ω5	2.3 (+70)	2.2 (+68)	2997, 3009	CH str
	ω6	-0.9 (-12)	-1.7 (-22)	1299, 1288	CH_2 rock
C_2H_4	ω1	1.7 (+54)	1.7 (+51)	3153	CH_2 s-s
	ω2	2.5 (+41)	2.2 (+36)	1655	CC str
	ω3	1.7 (+24)	1.4 (+20)	1370	CH_2 sc.
	ω4	2.8 (+30)	3.2 (+34)	1044	CH_2 tw.
	ω5	1.5 (+50)	1.5 (+48)	3232	CH_2 a-s
	ω6	0.2 (+3)	0.5 (+6)	1245	CH_2 rock

ω7	1.1 (+10)	1.6 (+15)	969	CH_2 wag
ω8	-3.1 (-30)	-1.3 (-13)	959	CH_2 wag
ω9	2.2 (+73)	2.1 (+70)	3234	CH_2 a-s
ω10	0.0 (0)	-1.3 (-11)	843	CH_2 rock
ω11	1.3 (+42)	1.3 (+42)	3147	CH_2 s-s
ω12	1.8 (+26)	0.9 (+14)	1473	CH_2 sc.

(a)Experimental estimates for harmonic frequencies from refs. 23, 24, 25, 30, 31, 32, 33.

TABLE 3 Anharmonic corrections ($\nu_r - \omega_r$), calculated at the SCF (DZP basis set) level, compared with experimental values (frequencies in cm^{-1})

Molecule	mode	$\nu_r - \omega_r$ (calc)	$\nu_r - \omega_r$ [(a)] (expt)
H_2O	ν_1	-167	-175
	ν_2	-57	-54
	ν_3	-178	-187
NH_3	ν_1	-124	-170
	ν_2	-98	-72
	ν_3	-155	-133
	ν_4	-47	-65
CH_4	ν_1	-103	-105
	ν_2	-39	-50
	ν_3	-129	-134
	ν_4	-36	-59

(a) Experimental values from refs. 34, 25.

TABLE 4 (in cm^{-1}) Comparison of ab initio and experimental vibrational band origins

Molecule	Mode	ν_r (calc)$^{(b)}$		ν_r $^{(a)}$ (obs)
H_2O	ν_1	3653	3633	3657
	ν_2	1594	1582	1595
	ν_3	3767	3748	3755
NH_3	ν_1	3333	3326	3336
	ν_2	950	937	950
	ν_3	3444	3441	3444
	ν_4	1645	1628	1626
CH_4	ν_1	2929	2916	2920
	ν_2	1566	1557	1533
	ν_3	3040	3035	3023
	ν_4	1347	1326	1309
HCN	ν_1	3277	3289	3312
	ν_2	2018	2028	2097
	ν_3	637	707	713
H_2CO	ν_1	2792	2791	2782
	ν_2	1737	1741	1746
	ν_3	1528	1522	1500
	ν_4	1185	1190	1167
	ν_5	2857	2855	2843
	ν_6	1268	1258	1249
C_2H_4	ν_1	3020	3019	3026
	ν_2	1662	1656	1623
	ν_3	1372	1368	1342
	ν_4	1053	1057	1023
	ν_5	3092	3093	3086
	ν_6	1228	1231	1222
	ν_7	983	969	949
	ν_8	912	929	940
	ν_9	3110	3107	3105
	ν_{10}	842	831	826
	ν_{11}	2984	2984	2989
	ν_{12}	1467	1455	1444

(a) Experimental band origins from refs. 35, 36, 24, 25, 31, 32.

(b) Anharmonic corrections for HCN taken from ref. 40. Anharmonic corrections for H_2CO and C_2H_4 taken from ref. 14.

TABLE 5 Optimised geometrical parameters for H_2O_2, determined at the SCF (DZP) and MP2 (TZ2p and TZ2p+f levels).

	r_{OH} /Å	r_{OO} /Å	θ_{OOH} /°	φ/°
SCF (DZP)	0.947	1.390	102.8	114.1
MP2 (TZ2p)	0.964	1.462	99.3	113.3
MP2(TZ2p+f)	0.967	1.454	99.5	114.5
Range of experimental values: [21].				
	0.950 -0.965	1.464 -1.475	94.9 -99.4	111.7 -119.0

TABLE 6 Ab initio prediction for band origins for H_2O_2 compared with experimental values.

		ω(MP2, TZ2p)	ν-ω (SCF)	ν_{pred}	ν_{obs} (a)
ν_1	OHs.s.	3817.3	-178.8	3602	3599
ν_5	OHa.s.	3816.7	-177.0	3603	3608
ν_2	$\theta_1 + \theta_2$	1428.7	-34.0	1394	1393
ν_6	$\theta_1 - \theta_2$	1321.9	-55.0	1266	1266
ν_3	OOstr	906.1	-26.7	871	864

(a) Observed values from refs. 37 and 38.

REFERENCES

1. W.J. Hehre, L. Radom, P.v.R. Schleyer and J.A. Pople, *Ab Initio Molecular Orbital Theory*, 1986, Wiley, New York.
2. J.A. Pople, R. Krishnan, H.B. Schlegel and J.S. Binkley, Int. J. Quant. Chem. Symp., 1979, **13**, 225.
3. E.D. Simandiras, N.C. Handy and R.D. Amos, Chem. Phys. Lett., 1987, **133**, 324.
4. J. Gerratt and I. Mills, J. Chem. Phys., 1968, **49**, 1219.
5. N.C. Handy and H.F. Schaefer, J. Chem. Phys., 1984, **81**, 5031.
6. *Geometrical Derivatives of Energy Surfaces and Molecular Properties*, ed. P. Jorgensen and J. Simons, NATO ASI series 166, 1986, D. Reidel, Dordrecht.
7. N.C. Handy, R.D. Amos, J.F. Gaw, J.E. Rice and E.D. Simandiras, Chem. Phys. Lett., 1985, **120**, 151.
8. The Cambridge Analytic Derivatives Package, Issue 4.0., R.D. Amos and J.E. Rice, 1987.
9. E.D. Simandiras, J.E. Rice, T.J. Lee, R.D. Amos and N.C. Handy, J. Chem. Phys., 1988, **88**, 3187.
10. J.F. Gaw, Y. Yamaguchi and H.F. Schaefer, J. Chem.Phys., 1985, **81**, 6395.
11. N.C. Handy, J.F. Gaw and E.D. Simandiras, J. Chem. Soc. Far. Trans. 2, 1987, **83**, 1577.
12. E.D. Simandiras, Ph.D. Thesis, University of Cambridge, 1988.
13. I.L. Alberts and N.C. Handy, J. Chem. Phys., 1988, **89**, 2041.
14. D.A. Clabo, W.D. Allen, R.B. Remington, Y. Yamaguchi and H.F. Schaefer, 1988, **123**, 187.
15. P. Pulay, G. Fogarasi, G. Pongor, J.E. Boggs and A. Vargha, J. Am. Chem. Soc., 1983, **105**, 7037.
16. H.B. Schlegel, J. Chem. Phys., 1986, **84**, 4530; P.J. Knowles and N.C. Handy, J. Chem. Phys., 1988, **88**, 6991; L. Radom and J.A. Pople, J. Chem. Phys., 1988.
17. T.J. Lee, J.E. Rice, G.E. Scuseria and H.F. Schaefer, Theor. Chim. Acta, 1989.
18. M. Page, P. Saxe, G.F. Adams and B.H. Lengsfield, Chem. Phys. Lett., 1984, **107**, 587.
19. T.H. Dunning, J. Chem. Phys., 1970, **53**, 2823; S. Huzinaga, J. Chem. Phys., 1965, **42**, 1293.
20. SPECTRO, a theoretical spectroscopy package. A. Willetts and J.F. Gaw, Cambridge, 1988.
21. J. Koput, J. Molec. Spectrosc., 1986, **115**, 438.
22. S. Carter and N.C. Handy, Comp. Phys. Comm., 1988.
23. A.R. Hoy, F.M. Mills and G. Strey, Molec. Phys., 1972, **24**, 1265.
24. W.S. Benedict and E.K. Plyler, Can. J. Phys., 1957, **35**, 1235.
25. D.L. Gray and A.G. Robiette, Molec. Phys., 1979, **37**, 1901.
26. K. Yamada, T. Nakagawa, K. Kuchitsu and Y. Morino, J. Molec. Spectrosc., 1971, **38**, 70.
27. G. Winnewisser, A.G. Maki and D.R. Johnson, J. Molec. Spectrosc., 1971, **39**, 149.
28. A. Baldacci, S. Ghersetti, S.C. Hurlock and K. N. Rao, J. Mol. Struct., 1976, **59**, 116.
29. T. Nakanaga, S. Kondo and S. Saeki, J. Chem. Phys., 1982, **76**, 3860.
30. G. Strey and I. M. Mills, J. Mol. Struct., 1976, **59**, 103.
31. G. Strey and I. M. Mills, Molec. Phys., 1973, **26**, 129.
32. J.L. Duncan and P.D. Mallinson, Chem. Phys. Lett., 1973, **23**, 597; D.E. Reisner, R.W. Field, J.L. Kinsey and H.-L. Dai, J. Chem. Phys., 1984, **80**, 5968.
33. J.L. Duncan, D.C. McKean and P.D. Mallinson, J. Molec. Spectrosc., 1973, **45**, 221.
34. I. M. Mills and A.G. Robiette, Molec. Phys., 1985, **56**, 743.
35. J. Pliva, V. Spirko and D. Papousek, J. Mol. Spectrosc., 1967, **23**, 331.
36. D. Van Lerberghe, I.J. Wright and J.L. Duncan, J. Molec. Spectrosc., 1972, **42**, 251.
37. R.L. Redington, W. B. Olson and P.C. Cross, J. Chem. Phys., 1962, **36**, 1311.
38. P.A. Giguere and T.K.K. Srinivasan, J. Raman. Spectr., 1974, **2**, 125.
39. P. Pulay (comment at Jerusalem International Quantum Molecular Science meeting, 1988).
40. R.D. Amos, J.F. Gaw, N.C. Handy, E.D. Simandiras and K. Somasundram, Theor. Chim. Acta, 1987, **71**, 41.

POTENTIAL ENERGY SURFACES OF SEVERAL ELEMENTARY CHEMICAL REACTIONS

Keiji MOROKUMA, Koichi YAMASHITA, and Satoshi YABUSHITA
Institute for Molecular Science
Myodaiji, Okazaki 444, Japan

ABSTRACT. Potential energy surfaces (PESs) of several elementary chemical reactions are investigated by ab initio calculations. Topics included are (1) spectroscopy during the K + NaCl and Na + KCl reactions, (2) F^+ + CO charge transfer reaction, (3) CH_3I photodissociation and (4) two examples of mode-selective reactions. In connection to recent experiments on each subject, qualitative as well as quantitative discussions on the PESs and dynamics are given. The last section deals with supercomputer algorithms for spin-orbit interaction.

1. Potential Energy Surfaces and Dynamics for Spectroscopy During Chemical Reactions [1]

Pioneering works by Polanyi and his group and by Brooks, Curl and their coworkers have opened a new field of "Spectroscopy of Transition State" and have attracted increasing experimental as well as theoretical interests. The final target there is to probe experimentally the PESs and dynamics in the interaction region where chemical-bond forming and breaking take place. The field is extensively reviewed by Brooks recently [2].

Relating to this problem, we studied previously the photon absorption process during the reaction K + NaCl [3]. For this system, Maguire et al. have reported evidence for absorption from the intermediate configurations of the reaction [4]. The PESs of the ground ($1^2A'$) and the electronically excited states ($2^2A'$, $3^2A'$, and $1^2A''$) are determined by the MRSD-CI method [5] with Hay and Wadt's effective core potentials [6]. The absorption intensity is assumed to be proportional to the density of KClNa classical trajectories on the ab initio ground-state PES fitted to extended LEPS and to the square of ab initio transition dipole moments at the respective geometries. It has been found there that the spectra consist of two broad bands which originate from two types of transitions: 1A'-2A' and 1A'-3A'(as well as 1A'-1A"), respectively. Also a spectroscopy that would probe bound states on the 2A' and 3A' excited states of the complex is proposed.

A further step, however, would be needed for the theoretical simulation to compare with the experimental emission spectra obtained by Maguire et al. [4]. A scheme of the photoabsorption and emission processes is given in Fig. 1. If there is only one channel on the excited state, absorption spectra would directly correspond to emission spectra. However, due to the

A. Laganà (ed.), Supercomputer Algorithms for Reactivity, Dynamics and Kinetics of Small Molecules, 37–56.

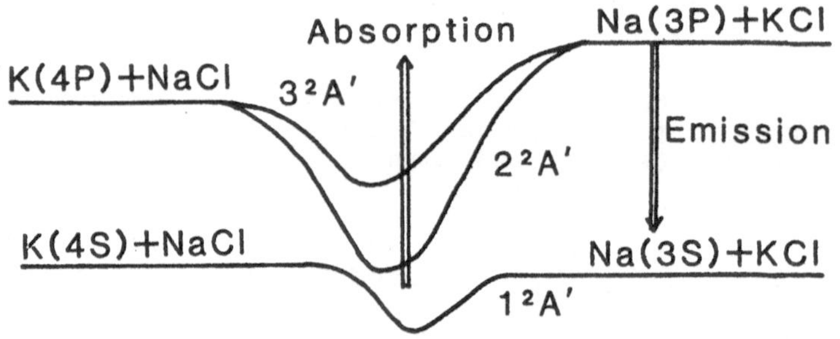

Figure 1. Scheme of absorption and emission processes during the K + NaCl reaction.

endothermicity (8 kcal/mol by experiment and 10 kcal/mol by calculation) of the excited-state reaction, in addition to this reactive channel, there is another possible channel which sends the system back to the reactant excited state. In the following we discuss the result of dynamics calculations on the excited states carried out to obtain the probability of forming K(4p) as well as Na(3p).

In order to treat the reaction in the laser field, we introduce an idea of "laser dressed" states [7]. Photoabsorption and photoemission processes are then modeled as the nonadiabatic transitions between the dressed ground and excited states. Under the assumptions of (1) two-state model, (2) single-photon excitation, and (3) rotating wave approximation, the laser dressed adiabatic PESs are given for two diabatic PESs W_1 and W_2 and the laser frequency ω as [7],

$$E_{1,2} = (W_1 + W_2 + \hbar\omega)/2 \pm [(W_2 - W_1 - \hbar\omega)^2 + 4|d_{12}|^2]^{1/2}/2,$$

where d_{12} is the nonadiabatic coupling due to the interaction between the laser field and the chemical system. In the case of dipole approximation d_{12} is given as [7],

$$d_{12} = \vec{E}\cdot\vec{\mu}_{12},$$

where $\vec{E}$ is the strength of laser field and $\vec{\mu}_{12}$ is the dipole transition moment between the states. The dynamics accompanied with nonadiabatic transitions between the laser dressed adiabatic PESs are then treated by surface hopping trajectory calculations [8]. Transition probability was calculated according to the one-dimensional Landau-Zener formula.

Fig. 2 shows the diabatic laser dressed PESs between the 1A' and 3A' states for the laser wavelength λ = 700 nm. The PESs are fitted to Murrell and Sorbie type analytical functions [9]. In

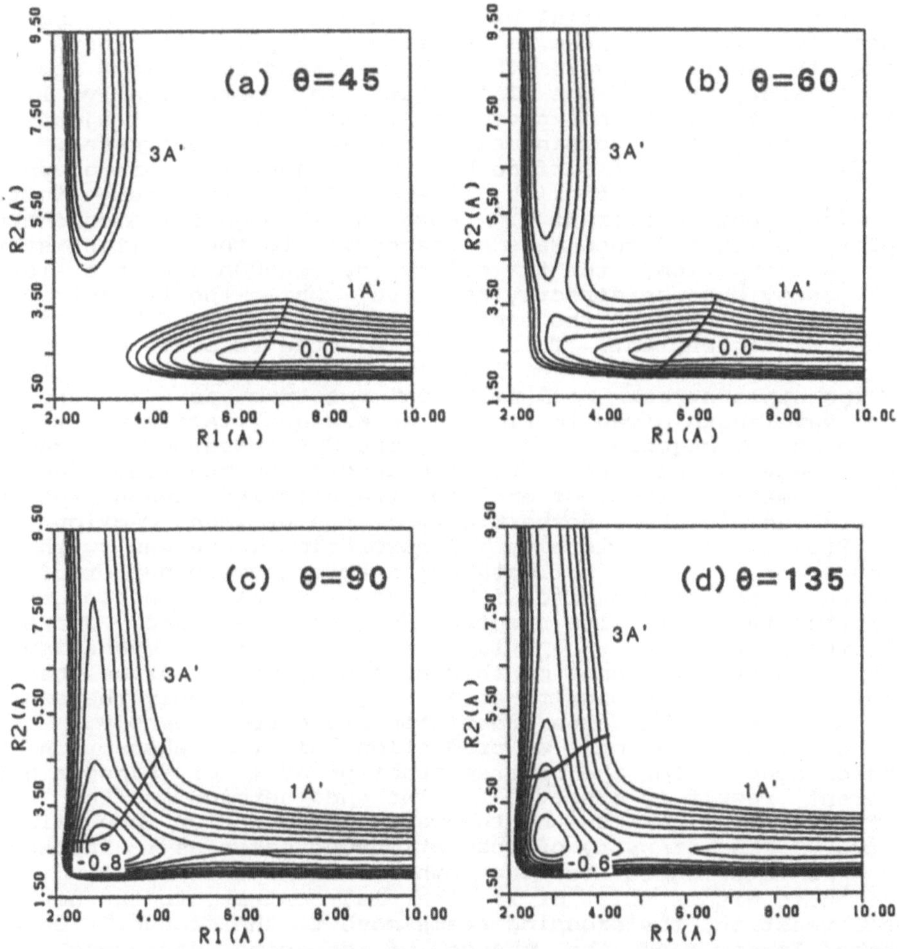

Figure 2. Laser-dressed PES for K + NaCl (λ = 700 nm). Contours are drawn for each 0.2 eV up to 1.2 eV above the ground state reactant. R1 and R2 denote the K-Cl and Na-Cl bond lengths, respectively and θ is the KClNa angle.

the case of smaller angles, 45° and 60°, the crossing occurs at around the shallow potential well on the reactant side. There exists a significant energy barrier at the reactant side of the seam. So trajectories can easily arrive at this crossing region, but have to have enough energy to overcome an energy barrier to be reactive. On the other hand, for larger angles, 90° and 135°, the crossing occurs at around the exit of the potential well located at the product side. Trajectories which cross this seam may stay at the potential well for a while and have a smaller chance to cross the seam. The important point is that if λ shifts to shorter wavelength, the crossing seams move to the product side, since the 1A' state is energetically lifted further. It is then expected that the number of trajectories which arrive at the crossing seams decrease as λ decreases.

Two examples of surface hopping trajectories are given in Fig. 3. Both trajectories (a) and (b) have the same initial condition, but different λ. Both go through the ground state complex region and then make a transition to the excited state. After a transition, the trajectory (a, λ=600nm) with a larger total energy becomes directly reactive, while the trajectory (b, λ=700nm) is trapped in the excited state complex and back to the excited state reactant. This latter example reflects the complex mechanism and the endothermicity of the excited state reaction.

The theoretical Na-D emission spectrum as a function of laser wavelength given in Fig. 4 agrees qualitatively with the experiment by Maguire et al. [4] The Na* emission spectrum is seen to be very different from the absorption spectrum, because only a small portion of excited trajectories reach the Na* product due to the endothermicity of the excited reaction (see Fig. 3(b)). The intensity of absorption increases with the increase of λ. This reflects a characteristic behavior of the crossing seams accompanied with change in λ as discussed before. The intensity of emission spectra, on the other hand, decreases with the increase of λ. This is due to a complex mechanism of the excited state. Based on this analysis, we may argue that the emission spectrum by Maguire et al. [4] reflects only the excited state dynamics but not the transition state spectroscopy.

We also have made a prediction of the absorption and emission spectra for the reverse reaction Na + KCl, since a quite different result is expected for the endothermic excited state reaction. Fig. 5 is the theoretical absorption and emission spectra. The intensity of absorption decreases as λ increases. This is attributed to the behavior of the crossing seam accompanied with a shift of λ; that is, contrary to the normal K + NaCl reaction, the crossing seams move to the product side as λ becomes longer and the chance of crossing decreases. The intensity of emission increases with the increase of λ from 650 to 750 nm. However at 800 and 850 nm, the intensity reaches an upper limit as a result of the decrease in the total number of trajectories excited, since almost all the trajectories that make a transition (corresponds to absorption) become excited-state reactive. Thus a quite different emission spectrum is predicted for the reverse reaction, and this emission spectrum reflects to a large extent the transition state spectroscopy.

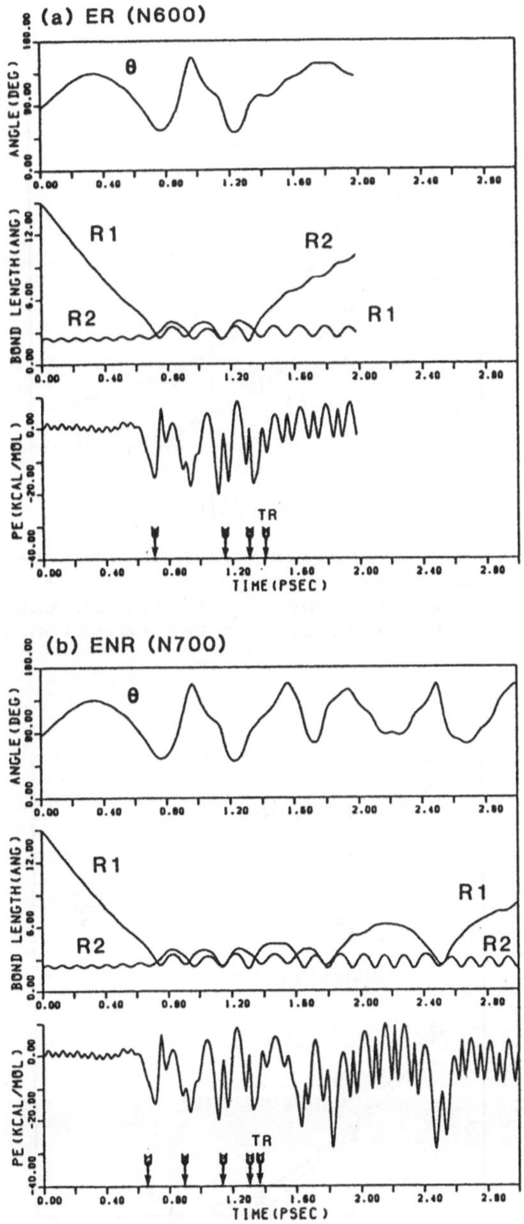

Figure 3. Plots for θ, R1, R2 and potential energy vs time for two trajectories with the same initial conditions. (a) for λ=600nm and (b) for λ=700nm.

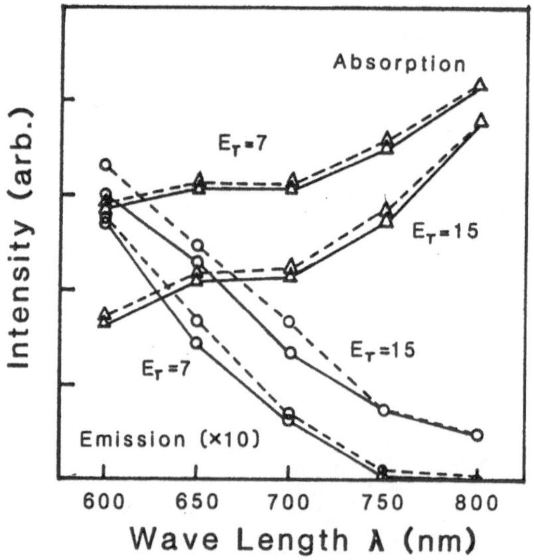

Figure 4. Theoretical absorption and emission spectra for K + NaCl. The solid lines and dotted ones are the results of 5000 and 2000 trajectory calculations. E_t is the initial translational energy in kcal/mol. The intensity of the emission is increased by 10 times.

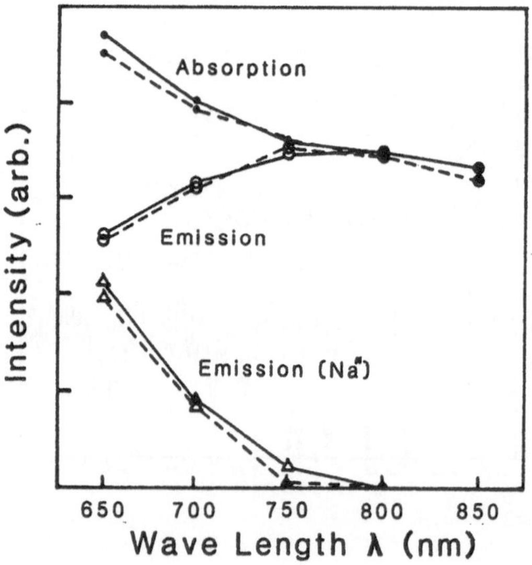

Figure 5. Theoretical absorption and emission spectra for Na + KCl.

2. Potential Energy Surfaces of Charge Transfer Reaction: $F^+ + CO \rightarrow F + CO^+$ [10]

Charge transfer reaction in a triatomic system, $A^+ + BC \rightarrow A + BC^+$, is a prototype chemical reaction involving nonadiabatic transitions between PESs. Recent experimental investigations using chemiluminescence or laser-induced fluorescence have revealed the vibrational and rotational distributions of some ion-molecule reactions and charge transfer reactions.

Kusunoki and Ishikawa [11] have investigated experimentally the chemiluminescent charge transfer reaction,

$$F^+(^3P) + CO(^1\Sigma^+) \rightarrow F(^2P) + CO^+(A^2\Pi, v').$$

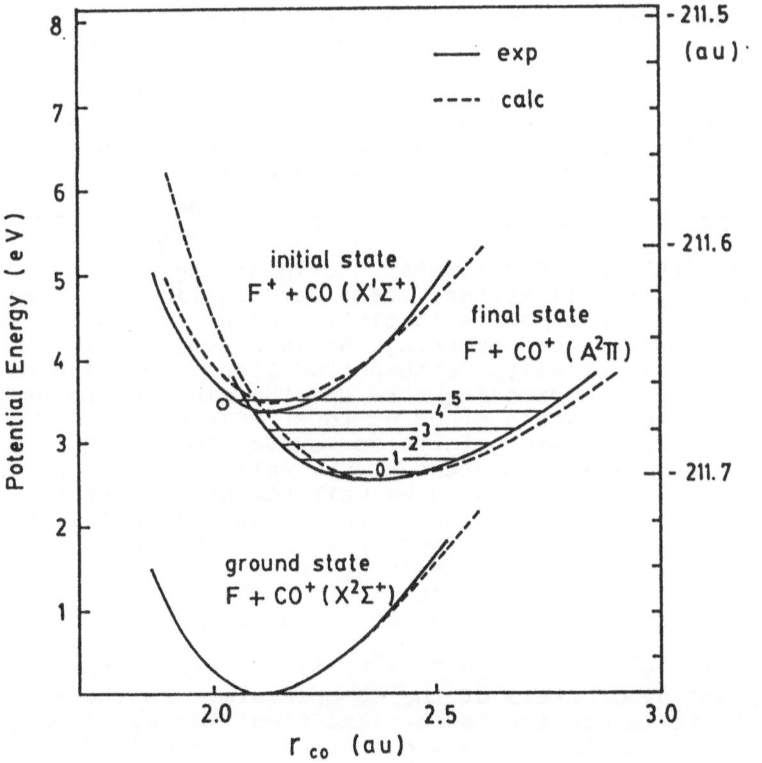

Figure 6. Potential energy curves for $(F + CO)^+$ at the infinite separation.

The dominant vibrational states of the electronically excited product CO^+ ion is v'=5. We discuss here the mechanism of (1) vibrational excitation of the product and (2) charge transfer; i.e. where the charge transfer occurs and which electronic states are involved.

The ab initio PESs are calculated by the MRSD-CI method using the DZP basis set. The low-lying six $^3A'$ and six $^3A''$ states, originated from the initial $F^+(^3P) + CO(X^1\Sigma^+)$ (6A', 5A" and 6A"), the final $F(^2P) + CO^+(A^2\Pi)$ (3, 4, and 5 A' and 2, 3, and 4A") and the ground state $F(^2P) + CO^+(X^2\Sigma^+)$ (1A' and 1A"), are investigated with respect to three internal coordinates between the F atom and the CO molecule.

The potential curves of the initial, final and ground states at the asymptotic regions are given in Fig. 6. The curvatures and the bond lengths of the diatoms are well reproduced by the calculations, though the energy difference between the initial and the ground states is slightly overestimated. It is noted that the potential curve of the initial state crosses that of the final state at r(C-O)=2.08 bohr near the v'=5 level of the final state. Therefore, the mechanism of the vibrational excitation of $CO^+(A)$ is well explained if this situation keeps up to the region where the charge transfer reaction takes place.

The potential contour maps of (a) the $2^3\Pi$ (3A' and 2A") and (b) the $3^3\Pi$ (6A' and 5A") states, originating respectively from the final and initial states, are drawn in Fig. 7. The minimum energy path of each state is plotted with broken lines. The crossing and avoided crossing seams between the states are indicated, respectively, with dotted lines and hatched region. In both states the crossing seam and the minimum energy path are almost parallel to each other up to R(F-C)= 8 bohr and then the avoided crossing seam starts to curve toward the potential minimum side. When R decreases, the valley becomes narrower and diminishes, and the repulsive wall stands against the F atom approach to the CO molecule. In the 2Π state the valley is broader and deeper than that of 3Π and the seam is located farther away from the minimum energy path.

Schematic pictures of atomic and molecular orbitals of the six states in the A' and A" symmetry are drawn for the perpendicular approach in the Fig. 8(a) and (b). The orbitals in the figures are the three 2p orbitals of the F atom and the two π's, 4σ, and 5σ orbitals of the CO molecule. In the A' symmetry, the 3A' state is constructed by 'exciting' two electrons from the initial reactant 6A' state and therefore the 6A'→3A' transition is expected to be forbidden. For the transitions 6A'→5A' and 6A'→4A', the former is expected to be stronger than the latter, since the phases of the orbitals involved in the charge transfer match in the former case. That is, the p orbital of the F atom is directed to the π of the CO molecule in the former, while that is parallel in the latter. Similarly, in the A" symmetry the transition 6A"→4A" is expected to be strong and the others are weak or forbidden. Actual calculations of the nonadiabatic coupling along the CO vibrational coordinate has supported these qualitative discussions. Furthermore, it has been found that at around R=8 bohr the coupling matrix elements for the 6A'→5A' and 6A"→4A" play an important role in charge transfer. At the

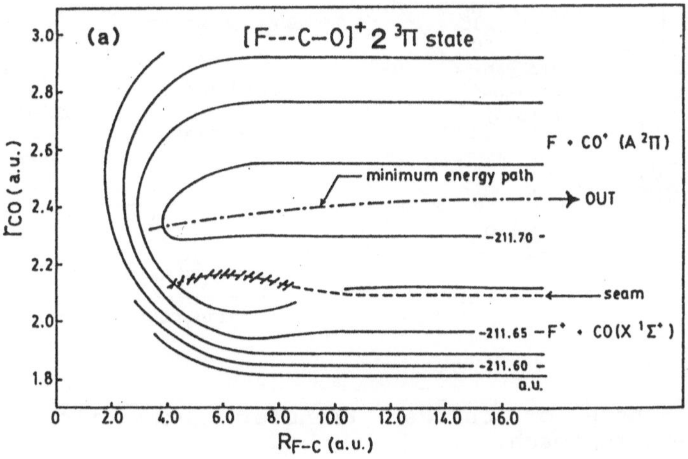

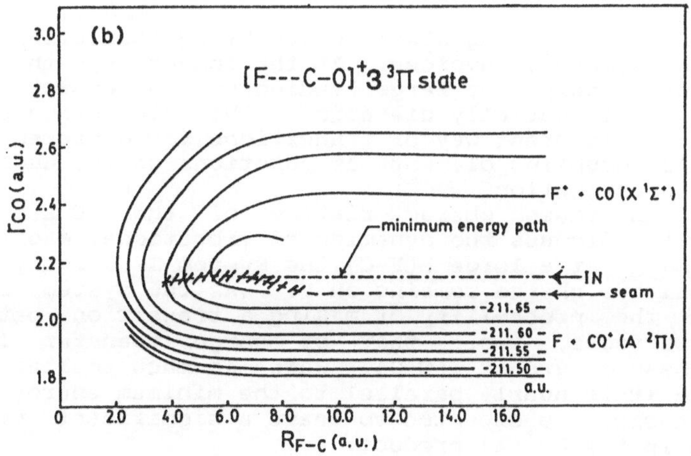

Figure 7. Contour maps (in hartree) for (a) $2^3\Pi$ (b) $3^3\Pi$ potential surfaces for θ=0.

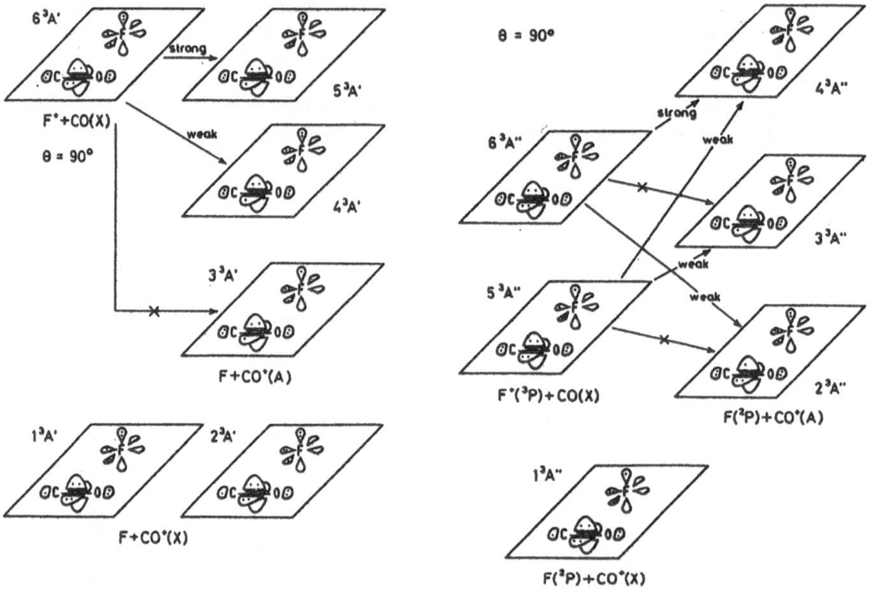

Figure 8. Scheme of charge transfer interaction for the perpendicular approach.

shorter R, the coupling elements are broad indicating that the states are strongly avoided. At the longer R, the coupling elements are sharp and large leading us to expect that the crossings are essentially diabatic. The interaction anisotropy may bring a different way of transitions for different angles. Nonadiabatic coupling elements as functions of the angle θ are now under investigation.

Based on these characteristics of the PESs, we may qualitatively discuss the dynamics of vibrational excitation of the $CO^+(A)$. At a large R(F-C) the system is largely diabatic, and the positive charge remains on F. As the system approaches R=8 bohr, the probability of making a transition between the diabatic surfaces, i.e., that of charge transfer increases. Since the seam occurs at r(C-O), which is much smaller than that in CO^+, and it is nearly parallel to the minimum energy path, the charge transfer is expected to leave a significant vibrational excitation in the $CO^+(A)$ product.

3. Potential Energy Surfaces for Rotational Excitation of CH_3 Product in Photodissociation of CH_3I [12]

Photodissociation dynamics of methyl iodine starting from the $A(n\rightarrow\sigma^*)$ band excitation has been subject to extensive experimental and theoretical studies. Recent experimental evidences indicate that photodissociation via the $A_1(^3Q_0$, parallel excitation) state which is correlated to CH_3 + excited $I^*(^2P_{1/2})$ is also responsible in part to the ground state $I(^2P_{3/2})$ product. Very recently Powis and Black [13] have found in the photodissociation of CD_3I that CD_3 produced in the I dissociation channel has rotational excitation, up to N=4 or larger, and that CD_3 is more probably rotating around an axis perpendicular to the CD_3 top axis. They have also found that CD_3 produced in the I* channel is rotationally cold with a propensity for N//K.

We have carried out calculations of PESs for symmetric(C_{3v}) and bent(C_s) dissociation of CH_3I. DZP basis functions with the iodine effective core potential were used and the molecular orbitals are determined by the Hartree-Fock calculation for an average of $n \rightarrow \sigma^*$ excited states. The spin-orbit interaction is taken into account by including an effective one-electron spin-orbit Hamiltonian

$$H^{SO} = (\alpha^2/2) \sum_i (Z_I^{eff}/r_{Ii}^3)\, L_{Ii} \cdot S_i$$

explicitly in the CI matrix evaluation. Here Z_I^{eff} is a parameter chosen to reproduce the experimental spin-orbit multiplet separation of the iodine atom. The configuration interaction includes all the singly and doubly excited singlet and triplet configurations, arising from the full valence (σ, σ^*, e_x, e_y)^6 reference configurations [14].

The conical intersection between $2A_1(^3Q_0)$ and $3E(^1Q_1)$ states occurs at R(C-I)=2.445 Å. Fig. 9 shows the angular dependency of the three potential surfaces (two A' states from $2A_1$ and 3E, and an A" state from 3E) at this distance, and at a slightly longer distance 2.646Å. The surfaces inside the intersection, not shown for brevity, reveal very little trend for bending. On the other hand, at the distance of intersection, one of the A' curves shows a minimum at a bent geometry, at θ=6.4° with $E(\theta=6.4°) - E(\theta=0°) = -5.3\times10^{-3}$ eV. The other A' curve has a smaller minimum at a geometry bent to the different direction. At the region outside the intersection, both 3E surfaces (A' and A") are very flat with respect to the bending angle. A schematic representation of relevant potential surfaces is shown in Fig. 10.

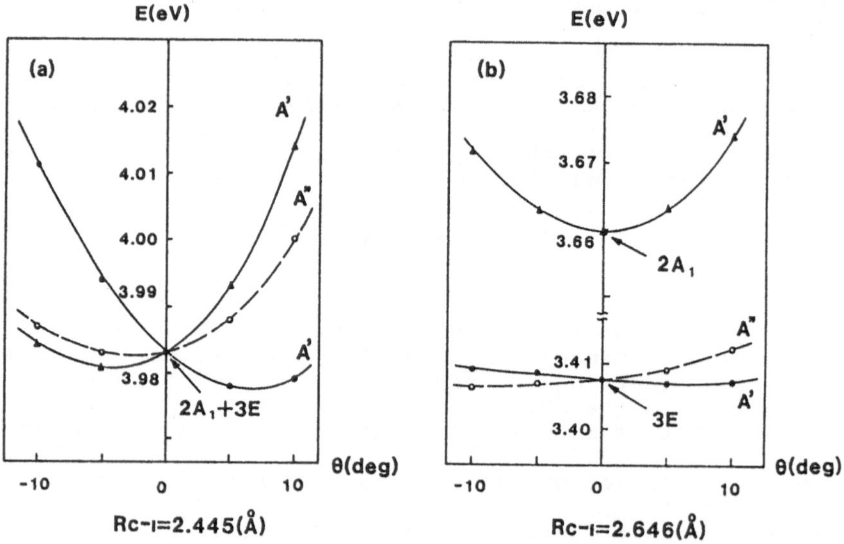

Figure 9. Angular dependency of the three potential surfaces at the conical intersection R(C-I)=2.445Å and at a longer distance 2.646Å.

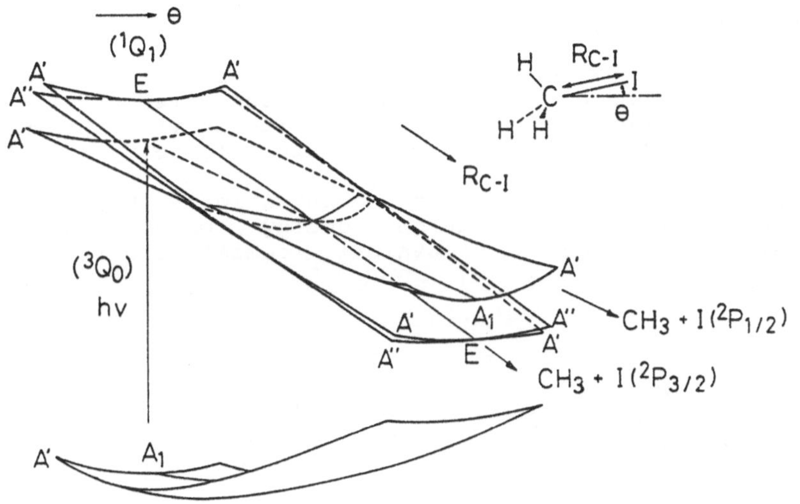

Figure 10. Schematic representation of potential energy surfaces for linear and bent dissociation.

The nonadiabatic coupling element was found to be sharply peaked in the vicinity of the avoided crossing between the two A' adiabatic surfaces. These results indicate that nonadiabatic transition from the originally excited $2A_1(^3Q_0)$ surface to the $3E(^1Q_1)$ surface takes place principally at the intersection region. The trajectory which "jumped the surface" in this region will receive a torsional force immediately and its torsional motion will not be disturbed as the system dissociates, because the lower 3E surfaces are now very flat after the intersection region. This will result in the rotational excitation of CH_3 produced in the ground state $I(^2P_{3/2})$ channel, with the angular momentum perpendicular to the dissociation axis. On the other hand, a trajectory which stayed on the $2A_1$ surface will see after the intersection region a potential function which has a minimum at $\theta=0°$ with a rather large bending force constant. This will make the CH_3 product in the $I^*(^2P_{1/2})$ channel rotationally cool.

4. Potential Energy Surface Characteristics and Mode-Selective Reactivity [15]

Mode-selective reactivity has been a subject of considerable interest due to recent developments of laser techniques. It is possible to prepare molecules in specific initial internal state and to study the detailed effect of internal energy and its role on chemical reactions.

For triatomic systems, one can investigate theoretically the detailed effect of internal energy based on dynamics calculations, based on classical, semiclassical or quantum

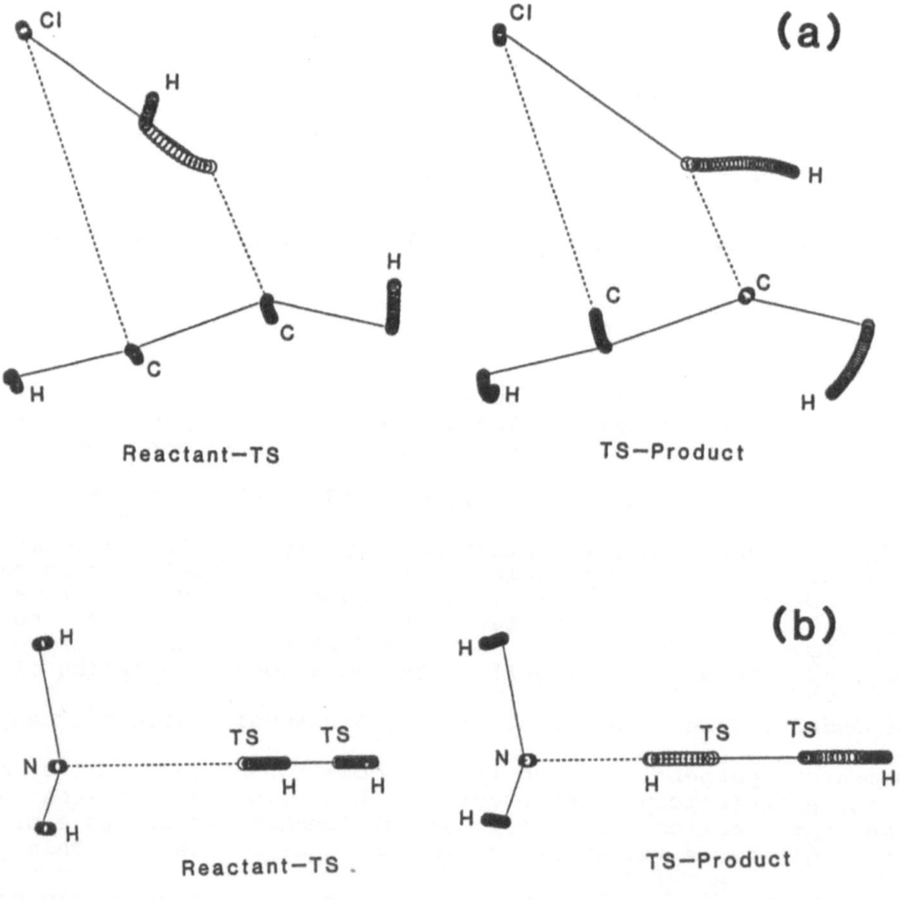

Figure 11. Geometrical change along IRC for (a) R1 and (b) R2.

scattering theory, if the PES is provided. For polyatomic reactions, Truhlar and Isaacson [16] have applied the variational transition-state theory for theoretical estimation of the effects of reagent vibrational excitation. They propose that the generalized transition state on the vibrationally adiabatic effective PES as "dynamical bottleneck" of reactions. However, how much a reaction is vibrationally adiabatic depends crucially on the location of the generalized transition state and the location of large reaction coordinate curvature. If the region of large curvature or large V-T coupling is locates earlier the barrier, then vibrational adiabacity is not a good approximation. We discuss this problem for the following reactions using the reaction coordinate model.

$$HCl + C_2H_2 \rightarrow C_2H_3Cl \quad (R1)$$

$$NH_3^+ + H2 \rightarrow NH_4^+ + H \quad (R2)$$

The effects of vibrational excitation of a reactant have been investigated experimentally by exciting selectively the HCl and CC stretching mode in R1 [17] and the umbrella bending mode of NH_3^+ ion in R2 [18].

The structures of the reactants, the transition states, and the products are optimized and the intrinsic reaction coordinates (IRC) connecting them are traced by the analytical gradient method at the 6-31G*/RHF (for R1) and MP2 (Møller-Plesset) (for R2) levels. The vibrational frequency analyses [19] along IRC are carried out to obtain the curvature of IRC and the coupling elements among the vibrational modes perpendicular to IRC.

The energetics of R1 and R2 are summarized in Table 1. Both R1 and R2 are the exothermic reactions. In R2, the inclusion of correlation effect is essential. Fig. 11 shows the geometrical change along the reaction coordinate for R1 and R2. In R1, the heavy chlorine atom moves little and the component of IRC is mainly comes from the motion of the hydrogen atom breaking the H-Cl bond and forming a new C-H bond. In R2, the motion of two hydrogen atoms of the reactant H_2 is a main component of IRC and the contribution from the part of NH_3 is not significant.

TABLE. 1. Energetics relative to the reactants of the reactions R1 and R2 (in kcal/mol).

		6-31G*/HF	6-31G*/MP2
R1	Transition State	50.99	52.66
	Product	-35.07	-36.20
R2	Transition State	14.33	4.59
	Product	-18.16	-31.83

Fig. 12 shows the curvature change along IRC. In R1 the curvature has a maximum at the reactant side (s=-1.4) due to the coupling between IRC and the mode Q_5 which relates to the HCl

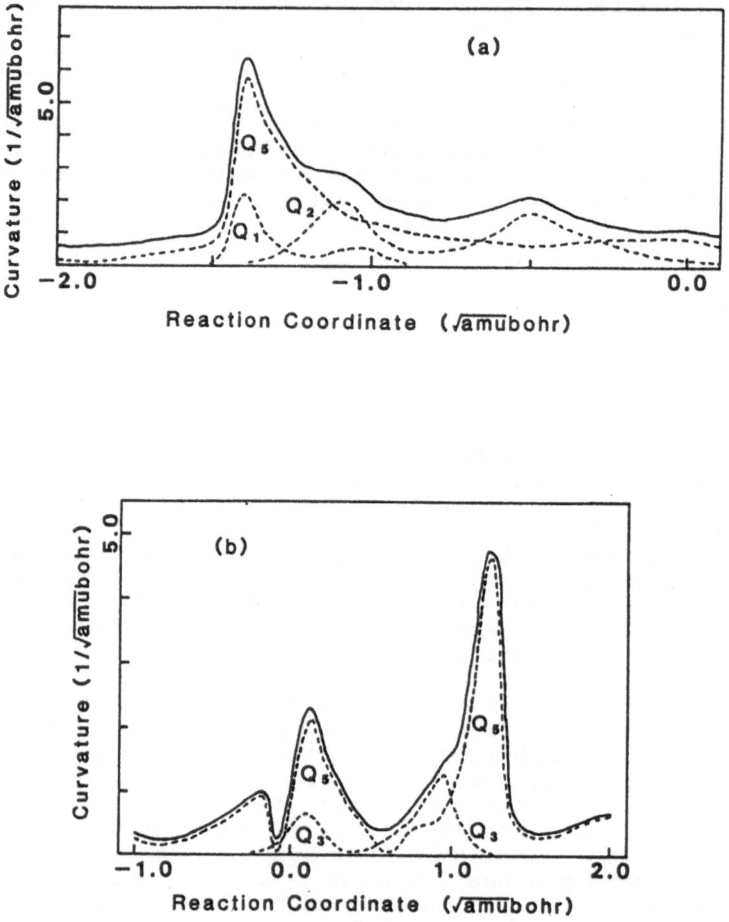

Figure 12. Curvature and its normal coordinate components of IRC for (a) R1 and (b) R2.

stretching mode at the reactant. Therefore vibrational adiabaticity of the mode does not hold at this region and the nonadiabatic energy transfer from Q_5 to the reaction coordinate is expected to occur. This indicates that R1 may be enhanced by the excitation of this HCl vibration at the reactant. However, a simple strong-coupling and statistical argument suggests that a substantial excess energy has to be provided in the Q_5 mode or in the translation before vibrational relaxation. On the other hand, in R2, the curvature has a maximum at around the transition state(s=0.0) and another on the product side (s=1.2). The modes Q_5 and Q_3, the main components of curvature, are the modes which correlate respectively to the H_2 stretching mode and the NH_3 umbrella mode at the reactant. Therefore in R1, the approximation of the vibrational adiabacity should hold up to the transition state. An analysis of the effective potential curve based on the vibrational adiabacity predicts a vibrational enhancement of the reactivity due to a significant change in the H_2 vibrational frequency. The NH_3 umbrella motion has a small effect.

We suggest qualitatively two different cases where the vibrational enhancement might be expected; (1) nonadiabatic vibrational energy transfer to the reaction coordinate, (2) characteristic change in vibrational frequency along the reaction coordinate. A question of which situation is more effective experimentally is reserved for further study. We are currently studying the dynamics of these systems classically, by reducing the degrees of freedom based on the reaction coordinate model.

5. An Implementation of Spin-Orbit Direct CI Method Based on the Spin-Dependent UGA [20]

In recent years, there has been an increasing interest in the inclusion of relativistic effects for molecules containing heavy atoms. One of the most practical yet reliable methods is to use relativistically derived effective core potentials. Major relativistic effects such as the Darwin and mass-velocity effects are easily taken into account in the form of a spin-free (SF) one-electron operator. The spin-orbit (SO) interaction is in general too strong to be considered as a small perturbation, and therefore should be treated explicitly as a part of the total Hamiltonian.

Although several computer programs have been implemented and developed in a similar line, little attention has been paid to the efficiency in actual computations, especially on vector-oriented computers. We have recently extended and reorganized the COLUMBUS direct CI program package to carry out large scale spin-orbit CI calculations. Numerical calculations reported in Section 3 have been carried out with this program.

In this method, we diagonalize the following Hamiltonian

$$H = H^{SF} + H^{SO} , \tag{1}$$

where $$H^{SF} = \sum_{i,j}^{n} \langle i|h|j\rangle E_{ij} + 1/2 \sum_{i,j,k,l} [ij;kl] e_{ij,kl} \quad (2)$$

and $$H^{SO} = \sum_{i,j,\sigma,\tau} \langle i_\sigma|h^{SO}|j_\tau\rangle E_{i_\sigma,j_\tau} \quad (3)$$

Here, $E_{i_\sigma,j_\tau}=a^+_{i_\sigma}a_{j_\tau}$ and $E_{ij}\ (=\sum_\sigma E_{i_\sigma,j_\sigma})$ are the infinitesimal generators of the unitary group U(2n) and U(n), respectively, and $e_{ij,kl}=E_{ij}E_{kl}-E_{il}\delta_{jk}$.

(a) Matrix elements of H in the spin-dependent UGA method. We use L-S coupled configuration state functions (CSFs) to diagonalize the spin-dependent Hamiltonian Eq.(1); in other words, we consider only in the subgroup U(n)×U(2) by separating out the space and spin parts of a system. In this manner, we can exploit the existing code with a minimum modification to obtain the contribution from the H^{SF} part to the so-called sigma vector,

$$\Delta\sigma^{SF}_{\alpha',S,Ms} = \sum_\alpha \langle\alpha',S,M_S|H^{SF}|\alpha,S,M_S\rangle C_{\alpha,S,Ms} \quad (4)$$

where α denotes the case number vector of each CSF.

Several forms of h^{SO} can be used in Eq.(3), for example,

"Empirical form" $$H^{SO} = \alpha^2/2 \sum_A^{Atom} (Z_A^{eff}/r_A^3)\vec{\ell}_A\cdot\vec{S} \quad (5)$$

"Ab-initio SO potential" $$H^{SO} = \sum_A^{Atom} \sum_{\ell_A} \xi_A(r_A)\vec{\ell}_A\cdot\vec{S}\sum|\ell_A,m\rangle\langle\ell_A,m| \quad (6)$$

The latter form was recently discussed by R. M. Pitzer et al. [21] in connection with the l- and j-dependent relativistic effective potentials. We proceed the discussion concentrating on a single term in (5) or (6) and expressing it as

$$f(\vec{r})\vec{\ell}\cdot\vec{S} = f(\vec{r}) \sum_{\gamma=-1}^{1} (-)^\gamma \ell_\gamma^{(1)} S_{-\gamma}^{(1)}$$

in the irreducible tensor operator form. Based on the pioneering works by Drake and Schlesinger [22], and Gould and Chandler [23], and using the segment values tabulated by Shavitt [24], we can readily derive the following compact expressions for the H^{SO} elements which are best suited to the actual coding.

$$\langle\alpha',S,M_S'|H^{SO}|\alpha,S,M_S\rangle=\langle a|f(\vec{r})\ell_{Ms-Ms'}^{(1)}|b\rangle(-)^{S-M_S}\begin{pmatrix} S & 1 & S \\ -M_S' & M_S'-M_S & M_S\end{pmatrix} \times\sqrt{(2S+1)/2}\ F_{ab} \quad (7)$$

$$\langle\alpha',S+1,M_s'|H^{SO}|\alpha,S,M_s\rangle=\langle a|f(\vec{r})\ell_{Ms-Ms'}^{(1)}|b\rangle(-)^{S-M_s}\begin{pmatrix}S & 1 & S+1\\ M_s & M_s'-M_s & -M_s'\end{pmatrix} \times\sqrt{S+1}\ \ F_{ab} \tag{8}$$

Here we assume that the bra and ket states have mismatching orbitals a and b, respectively ($1\leq a,b\leq n$), and F_{ab} are the segment value products defined as follows.

$$F_{ab}=\begin{cases} W_{\underline{R}}(a)\{\prod_{k=a+1}^{b-1} W_R(k)\}W_{R\underline{L}}^{(1)}(b)\{\prod_{k=b+1}^{n} W_{RL}^{(1)}(k)\} & (a<b) \quad (9)\\ W_{\underline{L}}(b)\{\prod_{k=b+1}^{a-1} W_L(k)\}W_{\underline{R}L}^{(1)}(a)\{\prod_{k=a+1}^{n} W_{RL}^{(1)}(k)\} & (a>b) \quad (10)\end{cases}$$

Here $W_A(1)$ is a one-body segment value for the i-level segment symbol A, and $W_{BC}^{(1)}(i)$ is a triplet coupled two-body segment value for the i-level segment symbol BC. We note here that Eqs.(9) and (10) are the same expressions as the segment value products required for the two-body SF operators $e_{a,n+1;n+1,b}$(we add one electron at the level n+1 in order to close the loop) except that there is no singlet coupled contribution $W_{BC}^{(0)}$. We also note that the segment value multiplications go from min(a,b) to the top level n showing a striking contrast to those for the one-body operator E_{ab}, in which the multiplications are just in the interval (a,b).

(b) <u>The use of the point group and time reversal symmetries.</u> These symmetries play a more important role in cutting down the computer time than in SF CI calculations. In an even number electron system with real MOs, the complex hermitian H matrix is changed to a completely real symmetric form by transforming the spin functions from the M_s eigenfunction form to the "real spherical" form and then multiplying a factor of i by the spin functions with S = 1,3,5,... . The real spherical form of the spin functions is also useful to make the CSFs symmetry adapted. In an odd number electron system, following the Wigner classification, we can divide all the Abelian point groups (which the program can handle) into three cases, (I) C_1,C_i ; (II) C_s,C_2,C_{2h} ; (III) C_{2v},D_2,D_{2h}, and devise the program to compute only one component of a Kramers pair.

(c) <u>External orbital treatment for the vectorization.</u> As in the original COLUMBUS CI program, we put all the external orbitals at the bottom of the DRT and treat them as symbolic orbitals. This feature is very important to vectorize the program using the DOT, AXPY, and MXM algorithms. In fact, we use the MXM algorithm for the treatment of two-external SO integrals in a similar fashion as for two-external two-electron integrals.

References

1. K. Yamashita and K. Morokuma, to be published.
2. P. R. Brooks, Chem. Rev. 88, 407 (1988).
3. K. Yamashita and K. Morokuma, J. Phys. Chem. 92, 3109 (1988).
4. T. C. Maguire, P. R. Brooks, R. F. Curl, J. H. Spence and S. J. Ulvick, J. Chem. Phys. 85, 844 (1986).
5. We used the program MELD, by E. R. Davidson, L. E. McMurcie, S. T. Elbert, S. R. Langhoff, D. Rawlings, and D. Feller, IMS Computer Center Library Program No. 030.
6. W. R. Wads and P. J. Hay, J. Chem. Phys. 82, 284 (1985).
7. T. F. George, J. Phys. Chem. 86, 10 (1982) and references therein.
8. J. R. Stine and J. T. Muckerman, J. Chem. Phys. 10, 3975 (1976).
9. J. N. Murrell, S. Carter, S. C. Farantos, P. Huxley and A. J. C. Varandas, in "Molecular Potential Energy Functions", (John Wiley & Sons, New York, 1984).
10. K. Yamashita, K. Morokuma, Y. Shiraishi, and I. Kusunoki, to be published.
11. I. Kusunoki and T. Ishikawa, J. Chem. Phys. 82, 4991 (1985).
12. S. Yabushita and K. Morokuma, to be published.
13. I. Powis and J. F. Black, private communication.
14. We have extended COLUMBUS program so that it can handle the spin-dependent Hamiltonian explicitly within the GUGACI algorithm.
15. K. Yamashita and K. Morokuma, to be published.
16. D. G. Truhlar and A. D. Isaacson, J. Chem. Phys. 77, 3516 (1982).
17. I. P. Herman and J. B. Marling, J. Chem. Phys. 71, 643 (1979).
18. R. J. S. Morrison, W. E. Conaway, T. Ebata, and R. N. Zare, J. Chem. Phys. 84 5527 (1986).
19. W. H. Miller, N. C. Handy, and J. E. Adams, J. Chem. Phys. 72, 99 (1980).
20. S. Yabushita and K. Morokuma, to be published.
21. R. M. Pitzer and N. W. Winter, J. Phys. Chem. 92, 3061 (1988).
22. G. W. F. Drake and M. Schlesinger, Phys. Rev. A15, 1990 (1977).
23. M. D. Gould and G. S. Chandler, Int. J. Quant. Chem. 26, 441 (1984). Note that these authors made several errors in obtaining the final equations which are corrected here.
24. I. Shavitt, in "The Unitary Group for the Evaluation of Electronic Energy Matrix Elements", J. Hinze, ed (Springer-Verlag, Berlin, 1981) 51.

CALCULATION AND CHARACTERIZATION OF REACTION VALLEYS FOR CHEMICAL REACTIONS

Thom H. Dunning, Jr., Lawrence B. Harding, and Elfi Kraka
Theoretical & Computational Chemistry Group
Chemistry Division, Argonne National Laboratory
Argonne, Illinois 60439 USA

ABSTRACT. The calculation and characterization of molecular potential energy surfaces for polyatomic molecules poses a daunting challenge even in the *Age of Supercomputers*. We have written a program, STEEP, which computes reaction paths (*IRC*s) for chemical reactions and characterizes the reaction valley centered on the *IRC*. This approach requires that only a swath of the potential surface be determined, a computationally tractable problem even for many-atom systems. We report *ab initio* reaction paths/valleys for two abstraction reactions: the OH + H_2 reaction, which is a simple, direct process; and the H + HCO reaction which can proceed along two distinct pathways, a direct pathway and an addition-elimination pathway. We find that the reaction path/valley method provides many insights into the detailed dynamics of chemical reactions.

1. Introduction

The concept of a reaction path which continuously describes the transform of reactants to products is firmly established in the *lore* of chemistry. A general definition of a reaction path for a polyatomic reaction has been advanced by Fukui (1), the *I*ntrinsic *R*eaction *C*oordinate (*IRC*) which is defined as the steepest descent path in mass-weighted cartesian coordinates from the transition state to reactants in one direction and products in the other (see also Ref. 2). The description of molecular potential energy surfaces and chemical reactions in terms of the *IRC* has been explored in detail by Fukui & coworkers (3) and many valuable insights into the mechanisms of chemical reactions have been gained (4). However, a detailed understanding of the dynamics of chemical reactions requires additional information on the molecular interactions. The potential energy surface, of course, embeds all of the information about a reacting system. Unfortunately, the potential surface for a molecular system is a function of (3N-6) variables, the calculation and representation of which provides a daunting challenge for all but the simpliest chemical systems even in the *Age of Supercomputers*.

A natural compromise is to characterize just the *reaction valley* leading from reactants through the transition state to products, *i.e.*, the (3N-7)-dimensional valley centered upon the reaction path. The idea of characterizing a reacting system by a coordinate set based on such a concept has frequently been used in chemistry; see, *e.g.*, the papers listed in Ref. 5. Miller, Handy & Adams (6) placed this approach on a firm theoretical foundation by deriving the nuclear Hamiltonian in a set of (3N-6) reaction path coordinates, namely, the *IRC* and the (3N-7) vibrational modes transverse to the reaction path. In this approach the reaction valley is

A. Laganà (ed.), Supercomputer Algorithms for Reactivity, Dynamics and Kinetics of Small Molecules, 57–71.

characterized by the path, energy profile and frequencies. In addition, there are dynamical terms which couple the translational and vibrational modes. From an analysis of all of the terms in the reaction path Hamiltonian which depend on the molecular potential energy surface, it is possible to provide a rationale for a number of the qualitative features of reaction dynamics.

2. Definition of the Reaction Valley

Let $x_s(s)$ denote the IRC, $\{Q_k(s)\}$ the transverse vibration modes and $(p_s(s), \{P_k(s)\})$ their conjugate momenta at the point s along the reaction path. In this coordinate system the molecular Hamiltonian $\mathcal{H}(x_s,p_s,\{Q_k\},\{P_k\})$ is given by:

$$\mathcal{H}(x_s,p_s,\{Q_k\},\{P_k\}) = \mathcal{T}(x_s,p_s,\{Q_k\},\{P_k\}) + \mathcal{V}(x_s,\{Q_k\}) \quad (1)$$

The potential energy at each point s on the reaction path is taken to be the classical energy at that point plus the potential energy for harmonic displacements perpendicular to the path (6):

$$\mathcal{V}(x_s,\{Q_k\}) = V_s(s) + \frac{1}{2}\sum_{k=1}^{3N-7} \omega_k(s)^2 Q_k(s)^2 \quad (2)$$

The kinetic energy is given exactly by (for zero total angular momentum) (6):

$$\mathcal{T}(x_s,p_s,\{Q_k\},\{P_k\}) = \frac{1}{2}\frac{[p_s - \sum_{k=1}^{3N-7} B_{kk'}(s)Q_k(s)P_{k'}(s)]^2}{[1 - \sum_{k=1}^{3N-7} B_{ks}(s)Q_k(s)]^2} + \frac{1}{2}\sum_{k=1}^{3N-7} P_k(s)^2 \quad (3)$$

There are two different types of terms in the Hamiltonian which explicitly depend on the potential energy surface:

(a) The shape terms: $x_s(s)$, $V_s(s)$ and $\{\omega_k(s)\}$. The shape of the reaction valley leading from reactants over the barrier to products is characterized by the reaction path, $x_s(s)$, which determines the meandering of the floor of the reaction valley in the (3N-6)-dimensional space, the classical potential energy along the reaction path, $V_s(s)$, which defines the steepness of the path, and the vibrational frequencies, $\{\omega_k(s)\}$, which define the width of the valley in the (3N-7)-dimensional subspace (low frequencies denote a wide valley, high frequencies a narrow valley).

(b) The coupling terms: $\{B_{ks}(s)\}$ and $\{B_{kk'}(s)\}$. These terms also play an important role in the description of the reaction dynamics. They specify the extent of the dynamical coupling between the translational and vibrational modes. It is through these terms that energy flows *non*-adiabatically from translation to vibration and vice-versa, $\{B_{ks}(s)\}$, and among the vibrations, $\{B_{kk'}(s)\}$.

Together these two types of terms fully define the (harmonic) reaction valley.

For the Li + HF reaction Dunning, Kraka & Eades (7) showed that the features of the reaction valley can be readily understood in terms of the changes in the electronic structure of the system as it evolves during the reaction. For the OH + H_2 reaction they showed that the terms in the reaction path Hamiltonian provide a rationale for many of the qualitative features of reaction dynamics, including such fine effects as the deposition of reactant vibrational excitation into product vibrational modes. The reaction valley approach thus provides a direct connection between the electronic structure of the system, the potential energy surface and the reaction dynamics.

3. Calculation of Reaction Valleys for Chemical Reactions

In order to determine the reaction valley the *IRC* first has to be calculated. Since the *IRC* is the steepest descent path it is the solution of the differential equation:

$$\frac{d\mathbf{x}_s(s)}{ds} = -\frac{\mathbf{g}}{g} \qquad (4)$$

where $\mathbf{g}$ is the gradient of the energy (g is its norm). Since gradients can be computed analytically for a variety of wave functions, including multireference CI wave functions (8), it would a first glance seem that Eq. (4) could be straightforwardly solved by the use of Euler-based methods or numerical predictor-corrector algorithms. Unfortunately, our experience and that of others (9) show that gradient-following techniques are very unstable, requiring extremely small step sizes if reliable frequencies and coupling terms are to be obtained. As such, it is not possible to solve Eq. (4) using gradients calculated by *ab initio* techniques, far too many gradient calculations would be required.

In order to cope with this problem a computer code for calculating reaction paths for polyatomic systems, STEEP, has been developed where the potential energy is defined in terms of a series of local force fields distributed along the reaction path and connected by switching functions. Each force field is expanded in internal coordinates and is determined from a grid of energies centered on a point near the reaction path; the expansions include up to quartic terms. In this representation gradients can be rapidly and economically calculated. Combining this with an integration technique developed for solving stiff differential equations (10), reaction paths for chemical reactions can be determined in a few minutes on a standard microcomputer. A minisupercomputer or supercomputer is, of course, needed to determine the series of local force fields.

For each point on the *IRC* STEEP calculates the gradient, the geometry and moments of inertia, the frequencies and normal modes for the vibrations perpendicular to the path, the distance from the saddle point (or other starting points), and the curvature and coriolis terms which describe the coupling between the path and the transverse vibrations. The use of the above representation of the potential energy surface leads to slight irregularities as a result of switching from one grid to a neighboring grid. For use in any dynamical studies these discontinuities would have to be suitably smoothed.

Recently, Jasien & Shepard (11) proposed a new approach for representing molecular potential energy surfaces. They expand the energy (and derivatives, if available) about a line which connects the reactants with the products through the saddle point. This line can be a simple interpolation between the geometries of the above points or it can be an approximate reaction path. This approach provides a smooth, continuous representation of the reaction valley from reactants through the transition state to products and will greatly facilitate the accurate calculation of reaction path parameters. Another promising approach has recently been put forward by Page & McIver (12). This method, based on a Taylor series expansion of the cartesian coordinate representation of the reaction path in terms of the reaction coordinate *s* appears to be far superior to techniques based solely on the gradient. However, second derivatives of the energy are required. This technique has been applied to a number of difficult reactions, *i.e.*, reactions with potential energy surfaces which are very flat in the region around the saddle point, by Koseki & Gordon with great success (13).

Truhlar & coworkers have recently made available a program, POLYRATE, which, among other things, calculates reaction paths for chemical reactions (14).

4. Reaction Valleys for Chemical Reactions: Two Examples

In this section we present two examples which illustrate the utility of the reaction valley model for describing the energetics, dynamics and mechanisms of chemical reactions. The first is a simple abstraction reaction with a single valley leading from reactants to products. Here we will focus on vibrational energy consumption and disposal in chemical reactions. The second is a not-so-simple abstraction reaction involving two radicals. For this reaction there are two valleys which lead to products: one based on a direct abstraction pathway, the other on an addition elimination pathway. Here we will focus on the relative features of the two valleys.

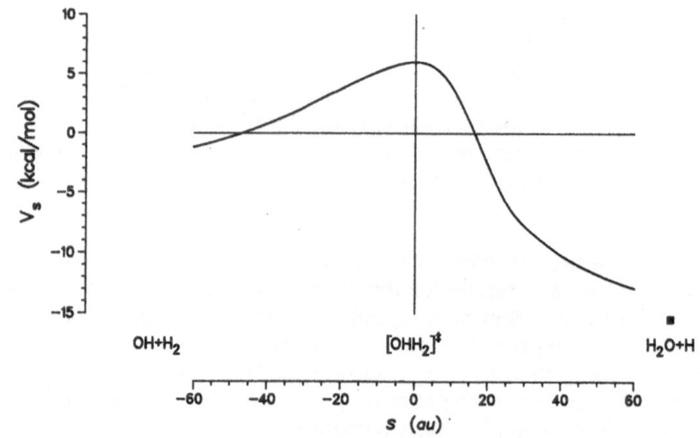

Figure 1. Calculated classical potential along the reaction path, $V_s(s)$, for the OH + H_2 reaction. The black square denotes the calculated energy defect of the reaction.

4.1. A SIMPLE ABSTRACTION REACTION: THE OH + H_2 REACTION

One polyatomic reaction which has received considerable attention is the reaction between hydroxyl radical and molecular hydrogen

$$OH + H_2 \rightarrow H_2O + H \qquad (5)$$

Calculations have been carried out on this system by Kraka & Dunning (15) (see also Ref. 16). These calculations used a polarized valence double zeta basis set and included all single and double excitations from the GVB wavefunction. The classical potential along the reaction path for reaction (5) is plotted in Figure 1. The height of the barrier is 6.0 kcal/mol. The calculated reaction energy defect is -15.5 kcal/mol which is just 0.6 kcal/mol above the experimental value (17). The OH + H_2 reaction is a simple hydrogen transfer reaction in which formation of the new bond occurs simultaneously with the breakage of the old bond and, thus, the barrier is expected to be only a small fraction of the H_2 bond energy (the bond being broken).

The hydroxyl radical can be considered a pseudo-halogen atom with the singly occupied π-orbital behaving much like the singly occupied p-orbital in the halogen atoms. The geometry of the saddle point is consistent with this, *i.e.*, the oxygen atom and the two hydrogen atoms in the H_2 moiety are nearly collinear ($\theta_{OHH} = 163°$). In addition, the saddle point is *early*, *i.e.*, the length of the OH bond at the saddle point relative to its length in the product (H_2O), 0.36 Å, is substantially larger than the change in the H_2 bond length from the reactants to the saddle point, 0.11 Å. Early saddle points are expected for an exoergic reaction.

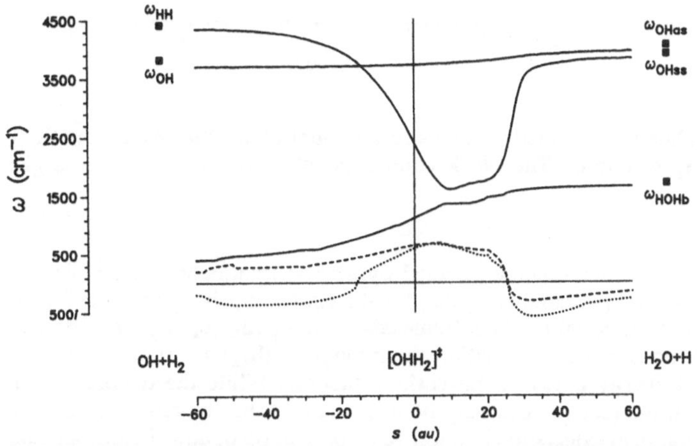

Figure 2. Calculated vibrational frequencies along the reaction path, $\{\omega_k(s)\}$, for the OH + H_2 reaction. The black square denotes the calculated frequencies in the reactants (H_2, OH) and products (H_2O).

The calculated frequencies for the OH + H_2 reaction are plotted in Figure 2. The vibrationally adiabatic potentials for the ground state, (n_{OH}=0, n_{H_2}=0), $V_s^{(0,0)}(s)$, and vibrationally excited

(n_{OH}=1, n_{H_2}=0) and (n_{OH}=0, n_{H_2}=1) states, $V_s^{(1,0)}(s)$ and $V_s^{(0,1)}(s)$, are plotted in Figure 3. We find that vibrational excitation of the OH molecule decreases the vibrationally adiabatic threshold by only 0.3 kcal/mol while excitation of the H_2 molecule decreases the threshold by 2.4 kcal/mol. Even in the latter case, vibrational excitation is relatively inefficient in overcoming the reaction barrier - the barrier is lowered by less than a quarter of the additional vibrational energy in the H_2. This is not unexpected given that the barrier to reaction is located in the entrance channel. These results are in line with the available experimental data. Zellner & Steinert (18) report a rate enhancement for $H_2(n_{H_2}=1)$ of a factor of 120±40 at 300°K while Glass & Chaturvedi (19) report a factor of 155±38. In contrast, Spencer, Endo & Glass (20) found that vibrational excitation of OH increased the rate the OH + H_2 reaction by less than a factor of two.

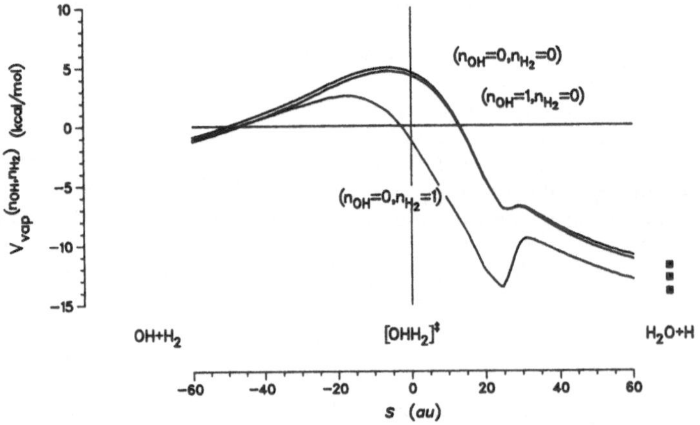

Figure 3. Calculated vibrationally adiabatic potential along the reaction path, $V_s^{(n_{OH},n_{H_2})}(s)$, for the OH + H_2 reaction. The black square denotes the calculated exoergicities of the reactions.

As noted in the previous section, a detailed dynamical treatment of the reaction requires, in addition to the classical potential and vibrational frequencies, the terms coupling motion along the path with the vibrational motions transverse to the path, $\{B_{ks}(s)\}$, and the terms coupling the transverse motions induced by motion along the path, $\{B_{kk'}(s)\}$. These terms govern the non-adiabatic flow of energy between the various modes. While the detailed influence of the coupling terms on the reaction can only be determined from dynamics calculations, valuable insights into vibrational effects in chemical reaction can be obtained from an examination of these terms alone. As can be seen in Figure 4, there is almost no coupling between the OH stretch and the reaction path in the entrance channel ($s < 0$). Thus, excitation of this bond will not enhance the rate of reaction - there is simply no way for the vibrational energy to flow into the reaction path. The OH bond is simply a *spectator* bond. The H_2 stretching mode, on the other hand, strongly couples with the reaction path in the entrance channel with $B_{HH,s}$ attaining its maximum value just before the saddle point. Energy can flow from the H_2 stretch into the reaction path via the $B_{HH,s}$ coupling term, and, hence, excitation of this bond has a significantly enhances the reaction rate. This is, of course, not unexpected as it is the H_2 bond which is broken in the reaction.

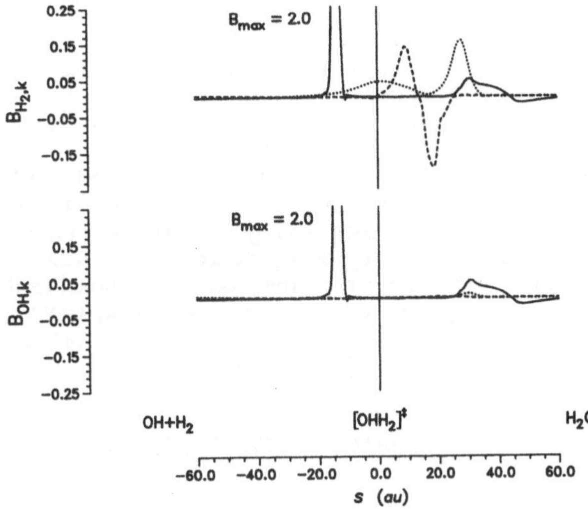

Figure 4. Calculated curvature and Coriolis coupling terms along the reaction path, $\{B_{ks}(s)\}$ and $\{B_{kk'}(s)\}$, for the OH + H_2 reaction.

Within the framework of the reaction path Hamiltonian approach, frequencies of the same symmetry do not cross. In the present case, the mode which correlates with the H_2 stretching mode drops in frequency as the saddle point is approached, while the mode which correlates with the OH stretching mode is largely unaffected. There is an avoided crossing of these two curves just before the saddle point. In Figure 4 we have also plotted the coriolis term coupling the HH and OH stretching modes. In the vicinity of the avoided crossing, this term is expected to be large and sharply peaked. In fact, in the entrance channel ($s < 0$), there is essentially no coupling between the OH and the H_2 stretch except near $s \approx -14$ au, the location of the avoided crossing. At that point there is a sharp peak in the $B_{OH,HH}$ term. This term will induce a transition between the two adiabatic levels as the crossing point is traversed. Hence, we conclude that it is more reasonable to assume that the OH and H_2 stretches preserve their character during the course of the reaction. Knowing this, we plotted the vibrational frequencies in Figure 2 as if the curves for $\omega_{OH}(s)$ and $\omega_{H_2}(s)$ do in fact cross. This same assumption was also used to compute the vibrationally adiabatic potentials plotted in Figure 3.

The influence of vibrational excitation on the product state distribution yields further insights into the dynamics of chemical reactions. In their quasiclassical trajectory study on an earlier semi-theoretical potential energy surface for OH + H_2 (21), Schatz & coworkers (22) found that for both vibrationally excited OH and H_2 the excess energy is deposited preferentially in the H_2O vibrational degrees of freedom. Essentially no energy is transferred to product translational motion and very little to the rotational degrees of freedom. The predicted distribution of the energy among the H_2O vibrational degrees of freedom was even more intriguing. For vibrationally excited H_2 Schatz *et al.* observed a nonspecific distribution of energy among the modes: with 42% going into the symmetric stretch mode, 35% into the asymmetric stretch mode and 22% into the bending mode. When the OH is initially excited, all of the excess energy goes into the OH stretching modes: 75% to the asymmetric stretch and 25% to the symmetric stretch. None of the excess energy flows into the bending mode.

While the trajectory calculations can determine the effect of vibrational excitation on the product state distribution, they do not provide a rationale for it. The reaction valley approach does so in a straightforward, conceptually appealing manner. The Coriolis coupling terms plotted in Figure 4 provide a framework in which to understand this observation. From Figure 2 we see that the H_2 mode evolves into the OH symmetric stretching mode (OH*ss*) in H_2O. In Figure 4 we see that there is a strong coupling between the ($H_2 \rightarrow$OH*ss*) mode and the OH*as* and HOH*b* modes in the exit channel. Thus, the excess energy in the H_2 mode can readily flow into both the stretching and bending modes in the water product. The net result is a nonspecific distribution of the excess vibrational energy among the product vibrational modes. For initial OH excitation, there is little coupling between the OH stretch and the HOH bending mode. The only important coupling is between the (OH$\rightarrow$OH*as*) and OH*ss* modes in the exit channel. In this case the excess energy is confined to the OH stretching modes in the products in line with the predictions of Schatz & coworkers (22).

4.2. A NOT-SO-SIMPLE ABSTRACTION REACTION: THE H + HCO REACTION

Reactions between radical species pose a number of challenges: not only are multiple surfaces often involved, but multiple pathways are likely on each surface. One such reaction that we have studied is that between atomic hydrogen and formyl radical. For this reaction two pathways lead to H_2 + CO:

$$H + HCO \rightarrow [H_2CO]^{\dagger} \rightarrow H_2 + CO \qquad (6)$$

$$H + HCO \rightarrow H_2 + CO \qquad (7)$$

The first pathway corresponds to an addition-elimination process, the second to a direct abstraction process.

The first of these reactions has been studied by a number of theorists. The energetics of the atomic decomposition channel ($H_2CO \rightarrow H + HCO$) in reaction (6) has been considered by Goddard & Schaefer (23) and by Palke & Kirtman (24). The former authors found that the reaction possessed no barrier beyond the endoergicity of the reaction. The latter authors carried out a thorough study of basis set effects on the threshold for reaction. They found that a large basis set was required in order to obtain an accurate reaction threshold. For the largest set considered, they reported a threshold of $85^1/_2$ kcal/mol (after correcting for zero point effects), just 1 kcal/mol below the most recent measurement (25).

The first reliable calculations on the $H_2CO \rightarrow H_2 + CO$ decomposition channel in reaction (6) were reported by Jaffe *et al* (26). They found a transition state having a planar, highly asymmetric structure in which both hydrogens are on the same side of the CO bond axis. Since then, a large number of calculations have been reported on this reaction examining the effects of electron correlation (27,28) and basis set convergence as well as higher order correlation corrections (29-32) on the transition state structure and/or reaction energetics. The latest heats of formation of H_2CO, CO and H_2 at 0° K (33,34) yield a $\Delta E(0\ K)$ for reaction (6) of -2.2 ± 0.15 kcal/mol. To compare to the calculations, this ΔE must be corrected for zero point energies (35,36). Applying these corrections gives $\Delta E_e = +5.0 \pm 0.2$ kcal/mol. The largest basis set calculations of Frisch *et al.* (31) predict $\Delta E_e = +3.2$ kcal/mol, implying a residual error of 1.8 kcal/mol.

The barrier height for reaction (6) is less accurately known from experiment. A recent review of the literature (37) concludes that the threshold for this reaction, including zero point effects, is approximately 84 kcal/mol. Using the frequencies of Frisch *et al.* (31) to correct for zero point energies, leads to an estimated experimental $\Delta E_e^{\dagger}$ of 89.5 kcal/mol. This result is in good accord with the results of several extensive calculations employing polarized double zeta basis sets, 91-92 kcal/mol (29,30). This agreement, though, is fortuitous since Frisch *et al.* (31) find that use of larger basis sets decreases the calculated barrier height to 86 kcal/mol. Most of this effect comes from expanding the valence s and p orbital basis set from a double zeta set to a triple zeta set. Dupuis *et al.* (30) have noticed a similar decrease in barrier height on expanding the basis set and have extrapolated a full CI-complete basis set barrier height of 86.5 ± 2.5 kcal/mol, slightly below the current experimental estimate.

One other important point to note here is that there are no low frequency vibrational modes at the transition state (30) for the molecular decomposition pathway - the lowest frequency is nearly 800 cm^{-1}. This is a direct result of the very ***constrained*** transition state for this pathway. The calculated high vibrational frequencies at the saddle point imply a narrow reaction channel at the transition state. The barrier is not only high but the peak is sharply pointed, *i.e.*, the reaction frequency is large, ω_s = 1840*i* cm^{-1} (Ref. 30).

The overall energetics of reactions (6) and (7) are summarized in Figure 5.

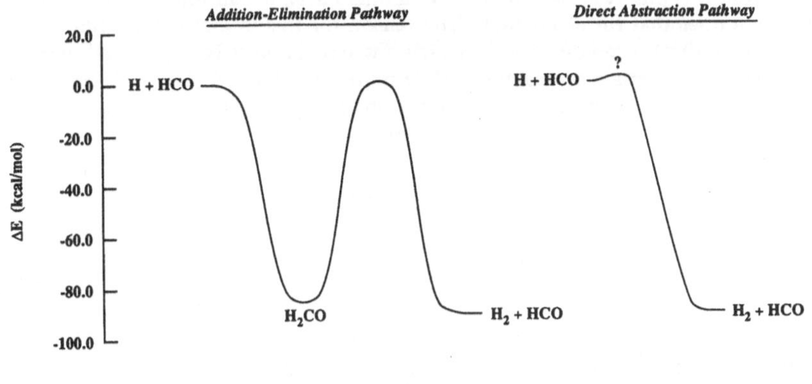

Figure 5. Calculated energetics for the addition-elimination and direct abstraction pathways in the H + HCO reaction.

Calculation of the reaction paths for both reactions (6) and (7) have been reported by Harding & Wagner (38). These calculations employed a polarized double zeta basis set and a four orbital - four electron CASSCF wavefunction. This wavefunction allows for the correlation of both CH bond pairs of formaldehyde and thus provides a qualitatively correct description of both reaction pathways. The angular dependence of the potential surface for a hydrogen atom moving in the field of a frozen formyl radical is shown in Figure 6. Two reactive channels are clearly seen in this plot. One, on the left side of the plot, leads to the addition product, formaldehyde, while the other, on the right side of the plot, leads directly to abstraction products, H_2 + CO. From this plot it can be seen that there is no barrier for either reaction.

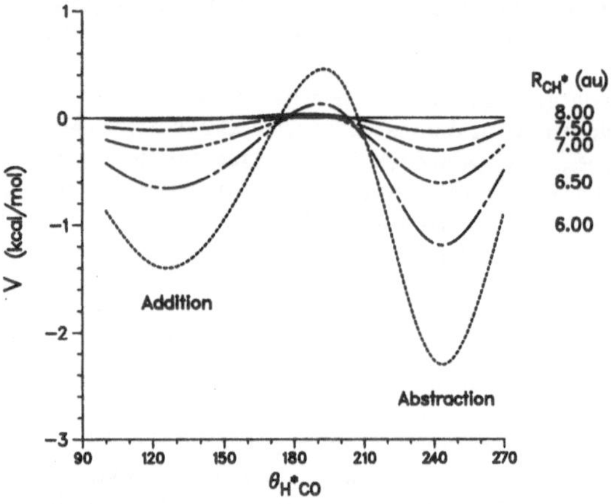

Figure 6. Calculated angular dependence of the potential energy surface for the attack of H on HCO. The formyl radical, HCO, has been frozen at its equilibrium geometry.

The energies and frequencies along the two reaction paths are plotted in Figures 7 & 8. In Figure 7, the classical energies along the reaction path are plotted as a function of the distance between the incoming hydrogen atom and the atom being attacked, *i.e.*, the carbon atom in reaction (6) and the hydrogen atom in reaction (7). It should be noted that the energies plotted here are from a steepest descent reaction path so all coordinates are changing, not just the one shown. From this plot it is clear that the potential curve for addition drops significantly more rapidly than does that for abstraction. The reason for this is that the addition reaction, (6), is a simple free radical recombination in which one new bond is being formed and no bonds are being broken. The abstraction reaction, (7), is more complex: a CH bond is being broken and a stronger, HH bond is being formed. It is this breaking of the CH bond that results in a flatter long range potential curve for the abstraction process.

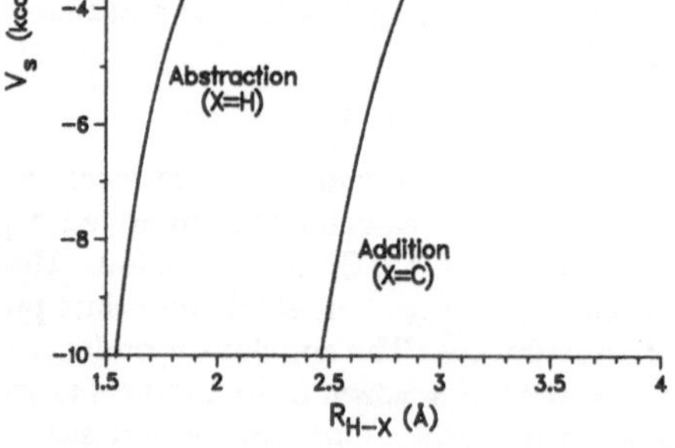

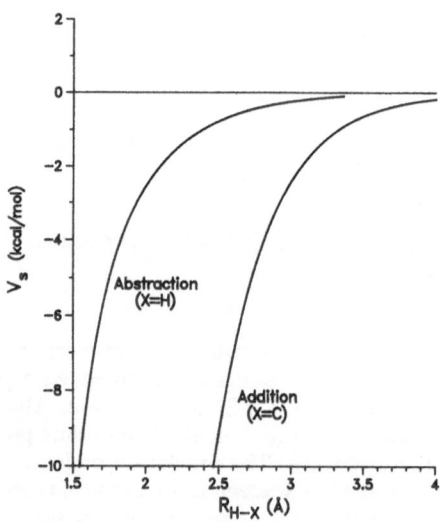

Figure 7. Calculated classical potentials along the reaction path, $V_s(s)$, for the H + HCO reaction.

The vibrational frequencies along the addition path are shown in Figure 8(a). Here it can be seen that the three modes of the formyl radical are only slightly affected by the approaching hydrogen atom, the most significant change being the increase in the CH stretching frequency. This is due to the strengthening of this "spectator" CH bond from just 15 kcal/mol in HCO to 86 kcal/mol in formaldehyde. The three new frequencies corresponding to motions of the attacking hydrogen are all seen to grow monotonically as the reaction progresses (in fact, it is found that they increase exponentially with decreasing distance).

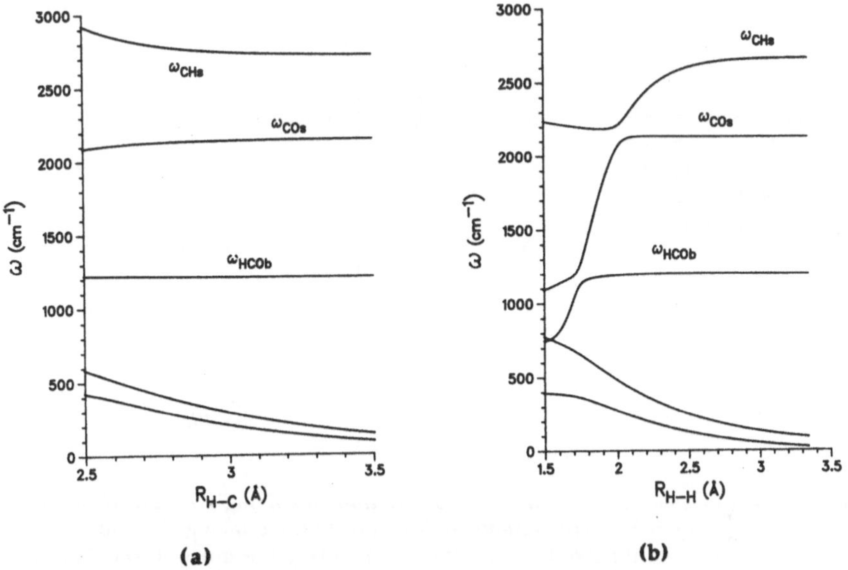

Figure 8. Calculated vibrational frequencies, $\{\omega_k(s)\}$, along the reaction paths for the (a) addition channel and (b) abstraction channel in the H + HCO reaction.

The frequencies variations along the abstraction path, Figure 8(b), are more complicated. At the reactant asymptote, the highest frequency corresponds to the CH stretch of the formyl radical. Since this bond is broken in the reaction, this frequency drops rapidly as the reaction progresses, crossing first the CO stretch and then the HCO bend. If this path were followed all the way to the product asymptote, this frequency would become zero as would the two frequencies which "grow-in" as the hydrogen atom approaches.

The calculated rate constants for the H + HCO reaction (38) are plotted in Figure 9, along with the measured (39,40). The rates for the direct abstraction and addition-elimination pathways are plotted separately in the figure. We find that the rate constant for this reaction is dominated by the direct abstraction pathway. At higher temperatures abstraction is more than twice as likely as addition-elimination. At room temperature the two rates are closer with the abstraction rate being about 20% larger than the addition-elimination rate. The small rate for the addition-elimination pathway is a direct result of the constrained transition state along the molecular elimination route, *i.e.*, while addition of the hydrogen atom to form H_2CO is rapid, the reaction valley leading back to reactants is much wider than that leading to products and so the complex eliminates a hydrogen atom (reforming reactants) rather than eliminating molecular hydrogen (to yield products). The increase in the addition-elimination rate relative to the abstraction rate at low temperature reflects the increasing importance of tunneling throught the sharply peaked elimination barrier.

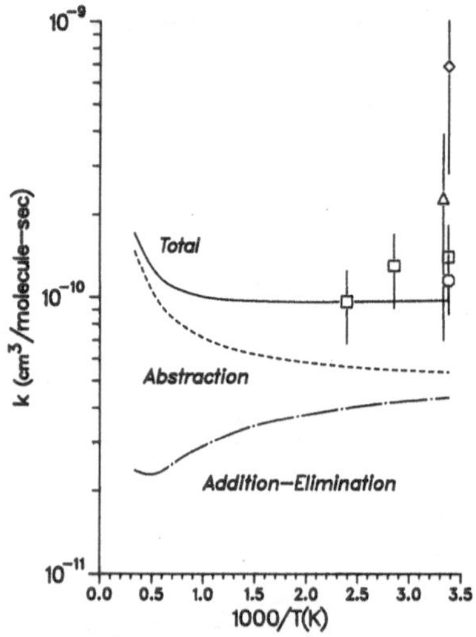

Figure 9. Calculated (curves) and experimental (polygons) rate constants for the H + HCO reaction. The diamond, triangle and circle are from Ref 39; the squares are from Ref. 40.

Experimental measurements of the rates of reactions involving two radical species, each of which must be cleanly formed and monitored, is a difficult undertaking. It is not surprising then that the calculated rates for the direct abstraction pathway are much lower than the earliest experimental result (39a). However, the two most recent experimental measurements confirm the essential correctness of the predicted rate constant. Finally, it should be noted that, at the time the first calculations were carried out, there were no measurements of the temperature dependence of the rate of this reaction. The calculations indicated that it was only weakly dependent on the temperature, a prediction subsequently borne out by the measurements of Gutman *et al.* (40).

ACKNOWLEDGEMENT

This work was supported by the Division of Chemical Science, Office of Basic Energy Research, U.S. Department of Energy under contract W-31-109-ENG-38

REFERENCES

1. Fukui, K., 1970, *J. Phys. Chem.* **74**, pp. 4161-4163; Fukui, K., 1979, *Recl. Trav. Chim. Pays-Bas* **98**, 75-77; Fukui, K., 1981, *Accts. Chem. Res.* **14**, pp. 363-368.

2. Eliason, M. A. and Hirschfelder, J. O., 1959, *J. Chem. Phys.* **30**, 1426-1436; Shavitt, I., 1959, Theoretical Chemistry Laboratory Report No. WIS-AEC-23 (University of Wisconsin, Madison, WI); Truhlar, D. G. and Kuppermann, A., 1971, *J. Am. Chem. Soc.* **93**, 1840-1851.

3. See, *e.g.*, Tachibana, A. and Fukui, K., 1978, *Theoret. Chim. Acta* **49**, pp. 321-347; Tachibana, T. and Fukui, K., 1980, *Theoret. Chim. Acta* **57**, pp. 81-94; Fukui, K., Tachibana, A. and Yamashita, K., 1981, *Intern. J. Quantum Chem.* **S15**, pp. 621-632; Fukui, K., *ibid.*, pp. 633-642.

4. See, *e.g.*, Fukui, K., Kato, S. and Fujimoto, H., 1975, *J. Am. Chem. Soc.* **97**, pp. 1-7; Ishida, K., Morokuma, K. and Komornicki, A., 1977, *J. Chem. Phys.* **66**, pp. 2153-2156; Kato, S. and Morokuma, K., 1980, *J. Chem. Phys.* **73**, pp. 3900-3914; Kato, S. and Morokuma, K., 1981, *J. Chem. Phys.* **74**, pp. 6285-6297; Yamashita, K., Yamabe, T. and Fukui, K., 1981, *Chem. Phys. Lett.* **84**, pp. 123-126; Yamashita, K., Yamabe, T. and Fukui, K., 1982, *Theoret. Chim. Acta* **60**, pp. 523-533; Yamashita, K. and Yamabe, T., 1983, *Intern. J. Quantum. Chem.* **S17**, pp. 177-189; Yamabe, T., Koizumi, M., Yamashita, K. and Tachibana, A., 1984, *J. Am. Chem. Soc.* **106**, pp. 2255-2260; Tachibana, A., Okazaki, I., Koizumi, M., Hori, K. and Yamabe, T., 1985, *J. Am. Chem. Soc.* **107**, pp. 1190-1196; Tachibana, A., Fueno, H. and Yamabe, T., 1986, *J. Am. Chem. Soc.* **108**, pp. 4346-4352.

5. See, *e.g.*, Hofacker, G. L., 1963, *Z. Naturforsch.* **18a**, pp. 607-619; Fischer, S. F., Hofacker, G. L. and Seiler, R., 1969, *J. Chem. Phys.* **51**, pp. 3951-3966; Hofacker, G. L. and Levine, R. D., 1971, *Chem. Phys. Lett.* **9**, pp. 617-620; Fischer, G. L. and Ratner, M. A., 1972, *J. Chem. Phys.* **57**, pp. 2769-2776; Marcus, R. A., 1966, *J. Chem. Phys.* **45**, pp. 4493-4499; 1966, *ibid.* **45**, 4500-4504; 1969 *ibid.* **49**, pp. 2610-2616.

6. Miller, W. H., Handy, N. C. and Adams, J. E., 1980, *J. Chem. Phys.* **72**, 99-112.

7. Dunning, Jr., T. H., Kraka, E. and Eades, R. A., 1987, *Faraday Discuss. Chem. Soc.* **84**, 427-440.

8. Shepard, R., 1987, *Intern. J. Quantum Chem.* **S31**, pp. 33-44; Shepard, R., Shavitt, I., Pitzer, R. M., Comeau, D. C., Pepper, M., Lischka, H., Szalay, P. G., Ahlrichs, R., Brown, F. B. and Zhao, J-G., *Intern. J. Quantum. Chem.* (to be published).

9. Garrett, B. C., Redmon, M. J., Steckler, R., Truhlar, D. G., Baldridge, K. K., Bartol, D., Schmidt, M. W. and Gordon, M. S., 1988, *J. Phys. Chem.* **92**, pp. 1476-1488.

10. Petzold, L. R. and Hindmarsch, A.C., *LSODA - Livermore Solver for Ordinary Differential Equations*, (Sandia National Laboratory, Livermore, CA, 1985).

11. Jasien, P. G. and Shepard R., 1988, *Intern. J. Quantum Chem.* (to be published).

12. Page, M. and McIver, J. W., Jr., 1988, *J. Chem. Phys.* **88**, pp. 922-935.

13. Koseki, S. and Gordon, M. S., *J. Phys. Chem.* (in press).

14. For a description of the program see: Isaacson, I., Truhlar, D. G., Rai, S. N., Steckler, R., Garrett, B. C. and Redmon, M. J., 1987, *Computer Phys. Comm.* **47**, 91-102; this program is available from the CPC Program Library, Queen's University of Belfast, Belfast, N. Ireland.

15. Kraka, E. and Dunning, Jr., T. H., (to be published)

16. Walch, S. P. and Dunning, Jr., T. H., 1980, *J. Chem. Phys.* **72**, pp. 1303-1311; Kochanski, E. and Flower, D. R., 1981, *Chem. Phys.* **57**, 217-225; Schlegel, H. B. and Sosa, C., 1988, *Chem. Phys. Lett.* **145**, pp. 329-333.

17. JANAF Thermochemical Tables, Natl. Stand. Ref. Data Ser., Natl. Bur. Stand. 37, 1970.

18. Zellner, R. and Steinert, W., 1981, *Chem. Phys. Lett.* **81**, pp. 568-572.

19. Glass, G. P. and Chaturvedi, B. K., 1981, *J. Chem. Phys.* **75**, pp. 2749-2752.

20. Spencer, J. E., Endo, H. and Glass, G. P., *16th Int. Symp. Combust.* (Combustion Institute, Pittsburgh, 1977), pp. 829-837.

21. Schatz, G. C., 1981, *J. Chem. Phys.* **74**, pp. 1133-1139.

22. Schatz, G. C. and Elgersma, H., 1980, *Chem. Phys. Lett.* **73**, pp. 21-25.

23. Goddard, J. D. and Schaefer, H. F., III, 1979, *J. Chem. Phys.* **70**, pp. 5117-5134.

24. Palke, W. E. and Kirtman, B., 1988, *Chem. Phys. Lett.* **148**, pp. 202-204.

25. Chuang, M. C., Foltz, M. F. and Mooré, C. B., 1987, *J. Chem. Phys.* **87**, pp. 3855-3864.

26. Jaffe, R. L., Hayes, D. M. and Morokuma, K., 1974, *J. Chem. Phys.* **60**, pp. 5108-5109; Jaffe, R. L. and Morokuma, K., 1976, *J. Chem. Phys.* **64**, pp. 4881-4886.

27. Harding, L. B., Schlegel, H. B., Krishnan, R. and Pople, J. A., 1980, *J. Phys. Chem.* **84**, pp. 3394-3401.

28. Goddard, J. D., Yamaguchi, Y. and Schaefer, H. F., III, 1981, *J. Chem. Phys.* **75**, pp. 3459-3465.

29. Frisch, M. J., Krishnan, R. and Pople, J. A., 1981, *J. Chem. Phys.* **85**, pp. 1467-1468.

30. Dupuis, M., Lester, W. A., Jr., Lengsfield, B. H., III and Liu, B., 1983, *J. Chem. Phys.* **79**, pp. 6167-6173.

31. Frisch, M. J., Binkley, J. S. and Schaefer, H. F., III, 1984, *J. Chem. Phys.* **81**, 1882-1893.

32. Adams, G. F., Bent, G. D., Bartlett, R. J. and Purvis, G. D., 1981, *J. Chem. Phys.* **75**, pp. 834-842.

33. Fletcher, R. A. and Pilcher, G., 1970, *Trans. Faraday Soc.* **66**, pp. 749-799.

34. Chase, M. W., Jr., Davies, C. A., Downey, J. R., Jr., Frurip, D. J., McDonald, R. A. and Syverud, A. N., 1985, *J. Phys. Chem. Ref. Data* **14**, Supplement No. 1.

35. B. Rosen, *Spectroscopic Data Relative to Diatomic Molecules*, (Pergamon, New York, 1970).

36. Romanowski, H. Bowman, J. M. and Harding, L. B., 1985, *J. Chem. Phys.* **82**, 4155-4165.

37. Moore, C. B. and Weisshaar, J. C., 1983, *Ann. Rev. Phys. Chem.* **34**, pp. 525-555.

38. Harding, L. B. and Wagner, A. F., *22nd Int. Symp. Combust.* (Combustion Institute, Pittsburgh, 1977), pp. 721-728.

39. (a) Reilly, J. P., Clark, J. H., Moore, C. B. and Pimentel G. C., 1978, *J. Chem. Phys.* **69**, 4381-4394; Hochanadel, C. J., Sworski, T. J. and Ogren, P. J., 1980, *J. Phys. Chem.* **84**, 231-235; (c) Nadtochenko, V. A., Sarkisov, O. M. and Vedeneev, V. I., 1978, *Dokl. Akad. Nauk SSSR* **243**, pp. 418; Nadtochenko, V. A., Sarkisov, O. M. and Vedeneev, V. I., 1979, *ibid.* **244**, pp. 152.

40. Timonen, R. S., Ratajczak, E. and Gutman, D., 1987, *J. Phys. Chem.* **91**, 692-694.

COMPUTED POTENTIAL ENERGY SURFACES FOR CHEMICAL REACTIONS

Stephen P. Walch
Eloret Institute
Sunnyvale, CA 94087
and
Celeste McMichael Rohlfing
Sandia National Laboratories
Livermore, CA 94551-0969

ABSTRACT. The results of multireference singles and doubles CI calculations of potential energy surfaces for hydrogen atom addition to O_2, N_2, and NO and recombination of OH + O are discussed. The errors due to the use of externally contracted CI and due to the neglect of correlation of O 2s and N 2s electrons are analyzed. Similarities and differences between the surfaces for the addition reactions are discussed. The calculated HN_2 addition surface is used in a simple dynamical treatment (one-dimensional tunneling through an Eckart barrier) to estimate the lifetime of the HN_2 species. The OH + O recombination potential is found to exhibit complex features which require that electrostatic forces (dipole-quadrupole) and chemical forces be treated consistently.

I. INTRODUCTION

H_2 combustion is of current interest to NASA because plans call for the National Aerospace Plane to be powered by an air-breathing, hydrogen-burning supersonic combustion ramjet (SCRAM jet) engine. The supersonic flow in the combustor region of the SCRAM jet leads to very short residence times and finite rate chemistry is critical to the design of the engine. This interest in H_2 combustion has led to the computation of a number of potential energy surfaces for reactions involving hydrogen and oxygen [1,2,3]. Among these are the reaction [1]

$$O(^1D) + H_2 \rightarrow H_2O^* \rightarrow OH + H. \tag{1}$$

Reaction (1) is of interest as a possible ignition process for hydrogen/oxygen mixtures, with the $O(^1D)$ being formed by photolysis of O_2. Potential surfaces have also been studied for several radical chain propagation reactions among which are the hydrogen abstraction reaction [2]

$$O(^3P) + H_2 \rightarrow OH + H \tag{2}$$

and the very important chain branching reaction [3]

$$H + O_2 \rightarrow HO_2^* \rightarrow HO + O. \tag{3}$$

A. Laganà (ed.), Supercomputer Algorithms for Reactivity, Dynamics and Kinetics of Small Molecules, 73–83.

Although there have been previous studies of reaction (3) [4], they focused on the $H + O_2$ region of the surface and a detailed examination of the $OH + O$ region was needed.

The presence of molecular N_2 also leads to important nitrogen chemistry in combustion. Among systems which have been studied are the $H + N_2$ reaction [5].

$$H + N_2 \rightarrow HN_2 \tag{4}$$

The initial interest in the HN_2 species at NASA centered around its possible role as an intermediate in the three-body recombination of H atoms in the presence of N_2. This and other three-body recombination reactions are critical to the design of the SCRAM jet engine since they are important heat producing reactions. The HN_2 species is also believed to be important in thermal de-NO_x processes. A key concern here is the lifetime of the HN_2 species. One previous calculation of the lifetime of HN_2 [6] was based on a potential energy surface derived with Møller-Plesset perturbation theory which is shown to break down severely for HN_2 [5].

The HNO species is also important in combustion and calculations have been carried out [7] for H atom addition to the N end of NO.

$$H + NO \rightarrow HNO \tag{5}$$

Here the potential is obtained for each of the lowest $^1A'$, $^3A''$, and $^1A''$ surfaces. There have been a number of previous ab initio studies of the near equilibrium geometry region of these three surfaces [8]. Nomura et al. [9] and Adams et al. [10] have also considered the large r_{NH} region. These calculations were CASSCF with a small basis set and UHF/MBPT, respectively, and are not expected to be of high accuracy, but they do indicate the existence of barriers on the $^1A''$ and $^3A''$ surfaces. Previous dynamical studies [11] have assumed that the barriers to H atom addition are negligible on all three of these surfaces.

High temperature air chemistry is an additional area in which molecular dynamics data is required in order to describe accurately the shock layers accompanying hypersonic vehicles. Two reactions which are important here are the exchange reactions

$$N + O_2 \rightarrow NO + O \tag{6}$$

and

$$O + N_2 \rightarrow NO + N \tag{7}$$

An analytic representation of the computed potential energy surface for reaction (6) was used in trajectory calculations which are discussed by Jaffe, Pattengill, and Schwenke in another article in this volume.

II. QUALITATIVE FEATURES

Localized orbital descriptions of the ground states of O_2, N_2, and NO are given in Fig. 1 as Eqn.(8), Eqn.(9), and Eqn.(10), respectively. Eqn.(8) is actually only one of two equivalent structures and resonance interactions between these structures lead to a $^3\Sigma_g^-$ ground state for O_2. The resonance interaction is significant for reaction (3) because the localization of the O_2 orbitals to a single structure, as needed to form an OH bond, may lead to a small barrier if the forming OH bond is not strong enough to compensate for loss of resonance energy. Combining

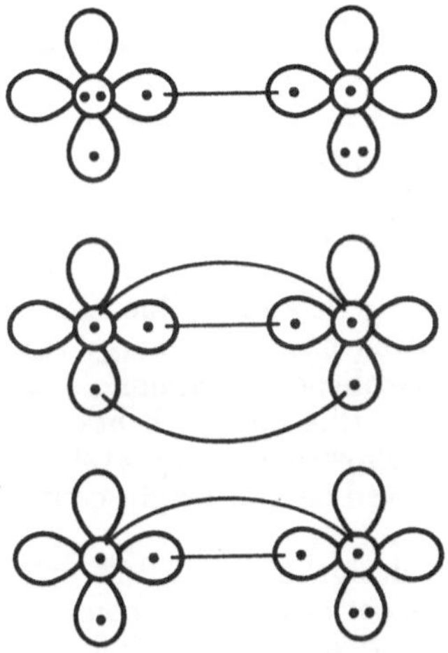

(8)

(9)

(10)

Fig. 1: Localized orbital description for O_2, N_2, and NO. The electronic structures of O_2, N_2, and NO are given by Eqn.(8), Eqn.(9), and Eqn.(10), respectively.

Eqn.(8) with a H atom leads to surfaces of ${}^{2,4}A''$ symmetry, but only the ${}^{2}A''$ surface connects to the ground state of O_2 and to the OH + O products. There is a low-lying ${}^{2}A'$ surface which correlates with the ${}^{1}\Delta$ state of O_2. Population of this surface in O + OH recombination could lead to chemiluminescence. From Eqn.(9) it is seen that N_2 is a triple bond and one expects a barrier to addition of H. Only one surface ${}^{2}A'$ connects H + N_2 to HN_2. From Eqn.(10) it is seen that NO has one two-electron π bond and one three-electron π bond. Adding H to the three-electron π bond leads to formation of the ground state of HNO (${}^{1}A'$). This process is expected to involve no barrier for formation of HNO. Adding H to the two-electron π bond of NO leads to ${}^{1,3}A''$ states. These surfaces are expected to involve significant barriers to H atom addition by analogy to HN_2. From Eqn.(10) it is evident that the A'' surfaces may also lead to the HON isomer, although this seems less favorable for the ${}^{1}A'$ surface. Recombination of HN + O or HO + N may populate the excited surfaces, leading to chemiluminescence.

III. CALCULATIONAL DETAILS.

The basis sets used in the present study are atomic natural orbital (ANO) basis sets[13]. These basis sets are optimal for describing the atomic correlation and thus have very small basis set superposition errors, but are sufficiently flexible to be used in molecular calculations at both the SCF and CI level. For most of the calculations described here the O and N basis sets were (13s8p6d4f)/[4s3p2d1f] and are described in detail in Ref. 13. The H basis set was (8s6p4d)/[3s2p1d] and is that developed by Almlöf and Taylor[13]. While these basis sets are expected to describe molecular potential curves accurately, additional flexibility is required to describe one-electron properties to a high degree of precision. As discussed elsewhere [13], a suitable basis set for properties may be obtained by uncontracting the most diffuse function of each group of functions, leading to a [5s4p3d2f/4s3p2d] basis set. In the present work, this larger basis set was used to calibrate the long range OH + O interaction. A different [5s4p3d2f/4s3p2d] basis set was used to calibrate the computed HN_2 potential. This basis set is obtained from the same primitive set but with additional atomic natural orbitals. The larger HN_2 basis set thus adds primarily tight functions designed to improve the description of electron correlation, while the larger HO_2 basis set adds more diffuse character in order to improve the description of properties.

The calculations are CASSCF/multireference single and double excitation CI (MRCI). The use of the MRCI method for computing potential energy surfaces is supported by full CI studies as discussed by Bauschlicher, Langhoff, and Taylor in another article in this volume. The accuracy of this approach has also been demonstrated for reaction (2) by Walch [2], who showed that by systematically expanding the active space and basis set, the computed barrier height converged to the currently accepted value [14].

Aside from the choice of the basis set, the design of an MRCI calculation requires the choice of a CASSCF active space and a set of reference configurations. In general the reference configurations are selected in two ways. In the first method, denoted as second order CI, all the configurations in the CASSCF are included as reference configurations. In the second method, denoted as selected references, occupations with CI coefficients in the CASSCF wavefunction greater than a cutoff value (typically 0.05) are included. In the present work the reference occupations include all unique occupations which have coefficients greater than the threshhold in any region of the surface. The latter constraint requires a preliminary set of CASSCF calculations to determine the reference occupations. The second order CI method eliminates this problem but often leads to CI expansions which are too large for present computational capabilities. In the present work the size of the second order CI expansions were often reduced by not correlating the O 2s and/or N 2s electrons.

The active orbitals in these calculations in general were those derived from the H 1s and O and/or N 2p orbitals. In the H + O_2 case the O 1s and O 2s electrons were not correlated. The justification for this approximation comes from calculations [2] on $O(^3P)$ + H_2 where the barrier height for abstraction was found to be the same whether or not the O 2s electrons were correlated, at the level of basis set and correlation treatment in the present calculations. The remaining nine electrons were correlated at the CASSCF and CI levels. The CASSCF active space here included five a′ and two a″ orbitals (denoted (52)). As discussed in detail elsewhere [3], the calculation was actually carried out using orbitals derived from a (41) CASSCF calculation but with a CI which was second order with respect to a (52) CASSCF wavefunction. The second order CI described above was carried out using the externally contracted CI method(CCI) of Siegbahn [15]. Comparisons were also made to uncontracted CI results along the minimum energy path determined at the CCI level.

For H + N_2 only the N 1s electrons were uncorrelated. The active space in the CASSCF was (62) and included one N 2s-like orbital. The inclusion of this orbital in the active space led to better convergence of the CASSCF process but the 4a′ orbital had an occupation number near two so the CASSCF is essentially equivalent to a (52) CASSCF calculation. MRCI was carried out using a selected list of reference configurations.

The initial set of calculations for HNO consisted of a second order CCI correlating eight electrons for a (52) CASSCF active space. Additional calculations were also carried out, using selected reference configurations, in which ten and twelve electrons were correlated (N 2s and O 2s).

The calculations were carried out on the NASA Ames Cray X-MP/48, on the Sandia Cray X-MP/416, and on the NASA NAS Facility Cray 2. These calculations used the MOLECULE[16]-SWEDEN[17] system of programs.

IV. DISCUSSION.

Considering first the H + O_2 reaction, we examine the effect of the approximations which were made in the potential surface calculation on the OH and O_2 spectroscopic constants listed in Table I. The CCI+Q and CI+Q results are CCI and CI results plus the multireference Davidson's correction for the case where the O 2s electrons were not correlated, as was done in the H + O_2 calculations, while the CCI(2s)+Q also includes correlation of the O 2s electrons at the CCI level. From Table I it is seen that the agreement with experiment is quite good for r_e and ω_e (maximum errors of 0.006 a_0 and 78 cm^{-1}), but is poorer for D_0. Looking first at the CCI+Q values, the error in D_0 is 0.20 eV (4.52 kcal/mole) for O_2 and 0.05 eV (1.15 kcal/mole) for OH. Part of this error is due to use of the CCI method ; the error in D_0 for O_2 is reduced by about half to 0.09 eV (2.12 kcal/mole) for CI+Q. For OH, the MRCI result for D_0 is 0.01 eV (0.25 kcal/mole) larger than experiment. This arises from not correlating the O 2s electrons in these calculations. Correlating the O 2s electrons decreases the D_e for O_2 by 1.34 kcal/mole more than for OH leading to a larger error in the exoergicity. From the errors in D_0, the error in the heat of reaction is 3.37 kcal/mole for CCI+Q and 1.87 kcal/mole for CI+Q, with OH + O too low with respect to H + O_2. About 1.0 kcal/mole of this error is a basis set defect, since the H basis set is based on H_2 natural orbitals, which are not fully optimal for describing a free H atom. Thus, correcting for the basis set error, the exoergicity is within 1.0 kcal/mole of experiment for CI+Q, and the errors due to use of the CCI method and not correlating the O 2s electrons are in the range of $\approx$ 1.0 kcal/mole. Finally, from Table I it is seen that the POL-CI calculations for O_2 give a longer r_e (error of 0.055 a_0), a smaller D_e (error of 0.11 eV compared to the present CI+Q results) and slightly smaller ω_e. These results are typical of POL-CI compared to MRCI results.

Table II shows the computed potential surface properties for H + O_2 in the reactant, entrance channel saddle point, and HO_2 minimum regions. A more detailed discussion of this potential is given in Ref. 3. In that paper, the variation of r_{OO} and H-O-O angle (θ) with r_{OH} is discussed. It is observed that the H atom initially approaches the O_2 molecule at large ($\approx$ 115 °) values of θ, and that the angle gradually decreases to $\approx$ 105 ° while r_{OO} gradually increases as the HO bond forms. The minimum energy path (MEP) found in the present studies is very similar to what was found in an earlier POL-CI study[4]. The main difference between the CCI results and the POL-CI results are that all the bond lengths are shorter and the well depths are larger in the present studies, as was also seen in the comparison of the properties of the O_2 molecule given in Table I.

From Table II it is seen that, at the MRCI level, the present calculations show essentially no barrier to H atom addition (computed barrier 0.1 kcal/mole), in agreement with the conclusion drawn from the POL-CI study. The latter study actually gave a barrier of $\approx$ 0.4 kcal/mole in agreement with the CCI results in the present study. From Table I and Table III the vibrational zero-point correction is +0.44 kcal/mole. However, the tunneling correction is of opposite sign and hence the threshhold may be nearly zero.

The computed HO_2 D_e with respect to the H + O_2 limit is 53.1 kcal/mole

Table I. Computed Spectroscopic Constants for O_2 and OH[a].

	O_2			OH		
	r_e	ω_e	D_0	r_e	ω_e	D_0
POL-CI[b]	2.336	1533	4.91			
CCI +Q	2.277	1575(1564[c])	4.920	1.833	3816(3840[c])	4.342
CCI(2s) +Q			4.72			4.20
CI +Q	2.281	1611(1604[c])	5.024	1.833	3798(3739[c])	4.403
Experiment	2.283	1580.19	5.116	1.833	3737.76	4.392

[a] r_e in a_0, ω_e in cm^{-1}, and D_0 in eV.
[b] From Ref. 4.
[c] From a quadratic fit to the potential, whereas, the other values result from a cubic fit to the potential.

Table II. Computed Potential Surface Properties for Reaction (1)[a]

	POL-CI[c]	CCI	CI	Exp.
ΔE_b	0.4	0.5	0.1	< 0.2
D_e(H-O_2)	42.1	53.1	53.9	53.7-55.2
r_{OH}(HOO)	1.873	1.840		1.846
r_{OO}(HOO)	2.588	2.520		2.523
θ(HOO)	103.3°	104.4°		104.1°
ω_1	3655.	3531.		3414.
ω_2	1457.	1417.		1389.
ω_3	1181.	1220.		1101.
D_e(HO-O)		63.2		68.3 ± 0.6
ΔE^b_{rx}		10.2	11.2	13.6

[a] Energies are in kcal/mole, geometries are in a_0, and vibrational frequencies are in cm^{-1}.
[b] Exoergicity relative to H + O_2.
[c] Ref. 4.

for CCI and 53.9 kcal/mole for MRCI as compared to 53.7-55.2 kcal/mole from experiment [18] and 42.1 kcal/mole from POL-CI [4]. The calculations seem to obtain a better HO_2 D_e than would be expected. This is consistent with the low heat of reaction, which suggests that the calculations describe the H + O_2 limit more poorly than the HO_2 molecule. From Table II it is seen that the HO_2 computed geometry is in very good agreement with experiment. The vibrational frequencies have somewhat larger errors than for the diatomics (117, 28, and 119 cm^{-1} for ω_1, ω_2, and ω_3, respectively). From Tables I and II it is seen that the OO stretch frequency (ω_3 in Table II) decreases by 344 cm^{-1} as compared to free O_2.

We now consider the OH + O region of the potential. From simple valence bond considerations it would be expected that as r_{OO} increases θ should decrease from the equilibrium value for HO_2 ($\approx$ 104°) to $\approx$ 90°. However, the calculations show that θ decreases slowly to values less than 90 ° and decreases sharply to 0 ° for r_{OO} greater than 5.5 a_0. This behavior is consistent with a curve crossing between chemical bonding interactions which favor $\theta \approx 90$ ° at short r_{OO}, and electrostatic (dipole-quadrupole) interactions which favor a collinear OH-O geometry at large r_{OO}. The shape of the surface in this region is shown in Fig. 2. From Fig. 2 and Table III it is seen that the surface has a minimum for $r_{OO} \approx 6.0$ a_0 and two symmetrically located saddle points (leading to HO_2) for $r_{OO} \approx 5.5$ a_0. It is clear from Fig. 2 that the OH + O interaction in the intermediate separation region cannot be described by a purely electrostatic model, but rather electrostatic and chemical forces must be treated consistently. In the region of large r_{OO} the dipole-quadrupole interaction should dominate the potential.

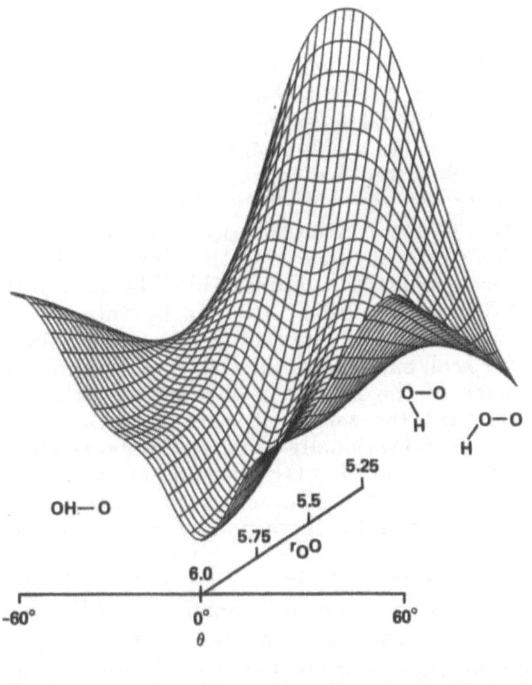

Fig. 2: Interaction potential for HO-O in the large r_{OO} region. For large r_{OO}, the electrostatic dipole-quadrupole interaction dominates leading to a collinear HO-O species, which is a minimum on the global surface. The O-O chemical bonding interaction becomes stronger as r_{OO} decreases, leading to a double minimum bending potential at intermediate r_{OO} and a single minimum for shorter r_{OO}. The surface has a two symmetrically located saddle points which lead to HO_2.

Table III. Stationary Points on the H + O_2 Surface.[a]

character	t.s.[b]	min.	t.s.[b]	min.
r_{OH}	4.135	1.840	1.839	1.839
r_{OO}	2.287	2.520	5.545	6.297
θ	116.4°	104.4°	50.3°	0.0°
ω_1	1515.	3531.	3807.	3780.
ω_2	354.	1417.	150.	224.
ω_3	412.*i*	1220.	149.*i*	96.
ΔE^c	0.5	-53.1	9.4	8.8

[a] Energies are in kcal/mole, geometries are in a_0, and vibrational frequencies are in cm^{-1}.
[b] A transition state (saddle point on the potential energy surface).
[c] Energy relative to H + O_2.

Additional calculations were carried out for the binding energy at the OH-O minimum using the [5s4p3d2f/4s3p2d] basis set. The resulting D_e is 1.5 kcal/mole for the [5s4p3d2f/4s3p2d] basis set compared to 1.2 kcal/mole for the [4s3p2d1f/3s2p1d] basis set. This result indicates that the computed potential energy surface is slightly too shallow in the electrostatic interaction region. This defect will probably not have significant consequences for the dynamics, since the bottleneck appears to occur at larger r_{OO} (vide infra).

The location of the bottleneck on the vibrationally adiabatic surface depends on zero-point vibrational effects. From Table III it is seen that including differential zero-point vibration effects, OH-O is bound by only 0.7 kcal/mole with respect to OH + O. Similarly including differential zero-point vibrational effects the OH-O saddle point is 0.1 kcal/mole above the OH-O minimum. Thus, zero-point vibration effects cancel much of the barrier in this region. For large r_{OO} the interaction energy is smaller than the zero-point energy in the bending mode leading to a small barrier on the vibrationally adiabatic potential for large r_{OO}. Tunneling should not be significant for O + OH recombination since both reactants are heavy and the potential is very flat. These results suggest that the bottleneck to OH + O recombination occurs for large r_{OO} where the interaction is predominantly electrostatic.

We now consider the calculations for HN_2. The features of the computed surface are summarized in Table IV. From Table IV it is seen that the potential for HN_2 has a shallow well $\approx$ 3 kcal/mole above the H + N_2 asymptote and separated from that asymptote by a barrier. The dissociation process is primarily controlled by quantum mechanical tunneling. Enough points on the HN_2 potential have been obtained to characterize the MEP. Approximate calculations (one-dimensional tunneling through an Eckart barrier) have been carried out to determine the rate of dissociation via tunneling. Additional electronic structure calculations with the [5s4p3d2f/4s3p2d] basis set were carried out at the stationary point geometries obtained with the smaller basis set. The larger basis set calculations indicate that the potential is converged to $\pm$ 1.0 kcal/mol. Based on this, the lifetime of HN_2 is estimated to be significantly smaller than that used in current models of the thermal de-NO_x process.

Table IV. Stationary Points on the H + N_2 Surface.[a]

character	min.	t.s.[b]	min.
r_{NH}	20.0	2.753	2.007
r_{NN}	2.095	2.146	2.262
θ		118.6°	116.3°
ω_1		1662i	2744
ω_2	2303	2072	1583
ω_3		771	1070
ΔE^c	0.0	15.16	3.01
zero point	3.29	4.06	7.72
$\Delta E_{corr.}$	0.0	15.93	7.44

[a] Energies are in kcal/mole, geometries are in a_0, and vibrational frequencies are in cm^{-1}.
[b] A transition state (saddle point on the potential energy surface).
[c] Energy relative to H + N_2.

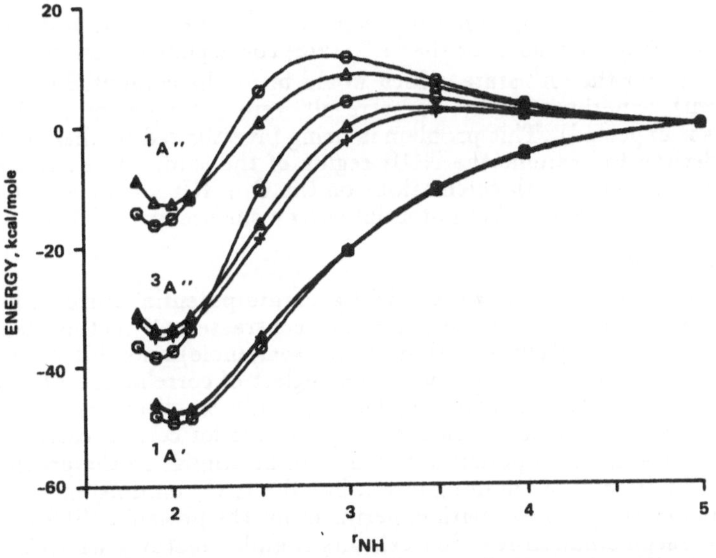

Fig. 3. The energy as a function of r_{NH} for H atom addition to the N end of NO for the lowest $^1A'$, $^3A''$, and $^1A''$ surfaces of HNO. The particular curves shown are for a cut in which r_{NO} is fixed at 2.3 a_0 and θ is fixed at 120 °. The curves with computed energies denoted by circles, triangles, and crosses are for CCI calculations in which eight, ten, and twelve electrons are correlated, respectively.

Fig. 3 shows one cut through the potential for H atom addition to the N end of NO for the $^1A'$, $^3A''$, and $^1A''$ surfaces with eight, ten, and twelve electrons correlated. The lowest surface ($^1A'$), which involves addition to the radical orbital of NO, has no barrier, while the other two surfaces ($^3A''$ and $^1A''$), which involve addition to the π bond of NO, have barriers. The minimum energy path for H atom addition leading to HNO has been characterized at the level of a second order CCI correlating eight electrons. The eight electron results give a good geometry for the ground state and a good excitation energy to the $^1A''$ state, but the geometry for the $^1A''$ state is in poor agreement with experiment(r_{NH} too short by $\approx 0.04\ a_0$). Fig. 3 shows the effect of correlating more electrons. For the $^1A'$ state the ten electron results are quite similar to the eight electron results, except the well depth is slightly ($\approx$ 2.0 kcal/mole) smaller, while the twelve electron results are indistinguishable from the ten electron results (maximum difference between the two potentials of $\approx$ 0.2 kcal/mole). Thus, for the ground state, correlation of the O 2s electrons has very little effect on the potential, while correlation of the N 2s electrons makes relatively small changes in the potential. For the excited states, on the other hand, there are large differences between the eight and ten electron results, i.e., correlation of the N 2s electrons makes significant changes in the potential, but the similarity of the ten and twelve electron results for the $^3A''$ state indicates that correlation of the O 2s electrons does not significantly change the potential, in agreement with the results for the exoergicity of the H + O_2 reaction discussed above. One encouraging result from Fig. 3 is that the minimum in r_{NH} along the cut given(which corresponds to varying r_{NH} with r_{NO} and θ fixed at 2.3 a_0 and 120 °, which are near the equilibrium values for the $^1A''$ state) corresponds to an increase in r_{NH} of about 0.04 a_0 for the $^1A''$ state, which would bring the computed r_{NH} into good agreement with experiment if the same result were obtained for a full geometry variation (as is expected). This problem is being investigated further. Calculations are also underway to examine the NOH region of the same three surfaces. These results will be combined with calculations on the O + NH and N + OH regions of the potential to produce a global potential energy surface for HNO [7].

V. CONCLUSIONS

The MRCI method is shown to yield accurate potential surfaces for the systems studied here. The use of the externally contracted CI method leads to relatively small errors(typically less than $\approx$ 1.0 kcal/mole) and this method is well suited to potential surface calculations. The neglect of correlation of the O 2s electrons (as in the H + O_2 calculations) leads to only small errors, but neglect of correlation of the N 2s electrons leads to larger errors for certain cases.

The H + O_2 addition potential is found to be similar to the previously published POL-CI potential of Dunning et al., but the H-O_2 binding energy and bond lengths are in better agreement with experiment for the present MRCI calculations. The OH + O recombination reaction exhibits complex features which require that electrostatic forces and chemical forces be treated equivalently.

Simple dynamical calculations based on the computed H + N_2 surface indicate a that the lifetime for the HN_2 species is significantly shorter than that used in current models of the thermal de-NO_x process.

ACKNOWLEDGEMENTS. S.P. Walch was supported by NASA Grant No. NCC2-478. The calculations on the CRAY 2 were made with a grant of computer time from the NAS facility. This work was also supported in part by the U. S. Department of Energy, Office of Basic Energy Sciences.

REFERENCES

1. S.P. Walch and L.B. Harding, J. Chem. Phys. **88**, 7653(1988).
2. S.P. Walch, J. Chem. Phys. **86**, 5670(1987).
3. S.P. Walch, C.M. Rohlfing, C.F. Melius, and C.W. Bauschlicher, Jr., J. Chem. Phys. **88**, 6273(1988).
4. T.H. Dunning,Jr, S.P. Walch, and M.M. Goodgame, J. Chem. Phys. **74**, 3482(1981) and references therein.
5. S.P. Walch, R.J. Duchovic, and C.M. Rohlfing, J. Chem. Phys. submitted.
6. L.A. Curtiss, D.L. Drapcho, and J.A. Pople, Chem. Phys. Lett. **103**, 437(1984).
7. S.P. Walch, C.M. Rohlfing, and C.F. Melius, J. Chem. Phys. to be published.
8. A. Heiberg and J. Almlöf, Chem. Phys. Lett. **85**, 542(1982) and references therein.
9. O. Nomura, S. Ikuta, and A. Iguwa, in "Applied Quantum Chemistry", edited V.H. Smith,Jr., H.F. Schaefer III, and K. Morokuma(Reidel, 1984), p 243.
10. G.F. Adams and G.D. Bent, U.S. Army Ballistic Research Laboratory Technical Report(BRL-TR-2737), June 1986.
11. M.C. Colton and G.C. Schatz, J. Chem. Phys. **83**, 3413(1985).
12. S.P. Walch and R.L. Jaffe, J. Chem. Phys. **86**, 6946(1987).
13. J. Almlöf and P.R. Taylor, J. Chem. Phys. **86**, 4070(1987).
14. T. Joseph, D.G. Truhlar, and B.C. Garrett, J. Chem. Phys. **88**, 6982(1988)
15. P.E.M. Siegbahn, Int. J. Quantum Chem. **23**, 1869(1983).
16. J. Almlöf, Molecule, a vectorized Gaussian integral program.
17. SWEDEN is a vectorized SCF-MCSCF-direct CI-conventional CI-CPF-MCPF program written by P.E.M. Siegbahn, C.W. Bauschlicher,Jr., B. Roos, P.R. Taylor, A. Heiberg, J. Almlöf, S.R. Langhoff, and D.P. Chong.
18. L.G.S. Shum and S.W. Benson, J. Phys. Chem. **87**, 3478(1983).

AN AB INITIO STUDY ON THE COORDINATION OF FORMALDEHYDE, CARBON DIOXIDE, DINITROGEN, AND RELATED MOLECULES TO IRON(0) AND NICKEL(0) FRAGMENTS

M. ROSI, A. SGAMELLOTTI, F. TARANTELLI,
Dipartimento di Chimica
Università di Perugia
Italy

C. FLORIANI,
ICMA
Universitè de Lausanne
Switzerland

ABSTRACT. 'Ab initio' calculations have been performed on the model systems $Fe(CO)_n(PH_3)_{4-n}$ $(\eta^2 - L)$ $L = CH_2O, CH_2S, CMe_2O$; n = 0, 2, 4; $Fe(CO)_2(PH_3)_2(\eta^2 - OCX)$, $Fe(CO)_2(PH_3)_2$ $(\eta^2 - SCX)$, $X = O, S, NH, CH_2$; $Fe(PH_3)_4(N_2)$, $Ni(PH_3)_2(L)$, $L = N_2$, N_2CH_2, N_2H_2; to investigate the nature and the energetics of the interaction between the metal and the unsaturated ligand. The results allow a complete description of the electronic structure of the model compounds: the bonding between the metallic fragment and the unsaturated molecule lies at the very extreme of the Chatt-Dewar-Duncanson model, being dominated by the π-back-donation. The coordinate bond of CH_2O with an iron(0) substrate and that of N_2 with a nickel(0) fragment are described in details, paying particular attention to the role of correlation effects.

1. Introduction

Fixation of carbon dioxide and formaldehyde in their intact form on a metal centre is a primary goal in metal-promoted transformations of a C_1 molecule, provided it forms metal-carbon bonds. Formation of formaldehyde and carbon dioxide complexes is, however, a quite rare reaction, in spite of the various strategies applied so far. On the contrary, coordination of dinitrogen has been found in a number of complexes: all the mononuclear compounds so far identified prefer the end-on bonding mode. Activation of dinitrogen, however, can be much more pronounced in case N_2 binds the metal in a side-on fashion.

We have recently started an extensive theoretical study on the nature of the interaction between transition metal fragments and unsaturated ligands, such as CH_2O, CO_2, N_2, and related molecules [1]. The purpose is to provide a theoretical ab initio interpretation of the coordinate bond between the transition metal (iron and nickel in the present study) and the unsaturated ligand in order to understand the nature and the energetics of the interaction of the ligand with the metal and the requirements for its activation.

We have analyzed the following systems:

A. Laganà (ed.), Supercomputer Algorithms for Reactivity, Dynamics and Kinetics of Small Molecules, 85–94.

$Fe(CO)_n(PH_3)_{4-n}(\eta^2 - L)$ $L = CH_2O, CH_2S, CMe_2O$; n = 0, 2, 4
$Fe(CO)_2(PH_3)_2(\eta^2 - OCX)$, $Fe(CO)_2(PH_3)_2(\eta^2 - SCX)$, $X = O, S, NH, CH_2$;
$Fe(PH_3)_4(N_2)$, $Ni(PH_3)_2(L)$, $L = N_2, N_2CH_2, N_2H_2$.

The results indicate that the bonding between the metallic fragment and the unsaturated molecule lies at the very extreme of the Chatt-Dewar-Duncanson model [2], being dominated by the π-back-donation. Dinitrogen prefers to bind to a low valent metallic fragment in a side-on fashion rather than in an end-on one, since the former mode implies a more pronounced π-back-donation than the latter.

In this report we will present a description of the coordinate bond of formaldehyde with an iron(0) substrate and that of dinitrogen with a nickel(0) fragment, paying particular attention to the role of the correlation effects.

2. Computation details

Two levels of theory have been employed in studying the ground states of the investigated species. 'Ab initio' spin restricted Hartree-Fock gradient calculations were used in partial geometry optimizations of the analyzed complexes and in deriving estimates of the binding energies of all complexes with respect to free unsaturated ligand and metallic fragment species. Configuration interaction calculations were subsequently performed on the $Fe(CO)_4(\eta^2 - CH_2O)$, $Ni(PH_3)_2(\eta^2 - N_2)$, $Ni(PH_3)_2(\sigma - N_2)$ complexes and associated fragments, including single and double excitations from the upper valence orbitals using the Direct-CI method [3]. All computations were performed using the GAMESS program package [4] implemented on FPS-164 and CRAY XMP/48 computers. The Gaussian basis sets employed are described in details in ref. [1]. Two different basis sets were employed for the iron systems. The first one, of double ζ quality for the iron and the main ligand and single ζ for the other ligands, was used in partial geometry optimization calculations. Subsequent single point SCF and CI calculations were performed at the optimized geometries, using the second basis set of double ζ quality. A double ζ expansion was used in all the calculations perfomed on the nickel systems.

All the geometrical parameters involving the metal centre and the atoms of the unsaturated ligand have been optimized, by means of gradient calculations, in all the analyzed systems.

3. Interaction between iron(0) and formaldehyde

The partially optimized geometries of the analyzed iron-formaldehyde systems are reported in Table 1, together with a comparison of the optimized geometry of $[Fe(CO)_2(PH_3)_2(\eta^2 - CH_2O)]$ with the experimental one of $[Fe(CO)_2P(OMe)_{32}(\eta^2 - CH_2O)]$ [5]. The distortion of the formaldehyde molecule upon coordination is seen to be satisfactorily reproduced in the optimized structures. In particular, the elongation of the C-O distance on bonding, from 1.21 to 1.32 A, is accurately predicted (1.311 A in the optimized structure), while the distortion angle, defined as the angle between the CH_2 plane and the $C - O$ bond, is calculated to be 29.1°. This value is in line with the experimental angle of 26.6° found in the complex $[Ni(PEt_3)_2(\eta^2 - CPh_2O)]$ [6]. The iron-formaldehyde bond is experimentally

	$Fe(CO)_4(CH_2O)$	$Fe(CO)_2(PH_3)_2(CH_2O)$	$Fe(PH_3)_4(CH_2O)$
$Fe-C$	2.097	1.952 (2.03)	1.900
$Fe-O$	2.055	1.971 (2.00)	1.917
$C-O$	1.262	1.311 (1.32)	1.354
$C-Fe-L$	104.6	102.1 (108.5)	110.1
$C-H$	1.080	1.082	1.087
$H-C-H$	115.1	112.4	110.5
distortion angle	16.1	29.1	37.3
binding energy	-22.6	-32.8	-70.4

Table 1: Optimized geometries and basis II binding energies of the systems under investigation. Bond lengths in angstroms, angles in degrees, energies in $kcal.mol^{-1}$. Experimental values in parentheses. L is CO or PH_3.

found to be very slightly asymmetric, with the $Fe-C$ distance (2.03 A) longer than the $Fe-O$ distance (2.00 A). Our partially optimized structure satisfactorily reproduces the absolute values of these bond distances (1.952 A for $Fe-C$ and 1.971 A for $Fe-O$), although the slightly greater deviation from experiment of the $Fe-C$ distance causes an inversion in their relative magnitude. The calculated $C-Fe-O$ angle of 39° is in excellent agreement with the experimental estimate of 38.2°. The distortion of the formaldehyde molecule increases with the substitution of carbonyl ligands with phosphines, suggesting the presence of a stronger interaction with the transition metal. In particular, the $C-O$ distance, which is 1.262 A in the tetracarbonyl compound, becomes 1.311 A in the dicarbonyldiphosphine system and 1.354 A in the tetrakis(phosphine) compound, while the distortion angle is calculated to be 16.1, 29.1, and 37.3° for the three compounds analyzed. Moreover, the distance between the iron fragment and the formaldehyde molecule decreases upon replacement of carbonyls with phosphines, suggesting the presence of a stronger bond in the tetrakis(phosphine) compound.

To understand qualitatively the nature of the coordinate bond between an iron fragment and CH_2O, it is useful to analyse the correlation between the molecular orbitals of the complex and those of the free fragments. Figure 1 shows such a correlation diagram for the orbitals of $Fe(CO)_2(PH_3)_2(\eta^2-CH_2O)$ and those of the singlet fragments, where only the main correlations are reported. The orbitals mainly involved in the iron-formaldehyde bonding are $33a'$, $35a'$, and $36a'$. The $33a'$ and $35a'$ molecular orbitals originate from the interaction of the filled orbitals $19a_1$ from $Fe(CO)_2(PH_3)_2$ and $1b_1$ from CH_2O. This is a four-electron destabilizing interaction. Effective bonding, therefore, is concentrated in the $36a'$ orbital, which is the bonding combination of $11b_1$ of $Fe(CO)_2(PH_3)_2$ (hereafter denoted as d_π), which is predominantly Fed_{xz} in character (the $Fe-\eta^2-C,O$ moiety lies in the xz plane), and the virtual orbital $2b_1$ of CH_2O (hereafter denoted as π^*). Clearly the main bonding interaction between iron and formaldehyde is the π-back-donation from the transition metal to the ligand, while the $CH_2O \rightarrow Fe$ σ-donation is almost absent. This picture is substantiated on performing a localization of the molecular orbitals by using the Foster-Boys algorithm [7]. The resulting localized molecular orbitals reveal that effective bonding between iron and formaldehyde is concentrated in a single localized molecular orbital, featuring overlap of the metal $3d_{xz}$ and the formaldehyde π^* orbitals.

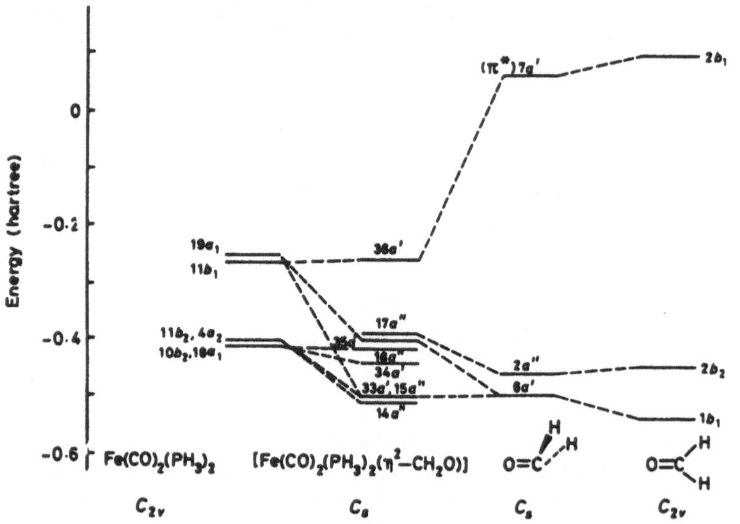

Figure 1: Molecular-orbital correlation diagram of $[Fe(CO)_2(PH_3)_2(\eta^2-CH_2O)]$

Those facts lead to the conclusion that the interaction of CH_2O with iron(0) in fragment $Fe(CO)_2(PH_3)_2$ can be described neither by the classic Chatt-Dewar-Duncanson model [2] nor by a metallacyclopropane structure [8], since both models require the utilization of two orbitals from the metal for bonding CH_2O (Figure 2). The chemical conclusion is that we need to transfer electrons to CH_2O (i.e. to reduce CH_2O) in order to bind it, and this occurs via a single metal orbital.

Also for the tetracarbonyl and the tetrakis(phosphine) compounds, the main bonding interaction is the π-back-donation from the metal to the unsaturated ligand. The binding energies (reported in Table 1) are computed with basis II to be -22.6, -32.8, and -70.4 $kcal.mol^{-1}$, for $Fe(CO)_4(\eta^2-CH_2O)$, $Fe(CO)_2(PH_3)_2(\eta^2-CH_2O)$, and $Fe(PH_3)_4(\eta^2-CH_2O)$, respectively. The replacement of carbonyl ligands with phosphines implies therefore an increase in the binding energy. The presence of electron-acceptor ligands such as carbonyl decreases the electron density at the iron atom, hence the back donation and therefore the strength of the interaction between Fe and CH_2O. Indeed, the increase in formaldehyde Mulliken population upon co-ordination is 0.21 e for the tetracarbonyl compound, 0.47 e for the dicarbonyldiphosphine system, and 0.76 e for the tetrakis(phosphine) complex. This shows that the π-back-donation increases upon replacing the carbonyls with phosphines. This increases the strength of the iron-formaldehyde bond. as is suggested by the binding energies.

To check the influence of correlation energy on the analysis performed, single reference configuration interaction calculations [hereafter referred to as Single plus Double Configuration Interaction (SDCI)] have been carried out on the compound $Fe(CO)_4(\eta^2-CH_2O)$, by using the direct-CI method [3]. 58 electrons have been frozen because of the dimensions

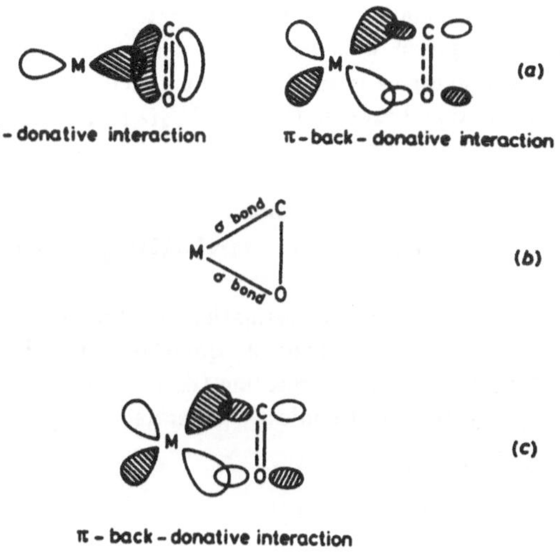

Figure 2: Bonding models for the side-on co-ordination of CH_2O to a metal fragment: (a) Chatt-Dewar-Duncanson; (b) metallacyclopropane; (c) present.

of the problem (49 doubly occupied orbitals and 142 basis functions using the bigger basis set). In the evaluation of the energies of $Fe(CO)_4$ and CH_2O, the orbitals correlating with those frozen and discarded in the complex have themselves been frozen and discarded. The SCF ground-state configuration was chosen as the reference function and all possible single and double excitations were included, except those from the frozen core orbitals.

Table 2 shows the details and results of the CI calculations on the complex and the separated fragments. The Davidson correction [9] was always added to correct for the lack of size consistency of the wavefunction. The binding energy is computed to be -15.6 and -11.2 $kcal.mol^{-1}$ at the optimized geometry, using basis I and II, respectively. These results show that correlation effects are particularly relevant in the basis II calculation and they seem to be more important for the fragment species than for the complex. Indeed the basis

	$Fe(CO)_4$		CH_2O	$[Fe(CO)_4(\eta^2 - CH_2O)]$	
	Basis I	Basis II		Basis I	Basis II
No. of active electrons	32	32	8	40	40
No. of active orbitals	38	79	18	56	97
No. of configurations	15825	128145	471	145433	595091
SCF Binding energy				-15.9	-22.6
CI Binding energy				-15.6	-11.2

Table 2: Details and results of the CI calculations on $[Fe(CO)_4(\eta^2 - CH_2O)]$. Binding energies in $kcal.mol^{-1}$

II binding energy at the CI level is smaller than that computed at the HF level, while the CI and SCF binding energies computed with basis I are comparable. However, since the difference between these binding energies is small, we can conclude with some confidence that the energetics of bond formation between Fe and CH_2O is not markedly affected by correlations effects.

4. Interaction between nickel(0) and dinitrogen

The binding energies of the complexes under investigation together with the results of the Mulliken population analysis are reported in Table 3. Bond formation between nickel and dinitrogen can result either from a substitution reaction (eq.1), or from addition to a "hot" metallic fragment (eq.3) thermally or photochemically generated in a previous stage (eq.2).

$$Ni(PH_3)_2L' + L \longrightarrow Ni(PH_3)_2L + L' \quad (1)$$
$$Ni(PH_3)_2L' \longrightarrow "Ni(PH_3)_2" + L' \quad (2)$$
$$"Ni(PH_3)_2" + L \longrightarrow Ni(PH_3)_2L \quad (3)$$

The binding energy between the metal and the ligand L is that related to reaction 3. Since we use frozen-geometry fragments and partially optimized complex geometries, we expect our calculated binding energies to overestimate somewhat (in absolute value) the true fragmentation energies of the complexes, but this should not unduly affect our comparative analysis. In our model the binding energy (BE) can be thought of as the sum of a positive deformation energy (DEF) arising from the distortion of an unsaturated ligand and a negative interaction energy (INT) resulting from the interaction between the nickel fragment and the distorted ligand [10]. The deformation and interaction energies are also reported in Table 3.

The SCF binding energies are calculated to be -15.1 and -12.6 $kcal.mol^{-1}$ for side-on and end-on dinitrogen complexes, respectively. To attempt an interpretation of these data we must first analyse the bond structure of the complexes. The binding of an unsaturated ligand, such as the dinitrogen, to a carbenoid metallic fragment is usually described by the Chatt-Dewar-Duncanson model with σ-donation from the ligand lone pair or π orbital to the metal atom and π-back-donation from the metal to the π^* orbital of the ligand [2]. This picture holds also for $Ni(PH_3)_2(N_2)$ with the dinitrogen bonded both side-on or end-on, although π-back-donation is more relevant than σ-donation. Indeed, the analysis of the molecular orbitals of $Ni(PH_3)_2(N_2)$ shows that the main bonding orbital between nickel and dinitrogen is the highest occupied one (HOMO) for both coordination modes of N_2. This orbital can be viewed as the overlap between the occupied $3d_{yz}$ orbital of the nickel (hereafter denoted as d_π) and one of the two components of the virtual $1\pi_g$ orbital of N_2 (hereafter denoted as π^*). Thus the HOMO describes π-back-donation from the nickel to the dinitrogen. σ-donation takes place from one component of $1\pi_u$ orbital of N_2 for side-on coordination and from $3\sigma_g$ orbital for end-on coordination to an hybrid orbital of nickel (hereafter denoted as d_σ), formed mainly by $4s$, $4p_z$. The predominant role played by π-back-donation, with respect to σ-donation is confirmed by the Mulliken population analysis, whose results are reported in Table 3 as population changes, i.e. Mulliken population of the complex minus Mulliken population of the fragments. We can notice a strong

	$Ni(PH_3)_2(N_2)$ side-on	end-on
BE	-15.1	-12.6
DEF	9.3	0.06
INT	-24.4	-12.7
$e_{\pi}*$	0.1067	0.1394
Nis	+0.11	+0.09
Nip	+0.22	+0.12
p_x	+0.02	-0.01
P_y	+0.02	-0.01
p_z	+0.18	+0.14
Nid	-0.43	-0.19
$d_x2_{-y}2$	+0.05	+0.03
d_z2	+0.03	-0.07
d_{xy}	+0.01	+0.01
d_{xz}	+0.04	-0.03
d_{yz}	-0.56	-0.13
$2PH_3$	-0.20	-0.04
N_2	+0.30	+0.02

Table 3: Energies ($kcal.mol^{-1}$) and Mulliken analysis of the $Ni(PH_3)_2(N_2)$ complexes. $e_{\pi}*$ is the π^* orbital energy (Hartree) of deformed N_2.

decrease of the population of the d_π orbital of the nickel (0.56 e for side-on; 0.13 e for end-on coordination) and, consequently, an increase of the population of the π^* orbital of dinitrogen. The increase of the population of the d_σ orbital of the nickel is also substantial, but, for side-on coordination, it is mainly due to the electron-donor character of the phosphine ligands. The results of Table 3 show that there is a stronger interaction, and, particularly, a more pronounced π-back- donation, for side-on coordination of dinitrogen, in agreement with the binding energies. Instead, the σ and π interactions are almost comparable for end-on coordination. The presence of a higher π-back-donation in the side-on complex is also confirmed by the energy of the π^* orbital of N_2 at the distorted geometry of the complexes, reported in Table 3. This orbital overlaps with the d_π orbital of the nickel fragment whose energy is -0.2438 hartree: the side-on coordination implies a better overlap and, therefore, an increased interaction. The energies of the π^* orbital are related to the degree of deformation of dinitrogen upon coordination; from Figure 3, where we have reported the partially optimized structures of the complexes, we can see that the N-N bond is elongated in the side-on complex (1.170 A), but not in the end-on one (1.106 A) (the experimental N-N bond distance is 1.098 A). The deformation energy of the side-on complex (9.3 $kcal.mol^{-1}$) is much higher than that of the end-on one (0.06 $kcal.mol^{-1}$), but it is more than compensated by the increased interaction energy (-24.4 for side-on; -12.7 $kcal.mol^{-1}$ for end-on).

The preference for the side-on coordination, with respect to the end-on one, for the dinitrogen interacting with the $Ni(PH_3)_2$ fragment is a rather unexpected result, since it is well known that the dinitrogen interacting with an isolated nickel atom prefers the end-on

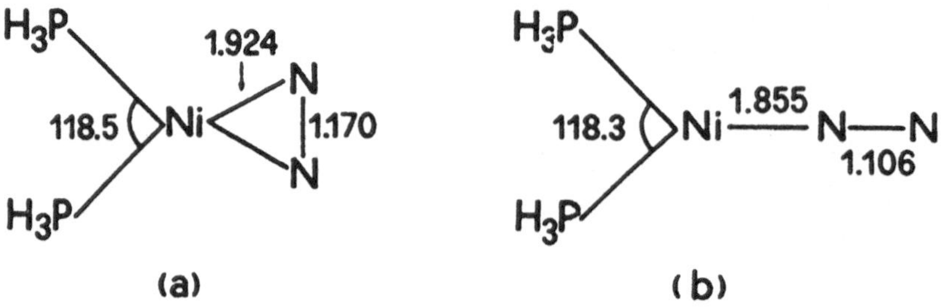

Figure 3: Optimized geometries of the $\eta^2 - N_2$ (a) and $\sigma - N_2$ (b) compounds

	$Ni(PH_3)_2$	N_2	$[Ni(PH_3)_2(N_2)$ side-on	end-on
No. of active orbitals	39	16	55	55
No. of active electrons	26	10	36	36
No. of SDCI configurations	14867	260	57797	58029
No. of MRSDCI configurations		2225	298247	115363
SCF Binding energy			-15.1	-12.6
SDCI Binding energy			-27.4	-21.8
MRSDCI Binding energy			-33.9	-19.8

Table 4: Details and results of the CI calculations on $[Ni(PH_3)_2(N_2)]$ and the separated fragments. Binding energies in $kcal.mol^{-1}$

coordination [11]. Since correlation effects may play a fundamental role and even reverse the relative stabilities of the analysed systems because of the small energy difference between them, configuration interaction calculations have been performed on the two dinitrogen complexes and relative fragments, by using the direct-CI method [3]. To reduce the size of the CI problem (39 doubly occupied orbitals and 88 basis functions) for the $Ni(PH_3)_2(N_2)$ complexes, 42 electrons have been frozen. In the evaluation of the energies of $Ni(PH_3)_2$ and N_2, the orbitals correlating with those frozen in the complex have also been frozen. For the $Ni(PH_3)_2$ fragment only single reference CI (SDCI) calculations have been performed since no configurations with a coefficient greater than 0.05 were present in the SDCI wavefunction. The preference for the side-on coordination of dinitrogen, with respect to the end-on one, in the case of the $Ni(PH_3)_2$ fragment is confirmed by the CI calculations, whose results are reported in Table 4. We can notice an increase of the binding energy for both coordination modes of dinitrogen; this increase, however, is more pronounced for side-on coordination and it is particularly relevant in the multi-reference CI [hereafter referred to as Multi-Reference Single plus Double Configuration Interaction (MRSDCI)] calculations.

At this point we can suggest with some confidence that the dinitrogen molecule prefers to bind to $Ni(PH_3)_2$ fragment in a side-on fashion, rather than in an end-on mode. This point is particularly interesting for its implications in the problem of nitrogen fixation. In the side-on bonded dinitrogen complex we have an higher electron transfer from the metal to N_2: the negative charge in the dinitrogen molecule is 0.30 e in the side-on complex and just 0.02 e in the end-on one. Moreover, the N-N bond is stretched only in the side-on complex. Therefore, we can conclude that only the side-on coordination leads to an activation of dinitrogen [12]. Such a coordination seems to be possible in the presence of appropriate ligands. The preference of Ni(0) in $Ni(PCy_3)_2N_2$ and $[Ni(PCy_3)_2]_2N_2$ [13] (PCy_3 = tricyclohexylphosphine) may be associated to a more complex energetic balance. The end-on bonding mode allows the coordination of a fourth ligand [14], i.e. solvent etc., or the formation of a dimer [13].

5. Acknowledgement

One of us (MR) thanks the Ente Nazionale Idrocarburi (ENI) for providing a fellowship.

References

1 M.Rosi, A.Sgamellotti, F.Tarantelli, and C.Floriani, Inorg. Chem. **26**, 3805 (1987); **27**, 69 (1988); J. Organomet. Chem. **332**, 153 (1987); **348**, C27 (1988); J. Chem. Soc., Dalton Trans., 249 (1988); M.Rosi, A.Sgamellotti, F.Tarantelli, C.Floriani, and M.F.Guest, J. Chem. Soc., Dalton Trans., 321 (1988); M.Rosi, A.Sgamellotti, F.Tarantelli, C.Floriani, and L.S.Cederbaum, J. Chem. Soc., Dalton Trans., in press.

2 M.J.S.Dewar, Bull. Soc. Chim. Fr. **18c**, 71 (1951); J.Chatt and J.A.Duncanson, J. Chem. Soc., 2939 (1953).

3 V.R.Saunders and J.H.Van Lenthe, Mol. Phys. **48**, 923 (1983).

4 M.Dupuis, D.Spangler, and J.Wendolowski, NRCC Software Catalog, National Resource for Computation in Chemistry, Lawrence Berkeley Laboratory, University of California, Berkeley, CA, 1980, Vol.1, Program No.QG01 (GAMESS); M.F.Guest and J.Kendrick, GAMESS User Manual, Daresbury Technical Memorandum, Daresbury Laboratory, UK, 1985.

5 H.Berke, W.Bankhardt, G.Huttner, J.v.Seyerl, and L.Zsolnai, Chem. Ber. **114**, 2754 (1981); H.Berke, G.Huttner, G.Weiler, and L.Zsolnai, J. Organomet. Chem. **219**, 353 (1981).

6 T.T.Tsou, J.C.Huffman, and J.K.Kochi, Inorg. Chem. **18**, 2311 (1979).

7 J.M.Foster and S.F.Boys, Rev. Mod. Phys. **32**, 300 (1960).

8 P.Conway, S.M.Grant, and A.R.Manning, J. Chem. Soc., Dalton Trans., 1920 (1979).

9 E.R.Davidson, in 'The World of Quantum Chemistry', eds. R.Daudel and B.Pullman, Reidel, Dordrecht, 1974.

10 S.Sakaki, K.Kitaura, and K.Morokuma, Inorg. Chem. **21**, 760 (1982).

[11] P.E.M.Siegbahn and M.R.A.Blomberg, Chem. Phys. **87**, 189 (1984).

[12] C.W.Bauschlicher, jr., L.G.M.Pettersson, and P.E.M.Siegbahn, J. Chem. Phys. **87**, 2129 (1987).

[13] P.W.Jolly, K.Jonas, C.Krüger, and Y-H.Tsay, J. organomet. Chem. **33**, 109 (1971); M.Aresta, C.F.Nobile, and A.Sacco, Inorg. Chim. Acta **12**, 167 (1975).

[14] C.A.Tolman, D.H.Gerlach, J.P.Jesson, and R.A.Schunn, J. Organomet. Chem. **65**, C23 (1974).

KINETIC PATHS FROM THE HYPERSPHERICAL PERSPECTIVE: *AB INITIO* POTENTIAL ENERGY SURFACE FOR THE $O(^3P) + H_2$ REACTION

Vincenzo Aquilanti, Simonetta Cavalli, Gaia Grossi,
Marzio Rosi, Antonio Sgamellotti, Francesco Tarantelli
Dipartimento di Chimica dell'Universitá,
06100 *Perugia, Italy*

ABSTRACT. The direct *ab initio* generation of potential energy surfaces for an elementary chemical reaction is discussed from the viewpoint of its use in dynamical calculations within the hyperspherical coordinate framework. The example is given of the reaction $O(^3P) + H_2 \longrightarrow OH + H$, for which *kinetic paths* (valley bottoms and ridge) as a function of the kinetic radius are computed at the Complete Active Space SCF level. Along the kinetic paths energies are refined using extensive Multi-Reference Configuration Interaction calculations.

1. Introduction

In the theory of elementary chemical reactions, the hyperspherical coordinate approach is providing an important alternative to the more conventional treatments based on the reaction path concept, both for the qualitative understanding of the dynamics [1] from features of the potential energy surfaces and for the quantitative implementation of advanced quantum mechanical treatments [2]. It seems therefore desirable to describe potential energy surfaces directly in hyperspherical coordinates and in particular to represent most accurately those features which are more relevant for an hyperspherical approach to the dynamics: these features include *kinetic paths* [3], *i.e.* extremal points as a function of the hyperradius and can be distinguished as *valley bottom* lines (describing asymptotically properties of reactants and products) and *ridge* lines (where strong coupling among channels localizes [4]).

As the first exploratory study in the direct *ab initio* generation of potential energy surfaces from the hyperspherical perspective the reaction

$$O(^3P) + H_2 \longrightarrow OH + H$$

was chosen as one of the most appropriate for comparison with extensive investigations by conventional approaches, being this system a prototype for asymmetric elementary reactions with a relatively small number of electrons. In particular, we have studied this system because recent experimental results have been obtained [5] that describe the long range interaction for this system, these making it possible a comparison between theory and experiments. Computational techniques are described in Sec. 2. Hyperspherical coordinates and related topological features of potential energy surfaces are presented in Sec.3.

A. Laganà (ed.), Supercomputer Algorithms for Reactivity, Dynamics and Kinetics of Small Molecules, 95–103.

As shown in Sec. 4, Complete Active Space SCF (CASSCF) methods including analytic gradient calculations are found to be useful for localizing the saddle point of the surface and some characteristics of the *valleys* and *ridges*. Multi-Reference Configuration Interaction (MRCI) calculations are necessary for improving the calculated energies: They are reported in Sec. 5 for the saddle point, the reagents, and the products, and shown to improve the results for the barrier height and the exoergicity.

As sketched in Sec.6, the final result of this investigation is a functional form for the accurate fitting of the dynamically relevant features of the surface, to be used in a quantum mechanical treatment of the reaction dynamics.

2. Computational details

Two different basis sets have been employed in our calculations. The first one (Basis I) is the double zeta valence quality set of Dunning and Hay [6], augmented with a d-type polarization function, whose exponent was 0.85 [7], on the oxygen atom and a p-type polarization function, whose exponent was 1.00 [7], on the hydrogen atoms. The second one (Basis II) is the triple zeta valence quality set of Dunning [8], augmented with two polarization functions. The exponents of the d-type polarization functions on the oxygen atom were 1.614 and 0.478 [7], while those of the p-type polarization functions on the hydrogen atoms were 1.591 and 0.3702 [7]. Our basis sets are comparable to basis sets II and IV, respectively, of ref.[7]

The first basis set was employed in most of the calculations, and, particularly, in the determination of the saddle point geometry and the minimun energy path (see Sec.4). The second basis set was used mainly to improve the calculated barrier height and the exoergicity of the reaction (see Sec.5).

The Complete Active Space SCF (CASSCF) method [9] was used in most of the calculations. For $C_{\infty v}$ symmetry, the $^3\Pi$ surface was obtained considering as active the orbitals 2σ, 3σ, 4σ, 1π, and 5σ, while the 1σ orbital was kept frozen: therefore, there are eight active electrons in six active orbitals. Analytical gradient calculations using the CASSCF wavefunction were performed in order to evaluate the saddle point geometry and the minimum energy path.

Multi-Reference Configuration Interaction (MRCI) calculations were subsequently performed on the saddle point, the reagents, and the products, in order to improve the determination of the barrier height and the exoergicity. These calculations, which include single and double excitations from the active orbitals, have been performed by using the direct-CI method [10]. Davidson's correction [11] was also added to account for the effect of unlinked clusters: this will be denoted MRCI+Q in Table 1.

All computations were performed by using the GAMESS [12] program package, implemented on a CRAY XMP/48 computer.

3. Coordinates and topology

In this section we describe those features of the potential energy surfaces in hyperspherical coordinates which are important from the point of view of reaction dynamics. We look for the stationary points of the potential energy surface at fixed values of the hyperradius (ρ) and follow their evolution along such a kinetic radius. To make clearer the description of our method it is useful to review the definition of the hyperspherical coordinates.

The kinetic radius ρ for a three body problem is defined as follows:

$$\rho = (R^2 + r^2)^{1/2}$$

where R and r are the mass scaled Jacobi coordinates

$$R = (\mu_{H_2,O}/\mu_{H_2})^{1/4} R_{H_2,O}$$

$$r = (\mu_{H_2}/\mu_{H_2,O})^{1/4} r_{H_2}$$

where r_{H_2} is the distance between the two H atoms and $R_{H_2,O}$ is the distance between the O atom and the center of mass of the molecule H_2, μ_{H_2} is the reduced mass of the molecule H_2 and $\mu_{H_2,O}$ is the reduced mass of the O atom with the molecule H_2.

The other two coordinates necessary to parametrize the internal modes and the interaction of the three bodies are the angles Θ and Φ of Smith's symmetric parametrization [13]. Θ is related to the bending mode and measures the area of the triangle: it is zero for the collinear configuration and $\pi/4$ for the equilateral one. The kinematic angle Φ describes the arrangements of the three atoms. In the collinear configuration it is defined as follows:

$$\Phi = \arctan(R/r)$$

and goes from zero to Φ_{max} which is the skewing angle

$$\Phi_{max} = \arctan[m_H(m_H + m_H + m_O)/(m_H m_O)]^{1/2}$$

46.69° in this case.

The potential energy surface is symmetric with respect to a change in sign of Θ [13]; in general, therefore, the first derivative of the potential with respect to Θ is equal to zero for $\Theta = 0$, and kinetic paths are to be searched by studying the collinear configuration. The evolution along ρ of the stationary points (maxima and minima), exhibited by the potential as a function of Φ identify the kinetic paths: in particular, the evolution of mimima defines valley bottoms and the evolution of maxima the ridges of the potential.

In general, for the three body problem the constant ρ cuts through the potential energy surface as a function of the angle Φ show a behaviour as ρ varies that is typical of the cusp catastrophe A_3 (see Fig.1) [14], Φ being the variable and ρ governing the control parameters. The critical points in the cusp catastrophe can be doubly and triply degenerate points or non degenerate. The triply degenerate point that occurs at the origin of the space of control parameters and the double degenerate points that determine the fold curve in this space, constitute the separatrix of the cusp. The space of the control parameters is divided into two regions by the separatrix; the function, the potential in our application, shows a double minimum for values of the parameters inside the area delimited by the two fold curves and the triply degenerate critical point; outside there is a single well and along the separatrix there is a minimum and an inflection point. Only symmetrical reactions had been discussed from this point of view before [3]: when the system is symmetric the parameter b is zero and

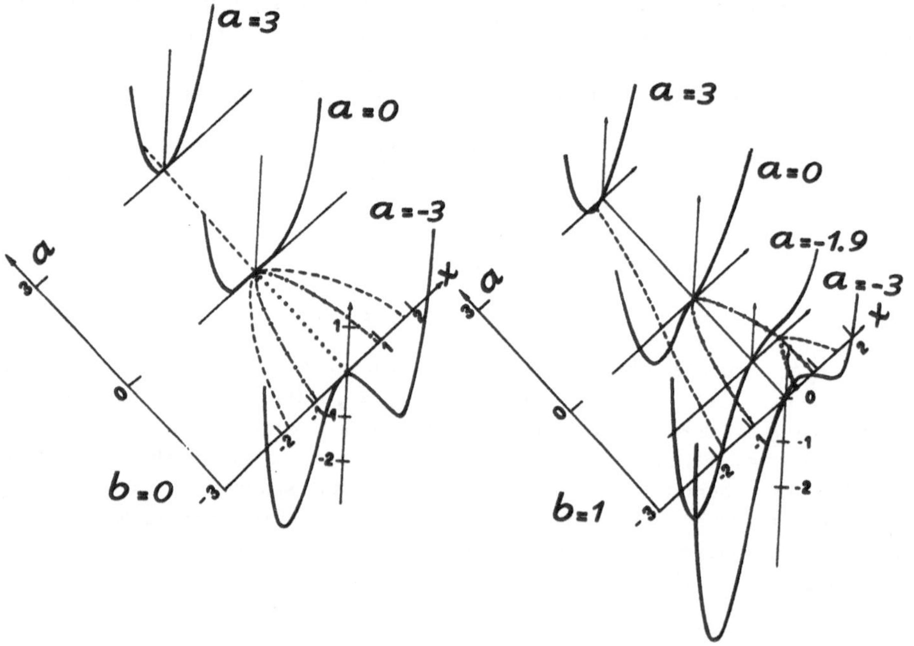

Figure 1: The canonical cusp catastrophe function, $A_3 = \frac{1}{4}x^4 + \frac{1}{2}ax^2 + bx$, at different values of the parameter a. The left panel illustrates the symmetric behavior for $b = 0$; the right panel for $b = 1$ illustrates the asymmetric behavior which occurs whenever b is different from zero. In the horizontal plane (a, x) we have drawn the locus of maxima as dotted lines, of minima as dashed lines and of inflections as dashed-dotted lines. The triply degenerate (or catastrophe) point occurs at $a = 0$ and $b = 0$ and bifurcation between single and double well modes occurs there for symmetric systems. For asymmetric systems, the bifurcation occurs on the fold line (at $a = -1.9$ for $b = 1$) where the cusp function has a doubly degenerate horizontal inflection point at $x = 0.8$.

the potential never shows a doubly degenerate inflection point. The only catastrophe point is the triply degenerate one, where the cut through the potential [3] exhibits the transition from a single well to a double well. A system like $O(^3P) + H_2$, as we will see, exhibits the characteristics of the cusp catastrophe which are generic for asymmetric systems. For such systems a minimum and a ridge arise at a bifurcation point that does not coincide with the catastrophe point.

4. Kinetic paths

Fig.2 shows that usual techniques may be misleading in identifying kinetic paths. The continuous curves are calculated as follows: By keeping r_{H_2} fixed and searching for energy minima as R_{O,H_2} was automatically varied, the branch for the entrance channel, *i.e* from $O + H_2$ to the saddle of the surface were obtained; similarly, to obtain the branch for the exit channel, *i.e* from saddle to $OH + H$, R_{O,H_2} was held fixed and the energy minimized as a function of r_{H_2}. These two branches meet at the saddle, as they should. They coincide only there, and asymptotically, with the kinetic paths obtained as described below. Coordinates for the saddle by the CASSCF technique are, for basis sets I and II, respectively, 1.860 and 1.825 a_o for the $H-H$ distance and 2.218 and 2.238 a_o for the $O-H$ distance, corresponding to $\rho = 4.530$ and 4.526 a_o and $\Phi = 72.60°$ and 72.92°. This agreement between basis sets is taken as support for the use of the simpler basis set I for a search of the kinetic paths.

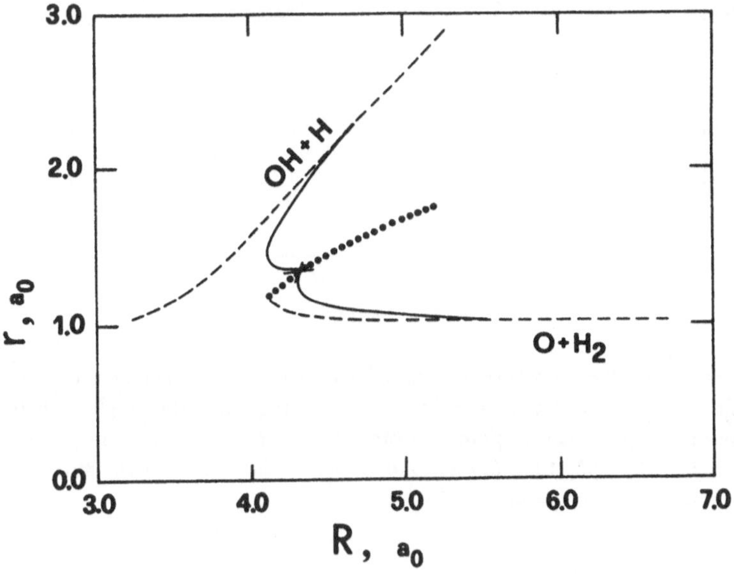

Figure 2: Kinetic paths (broken curves show the two valley bottoms and dots the ridge) and the minimum energy path (solid curve) as a function of the mass scaled Jacobi coordinates, obtained as described in Sec.4. The saddle of the surface is marked by a cross.

In Fig.3 we have drawn some typical computed cuts of the potential energy at several ρ values for this reaction: they show a single minimum and a double minimum, the transition occuring at the ρ value for which the potential has an inflection point. This is a point on the separatrix of the cusp, as stressed above. At short ρ values, there is only one kinetic path (Fig.4) which is the locus of the energy minima of the single well potential; at $\rho = 4.32a_o$, when the separatrix of the cusp is crossed in the control parameter space, the left hand side minimum and the maximum are degenerate and the potential has an inflection point. Increasing ρ the system moves into the double well region corresponding to separated reactant and product channels: another valley bottom, corresponding to the second minimum of the potential, and the separating ridge appear.

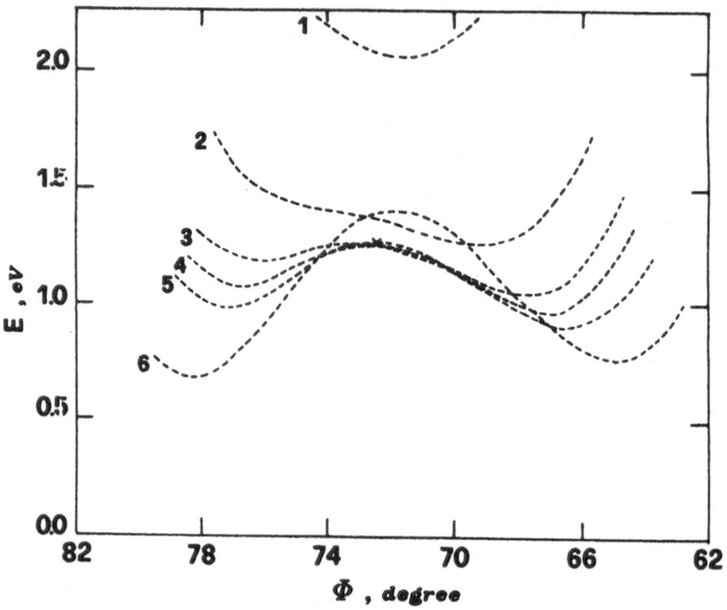

Figure 3: Typical computed cuts through the potential energy surface as a function of the kinematic angle Φ at values of the hyperradius ρ, ranging from the single well to double well mode, going through inflection points. Cuts marked as $1, 2, \cdots, 6$ corresponds to the ρ values of 3.75, 4.2, 4.45, 4.53, 4.65, 5.0 a_o respectively. The saddle of the surface, marked by a cross, corresponds to the value $\rho = 4.53$ a_o.

5. Potential energies along the kinetic paths

Fig.4 shows the energy profiles of the surface at the CASSCF level along the kinetic paths. While this level is expected to be reliable in providing geometric features of surfaces, such as the localization of kinetic paths, more refined computational techniques are needed for energies. The methods outlined in Sec.2 must therefore be used , particularly at those values of interatomic distances which correspond to critical configurations along the kinetic paths.

Typical results are shown in the table for the saddle height with respect to the incoming channel $O + H_2$, and for the asymptotic difference between the incoming channel and the exit channel $OH + H$. The table indicates that the computed values improve substantially in the CI treatment with respect to the experimental values for those quantities, the resulting error for the more refined Basis II being comparable to those for the best computed surfaces avalaible in the literature for this reaction [15]. Detailed comparisons and further results will be presented elsewhere.

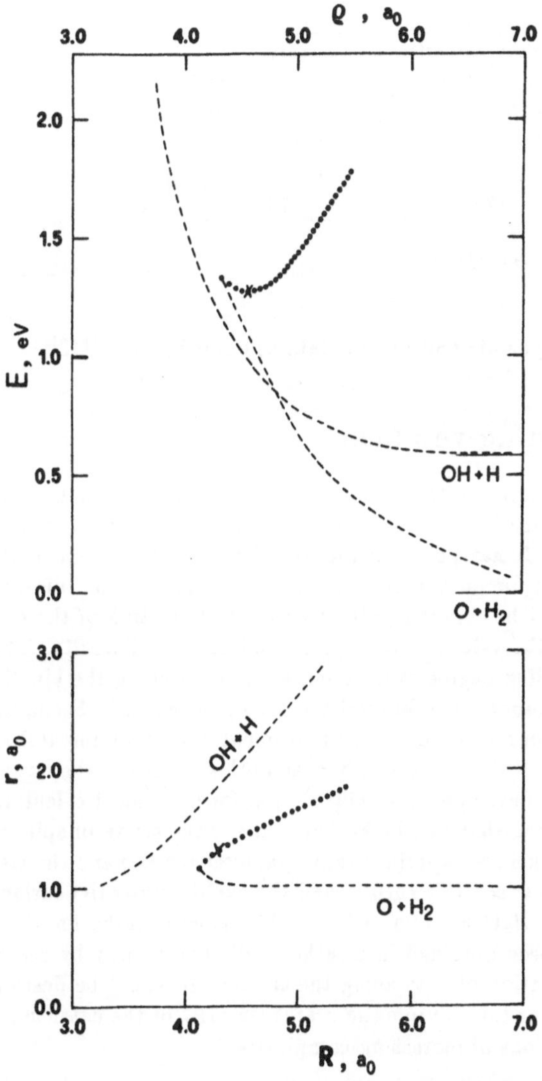

Figure 4: The upper panel shows the energy dependence of the kinetic paths as a function of ρ and the lower panel the evolution of the position of minima and maxima as a function of the mass scaled Jacobi coordinates. The broken curves identify valley bottoms and dots the ridge. The saddle point is marked by crosses.

Table 1: Calculated versus experimental energies[a].

Method[b]	*Saddle Height*[c]		*Exoergicity*[c]	
	Basis I	*Basis II*	*Basis I*	*Basis II*
CASSCF	29.5	27.9	13.3	10.9
MRCI	20.2	17.4	8.8	6.2
MRCI+Q	18.2	14.5	8.0	5.5
EXP.[d]	12.5		2.9	

[a] Energies are in $Kcal/mole$.
[b] See Sec.2
[c] See Sec.5
[d] From thermodynamic and kinetic data as quoted in Refs.[15]

6. Concluding remarks

This exploratory study of the potential energy surface for the $O + H_2$ system from the hyperspherical perspective has been limited to the collinear configuration, which is dominant for this reaction. It has provided the first detailed illustration of the topological features which are generic for an asymmetric process. The present information on the collinear configuration must be complemented by a characterization of the surface as the area angle Θ varies around the values $\Theta = 0$ corresponding to collinearity. This essentially amounts to study the bending degree of freedom, as it varies along the kinetic paths. Furthermore, refined energy computations beyond CASSCF, as shown in Sec.5, are essential to provide the necessary scaling to improve the agreement with experimental saddle height and exoergicity. A fitting of this surface, which smoothly describes the cusp catastrophe topology, is currently being developed and will be used for dynamical calculations.

Future extensions that can be anticipated are the study of spin orbit effects (for which the entrance channel are experimentally characterized [5]) and the treatment of the nonadiabatic coupling between this surface and the excited potential surface for $O(^1D) + H_2 \longrightarrow OH + H$, having H_2O as intermediate. The kinetic paths for the example discussed in this paper have been obtained in a rather rudimentary way by searching minima through pointwise computation of cuts along the surface: It would be desirable to implement gradient techniques for their automatic search, in view of the extension of the hyperspherical approach to reactions of increasing complexity.

References

1 V.Aquilanti, in: Theory of chemical reaction dynamics, ed. D.C.Clary (Reidel, Dordrecht, 1986) p.383

2 A.Kuppermann and P.G.Hipes, J. Chem. Phys. **84** 5962 (1986);
G.A.Parker, R.T.Pack, B.J.Archer and R.B.Walker, Chem. Phys. Lett. **137** 564 (1987);
J.Linderberg, Intern. J. Quantum Chem. Symp. **19** 467 (1986); J.Linderberg and B.Vessal, Intern. J. Quantum Chem. **31** 65 (1987);

J.Manz, Comments At. Mol. Phys. **17** 91 (1985);
B.Lepetit, J.M.Launay and M. Le Dourneuf, Chem.Phys. **106** 103 (1986); **106** 111 (1986)

[3] V.Aquilanti and S.Cavalli, Chem. Phys. Lett. **141** 309 (1987)

[4] V.Aquilanti, S.Cavalli, G.Grossi and A.Laganà, J.Mol.Struct. Theochem. **93** 319 (1983); **107** 95 (1984);
V.Aquilanti, G.Grossi and A.Laganà, Chem.Phys.Lett. **93** 179 (1982);
V.Aquilanti, S.Cavalli and G.Grossi, Chem.Phys.Lett. **110** 43 (1984)

[5] V.Aquilanti, R.Candori, L.Mariani, F.Pirani and G.Liuti, J.Phys.Chem. in press (1988)

[6] T. H. Dunning, Jr. and P. J. Hay, Modern Theoretical Chemistry, Vol.4, p.1, ed. H. F. Schaefer III, Plenum, New York, (1977).

[7] S. P. Walch, T. H. Dunning, Jr., R. C. Raffenetti, and F. W. Bobrowicz, J. Chem. Phys. **72** 406 (1980).

[8] T. H. Dunning, Jr., J. Chem. Phys. **55** 716 (1971).

[9] P. E. M. Siegbahn, J. Almlöf, A. Heiberg, and B. O. Roos, J. Chem. Phys. **74** 2384 (1981); B. O. Roos, P. R. Taylor, and P. E. M. Siegbahn, Chem. Phys. **48** 157 (1980).

[10] V. R. Saunders and J. H. Van Lenthe, Molec. Phys. **48** 923 (1983).

[11] E. R. Davidson, in The World of Quantum Chemistry, eds. R. Daudel and B. Pullman, Reidel, Dordrecht, (1974).

[12] M. Dupuis, D. Spangler, and J. Wendolowski, NRCC Software Catalog, National Resource for Computation in Chemistry, Vol.1, Program N. QG01 (GAMESS), Lawrence Berkeley Laboratory, University of California, Berkeley, CA, 1980); M. F. Guest and J. Kendrick, GAMESS User Manual, Daresbury Technical Memorandum, Daresbury Laboratory, UK, (1985).

[13] F.T. Smith, J. Math. Phys. **3** 735 (1962)

[14] R.Gilmore, Catatrophe theory for scientists and engineers (Wiley, New York, 1981);
J.N.L.Connor, Mol. Phys. **31** 33 (1976);
R.Gilmore, S.Kais and R.D.Levine Phys. Rev. A **34** 2442 (1986)

[15] Recent surfaces for this system include:
B.R.Johnson and N.W.Winter, J.Chem.Phys. **66** 4116 (1977);
R.E.Howard, A.D.McLean and W.A.Lester, Jr., J.Chem.Phys. **71** 2412 (1979);
S.P.Walch, T.Dunning, Jr., F.W.Bobrowicz and R.Raffenetti, J.Chem.Phys. **72** 406 (1980);
R.Jaquet and V.Staemmler, Chem.Phys. **59** 373 (1981);
J.S.Wright, D.J.Donaldson and R.J. Williams, J.Chem.Phys. **81** 397 (1984);
G.Durand and X.Chapuisat, Chem.Phys. **96** 381 (1985)

EXACT QUANTUM RESULTS FOR REACTIVE SCATTERING USING HYPERSPHERICAL (APH) COORDINATES.

G. A. Parker[a], R. T Pack, A. Laganà[b], B. J. Archer[c], J. D. Kress and Z. Bačić[d].
Group T-12, MS J569
Los Alamos National Laboratory
Los Alamos, New Mexico 87545

ABSTRACT. Adiabatically adjusting Principal-axis Hyperspherical(APH) coordinates are used in a fully 3-dimensional quantum mechanical formulation of reactive scattering. Exact results will be presented for the following systems:

a) $Li + FH \longrightarrow LiF + H$
b) $F + H_2 \longrightarrow HF + H$
c) $e^+ + H \longrightarrow Ps + p^+$
d) $d^+ + t^+\mu^- \longrightarrow t^+ + d^+\mu^-$
e) $H_2O + h\nu \longrightarrow OH + H$.

These hyperspherical coordinates consist of a hyperradius(ρ), three Euler angles (α, β, γ) which specify the rotation and tumbling motion of the triatomic system, and two internal angles (θ, χ) which were variationally determined to minimize the Coriolis coupling. The exact wavefunction is expanded in a product of Wigner rotation matrices and surface functions. The surface functions are determined from numerically solving a partial differential equation in the two internal angular coordinates (θ, χ) which we refer to as the surface function Hamiltonian. This surface function Hamiltonian parametrically depends on the hyperradius giving sector adiabatic basis functions. We will report on 4 distinct methods for solving this surface function Hamilitonian:

a) Finite Element
b) Improved Finite Element
c) Discrete Variable Representation
d) Analytic basis set.

---

[a] *Present Address: Department of Physics and Astronomy, University of Oklahoma, Norman, Oklahoma 73019*

[b] *Present Address: Dipartimento di Chimica dell'Universitá, Via Elce di Sotto 8, I-06100 Perugia, Italy*

[c] *Present Address: Department of Physics, Rice University, Houston, Texas 77251-1892*

[d] *Present Address: Department of Chemistry, New York University, 4 Washington Place, New York, New York 10003*

A. Laganà (ed.), Supercomputer Algorithms for Reactivity, Dynamics and Kinetics of Small Molecules, 105–129.

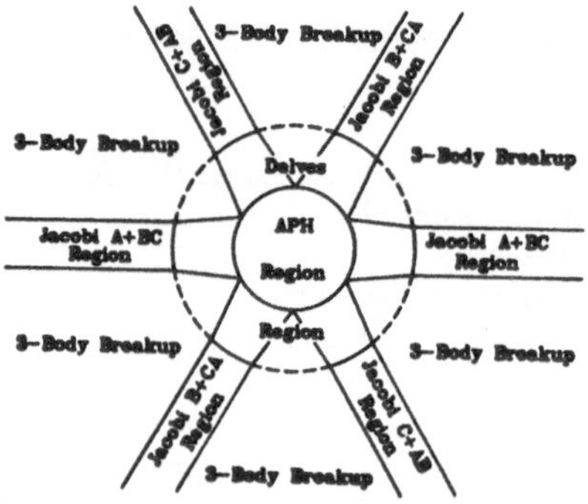

Figure 1. Schematic showing the two hyperspherical regions (APH and Delves) and the Jacobi region. There are six arrangement channels as a result of covering configuration space twice.

1. INTRODUCTION

In this paper we present an accurate quantum theory for reactive (exchange or rearrangement) scattering processes of the type

$$A + BC \rightleftharpoons \begin{cases} AB + C \\ AC + B \end{cases} \tag{1}$$

where A, B and C are atoms or simple particles.

In 1975, after great programming and computational effort converged, essentially exact Coupled-Channel (CC) scattering results were reported for the simplest chemical reaction $H + H_2 \rightleftharpoons H_2 + H$ by Schatz and Kuppermann[1,2], and Elkowitz and Wyatt[3]. This was indeed a real advance! These results were immediately extended to higher energies and better potential energy surfaces. Then, there was a lapse of over 10 years before calculations of equivalent accuracy were published. However we are now seeing a plethora of new and innovative methods for obtaining accurate reactive scattering cross sections. The next few years will be rewarding as we gain an indepth understanding of reactive processes.

In Section 2 we review the APH coordinate theory of reactive scattering. In Section 3 we present several methods used to solve the surface function Hamiltonian which is currently the most computationally time consuming part of the calculation. In Section 4 we present some results for variety of systems. Finally, in Section 5 we offer a few concluding remarks.

2. THEORY

A detailed description of the APH theory has been given elsewhere[4] and we will only briefly discuss the theory, listing the equations necessary for an accurate treatment of reactive scattering. We have divided configuration space into three separate regions, each region being labeled by coordinates used in that region. In Fig. 1 we give a schematic portraying these regions. The inner strong interaction region where reactive process occur is the APH region. Outside of the APH region we have a Delves region where there is strong coupling within each arrangement channel but negligible coupling between arrangement channels. Just beyond the Delves region we have the Jacobi region where the coupling within each arrangement channel varies from moderate to zero. When there is zero coupling, asymptotic boundary conditions can be applied. Since Jacobi coordinates are the usual coordinates used in reactive scattering and since the APH and Delves coordinates are defined in terms of Jacobi coordinates we will discuss these regions in the reverse order.

2.1 Jacobi Regions

2.1.1 *Mass-scaled Jacobi Coordinates*

Let A, B, and C be the three atoms or particles of interest with $m_\tau(\tau = A, B, C)$ being their masses and $\mathbf{x}_\tau$ being the column vectors of their coordinates relative to an origin fixed in the laboratory. Then, after separation of the center of mass motion, the well-known Jacobi coordinates for relative motion are

$$\mathbf{R}_\tau = \mathbf{x}_\tau - \frac{m_{\tau+1}\mathbf{x}_{\tau+1} + m_{\tau+2}\mathbf{x}_{\tau+2}}{m_{\tau+1} + m_{\tau+2}}, \qquad \mathbf{r}_\tau = \mathbf{x}_{\tau+2} - \mathbf{x}_{\tau+1}, \tag{2}$$

where $\tau, \tau+1$, and $\tau+2$ are any cyclic permutation of A, B, and C. The corresponding *mass-scaled* Jacobi coordinates are simply[4-13]

$$\mathbf{S}_\tau = d_\tau \mathbf{R}_\tau, \qquad \mathbf{s}_\tau = d_\tau^{-1}\mathbf{r}_\tau \tag{3}$$

where the S and s symbols imply "scaled", and the d_τ are the dimensionless scaling factors

$$d_\tau = \left[\frac{m_\tau}{\mu}\left(1 - \frac{m_\tau}{M}\right)\right]^{\frac{1}{2}}. \tag{4}$$

The three-body reduced mass μ, and the total mass of the system are

$$\mu = \left[\frac{m_A m_B m_C}{M}\right]^{\frac{1}{2}}, \qquad M = m_A + m_B + m_C. \tag{5}$$

The six mass-scaled Jacobi coordinates $(S_\tau, \hat{S}_\tau, s_\tau, \hat{s}_\tau)$ are the space-fixed (SF) Jacobi coordinates whereas the coordinates $(S_\tau, s_\tau, \Theta_\tau, \alpha_\tau, \beta_\tau, \gamma_\tau)$ are the body-fixed (BF) Jacobi coordinates. The angles $\hat{S}_\tau$ and $\hat{s}_\tau$ are the two sets of polar and azimuthal angles associated with the vectors $\mathbf{S}_\tau$ and $\mathbf{s}_\tau$ respectively, and the angle between these vectors is

$$\Theta_\tau = \cos^{-1}\left[\frac{\mathbf{S}_\tau \cdot \mathbf{s}_\tau}{S_\tau s_\tau}\right].$$

The three Euler angles $(\alpha_\tau, \beta_\tau, \gamma_\tau)$ of the BF system, are usually chosen to orient the body-fixed z-axis along $\mathbf{S}_\tau$ or $\mathbf{s}_\tau$ depending on the relative masses of the particles involved.

2.1.2 *Jacobi Coordinate Hamiltonians*

The kinetic energy operator in SF_τ or BF_τ coordinates is easily derived[13] by writing the Laplacian operators in polar coordinates

$$T = -\frac{\hbar^2}{2\mu}\left[\frac{1}{S_\tau}\frac{\partial^2}{\partial S_\tau^2}S_\tau + \frac{1}{s_\tau}\frac{\partial^2}{\partial s_\tau^2}s_\tau\right] + \frac{\mathbf{L}_\tau^2}{2\mu S_\tau^2} + \frac{\mathbf{J}_\tau^2}{2\mu s_\tau^2}. \tag{6}$$

Here $\mathbf{J}_\tau$ is the rotor angular momentum and $\mathbf{L}_\tau$ the orbital angular momentum of atom τ about the diatom of the τ system. The BF_τ kinetic energy operator is obtained from Eq. (6) by using the fact that $\mathbf{J} = \mathbf{L}_\tau + \mathbf{J}_\tau$ is the total angular momentum to eliminate $\mathbf{L}_\tau$ in (6).

2.1.3 *Jacobi Coordinate Wavefunctions*

In SF Jacobi coordinates Ψ can be written as

$$\Psi^{JM\tau_i\nu_ij_i\ell_i} = \sum_{\tau_f\nu_fj_f\ell_f} \frac{1}{s_{\tau_f}S_{\tau_f}} G^{J\tau_i\nu_ij_i\ell_i}_{\tau_f\nu_fj_f\ell_f}(S_{\tau_f})\mathcal{X}_{\nu_fj_f}(s_{\tau_f})\mathcal{Y}^{JM}_{j_f\ell_f}(\hat{s}_{\tau_f},\hat{S}_{\tau_f}), \tag{7}$$

where, as usual,[14]

$$\mathcal{Y}^{JM}_{j_f\ell_f} = \sum_{m_f} C(j_f\ell_fJ;m_fM-m_f,M)Y_{j_fm_f}(\hat{s}_{\tau_f})Y_{\ell_f,M-m_f}(\hat{S}_{\tau_f}). \tag{8}$$

The $C(j,\ell,J;m_j,m_\ell,M)$ are Clebsch-Gordan coefficients and the Y's are spherical harmonics. In Jacobi coordinates it is convenient to use asymptotic vibrational wavefunctions. They satisfy

$$\left[-\frac{\hbar^2}{2\mu}\frac{\partial^2}{\partial s_f^2} + \frac{\hbar^2 j_f(j_f+1)}{2\mu s_f^2} + v_f(s_f) - \epsilon_{\nu_fj_f}\right]\mathcal{X}_{\nu_fj_f}(s_f) = 0, \tag{9}$$

where $s_f = s_{\tau_f}$ and v_f is the asymptotic diatomic potential of the diatomic molecule of the τ_f arrangement channel.

2.1.4 *Jacobi CC Equations*

The Jacobi coordinates are only used at distances outside the exchange region, so that from Eqs. (6) and (7), we get the usual[13-15] SF Coupled Channel (CC) equations,

$$\left[\frac{d^2}{dS_f^2} + k^2_{\nu_fj_f} - \frac{\ell_f(\ell_f+1)}{S_f^2}\right] G^{J\tau_i\nu_ij_i\ell_i}_{\tau_f\nu_fj_f\ell_f}(S_f)$$
$$= \frac{2\mu}{\hbar^2}\sum_{\nu_nj_n\ell_n}\langle\tau_f\nu_fj_f\ell_f|V_f|\tau_f\nu_nj_n\ell_n\rangle G^{J\tau_i\nu_ij_i\ell_i}_{\tau_f\nu_nj_n\ell_n}(S_f), \tag{10}$$

where $k^2_{\nu j} = \frac{2\mu}{\hbar^2}[E-\epsilon_{\nu j}]$, and $V_f = V - v_f$ with V the total potential, so that V_f is the atom-diatom potential in arrangement channel f, and the integrals involve the $\mathcal{X}_{\nu_nj_n}$ and $\mathcal{Y}^{JM}_{j_n\ell_n}$ basis functions.

In the present context, the Jacobi CC equations are used only to propagate an **R** matrix in the weak interaction region where there is negligible coupling between different arrangement channels, so that the boundary conditions can be applied.

We note that propagation of the CC equations in Jacobi coordinates in this region is faster than propagation in Delves or APH coordinates not only because the different arrangements can be propagated separately with fewer coupled equations but also because the basis functions $\mathcal{X}$ are independent of the propagation variable and need not be redetermined at each step.

2.1.5 *Jacobi Coordinate Boundary Conditions*

The present calculations generate a Wigner $\mathbf{R}^J$ matrix,

$$\mathbf{R}^J = \mathbf{G}^J(\mathbf{G}^{J\prime})^{-1}, \tag{11}$$

at the sector boundaries where we have used matrix notation $\mathbf{G}^J = \{G^J_{fi}\}$ and composite indices for simplicity. The prime implies the derivative with respect to the appropriate propagation variable. At large distances the propagated solutions can be expressed as a linear combination of spherical Riccati–Bessel functions giving the "reactance" (which has no special relation to reactive scattering) matrix $\mathbf{K}^J$, scattering matrix $\mathbf{S}^J$ and transition matrix $\mathbf{T}^J$:

$$\mathbf{K}^J = [\mathbf{R}^J\mathbf{b}' - \mathbf{b}]^{-1}[\mathbf{R}^J\mathbf{a}' - \mathbf{a}], \qquad \mathbf{S}^J = (\mathbf{1} + i\mathbf{K}^J)(\mathbf{1} - i\mathbf{K}^J)^{-1}, \qquad \mathbf{T}^J = \mathbf{1} - \mathbf{S}^J. \quad (12)$$

For the open (asymptotically allowed) channels, the elements of $\mathbf{a}$ and $\mathbf{b}$ are simply proportional to the spherical Riccati--Bessel functions,[16]

$$a_{fi} = \delta_{fi} k_f^{\frac{1}{2}} S_f j_{\ell_f}(k_f S_f) \qquad \text{and} \qquad b_{fi} = \delta_{fi} k_f^{\frac{1}{2}} S_f y_{\ell_f}(k_f S_f). \quad (13)$$

For asymptotically closed channels, we let $k_f = i\kappa_f$ where $\kappa_f = |k_f|$, and we follow McLenithan and Secrest[17] in choosing

$$a_{fi} = \delta_{fi} \kappa_f^{\frac{1}{2}} S_f i^{-\ell_f} j_{\ell_f}(i\kappa_f S_f) \qquad \text{and} \qquad b_{fi} = \delta_{fi} \kappa_f^{\frac{1}{2}} S_f i^{\ell_f+2} h_{\ell_f}^{(1)}(i\kappa_f S_f). \quad (14)$$

The a_{fi} are real and closely related to the modified spherical Bessel functions $I_{\ell_f+\frac{1}{2}}$ which are regular at the origin, and the b_{fi} are also real and closely related to the modified spherical Bessel functions $K_{\ell_f+\frac{1}{2}}$ which die exponentially at large distances.

The scattering amplitude is then given by

$$\begin{aligned} &\mathrm{f}(\tau_f \nu_f j_f m_f \leftarrow \tau_i \nu_i j_i m_i | \mathbf{k}_i, k_f, \hat{S}_{\tau_f}) \\ &= \frac{2\pi}{(k_i k_f)^{\frac{1}{2}}} \sum_{JM\ell_i\ell_f} i^{\ell_i - \ell_f + 1} C(j_i \ell_i J; m_i, M - m_i, M) C(j_f \ell_f J; m_f, M - m_f, M) \\ &\times Y^*_{\ell_i, M-m_i}(\hat{k}_i) Y_{\ell_f, M-m_f}(\hat{S}_{\tau_f}) T^J(\tau_f \nu_f j_f \ell_f | \tau_i \nu_i j_i \ell_i). \end{aligned} \quad (15)$$

The integral cross section has the usual form and all averaged cross sections, which are obtained by summing over certain final quantum numbers and averaging over certain initial quantum numbers, show the usual[1,3,14,15,18] simplifications and will not be detailed here.

2.2 Delves Region

2.2.1 *Delves Hyperspherical Coordinates*

The Delves[9] coordinates of arrangement τ are very simply obtained from the scaled Jacobi coordinates *via* the definitions

$$\rho = (S_\tau^2 + s_\tau^2)^{\frac{1}{2}}, \quad (16)$$

and

$$\theta_{D_\tau} = \tan^{-1}\left(\frac{s_\tau}{S_\tau}\right). \quad (17)$$

Here ρ is the hyperradius. No subscript is put on ρ because ρ is independent of arrangement channel. However, the Delves hyperangle θ_{D_τ} does depend on the arrangement channel τ. In addition to ρ and θ_{D_τ}, the four SF or four BF angles of the τ arrangement complete the Delves coordinate set.

For some rearrangement scattering processes, it may be convenient to use the three sets of Delves coordinates everywhere in the strong interaction region and then project onto functions of the three Jacobi sets at large ρ.

At this point we note that Kuppermann[1a,19-21] and co-workers use what we shall call Doubled Delves hyperspherical (DDH) coordinates. They are the same as the Delves coordinates just discussed except that the angle θ_{D_τ} is replaced by $\theta_{K_\tau} = 2\theta_{D_\tau}$.

2.2.2 *Delves Coordinate Hamiltonians*

The SF Delves Hamiltonian operator has the simple form,

$$H = -\frac{\hbar^2}{2\mu\rho^5}\frac{\partial}{\partial\rho}\rho^5\frac{\partial}{\partial\rho} + \frac{\Delta^2}{2\mu\rho^2} + V(\rho,\theta_{D_\tau},\Theta_\tau), \tag{18}$$

where Δ^2, given here by

$$\Delta^2 = -\frac{\hbar^2}{\sin^2 2\theta_{D_\tau}}\frac{\partial}{\partial\theta_{D_\tau}}\sin^2 2\theta_{D_\tau}\frac{\partial}{\partial\theta_{D_\tau}} + \frac{\mathbf{L}_\tau^2}{\cos^2\theta_{D_\tau}} + \frac{\mathbf{J}_\tau^2}{\sin^2\theta_{D_\tau}}, \tag{19}$$

is the square of Smith's grand angular momentum[5] operator. As before the BF Delves Hamiltonian operator is obtained by replacing $\mathbf{L}_\tau$ by $\mathbf{J} - \mathbf{J}_\tau$.

2.2.3 *Delves Coordinate Wavefunctions*

The Delves coordinate wavefunctions are

$$\Psi^{JM\tau_i v_i j_i \ell_i} = 2\sum_{\tau_f v_f j_f \ell_f} \frac{\Gamma^{J\tau_i v_i j_i \ell_i}_{\tau_f v_f j_f \ell_f}(\rho)}{\rho^{5/2}}\frac{\Upsilon_{v_f j_f}(\theta_{D_{\tau_f}};\rho)}{\sin(2\theta_{D_{\tau_f}})}\mathcal{Y}^{JM}_{j_f\ell_f}(\hat{s}_{\tau_f},\hat{S}_{\tau_f}). \tag{20}$$

Here the three possible arrangement channels τ_f are distinctly included and the angular functions $\mathcal{Y}$ are exactly the same as those used in the Jacobi wavefunctions of Eqs. (7) and (8). The $\Upsilon(\theta_{D_{\tau_f}};\rho)$ are vibrational wavefunctions which depend parametrically upon ρ. We use "sector-adiabatic" basis functions of the Delves coordinates. That is, the basis functions change from sector to sector but not within a sector. On the ξth sector we choose the vibrational functions Υ to satisfy

$$\left\{\frac{\hbar^2}{2\mu\rho_\xi^2}\left[-\frac{\partial^2}{\partial\theta_{D_f}} + \frac{j_f(j_f+1)}{\sin^2\theta_{D_f}}\right] + v_f - \mathcal{E}_{v_f j_f}(\rho_\xi)\right\}\Upsilon_{v_f j_f}(\theta_{D_f};\rho_\xi) = 0, \tag{21}$$

where $v_f(s_f) = v_f(\rho_\xi \sin\theta_{D_f})$ is the asymptotic diatomic potential of the τ_f arrangement as before.

2.2.4 *Delves CC Equations*

Currently we are using Delves coordinates only in the moderate interaction region where there is strong rovibrational coupling but negligible coupling between different arrangement channels.

The Υ's are used in Eq. (18) together with Eqs. (20) and (21), to obtain the SF CC equations

$$\begin{aligned}\left\{\frac{\partial^2}{\partial\rho^2} + \frac{2\mu}{\hbar^2}\left[E - \frac{\rho_\xi^2}{\rho^2}\mathcal{E}_{v_f j_f}(\rho_\xi)\right] - \frac{15}{4\rho^2}\right\}\Gamma^{J\tau_i v_i j_i \ell_i}_{\tau_f v_f j_f \ell_f}(\rho) \\ = \frac{2\mu}{\hbar^2}\sum_{v_n j_n \ell_n}\langle\Upsilon_{v_f j_f}\mathcal{Y}^{JM}_{j_f\ell_f}|V - \frac{\rho_\xi^2}{\rho^2}v_f(\rho_\xi\sin\theta_{D_f}) \\ + \frac{\hbar^2\ell_n(\ell_n+1)}{2\mu\rho^2\cos^2\theta_{D_f}}|\Upsilon_{v_n j_n}\mathcal{Y}^{JM}_{j_n\ell_n}\rangle\Gamma^{J\tau_i v_i j_i \ell_i}_{\tau_f v_n j_n \ell_n}(\rho)\end{aligned} \tag{27}$$

All the terms on the right side except V are diagonal in j and ℓ.

2.2.5 *Delves Coordinate Boundary Conditions*

For systems with long range potentials it is computationally faster to project the Delves solutions onto Jacobi solutions, and propagate those solutions on out to where the boundary conditions can be applied. However, it is conceptually simpler and often practicable to propagate the Delves solutions out to the asymptotic region and apply the boundary conditions directly in Delves coordinates. To allow that, we transform the boundary conditions into Delves coordinates. We note that the Υ are obtained at a finite set of ρ values, which are the centers of the propagation sectors.

At large hyperradii we can write

$$\Gamma^J = \mathcal{A} - \mathcal{B}\mathbf{K}^J, \qquad \text{and} \qquad \frac{\partial \Gamma^J}{\partial \rho} = \mathcal{E} - \mathcal{F}\mathbf{K}^J, \tag{23}$$

where

$$\mathcal{A}_{fi} = \delta_{\tau_f\tau_i}\delta_{j_fj_i}\delta_{\ell_f\ell_i}\rho^{\frac{1}{2}} \int_0^{\pi/2} d\theta_{D_f} \Upsilon_f^*(\theta_{D_f};\rho_\xi) a_{ii}(S_f)\mathcal{X}_i(s_f), \tag{24}$$

$$\mathcal{B}_{fi} = \delta_{\tau_f\tau_i}\delta_{j_fj_i}\delta_{\ell_f\ell_i}\rho^{\frac{1}{2}} \int_0^{\pi/2} d\theta_{D_f} \Upsilon_f^*(\theta_{D_f};\rho_\xi) b_{ii}(S_f)\mathcal{X}_i(s_f), \tag{25}$$

$$\mathcal{E} = \frac{1}{2\rho}\mathcal{A} + \mathcal{C}, \qquad \mathcal{F} = \frac{1}{2\rho}\mathcal{B} + \mathcal{D}, \tag{26}$$

$$\mathcal{C}_{fi} = \delta_{\tau_f\tau_i}\delta_{j_fj_i}\delta_{\ell_f\ell_i}\rho^{\frac{1}{2}} \int_0^{\pi/2} d\theta_{D_f} \Upsilon_f^* \left[\cos\theta_{D_f}\mathcal{X}_i \frac{\partial a_{ii}}{\partial S_f} + \sin\theta_{D_f} a_{ii}\frac{\partial \mathcal{X}_i}{\partial s_f}\right], \tag{27}$$

and

$$\mathcal{D}_{fi} = \delta_{\tau_f\tau_i}\delta_{j_fj_i}\delta_{\ell_f\ell_i}\rho^{\frac{1}{2}} \int_0^{\pi/2} d\theta_{D_f} \Upsilon_f^* \left[\cos\theta_{D_f}\mathcal{X}_i \frac{\partial b_{ii}}{\partial S_f} + \sin\theta_{D_f} b_{ii}\frac{\partial \mathcal{X}_i}{\partial s_f}\right]. \tag{28}$$

These integrals all consist of known functions and can be performed by quadrature. It is also convenient in these formulas to now set ρ equal to ρ_ξ everywhere ρ_ξ occurs. Then, from Eq. (23) the reactance matrix is

$$\mathbf{K}^J = (\mathbf{R}_\Gamma\mathcal{F} - \mathcal{B})^{-1}(\mathbf{R}_\Gamma\mathcal{E} - \mathcal{A}), \tag{29}$$

where $\mathbf{R}_\Gamma = \Gamma^J(\partial\Gamma^J/\partial\rho)^{-1}$ is the Wigner R matrix in the Delves coordinates.

Thus, the $\mathbf{K}^J$ (and hence $\mathbf{S}^J$) matrices can be obtained directly from the Delves wavefunctions.

2.3 APH Region

2.3.1 *APH Coordinates*

The APH coordinates[4,22] are used in the strong reactive region. As with Delves coordinates they are simply obtained from the scaled Jacobi coordinates

$$\rho = (S_\tau^2 + s_\tau^2)^{\frac{1}{2}}, \tag{30}$$

$$\tan\theta = \frac{[(S_\tau^2 - s_\tau^2)^2 + (2\mathbf{S}_\tau\cdot\mathbf{s}_\tau)^2]^{\frac{1}{2}}}{2S_\tau s_\tau \sin\Theta_\tau}, \tag{31}$$

$$\sin(2\chi_\tau) = \frac{2\mathbf{S}_\tau \cdot \mathbf{s}_\tau}{[(S_\tau^2 - s_\tau^2)^2 + (2\mathbf{S}_\tau \cdot \mathbf{s}_\tau)^2]^{1/2}} \quad \text{and} \quad \cos(2\chi_\tau) = \frac{(S_\tau^2 - s_\tau^2)}{[(S_\tau^2 - s_\tau^2)^2 + (2\mathbf{S}_\tau \cdot \mathbf{s}_\tau)^2]^{1/2}}. \tag{32}$$

To complete the APH coordinate specification we choose the Euler angles $(\alpha_Q, \beta_Q, \gamma_Q)$ to orient the body-fixed z-axis along the smallest principal moment of inertia. As shown elsewhere[22] these coordinates minimize the coriolis coupling for reactions that are collinearly dominated.

We note that the APH and Delves hyperradius are identical but all of the other coordinates are distinct.

2.3.2 *APH Coordinate Hamiltonians*

Using the procedure of Podolsky[23], Johnson[12] has clearly detailed the procedure for transforming the kinetic energy operator into the APH system, and we have verified his results. The resulting APH T is

$$\begin{aligned} T = &-\frac{\hbar^2}{2\mu\rho^5}\frac{\partial}{\partial\rho}\rho^5\frac{\partial}{\partial\rho} - \frac{\hbar^2}{2\mu\rho^2}\left\{\frac{4}{\sin 2\theta}\frac{\partial}{\partial\theta}\sin 2\theta\frac{\partial}{\partial\theta} + \frac{1}{\sin^2\theta}\frac{\partial^2}{\partial\chi_i^2}\right\} \\ &+ \frac{J_x^2}{\mu\rho^2(1+\sin\theta)} + \frac{J_y^2}{2\mu\rho^2\sin^2\theta} + \frac{J_z^2}{\mu\rho^2(1-\sin\theta)} - \frac{i\hbar\cos\theta}{\mu\rho^2\sin^2\theta}J_y\frac{\partial}{\partial\chi_i}, \end{aligned} \tag{33}$$

where the J_i are the BF_Q components of the total angular momentum.

It is often convenient to write Eq. (33) as

$$T = T_\rho + T_h + T_r + T_c, \tag{34}$$

where h implies the θ, χ motion on the surface of the "hypersphere" (internal coordinate sphere), r stands for rotational, and c for Coriolis. The rotational kinetic energy can be written as

$$T_r = AJ_x^2 + BJ_y^2 + CJ_z^2, \tag{35}$$

where

$$A = \frac{1}{\mu\rho^2(1+\sin\theta)}, \qquad B = \frac{1}{2\mu\rho^2\sin^2\theta}, \qquad \text{and} \qquad C = \frac{1}{\mu\rho^2(1-\sin\theta)}. \tag{36}$$

2.3.3 *APH Coordinate Wavefunctions*

On the sector centered at ρ_ξ, the APH coordinate wavefunctions can be written as

$$\Psi^{JMpn} = 4\sum_{t,\Lambda}\rho^{-5/2}\psi_{t\Lambda}^{Jpn}(\rho)\Phi_{t\Lambda}^{Jp}(\theta,\chi_i;\rho_\xi)\hat{D}_{\Lambda M}^{Jp}(\alpha_Q,\beta_Q,\gamma_Q), \tag{37}$$

where only $\Lambda \geq 0$ occurs.

The basis functions Φ satisfy a two dimension Schrödinger equation on the surface of the upper half of the sphere (or "hypersphere") made up of V, T_h of Eq. (33) and the symmetric top part of T_r [Eq. (35)],

$$\begin{aligned} \Bigg\{ &-\frac{\hbar^2}{2\mu\rho_\xi^2}\left[\frac{4}{\sin 2\theta}\frac{\partial}{\partial\theta}\sin 2\theta\frac{\partial}{\partial\theta} + \frac{1}{\sin^2\theta}\frac{\partial^2}{\partial\chi_i^2}\right] + \frac{A+B}{2}\hbar^2 J(J+1) + \frac{15\hbar^2}{8\mu\rho_\xi^2} \\ &+ \left[C - \frac{(A+B)}{2}\right]\hbar^2\Lambda^2 + V(\rho_\xi,\theta,\chi_i) - \mathcal{E}_{t\Lambda}^{Jp}(\rho_\xi)\Bigg\}\Phi_{t\Lambda}^{Jp}(\theta,\chi_i;\rho_\xi) = 0. \end{aligned} \tag{38}$$

Here A, B, and C are the functions defined by Eq. (36), and ρ is fixed at the center of the ξth sector.

2.3.4 ***APH Coordinate CC Equations***

The APH surface functions Φ are also "sector adiabatic". When the APH wavefunction of Eq. (37) is substituted into the Schrödinger equation, the resulting exact Coupled Channel or Close Coupling (CC) equations are of the form

$$\left[\frac{\partial^2}{\partial\rho^2} + \frac{2\mu E}{\hbar^2}\right]\psi_{t\Lambda}^{Jpn}(\rho) = \frac{2\mu}{\hbar^2}\sum_{t'\Lambda'}\langle\Phi_{t\Lambda}^{Jp}\hat{D}_{\Lambda M}^{Jp}|H_i|\Phi_{t'\Lambda'}^{Jp}\hat{D}_{\Lambda' M}^{Jp}\rangle\psi_{t'\Lambda'}^{Jpn}(\rho), \tag{39}$$

where, with terms as at Eqs. (34) and (35),

$$H_i = T_h + T_r + T_c + \frac{15\hbar^2}{8\mu\rho^2} + V(\rho,\theta,\chi_i). \tag{40}$$

Using Eq. (38) and the ρ^{-2} dependence of A,B, and C, we obtain

$$\langle\Phi_{t\Lambda}^{Jp}\hat{D}_{\Lambda M}^{Jp}|H_i|\Phi_{t'\Lambda'}^{Jp}\hat{D}_{\Lambda' M}^{Jp}\rangle = \langle\Phi_{t\Lambda}^{Jp}\hat{D}_{\Lambda M}^{Jp}|\frac{\rho_\xi^2}{\rho^2}\mathcal{E}_{t'\Lambda'}(\rho_\xi) + V(\rho,\theta,\chi_i)$$

$$- \frac{\rho_\xi^2}{\rho^2}V(\rho_\xi,\theta,\chi_i) + \frac{A-B}{2}(J_x^2 - J_y^2) + T_c|\Phi_{t'\Lambda'}^{Jp}\hat{D}_{\Lambda' M}\rangle, \tag{41}$$

$$= \frac{\rho_\xi^2}{\rho^2}\mathcal{E}_{t\Lambda}(\rho_\xi)\delta_{tt'}\delta_{\Lambda\Lambda'} + \delta_{\Lambda\Lambda'}\langle\Phi_{t\Lambda}^{Jp}|V(\rho,\theta,\chi_i) - \frac{\rho_\xi^2}{\rho^2}V(\rho_\xi,\theta,\chi_i)|\Phi_{t'\Lambda}^{Jp}\rangle$$

$$+ \langle\Phi_{t\Lambda}^{Jp}\hat{D}_{\Lambda M}^{Jp}|\frac{A-B}{2}(J_x^2 - J_y^2) + T_c|\Phi_{t'\Lambda'}^{Jp}\hat{D}_{\Lambda' M}^{Jp}\rangle. \tag{42}$$

2.3.5 ***APH Coordinate Boundary Conditions***

The BF Delves **R** matrix is obtained from the APH **R** matrix via

$$\mathbf{R}^J{}_\Gamma(BF) = \mathbf{U}^J\mathbf{R}^J{}_\psi\tilde{\mathbf{U}}^J, \tag{43}$$

where

$$U_{\tau_f\nu_f j_f\Lambda_f,t\Lambda}^J = 2\int_0^{\pi/2}\sin(2\theta_{D_f})d\theta_{D_f}$$

$$\times\int d\hat{s}_f d\hat{S}_f \Upsilon_f P_f \hat{D}_{\Lambda_f M}^{J\,*}(\alpha_f,\beta_f,\gamma_f)\Phi_{t\Lambda}^{Jp}\hat{D}_{\Lambda M}^{Jp}(\alpha_Q,\beta_Q,\gamma_Q) \tag{44}$$

Hence, the transformation is straightforward once **U** is evaluated. Thus, the Delves $\mathbf{R}^J$ matrix can be obtained from the APH $\mathbf{R}^J$ matrix. If ρ is large enough, the boundary conditions can be applied.

3. SURFACE FUNCTIONS

The amount of computer time required to generate sector adiabatic surface functions at many ρ's can be substantial. Hence, we have used several methods for solving the 2−D surface function equation. We will now describe each of these methods.

3.1 Finite Element Method

The finite element method (FEM) is a powerful numerical technique often used by the engineering community for solving partial differential equations. The method is based on the variational equation

$$\int_0^{2\pi} d\chi \int_0^{\pi/2} f d\theta,$$

where

$$f = \frac{1}{2}(\Phi_\theta, \Phi_\chi, \Phi) \begin{pmatrix} K_{11} & 0 & 0 \\ 0 & K_{33} & 0 \\ 0 & 0 & K_{44} \end{pmatrix} \begin{pmatrix} \Phi_\theta \\ \Phi_\chi \\ \Phi \end{pmatrix} \tag{55}$$

and

$$K_{11} = 4\sin 2\theta, \qquad K_{33} = \frac{2\cos\theta}{\sin\theta}, \qquad \text{and} \qquad K_{44} = \frac{2\mu\rho^2}{\hbar^2}(V - E)\sin 2\theta. \tag{56}$$

Finding the surface functions Φ which minimize the variational equation is equivalent to solving the surface function hamiltonian Eq. (38). In our implementation of the finite element method one divides the (θ, χ) domain into quadrilateral subdomains called elements. The (θ_i, χ_i) coordinates of the corners, the midpoint of each side, and the center of the quadrilateral elements are called nodal points, giving 9 nodal points per element. Polynomial basis functions localized within each element are used to approximate the surface function Φ. These basis functions are zero at 8 of the 9 nodal points and 1 at the remaining nodal point. Hence, there are 9 polynomial basis functions per element. The resulting surface functions are continuous everywhere. However, the derivatives of the surface functions need not be continuous across the element boundaries.

A major advantage of the finite element method is that the size of the elements can be varied across the domain. Hundred of non-overlapping elements of different sizes make a mesh. In regions where the amplitudes of the surface functions are large it is desirable to use as fine as a mesh as one can afford. However, in regions where the surface functions are nearly zero one can use fairly large elements. Ideally, the mesh would be very dense in the strongly attractive regions, very sparse in the highly repulsive regions and have a smooth transition between the two regions. To obtain this desired result we start with a very sparse mesh and then subdivide it many times, concentrating each time on the regions where the root mean square (RMS) of the surface functions calculated at the previous value of ρ is large. Basing the mesh on the RMS of the surface functions places nodal points in all regions where they are needed to accurately represent the surface function. An optimized mesh has about 3,000 to 5,000 nodal points instead of $\approx$50,000 nodal points of a uniform mesh. This procedure allows one to obtain accurate surface functions with minimal cost.

This finite element basis yields the generalized eigenvalue problem

$$\mathbf{H\Phi} = \mathbf{S\Phi E} \tag{57}$$

where **S** is the overlap matrix and **H** is the Hamiltonian matrix. Currently, these matrices are large (order n =3,000-5,000, with a half band width m =100-200). Hence, we choose to obtain the lowest eigenvalues using iterative methods rather than solving for the complete spectrum. We have used both the subspace iteration method and the Lanczos method. Since we have a good guess to the initial subspace constructed from the eigenvectors calculated at the previous ρ, the subspace iteration method is only a factor of 2 slower than the Lanczos method. Also, because we have fewer numerical problems with the subspace iteration method we choose to use it for production runs.

It should be noted that because of the way in which the finite element basis is chosen the elements of the eigenvectors are actually the values of the surface function at the nodal points.

3.2 Improved Finite Element Method

At very large hyperradii the surface functions are highly localized into small regions of the (θ,χ) domain. Also, for molecular problems the vibrational motion of the surface function can be approximated as a harmonic oscillator (a Gaussian centered at the equilbrium position of the isolated diatom times a Hermite polynomial). Hence, it becomes difficult to represent this exponential function accurately with a polynomial basis. By factoring out this non-polynomial behavior we can obtain accurate solutions with less computational expense. To do this we substitute $\Phi = g\bar{\Phi}$ into Eq. (55) giving

$$f = \frac{1}{2}\left(\bar{\Phi}_\theta, \bar{\Phi}_\chi, \bar{\Phi}\right)\begin{pmatrix}\bar{K}_{11} & 0 & \bar{K}_{14}\\ 0 & \bar{K}_{33} & \bar{K}_{34}\\ \bar{K}_{41} & \bar{K}_{43} & \bar{K}_{44}\end{pmatrix}\begin{pmatrix}\bar{\Phi}_\theta\\ \bar{\Phi}_\chi\\ \bar{\Phi}\end{pmatrix}, \tag{58}$$

where the diagonal matrix elements are

$$\bar{K}_{11} = g^2 K_{11}, \qquad \bar{K}_{33} = g^2 K_{33}, \qquad \bar{K}_{44} = g_\theta^2 K_{11} + g_\chi^2 K_{33} + g^2 K_{44}, \tag{59}$$

and the off-diagaonal matrix elements are

$$\bar{K}_{14} = \bar{K}_{41} = g g_\theta K_{11}, \qquad \text{and} \qquad \bar{K}_{34} = \bar{K}_{43} = g g_\chi K_{33} \tag{60}$$

This new variational equation will converge faster if the scaling function g correctly factors out of the surface functions the non-polynomial behaivor. A good guess for g would be

$$g = e^{-\beta_A(S_A - S_A^{eq})^2} + e^{-\beta_B(S_B - S_B^{eq})^2} + e^{-\beta_C(S_C - S_C^{eq})^2}, \tag{61}$$

where S_A^{eq} is the vibrational equilibrium distance.

Again we must solve a large generalized eigenvalue problem and use the subspace iteration method to obtain a partial spectrum.

3.3 Discrete Variable Method

Determining the surface functions amounts to finding the bound eigenstates of the two-dimensional surface function Hamiltonian in Eq. (38). The discrete variable representation (DVR)[24] provides a very general and powerful approach for solving bound state problems. In contrast to the standard variational approaches[25], which expand the eigenstates in terms of basis functions, the DVR yields a localized basis for one or more internal degrees of freedom. This localized basis consists of the points of the Gaussian quadrature defined in an orthogonal polynomial basis, to which the DVR is related by a unitary transformation[24]. The DVR-based formulation of the bound state problem (possible in many coordinate systems) allows development of a variety of computational schemes, some of which treat all internal coordinates in the DVR[26], while others[27] combine the DVR with the distributed gaussian basis (DGB)[28] for some of the degrees of freedom. As calculations on LiCN/LiNC[27(a,c)], HCN/HNC[27(b,c)] and H_2O[27(d)] have demonstrated, the DVR-based approaches are very versatile and permit exceptionally efficient calculation of highly excited, large amplitude vibrational levels in many-mode, strongly coupled molecular systems, such as isomerizing molecules, van der Waals complexes and other floppy molecules, with arbitrarily shaped potential surfaces.

Several factors contribute to the efficiency of the DVR-based bound state treatments. The DVR basis is readily tailored to the potential, i.e., only those DVR points lying in the energetically accessible regions of the potential surface need to be kept in the basis[27]. The potential marix is diagonal on the DVR points, thus reducing the dimension of the integrals, appearing in the potential matrix elements, which have to be evaluated numerically[24,27]. What is even more important, DVR allows us to define, at each of the discretized values of one coordinate, a lower dimensional vibrational problem in the other coordinates[27]. A truncated set of eigenvectors obtained by solving these reduced vibrational problems provides a contracted, very compact basis for the final matrix representation of the full Hamiltonian, whose diagonalization gives the desired eigenvalues and eigenfunctions[27].

In the present calculations of the surface functions, DVR is applied to both the θ and χ coordinate. The θ coordinate is discretized in the DVR consisting of the Gauss-Legendre quadrature points, while χ is treated in the DVR defined in the properly symmetrized set of functions $\cos(n-1)\chi, \sin n\chi$[26]. For the discrete values of θ (only those for which the potential is less than some maximum value V_{max}), we diagonalize one-dimensional vibrational problems in χ[26,27(e)]. The full surface Hamiltonian matrix is formed in the truncated set of these 1-D eigenvectors and subsequently diagonalized, yielding the surface functions. In the surface function calculations for the H_3 system, the DVR Hamiltonian matrices are quite small, their dimension ranging from 470 to 630. Consequently we can solve efficiently for the full spectrum, out of which some 90-100 lowest surface eigenstates are highly converged (to four significant figures or better). These calculations also show that determining reasonable parameters which define the DVR basis is rather straightforward, and involves little preliminary calculation.

3.4 Analytic Basis Set Method

In our analytic basis set approach for solving the APH surface function Hamiltonian Eq. (38) we use a linear combination of harmonic oscillators in each arrangement channel times Legendre polynomials.

$$\begin{aligned}\Phi_{t\Lambda}^{Jp} = &\sum_{v_A j_A} c_{v_A j_A}\phi_{v_A j_A}(\theta_A)P_{j_A\Lambda}(\Theta_A)\begin{cases}1 & \text{if} \quad p=0\\ \cos\chi_A & \text{if} \quad p=1\end{cases}\\ &+\sum_{v_B j_B} c_{v_B j_B}\phi_{v_B j_B}(\theta_B)P_{j_B\Lambda}(\Theta_B)\begin{cases}1 & \text{if} \quad p=0\\ \cos\chi_B & \text{if} \quad p=1\end{cases}\\ &+\sum_{v_C j_C} c_{v_C j_C}\phi_{v_C j_C}(\theta_C)P_{j_C\Lambda}(\Theta_C)\begin{cases}1 & \text{if} \quad p=0\\ \cos\chi_C & \text{if} \quad p=1\end{cases}\end{aligned} \tag{62}$$

where $\phi_{v_f j_f}(\theta_f)$ is a harmonic oscillator. Quadratures appropriate for each arrangement channel are used to calculate overlap and hamiltonian matrix elements. This basis is non-orthogonal giving a generalized eigenvalue problem. Since the matrices are small (≈100-300), we choose to solve for the complete spectrum.

3.5 Comparison of Methods for Generating Surface Functions

The improved finite element method gives accurate ground state eigenvalues with only a few elements. However, it still requires a fairly large mesh to accurately produce the highly excited rotational states.

As we will explain, the DVR approach to calculating surface functions has several additional advanages over the FEM. The DVR basis needs to be changed only a few times over the ρ range of interest (four times in the case of H_3), thus minimizing the possibility that surface functions at neighbouring ρ's may be calculated with appreciably different bases (which is manifested in visible discontinuities in the surface energy correlation diagram). Convergence studies, i.e., determining the DVR basis parameters adequate for the whole ρ range, are quite simple and require relatively little effort and time.

In the H_3 surface function calculation, the ρ range of interest (ρ = 1.6–9.1 a.u.) was divided in four segments, with end points at 5.1, 6.0, 7.0 and 9.1 a.u., respectively. The respective orders of the DVRs for θ and χ were 25 and 40 in the first segment, 30 and 50 in the second, 33 and 70 in the third, 36 and 90 in the fourth segment. The dimension of the surface function Hamiltonian matrices to be diagonalized varied as function of ρ, from 470 to 630. The surface functions were calculated at 179 ρ values. The complete calculation, which included calculating the potential coupling and overlap (between surface functions at neighbouring ρ's) matrix elements, took 2.79 hours of Cray XMP CPU time.

The above CPU time is about a factor of three smaller than the CPU time for a comparable calculation performed using the FEM method. However, from the FEM calculation, only 71 surface functions were obtained, out of which some 40-45 eigenstates were well converged (i.e., gave smooth energy correlation diagrams). In contrast, the DVR calculation generates the full surface eigenstate spectrum for every ρ; the lowest 90-100 eigenstates (up to total energies of 2.2-2.3 eV) are converged to at least four significant figures. Thus, the DVR vs. FEM comparison based on CPU times alone, is somewhat misleading. In fact, the DVR calculation of only the surface functions is at least 6-7 times faster than with the FEM; however, calculation of the necessary matrix elements begins to take almost the same amount of CPU time. Hence, decreasing the CPU time required for generating matrix elements will net another factor of 2.

Preliminary results from the analytic basis set method indicate it to be a factor of 60 faster than the finite element method on the $LiFH$ problem.

4. SAMPLE CALCULATIONS AND RESULTS

4.1 $Li + FH \longrightarrow LiF + H$

Chen and Schaefer[29] have calculated an accurate *ab initio* potential energy surface for $LiFH$. An analytic Bond-order fit to this potential energy surface was performed by Lagana.[30-35] Quasiclassical calculations of the $Li + FH \longrightarrow LiF + H$ reactive scattering properties were used to further refine the potential energy surface so as to reproduce experimental results. Three dimensional reactive scattering cross sections have also been carried out using reduced dimensionality techniques.

In Fig. 2 we show isoenergetic contours of the potential energy surface at two values of ρ. Fig. 2a shows that the LiF and HF channel surface functions will be strongly mixed. Also, one can see the barrier to reaction. Fig. 2b shows that the arrangement channels have separated and the HF is somewhat rotationally hindered.

In Figs. 3 and 4 we show the lowest surface functions at the same two values of ρ. Fig. 3a shows that the surface functions for different arrangement channels are strongly mixed. Fig. 3b shows that lowest states are simply rotational states of the LiF arrangement.

In Fig. 5 we show an energy correlation diagram for the lowest 80 states of $Li+FH \longrightarrow LiF + H$. In this figure we can clearly see the $\nu = 0 - 4$ vibrational manifolds of LiF and the lowest rotational states of HF for $\nu = 0$. We see that at distances larger than $\approx 6a_0$ the curves are basically diabatic. Whereas, for smaller distances we see strong adiabatic coupling. Reactivity will therefore be strongly dominated by short range interactions as one would expect. Long range interactions will then cause rotational vibrational mixing in each arrangement channel.

4.2 $F + H_2 \longrightarrow HF + H$

This reaction has been the focus of much experimental and theoretical effort. For example, *some* theoretical techniques with which the dynamics have been investigated include quasiclassical trajectories[36], variational transition state theory[37], collinear 1-D quantum scattering[38], and an approximate 3-D quantum treatment[39-40]. Recently, Zhang and Miller[41] reported the first accurate 3-D quantum reaction probabilities for this system

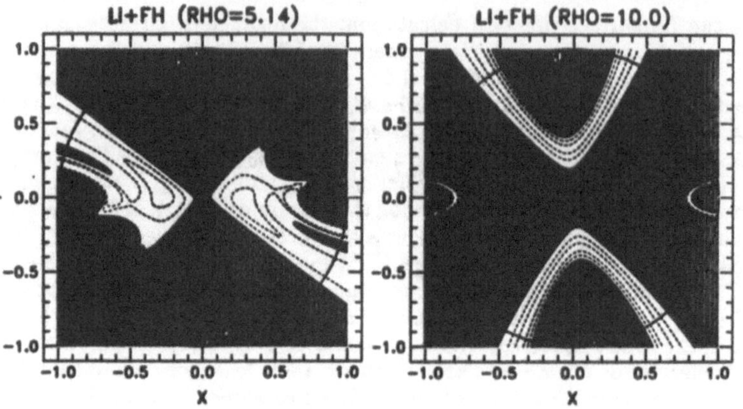

Figure 2. Isoenergetic contours of the $Li{+}FH \longrightarrow LiF{+}H$ Bond-Order potential energy surface. The Contours are drawn at 0.2, 0.5, 1.5 and 3.7 eV. The potential energy surface above 3.7 eV has been shaded. The distance from the center of the plot is a measure of θ whereas the azimuthal angle χ_A is measured from the positive x-axis. **a)** Stereographic projection at $\rho = 5.14a_0$. **b)** Stereographic projection at $\rho = 10.0a_0$.

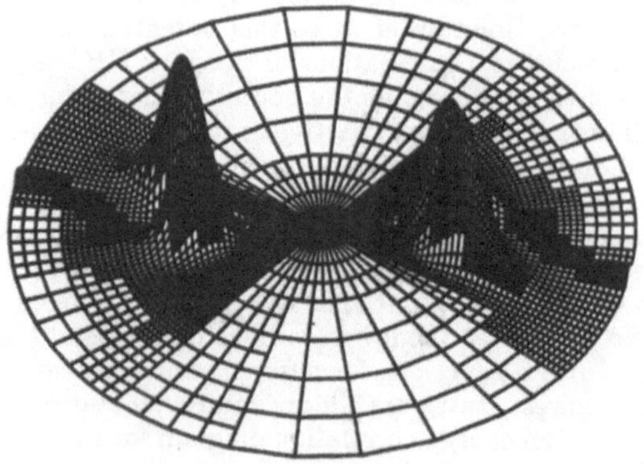

Figure 3. The 4 lowest surface functions for $Li + HF \rightleftharpoons LiF + H$ calculated at $\rho = 5.14a_0$.

using an S-matrix version of the Kohn variational principle and the Muckerman 5 ($M5$) potential energy surface[36].

Truhlar and coworkers[38] have generated a partly empirical, partly *ab-initio*, potential energy surface for FH_2, denoted surface No. 5a, which corrects some of the observed deficiencies in the $M5$ surface. In Fig. 6 we show isoenergetic contours of the No. 5a surface at two values of ρ. Fig. 6a shows that at $\rho = 7.3a_0$ the H_2 and HF channels are well separated with each channel resembling its asymptotic ($\rho \rightarrow \infty$) form. Fig. 6b shows that at $\rho = 5.2a_0$ the channel surface functions between the H_2 and HF channels will

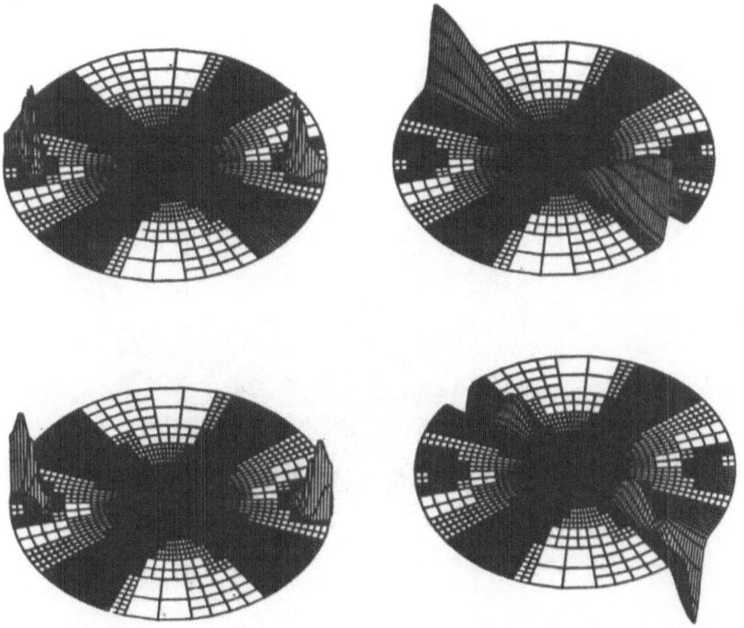

Figure 4. The 4 lowest surface functions for $Li + HF \rightleftharpoons LiF + H$ calculated at $\rho = 10.0a_0$.

be strongly mixed; this distance is near the FHH transition state. The barriers between channels seen in Fig. 1a have now decreased in height to yield transition states located in Fig. 1b along the equator (on the rim) near $\chi_A = \pi/6, -\pi/6, 5\pi/6$, and $-5\pi/6$.

In Figs. 6 and 7 we show two surface functions each for the same two values of ρ calculated using the finite element method. Fig. 7a shows the lowest energy surface function that correlates with an H_2 state asymptotically. Fig. 7b is an example of a surface function which correlates with a $v = 2$ state of HF asymptotically. Note the lack of surface function amplitude between the H_2 and HF channels in both Figs. 2a and 2b. Conversely, Figs. 2c and 2d show two different surface functions at $\rho = 5.2a_0$ which exhibit amplitude in *both* the H_2 and HF channels. These surface functions provide reaction flux to the total wavefunction.

In Fig. 8 we present an energy correlation diagram for $F + H_2 \rightleftharpoons HF + H$. The finite element method is too costly to converge for this system, however the DVR method is economical. To obtain all of the surface functions and eigenenergies at all ρ values only required 210 CPU minutes on the Cray-XMP. These results will allow us to obtain converged transition probabilities from threshold to more than 2.5 eV. Scattering calculations are presently underway.

4.3 $e^+ + H \longrightarrow Ps + p^+$

The APH method is applicable to any three particle rearrangement collision for which the potential is known. A good example of this is a problem from atomic physics, a positron scattering with a hydrogen atom. Positrons are antiparticles, positively charged electrons. Besides the usual elastic and inelastic scattering processes a rearrangement process also

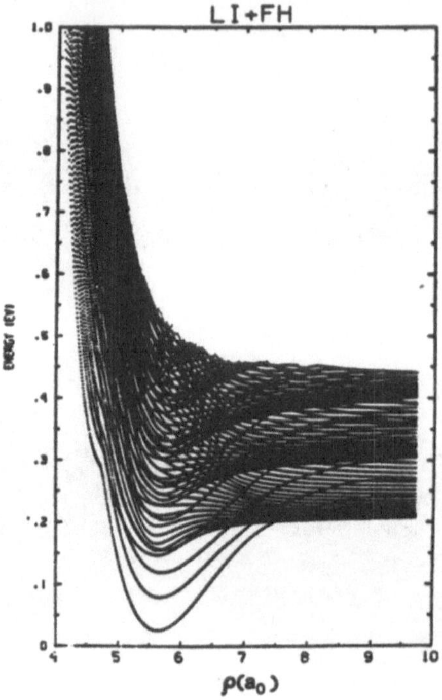

Figure 5. Energy correlation diagram for the lowest 80 vibrational-rotational energy levels of $Li + HF \rightleftharpoons LiF + H$ as a function of the hyperradius ρ.

exists. The electron can be stripped from the hydrogen atom and transferred to the positron, forming a positronium atom in either the ground state or in excited states. All four processes are shown in the following schematic:

$$e^+ + H \rightleftharpoons \begin{cases} e^+ + H & \text{elastic,} \\ e^+ + H^* & \text{inelastic,} \\ p^+ + Ps & \text{elastic rearrangement,} \\ p^+ + Ps^* & \text{inelastic rearrangement.} \end{cases}$$

Interestingly, positronium is not stable. If the spins are aligned Ps annihilates in approximately 1.4×10^{-7}sec by emitting 3 photons[42]. If the spins are opposed 2 photons are emitted[43] in about 1.25×10^{-10}sec. We assume that the collision process occurs on a shorter time scale than annihilation. Scattering energies will be kept below the three body break up, or ionization, energy and only the $J = 0$ partial wave will be examined. Further, since there are no identical particles spin will be ignored.

The primary physical interest in $e^+ + H$ comes from astrophysics where the cross sections are of use in analyzing positron annihilation radiation from the galactic center[44-46]. Theoretically this process is also interesting since it is the simplest scattering process

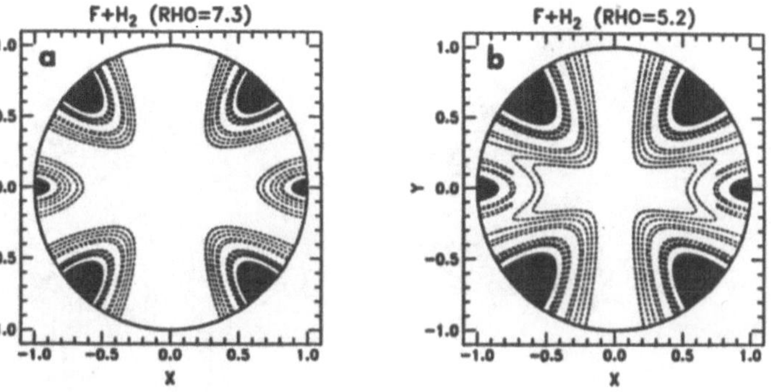

Figure 6. Isoenergetic contours of the $F + H_2 \rightleftharpoons HF + H$ No. 5a potential energy surface[38]. All contours are measured from the bottom of the asymptotic well for HF. The contours are drawn at 0.35, 0.75, 1.44 (a little above the bottom of the asymptotic well for H_2), 2.1, and 2.8 eV. The potential energy surface above 10.0 eV has been shaded. The distance from the center of the plot is a measure of θ whereas the azimuthal angle χ_A is measured from the positive x axis. **a)** Stereographic projection at $\rho = 7.3a_0$. **b)** Stereographic projection at $\rho = 5.2a_0$.

involving positrons. After 25 years of work on this system by many groups[47], only the elastic cross section[48,49] below the first excited state is well known. The rearrangement partial cross section into the Ps ground state has been accurately calculated[50] but not verified. Recently the elastic and rearrangement cross sections have been calculated[47,51] below the first inelastic threshold. As yet no experimental results exist for this system.

At this point our calculation is not quite finished, but we have been able to verify all the previous results. Figure 9 shows the energy correlation diagram for the 20 lowest surface functions. Bound states of wells in the adiabatic curves correspond to metastable scattering states and show up as resonances in the partial cross sections. Using these metastable state energies we have confirmed the lowest few resonances for the H(2)[52], Ps(2), H(3)[53], H(4), and Ps(3)[54] states.

Coulomb singularities occur at the atoms for $e^+ + H$ making this a particularly difficult potential. The ground state wavefunctions, H(1s) and Ps(1s), have cusps and are difficult to model using the finite element method in APH coordinates. A very sensitive test of the accuracy of these two states is the scattering phase shifts. At the lowest energy, k=0.1 (k is given in atomic units), our elastic phase shift is 44% higher than the accurate one[48,49]. As the energy increases our phase shifts rapidly approach the accurate ones, 17% high at k=0.2, 7% at k=0.4, and within 4% at energies higher than k=0.5. The Ps formation phase shift is 10% higher than Humberston[9] at k=0.71, 2.5% high at k=0.75, and within 0.5% at higher energies.

Figure 10 shows the $J = 0$ partial cross section from H(1s) to Ps(1s) for energies between the Ps(1) threshold and the H(2) threshold. Also plotted are the results of Humberston[50], with which we agree to within 15%, and the next two closest calculations. Our results and those of Humberston agree better than any other two previous calculations for this partial wave. Both the very abrupt Ps(1) threshold and onset of the H(2) threshold are shown in our cross section.

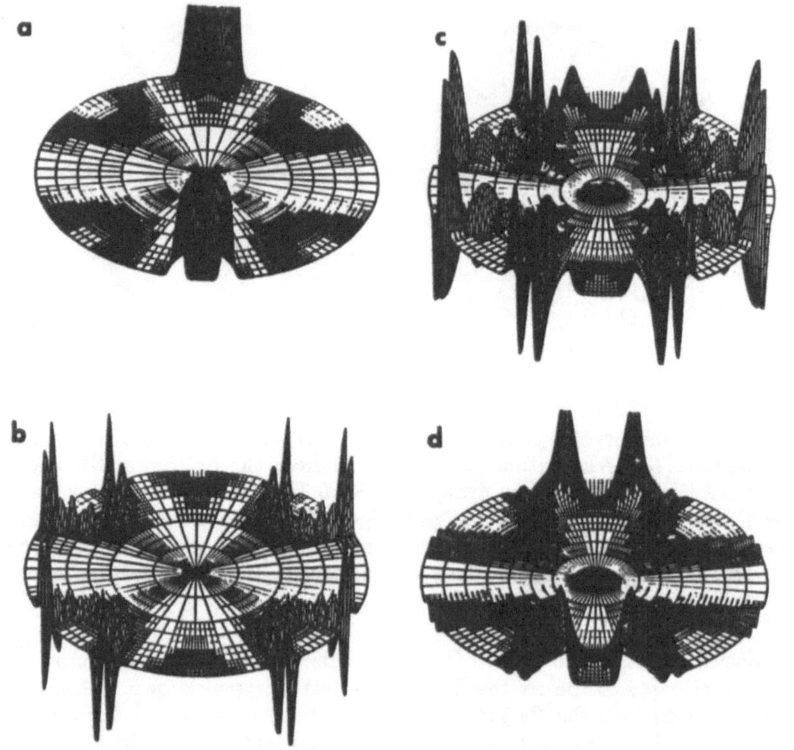

Figure 7. Surface functions for the No. 5a potential energy surface[38]. An H_2 channel, centered at $\chi_A = 0$, faces the reader. **a)** The surface function $\Phi^0_{59,0}(\theta, \chi_A; \rho)$ at $\rho = 7.3a_0$. **b)** The surface function $\Phi^0_{60,0}(\theta, \chi_A; \rho)$ at $\rho = 7.3a_0$. **c)** The surface function $\Phi^0_{63,0}(\theta, \chi_A; \rho)$ at $\rho = 5.2a_0$. **b)** The surface function $\Phi^0_{64,0}(\theta, \chi_A; \rho)$ at $\rho = 5.2a_0$.

Presently we can confirm all the theoretical predictions for $J = 0$ $e^+ + H$ scattering. Extensions to higher energies below the ionization threshold are under way. Higher total angular momenta will be worked on using the basis set approach for calculating surface functions.

4.4 $d^+ + t^+\mu^- \longrightarrow t^+ + d^+\mu^-$

Muon catalyzed fusion reactions are currently being investigated by many investigators for obvious reasons. The cross section for muon transfer between triton and deuteron atoms is one of the many reactions of interest.

For this system the interaction is simply a sum of coloumbic terms. In Fig. 11 we present two energy correlation diagrams. The energy levels at large ρ approach the energies of the isolated muonic atoms $d^+\mu^-$ and $t^+\mu^-$. We see that the states interact strongly. We also see that the density of states increases rapidly at higher energies.

In Fig. 12 we present reactive cross sections for muon transfer. We see no resonances.

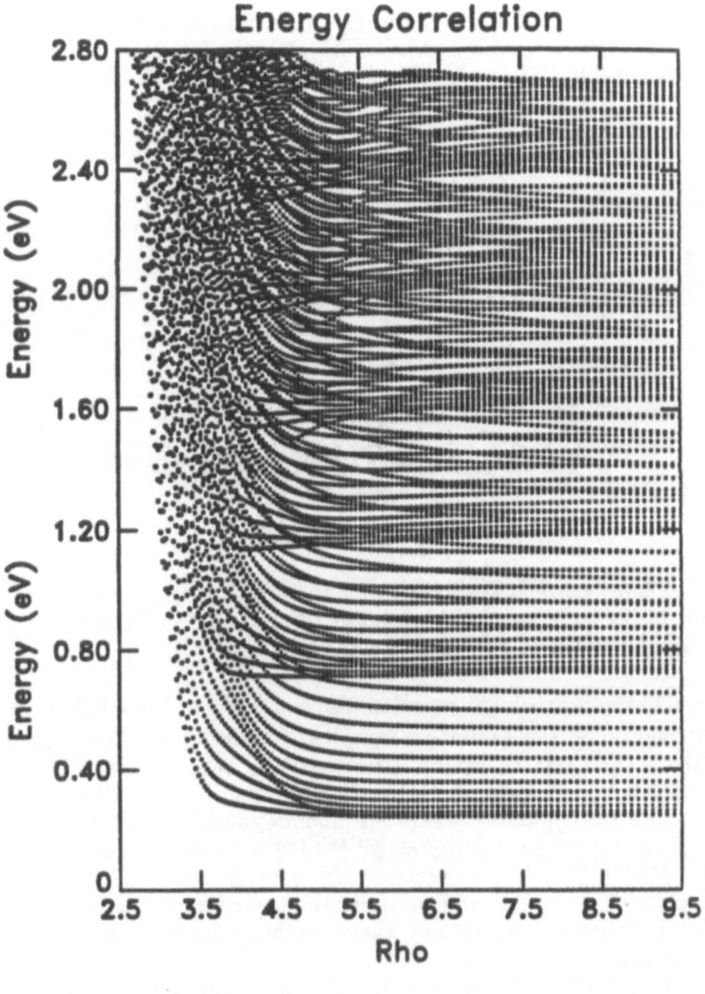

Figure 8. Energy correlation diagram for the lowest vibrational-rotational energy levels of $F + H_2 \rightleftharpoons HF + H$ as a function of the hyperradius ρ.

4.5 $H_2O + h\nu \longrightarrow OH + H$.

The photodisociation of H_2O in the first absorption band ($\lambda \sim 165$ nm) proceeds through a two step process[57]. First, a bound state of the ground ($\tilde{X}\ {}^1A_1$) electronic potential energy surface is excited to the $\tilde{A}\ {}^1B_1$ potential energy surface by a photon of energy $h\nu$. Second, since the $\tilde{A}$ surface is repulsive, the excited H_2O molecule dissociates into $OH + H$. This second step can be viewed as the final half of a reactive collision. To treat the exact quantum dynamics of this process requires the knowledge of the initial bound state wavefunction $\Psi^{JMp(0)}$ on the $\tilde{X}$ surface, as well as the appropriate scattering wavefunction $\Psi^{J'M'p'(+)}$ on the $\tilde{A}$ surface. $\Psi^{JMp(0)}$ is calculated using a linear combination

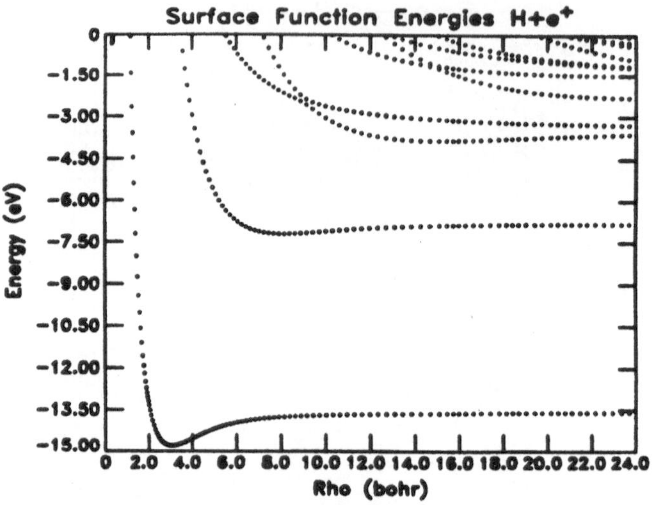

Figure 9. Energy correlation diagram for the lowest 20 energy levels of $e^+ + H$ as a function of the hyperradius ρ. Ionization occurs at 0 eV. Asymptotically these curves correspond to bound states of H and Ps.

of the n solutions to the Schrödinger equation [the expansion of Eq. (37)] subject to bound state boundary conditions. The excitation from the $\tilde{X}$ surface to and subsequent scattering on the $\tilde{A}$ surface is then treated with a distorted wave approximation[58] for the *dipole coupling* and with full close coupling for the nuclear motion. This results in a set of coupled *inhomogeneous* scattering equations, where the inhomogeneous driving term is proportional to the projection of the bound portion of $\Psi^{J'M'p'(+)}$ onto the transition dipole moment operator times $\Psi^{JMp(0)}$. The dipole selection rule connects the initial state $J = 0$ with the final scattering state $J' = 1$. The odd parity of the transition is carried by the *electronic* wavefunctions of the initial and final states, therefore the parity of the *nuclear* wavefunctions is conserved ($p = even$ to $p' = even$).

Surface functions calculated on both the $\tilde{X}$ and $\tilde{A}$ surfaces are shown in Fig. 13. We used the $\tilde{X}$ surface of Sorbie and Murrell[59], and the $\tilde{A}$ surface of Engel, Schinke, and Staemmler[57]. $\rho = 2.67a_0$ is the value derived from the equilibrium geometry of the ground vibrational state of the $\tilde{X}$ surface. Figs. 13a and 13b are the lowest and fourth lowest (in energy) surface functions on the $\tilde{X}$ surface for ($J = 0$, $p = even$) at $\rho = 2.67a_0$. Figs. 13c and 13d are the lowest and fourth lowest surface functions on the $\tilde{A}$ surface for ($J' = 1$, $p = even$) at $\rho = 2.67a_0$. Qualitatively the $\tilde{X}$ and $\tilde{A}$ surfaces are very similar in shape at this value of ρ as the lowest surface functions (Fig. 13a *vs.* 13c), as well as the fourth lowest surface functions (Fig. 13b *vs.* 13d), exhibit only small differences barely discernable to the eye. In a Franck-Condon description of the transition (vertical excitation at $\rho = 2.67a_0$), the surface functions in Figs. 13a and 13c would overlap strongly, as would the surface functions in Figs. 13b and 13d.

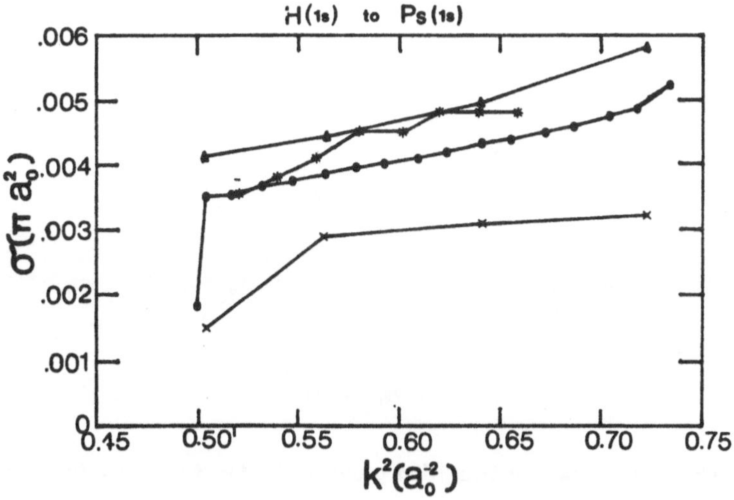

Figure 10. $J = 0$ partial cross section for H(1s) to Ps(1s). The curve marked with o are the present results and Δ are those of Humberston[50]. Results of Stein and Sternlicht[55] are marked by $*$, and $\times$ marks Chan and Fraser[56]. The Ps(1s) threshold occurs at $k^2 = 0.5$ and the H(2) threshold is at $k^2 = 0.75$.

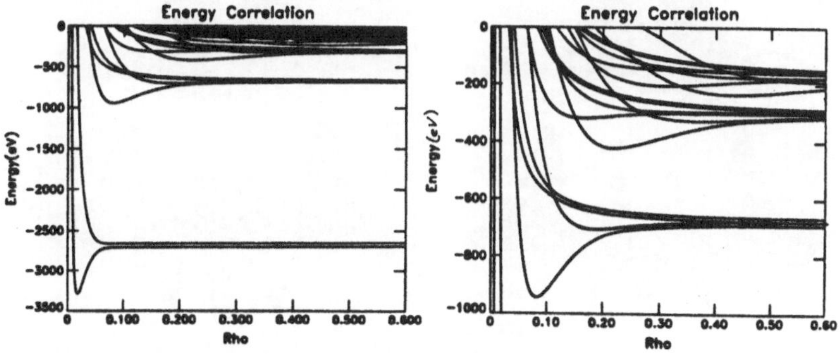

Figure 11. Energy correlation diagram of the lowest 56 energy levels of $d^+ + t^+\mu^- \longrightarrow t^+ + d^+\mu^-$ as a function of the hyperradius ρ. **a)** Ground and excited electronic states. **b)** Excited electronic states.

5. CONCLUSIONS

The APH coordinates are particularly well suited for studying reactive scattering processes. Some of the important characteristics of the APH coordinates are:

1. They are orthogonal.
2. The kinetic energy operator is fairly simple.
3. All of the arrangement channels are treated equivalently.
4. They allow for a single consistent CS_{APH} approximation.
5. They provide a convenient mapping to easily visualize potential energy surfaces.

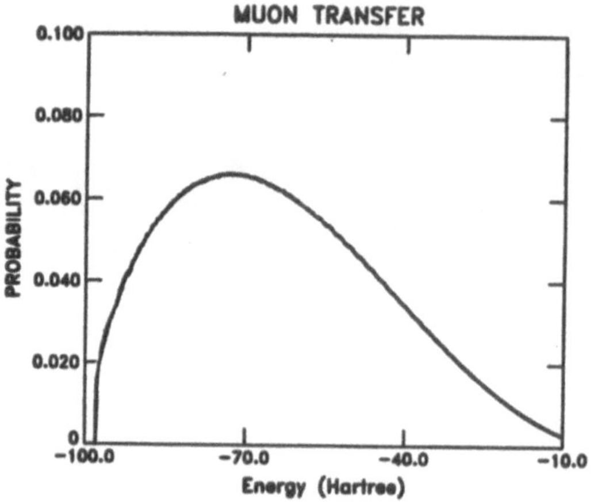

Figure 12. Muon transfer transition probabilities for $J = 0$ as a function of the incident energy.

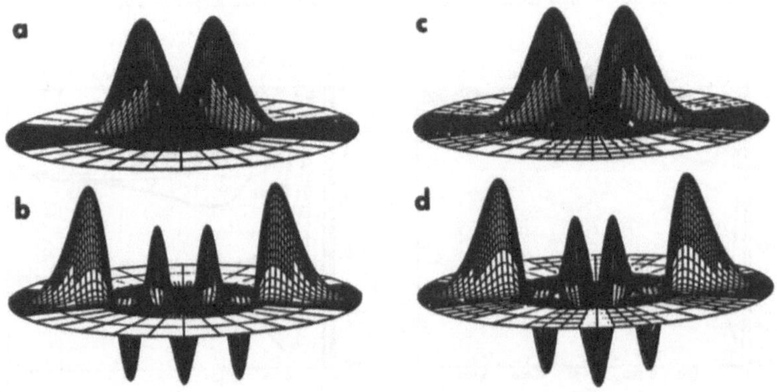

Figure 13. Surface functions for $H_2O + h\nu \longrightarrow H + OH$. The H_2 channels are centered in the two regions of densest mesh along the equator (rim). **a)** The surface function $\Phi^0_{1,0}(\theta, \chi_A; \rho)$ on the $\tilde{X}$ surface[58] at $\rho = 2.67a_0$. **b)** The surface function $\Phi^0_{4,0}(\theta, \chi_A; \rho)$ on the $\tilde{X}$ surface[58] at $\rho = 2.67a_0$. **a)** The surface function $\Phi^1_{1,1}(\theta, \chi_A; \rho)$ on the $\tilde{A}$ surface[57] at $\rho = 2.67a_0$. **b)** The surface function $\Phi^1_{4,1}(\theta, \chi_A; \rho)$ on the $\tilde{A}$ surface[57] at $\rho = 2.67a_0$.

6. They provide promising possibilities for accurately treating collision induced dissociation processes.
7. They are an instantaneous principal moment of inertia coordinate system and swing smoothly from reactants to products.
8. They are applicable to symmetric or asymmetric systems with light or heavy masses.
9. Only regular solutions are required.
10. Matching between differnent coordinate regions is done *via* simple projection.

The most time consuming part of the calculation, the determination of the surface functions, has been improved by almost two orders of magnitude. The DVR and analytic basis set methods are preferred over finite elements methods. We have designed an efficient numerical method for studying reactive scattering that can produce exact and approximate results for systems and energies which have heretofore been impossible to study.

6. ACKNOWLEDGMENTS

We thank Robert B. Walker for countless useful discussions and helpful suggestions, William A. Cook for much assistance with the finite element program, Susan B. Woodruff for much assistance with the implementation and optimization of the Lanczos code and Melvin L. Prueitt for assistance with computer graphics. This work was performed under the auspices of the U. S. Department of Energy and was partially supported by the National Science Foundation under Grant No. CHE-8706385.

7. REFERENCES

[1] For recent reviews, see a) A. Kuppermann, *Theor. Chem. Advances and Perspectives* **6A**, 79 (1981); b) R. B. Walker and J. C. Light, *Ann. Rev. Phys. Chem.* **31**, 401 (1980); c) D. G. Truhlar and R. E. Wyatt, *ibid.* **27**, 1(1976); d) M. Baer, *Adv. Chem. Phys.* **49**, 191 (1982); e) G. C. Schatz, in *The Theory of Chemical Reaction Dynamics*, edited by D. C. Clary (D. Reidel, Netherlands, 1986), pp. 1-26; and references therein.

[2] A. Kuppermann and G. C. Schatz, *J. Chem. Phys.* **62**, 2502 (1975); G.C. Schatz and A. Kuppermann, ibid. **65**, 4642 and 4668 (1976).

[3] A. B. Elkowitz and R. E. Wyatt, *J. Chem. Phys.* **62**, 2504 (1975); **63**, 702 (1975).

[4] R. T Pack and G. A. Parker, *J. Chem. Phys.* **87**, 3888(1987).

[5] F. T. Smith, *J. Chem. Phys.* **31**, 1352 (1959); *Phys. Rev.* **120**, 1058 (1960).

[6]J. O. Hirschfelder and J. S. Dahler, *Proc. Nat. Acad. Sci.* **42**, 363 (1956); D. W. Jepsen and J. O. Hirschfelder, *ibid.* **45**, 249 (1959).

[7] J. O. Hirschfelder, *Intern. J. Quantum Chem. Symp.* **3**, 17 (1969).

[8] K. H. Yang, J. O. Hirschfelder and B. R. Johnson, *J. Chem. Phys.* **75**, 2321 (1981); B. R. Johnson, J. O. Hirschfelder, and K. H. Yang, *Rev. Mod. Phys.* **55**, 109 (1983).

[9] F. T. Smith, *J. Math. Phys.* **3**, 735 (1962); R. C. Whitten and F. T. Smith, *ibid.* **9**, 1103 (1968); and R. C. Whitten, *ibid.* **10**, 1631 (1969).

[10] L. M. Delves, *Nucl. Phys.* **9**, 391 (1959); **20**, 275 (1960).

[11] B. R. Johnson, *J. Chem. Phys.* **73**, 5051 (1980).

[12] B. R. Johnson, *J. Chem. Phys.* **79**, 1906, 1916 (1983).

[13] R. T Pack, *J. Chem. Phys.* **60**, 633 (1974).

[14] A. M. Arthurs and A. Dalgarno, *Proc. R. Soc. London* **A256**, 540 (1960).

[15] P. McGuire and D. J. Kouri, *J. Chem. Phys.* **60**, 2488 (1974); D. J. Kouri, in *Atom-Molecule Collision Theory: a Guide for the Experimentalist*, edited by R. B. Bernstein,

(Plenum Press, New York, 1979), pp. 301-358; D. J. Kouri and D. E. Fitz, *J. Phys. Chem.* **86**, 2224 (1982); and references therein.

[16] M. Abramowitz and I. A. Stegun, editors, *Handbook of Mathematical Functions* (Dover, New York, 1968), pp. 437f.

[17] K. D. McLenithan, Ph. D. thesis, University of Illinois at Urbana-Champaign, 1982, Appendix C.

[18] Cf. C. F. Curtiss and F. T. Adler, *J. Chem. Phys.* **20**, 249 (1952).

[19] A. Kuppermann, *Chem. Phys. Lett.* **32**, 375 (1975). See also Ref. 1a.

[20] a) A. Kuppermann and P. G. Hipes, *J. Chem. Phys.* **84**, 5962 (1986); b) P. G. Hipes and A. Kuppermann, *Chem. Phys. Lett.* **133**, 1 (1987).

[21] D. M. Hood and A. Kuppermann, in *The Theory of Chemical Reaction Dynamics*, edited by D. C. Clary (D. Reidel, Netherlands, 1986) pp. 193-214.

[22] R. T Pack, *Chem. Phys. Lett.* **108**, 333 (1984).

[23] B. Podolsky, *Phys. Rev.* **32**, 812 (1928).

[24] (a) J. V. Lill, G. A. Parker, and J. C. Light, Chem. Phys. Lett. **89**, 483 (1982); (b) J. C. Light, I. P. Hamilton, and J. V. Lill, J. Chem. Phys. **82**, 1400 (1985).

[25] For example, (a) J. Tennyson, Comput. Phys. Rep. **4**, 1 (1986); (b) S. Carter and N. C. Handy, *ibid.* **5**, 115 (1986).

[26] R. M. Whitnell and J. C. Light, J. Chem. Phys., in press.

[27] (a) Z. Bačić and J. C. Light, J. Chem. Phys. **85**, 4594 (1986); (b) *ibid.* **86**, 3065 (1987); (c) J. C. Light and Z. Bačić, *ibid.* **87**, 4008 (1987); (d) Z. Bačić, D. Watt, and J. C. Light, *ibid.* **89**, 947 (1988); (e) Z. Bačić, R. M. Whitnell, D. Brown, and J. C. Light, Comput. Phys. Commun., in press.

[28] I. P. Hamilton and J. C. Light, J. Chem. Phys. **84**, 306 (1986).

[29] M. M. Chen and H. F. Schaefer, *J. Chem. Phys.* **72**, 4376 (1980).

[30] I. Noorbatcha, and Sathyamurthy, *J. Chem. Phys.* **76**, 6447 (1982); *J. Am. Chem. Soc.* **104**, 1766 (1982); *Chem. Phys. Lett.* **93**, 432 (1982).

[31] I. Noorbatcha, and Sathyamurthy, *Chem. Phys.* **77** 67 (1983).

[32] J. M. Alvariño, O. Gervasi, and A. Lagana, *Chem. Phys. Lett.* **87** 254 (1982).

[33] A. Laganà, M. L. Hernandez, and J. M. Alvariño, Chem. Phys. Lett. **106** 41 (1984).

[34] A. Laganà, and E. Garcia, *Chem. Phys. Lett.* **139**, 140 (1987).

[35] For more details see A. Laganà, E. Garcia, O. Gervasi, *Faraday Discuss. Chem. Soc.* **84** general discussion of paper 23 (1987).

[36] J. T. Muckerman, *Theor. Chem. Adv. Perspect.* **6A**, 1 (1981).

[37] D. G. Truhlar, B. C. Garrett, and N. C. Blais, *J. Chem. Phys.* **80**, 232 (1984).

[38] R. Steckler, D. G. Truhlar, and B. C. Garrett, *J. Chem. Phys.* **82**, 5499 (1985); F. B. Brown, R. Steckler, D. W. Schwenke, D. G. Truhlar, and B. C. Garrett, *ibid.*, 188 (1985).

[39] G. C. Schatz, J. M. Bowman, and A. Kupperman, *J. Chem. Phys.* **63**, 674 (1985).

[40] M. J. Redmon and R. E. Wyatt, *Intern. J. Quantum Chem.* **95**, 403 (1975); **115**, 343 (1977); *Chem. Phys. Lett.* **63**, 209 (1979).

[41] J. Z. H. Zhang and W. H. Miller, *J. Chem. Phys.* **88**, 4549 (1988).

[42] W. E. Caswell, G. P. Lepage, and J. Sapirstein, Phys. Rev. Lett. **38**, 488 (1977); T. C. Griffith, G. R. Heyland, K. S. Lines, and T. R. Twomey, J. Phys. B. **11**, L743 (1978); D. W. Gidley, A. Rich, P. W. Zitzewitz, and D. A. L. Paul, Phys. Rev. Lett. **40**, 737 (1978).

[43] H. J. Ache, *Positronium and Muonium Chemistry*, edited by H. J. Ache (American Chemical Society, Washington, 1979).

[44] M. Leventhal and C. J. MacCallum, *Positron Annihilation*, edited by P. C. Jain, R. M. Singru, and K. P. Gopinathan (World Scientific, Singapore, 1985), and references therein.

[45] B. L. Brown, M. Leventhal, and A. P. Miller, Jr., *Positron Annihilation*, edited by P. C. Jain, R. M. Singru, and K. P. Gopinathan (World Scientific, Singapore, 1985), and references therein.

[46] B. L. Brown, *Positron Studies of Solids, Surfaces, and Atoms*, edited by A. P. Mills, Jr., W. S. Crane, and K. F. Canter (World Scientific, Singapore, 1986), and references therein.

[47] J. W. Humberston, *Adv. At. Mol. Phys.* **15**, 101 (1979) and references therein; J. W. Humberston, *Positron (Electron)-Gas Scattering*, edited by W. E. Kauppila, T. S. Stein, and J. M. Wadehra (World Scientific, Singapore, 1985), and references therein.

[48] A. K. Bhatia, A. Temkin, R. J. Drachman, and H. Eiserike, *Phys. Rev. A* **3**, 1328 (1971); S. K. Houston and R. J. Drachman, *ibid.* **3**, 1335 (1971).

[49] M. Abdel-Raouf, *J. Phys. B* **12**, 3349 (1979); *Can. J. Phys.* **60**, 577 (1982).

[50] J. W. Humberston, *Can. J. Phys.* **60**, 591 (1982); J. Phys. B **17**, 2353 (1984).

[51] C. J. Brown and J. W. Humberston, *J. Phys. B* **17**, L423 (1984); **18**, L401 (1985).

[52] E. Pelikan and H. Klar, *Z. Phys. A* **310**, 153 (1983).

[53] Y. K. Ho and C. H. Greene, *Phys. Rev. A* **35**, 3169 (1987).

[54] Y. K. Ho, *Atomic Physics with Positrons* edited by J. W. Humberston and E. A. G. Armour (Plenum, New York, 1987).

[55] J. Stein and R. Sternlicht, *Phys. Rev. A* **6**, 2165 (1972).

[56] Y. F. Chan and P. A. Fraser, *J. Phys. B* **6**, 2504 (1973).

[57] V. Engel, R. Schinke, and V. Staemmler, *J. Chem. Phys.* **88**, 129 (1988).

[58] Y. B. Band, K. F. Freed, and D. J. Kouri, *J. Chem. Phys.* **74**, 4380 (1981).

[59] K. S. Sorbie and J. N. Murrell, *Mol. Phys.* **29**, 1387 (1975); **31**, 905 (1976).

COMPUTATIONAL STRATEGIES AND IMPROVEMENTS IN THE LINEAR ALGEBRAIC VARIATIONAL APPROACH TO REARRANGEMENT SCATTERING

David W. SCHWENKE
NASA Ames Research Center, Mail Stop 230–3, Moffett Field, CA 94036 USA

Mirjana MLADENOVIC, Meishan ZHAO, and Donald G. TRUHLAR
Department of Chemistry, Chemical Physics Program, and Supercomputer Institute, University of Minnesota, Minneapolis, MN 55455 USA

and Yan SUN and Donald J. KOURI
Department of Chemistry, University of Houston, Houston, TX 77204–5641 USA

ABSTRACT. We discuss the computational steps in calculating quantum mechanical reactive scattering amplitudes by the $\mathscr{L}^2$ generalized Newton variational principle with emphasis on computational strategies and recent improvements that make the calculations more efficient. We place special emphasis on quadrature techniques, storage management strategies, use of symmetry, and boundary conditions. We conclude that an efficient implementation of these procedures provides a powerful algorithm for the accurate solution of the Schroedinger equation for rearrangements.

1. Introduction

Rearrangement scattering is certainly one of the most difficult problems in few–body physics. The quantum mechanical theory of chemical reactions, as a subfield of the larger topic of rearrangement scattering, is similarly one of the most difficult subtopics in molecular collision theory. Successful and useful procedures for rearrangement scattering problems have been advanced from many different directions, including (references are perforce representative rather than exhaustive) the theory of compound nuclei and resonances,[1] formal many–body physics,[2] and multichannel inelastic scattering theory,[3–5] and the theory will certainly continue to benefit from further progress in those fields in which it has its roots.

A. Laganà (ed.), Supercomputer Algorithms for Reactivity, Dynamics and Kinetics of Small Molecules, 131–168.

The approach discussed here has its origins in both many–body theory and multichannel scattering theory. From many–body theory it builds on the method called "configuration interaction" or "superposition of configurations" (SOC), especially as that method is viewed as a systematic way to enlarge multiconfiguration Hartree–Fock basis sets until convergence is reached. This approach has its roots in bound–state electronic structure theory,[6] and applications to dynamics may be found in electron–atom scattering,[3,7–9] electron–molecule scattering,[10] and nuclear physics[11] (where it is sometimes called the "resonating group method"). In the work discussed here the SOC approach is used to expand the reactive amplitude density,[12–15] rather than the scattering wave function,[3,4,7–10,16–18] and the coefficients in the expansion are obtained variationally by a generalization[19,20] of a variational principle due to Newton.[21] Since the reactive amplitude density is vector of $\mathscr{L}^2$ (i.e., square integrable) functions in arrangement channel space, the configurations in our SOC may also be taken as $\mathscr{L}^2$, and this simplification of the basis set is one of the desirable features of this approach which motivates our work.

Another important element in our work is the use of a distortion potential which defines a partially decoupled scattering problem whose solutions are obtained as the first step of the dynamics calculation. Distortion potentials appear as the first step of many inelastic and rearrangement scattering calculations,[22–24] often in conjunction with a variational principle, a combination which is motivated by the expectation that as one increases the coupling in the distortion potential, the level of theory required to treat the remaining channel coupling accurately will be lower. We have recently[20] shown this to be true in our formalism, in particular as we increased the coupling in the distortion potential, the size of the basis set required to solve the full problem decreased. Our approach allows for *multichannel* single–arrangement distortion potentials, as applied previously[25–30] to rearrangement problems in various contexts. We have shown, however, that the combination of this approach with the $\mathscr{L}^2$ generalized Newton variational principle (GNVP) is a particularly powerful method for treating chemical reactions,[19,20,31,32] and it has allowed us to converge very large–scale problems, involving (so far) up to[33] 1035 coupled channels.

Our previous work relevant to this approach has been published in several papers. Our original calculations were based on an $\mathscr{L}^2$ SOC expansion of the reactive amplitude density with the coefficients obtained by the method of moments. We have published several applications,[13–15,34–36] full details of the numerical procedures,[14] a discussion of storage management,[37] and three derivations[14,15,38] of the final equations. In more recent work we have obtained the coefficients of the expansion by the GNVP. We have carried out new applications,[19,20,31–33,39] and we have presented full details of the method and its initial implementation.[20] The present paper presents a discussion of improved computational and storage management strategies for GNVP calculations, including recent improvements in the quadratures. We also give an overview of the method, including a reformulation in terms of the scattering matrix and a discussion of symmetry decoupling, which subjects are also treated elsewhere.[40,41]

2. Generalized Newton Variational Formalism

2.1. BASIC EQUATIONS

The theoretical developments leading up to the equations presented below are given in detail elsewhere,[20] thus only an overview of the formalism will be given, and the emphasis will be on selected details and improvements. Reference 20 plus the current paper should be considered as a pair; together they express the current state of our program.

Our methods are designed to be as efficient as possible for calculations at a single total energy on vector pipeline supercomputers. Thus quantities which are used only once or a few times for a single energy calculation are usually not saved but rather calculated when they are required, even if they are independent of energy. This strategy influences the choice of several computational details and will be discussed in more detail in Sect. 4.

We consider reactive scattering between different arrangements of the atoms A, B, and C. The label α specifies the asymptotic partitioning of the atoms: $\alpha = 1$ for A + BC, $\alpha = 2$ for B + CA, and $\alpha = 3$ for C + AB. We mass scale the Jacobi coordinates for each of the three arrangements to the single reduced mass μ defined by

$$\mu = [m_A m_B m_C/(m_A + m_B + m_C)]^{\frac{1}{2}} \tag{1}$$

where m_A is the mass of atom A, etc. For arrangement $\alpha = 1$, $\vec{R}_\alpha$ is the mass-scaled vector from the center of mass of BC to atom A, $\vec{r}_\alpha$ is the mass-scaled vector from atom B to atom C, and $\cos\gamma_\alpha = \hat{r}_\alpha \cdot \hat{R}_\alpha$. The coordinates of the other arrangements are defined by cyclic permutation of the three atoms. It will be also convenient to let $\mathbf{x}_\alpha$ denote the collection of all coordinates except the scalar length R_α.

The Hamiltonian operator is partitioned into arrangement-dependent components as

$$H = H^D_\alpha + V^C_\alpha, \tag{2}$$

where H^D_α is the Hamiltonian for non-reactive scattering in arrangement channel α with the distortion potential V^D_α, and V^C_α is the coupling potential responsible for connecting the various distortion potential blocks and various arrangements. We denote the Green's function corresponding to the potential H_α^D as G_α^D.

The index n is used to specify a particular α, v, j, ℓ channel where v, j, and ℓ are vibrational, rotational, and orbital quantum numbers; the index m is used to specify a particular transitional basis function, and β denotes a particular pair of n and m. Note that sometimes for clarity we will label quantities by both α and n although the specification of α is redundant. In addition we use $|\alpha_n n>$ to denote a product of an Arthurs–Dalgarno rotational–orbital function (depending on j, ℓ, and J) and a vibrational function (depending on v and j) in channel n. We use α_n to denote the value of α for channel n. Then we may write

$$H_\alpha^D = T + V_\alpha^{diat} + \sum_n \sum_{n'} \delta_{\alpha_n \alpha} \Delta_{nn'}^\alpha |\alpha_{n'} n'> V^{int,\alpha} <\alpha_n n| \tag{2a}$$

where T is the total kinetic energy, V_α^{diat} is the diatomic potential in channel α,

$$V^{int,\alpha} = V - V_\alpha^{diat}, \tag{2b}$$

V is the total potential, and $\Delta_{nn'}^\alpha$ is unity if channel n and channel n′ belong to the same distortion potential block and zero otherwise.

The GNVP may be applied to calculate the reactance matrix **K**, the scattering matrix S, or the transition matrix, **T**, from any of which one may calculate all physical observables for the collision processes by standard[42,43] formulas. In this section, as well as Sects. 3 and 4, we consider the application to the reactance matrix; in Sect. 5 we will consider the direct calculation of the scattering matrix. In all sections, we use a generalized reactance matrix with the form

$$\mathbf{K} = \begin{bmatrix} \mathbf{K}^{oo} & \mathbf{K}^{oc} \\ \mathbf{K}^{co} & \mathbf{K}^{cc} \end{bmatrix} \tag{3}$$

where o stands for open and c for closed. Although the physical results depend only on $\mathbf{K}^{oo}$, we retain the other parts in the calculation because it simplifies the code and adds negligibly to the computer time. (In addition we will use the closed channel parts to simplify the arithmetic in Sect. 5). In practice the channels are not actually stored in the order necessary to partition **K** as indicated in (3), but they are re–ordered this way prior to calculating eq. (53) below.

Using n to label the channels and β to label the basis functions, coupling the various arrangement components of the total system wave function by the Fock coupling scheme,[4,5,7,16,44] expanding the Fock reactive amplitude density[14,15] for each arrangement in an $\mathscr{L}^2$ basis set expressed in the coordinates of that arrangement, and solving for the coefficients by the GNVP, yields the following expression for the elements of the generalized full reactance matrix:[20]

$$K_{nn_0} = \delta_{\alpha\alpha_0}\, {}^oK_{nn_0} + \mathcal{K}_{nn_0}, \tag{3a}$$

where ${}^O\mathbf{K}$ is the collection of all nonzero reactance matrix elements for the decoupled distortion blocks, and

$$\begin{aligned}\mathcal{K}_{nn_0} \equiv\ & <\alpha_n nE|\,\mathcal{K}|\,\alpha_0 n_0 E> = <\alpha_n nE|\,\mathcal{U}|\,\alpha_0 n_0 E> + \\ & + \sum_\beta \sum_{\beta'} <\alpha_n nE|\,\mathcal{U}\mathcal{G}|\beta><\beta|\,\mathcal{G} - \\ & - (\mathcal{G}\mathcal{U}\mathcal{G})^{-1}|\beta'><\beta'|\,\mathcal{G}\mathcal{U}|\,\alpha_0 n_0 E>\end{aligned} \tag{4}$$

where $\alpha_0(\equiv \alpha_{n_0})$ is the initial arrangement, $|\alpha_n nE>$ is a regular standing–wave distorted–wave scattering state for the Hamiltonian $H^D_{\alpha_n}$,

$$\mathcal{U} = (-2\mu/\hbar^2)(H-E), \tag{5}$$

$$\mathcal{G} = \sum_\alpha G^D_\alpha \hat{P}_\alpha, \tag{5a}$$

$$\mathcal{G}\mathcal{U}\mathcal{G} = (-2\mu/\hbar^2) \sum_{\alpha'} \sum_\alpha \hat{P}_{\alpha'} G^D_{\alpha'} (H-H^D_\alpha) G^D_\alpha \hat{P}_\alpha, \tag{5b}$$

$\hat{P}_\alpha$ is a projector on arrangement α, and the basis set is taken to have the form

$$\begin{aligned}|\beta> &= |\alpha_\beta n_\beta m_\beta> \\ &= R^{-1}_{\alpha_\beta} \phi^{\alpha_\beta}_{n_\beta}(\mathbf{x}_{\alpha_\beta}) t_{m_\beta n_\beta}(R_{\alpha_\beta}).\end{aligned} \tag{6}$$

In Eq. (6), $t_{m_\beta n_\beta}$ is a translational basis function, and $\phi^{\alpha_\beta}_{n_\beta}$ is a vibrational–rotational–orbital function given by

$$\phi^\alpha_n(\mathbf{x}_\alpha) = \chi_{\alpha v_n j_n}(r_\alpha) \mathcal{Y}^{JM}_{j_n \ell_n}(\hat{r}_\alpha, \hat{R}_\alpha) \tag{6a}$$

where $\chi_{\alpha v_n j_n}$ is an asymptotic vibrational eigenfunction, v_n is a vibrational quantum number, j_n is a rotational quantum number, ℓ_n is an orbital angular momentum quantum number for relative translational motion, and $\mathcal{Y}^{JM}_{j\ell}$ is a laboratory–frame angular function[7] coupled to total angular momentum quantum numbers J and M. All equations are decoupled in J and in parity P, given by

$$P = (-1)^{j+\ell}, \tag{6b}$$

so we treat each JP block independently and set M = 0. In the rest of this article it will be understood that all equations refer to a single JP block.

In what follows we write Eq. (4) as

$$\mathscr{K} = \mathscr{K}^{B} + \mathbf{B}^{T}\mathbf{C}^{-1}\mathbf{B} \tag{7}$$

where T denotes a matrix transpose, and $\mathscr{K}^{B}$ is the distorted–wave Born approximation[22–30] to the reactance matrix. In practice, Eq. (7) is replaced by

$$\mathscr{K} = \mathscr{K}^{B} + \mathbf{B}^{T}\,\overline{\mathbf{B}} \tag{7a}$$

where

$$\mathbf{C}\,\overline{\mathbf{B}} = \mathbf{B}. \tag{7b}$$

We note that **C** is symmetric, which guarantees the symmetry of $\mathscr{K}$. The matrix elements diagonal in the arrangement label ($\alpha_n = \alpha_0 = \alpha$) are given by

$$\mathscr{K}^{B}_{nn_0} = \sum_{n'} \Delta^{\alpha}_{nn'} \int dR_{\alpha} \, {}^{(r)}f^{\alpha}_{n'n}(R_{\alpha}) \sum_{n''} V^{\alpha}_{n'n''}(R_{\alpha}) \Delta^{\alpha}_{n''n_0} \times$$
$$\times \, {}^{(r)}f_{n''n_0}(R_{\alpha}), \tag{8}$$

$$B_{\beta n_0} = \sum_{n'} \Delta^{\alpha_0}_{n_{\beta}n'} \int dR_{\alpha} \grave{g}^{N}_{n'\beta}(R_{\alpha}) \sum_{n''} V^{\alpha}_{n'n''}(R_{\alpha}) \Delta^{\alpha}_{n''n_0} \times$$
$$\times \, {}^{(r)}f^{\alpha}_{n''n_0}(R_{\alpha}), \tag{9}$$

and

$$C_{\beta\beta'} = \Delta^{\alpha}_{n_{\beta}n_{\beta'}} \int dR_{\alpha} \grave{g}^{N}_{n_{\beta'}\beta}(R_{\alpha}) t^{\alpha}_{m_{\beta'}n_{\beta'}}(R_{\alpha}) -$$
$$- \sum_{n'} \Delta^{\alpha}_{nn'} \int dR_{\alpha} \grave{g}^{N}_{n'\beta}(R_{\alpha}) \sum_{n''} \times$$
$$\times \, V^{\alpha}_{n'n''}(R_{\alpha}) \Delta^{\alpha}_{n''n_{\beta'}} \grave{g}^{N}_{n''\beta'}(R_{\alpha}), \tag{10}$$

while those off–diagonal in arrangement are given by

$$\mathscr{K}^{B}_{nn_0} = \sum_{n'} \Delta^{\alpha_n}_{nn'} \int dR_{\alpha_n} \int dR_{\alpha_0} \, {}^{(r)}f^{\alpha_n}_{n'n}(R_{\alpha_n}) \times$$
$$\times \sum_{n''} W^{\alpha_n \alpha_0}_{n'n''}(R_{\alpha_n}, R_{\alpha_0}) \Delta^{\alpha_0}_{n''n_0} \, {}^{(r)}f^{\alpha_0}_{n''n_0}(R_{\alpha_0}), \tag{11}$$

$$B_{\beta n_0} = \sum_{n'} \Delta^{\alpha_\beta}_{n_\beta n'} \int dR_{\alpha_\beta} \int dR_{\alpha_0} \grave{g}^{N}_{n'\beta}(R_{\alpha_\beta}) \sum_{n''} W^{\alpha_\beta \alpha_0}_{n'n''}(R_{\alpha_\beta}, R_{\alpha_0}) \times$$
$$\times \Delta^{\alpha_0}_{n''n_0} \, {}^{(r)}f^{\alpha_0}_{n''n_0}(R_{\alpha_0}), \tag{12}$$

and

$$C_{\beta\beta'} = \sum_{n'} \Delta^{\alpha_\beta}_{n'n_\beta} \int dR_{\alpha_\beta} \int dR_{\alpha_{\beta'}} \grave{g}^{N}_{n'\beta}(R_{\alpha_\beta}) \mathscr{B}^{\alpha_\beta \alpha_{\beta'}}_{n\,'n_{\beta'}}(R_{\alpha_\beta}, R_{\alpha_{\beta'}}) \times$$
$$\times t^{\alpha_{\beta'}}_{m_{\beta'} n_{\beta'}}(R_{\alpha_{\beta'}}) - \sum_{n'} \Delta^{\alpha_\beta}_{n_\beta n'} \int dR_{\alpha_\beta} \int dR_{\alpha_{\beta'}} \grave{g}^{N}_{n'\beta}(R_{\alpha_\beta}) \times$$
$$\times \sum_{n''} W^{\alpha_\beta \alpha_{\beta'}}_{n\,'n''}(R_{\alpha_\beta}, R_{\alpha_{\beta'}}) \Delta^{\alpha_{\beta'}}_{n''n_{\beta'}} \grave{g}^{N}_{n''\beta'}(R_{\alpha_{\beta'}}), \tag{13}$$

where ${}^{(r)}f^{\alpha}_{nn'}$ is a radial distorted wave function, $V^{\alpha}_{nn'}$ is an intra-arrangement matrix element of the coupling potential given by

$$V^{\alpha}_{nn'} = U^{\alpha}_{nn'} - U^{D\alpha}_{nn'}, \tag{14}$$

where

$$U^{D\alpha}_{nn'} = \Delta^{\alpha}_{nn'} U^{\alpha}_{nn'}, \tag{14a}$$
$$U^{\alpha}_{nn'} = (-2\mu/\hbar^2) \int dx_\alpha \phi^{\alpha *}_{n}(x_\alpha) V^{int,\alpha}(R_\alpha, x_\alpha) \phi^{\alpha}_{n'}(x_\alpha), \tag{15}$$

and $V^{int,\alpha}$ is the interaction potential for arrangement α. The function $\grave{g}^{N}_{n\beta}$ is a half-integrated Green's function; $W^{\alpha\alpha_0}_{nn_0}$ is given by

$$W^{\alpha\alpha_0}_{nn_0} = \mathscr{C}^{\alpha\alpha_0}_{nn_0}(R_\alpha, R_{\alpha_0}) - \sum_{n_0'} \mathscr{B}^{\alpha\alpha_0}_{nn_0'}(R_\alpha, R_{\alpha_0}) U^{D\alpha_0}_{n_0' n_0}(R_{\alpha_0}), \tag{16}$$

where $\mathscr{C}^{\alpha\alpha_0}_{nn_0'}$ is $-2\mu/\hbar^2$ times the inter-arrangement matrix element of the interaction potential; and $\mathscr{S}^{\alpha\alpha_0}_{nn_0'}$ is an inter–arrangement overlap matrix element. Further details of all these quantities are given in Ref. 20.

2.2. COMPUTATIONAL STEPS

The organization of the computational steps is as follows. First the parameters which specify the calculation are initialized. These include masses, the total energy, basis set parameters, and various other numerical parameters. Next, commonly used quantities are pre–calculated and stored. These include vibrational wavefunctions and weights and nodes for the various quadratures used. After that, the radial distorted wave functions and the radial half–integrated Green's functions are calculated and the values of these functions at the radial quadrature points are stored, either in memory or on disk, depending on options. Then the integrations to obtain the $\mathscr{K}^{B}_{nn'}$, $B_{\beta n}$, and $C_{\beta\beta'}$ matrix elements are carried out. Next, the correction $\mathscr{K}$ to the reactance matrix is evaluated by means of Eqs. (7a) and (7b). Using this matrix, the full reactance matrix is calculated from Eq. (3), followed by the evaluation of the scattering matrix and related quantities useful in the interpretation of the outcome of the collisions.

All of these computational steps involve special considerations in order that they and subsequent steps are performed as efficiently as possible. Thus we now discuss the steps in more detail.

2.2.1. *Asymptotic channel states.* We first consider the calculation of the vibrational functions and their associated eigenenergies. The vibrational functions $\chi_{\alpha vj}$ are expanded in terms of harmonic oscillator wave functions with coefficients $B^{\alpha vj}_{\gamma}$. The matrix elements of the diatomic Hamiltonian are evaluated by Gauss–Hermite quadrature, and the Rayleigh–Ritz variational principle is used to determine the expansion coefficients and eigenenergies. The harmonic basis for the vibrational eigenstates is chosen for computational convenience of later steps, namely the calculation of the inter–arrangement integrals. There the vibrational functions will have to be evaluated at many different bond lengths, and it is advantageous to avoid the necessity of interpolating a numerical wave function. As it is, the vibrational functions are evaluated by first recursively calculating the harmonic oscillator basis functions at many bond lengths simultaneously using vector operations and then transforming to the $\chi_{\alpha vj}$ by a call to the very efficient Cray matrix multiply library routine MXM or a new fast matrix multiplication routine[45] utilizing local memory and Strassen's[46] algorithm. In addition to these considerations, another reason for using analytic basis functions is that they provide a more compact representation than a numerical wave function, i.e., less storage space is required to save the expansion coefficients then would be required if one were to save the wave function on a grid.

One difficulty with using harmonic oscillator wave functions is that they behave in a non–physical manner as the bond length goes to zero. These functions go to zero in the limit as the bond length goes to minus infinity, rather than as the bond length goes to zero. However this is not a problem in practice because molecular potentials are very repulsive for small bond lengths, and so the variational principle causes the wave function to be small in that region. It is sometimes necessary to modify the vibrational potentials near the origin or at negative r_α so that the integrals involved in the Rayleigh–Ritz variational principle are all finite; however the final results are not sensitive to the exact nature of this modification provided that the potential for small and non–physical values of the bond length are sufficiently repulsive. These considerations apply as well to the centrifugal potential $j(j+1)\hbar^2/2\mu r_\alpha^2$, which is singular at the origin. We make this nonsingular by replacing r_α^2 with $r_\alpha^2 + r_0^2$, where r_0 is a small distance we have taken to be $1.0 \times 10^{-3}\, a_0$.

The integrals $U^\alpha_{nn'}$ are computed by optimized quadratures[47] with weights $w_{\alpha v'j'vj}$.

2.2.2. *Distorted waves and Green's functions for nonreactive scattering.* Next we turn to the evaluation of the functions ${}^{(r)}f^\alpha_{nn'}$ and $\grave{g}^N_{n\beta}$, where $\alpha_n = \alpha_\beta = \alpha$; these are respectively the scattering solution and half–integrated Green's functions governed by the last term in Eq. (2a). The distorted wave radial functions solve the homogeneous equations

$$[d^2/dR_\alpha^2 - \ell_n(\ell_n+1)/R_\alpha^2 + k_n^2]\; {}^{(r)}f^\alpha_{nn''}(R_\alpha) + \sum_{n'} U^{D\alpha}_{nn'}(R_\alpha)\; {}^{(r)}f^\alpha_{n'n''}(R_\alpha) = 0 \tag{17}$$

subject to the boundary conditions

$${}^{(r)}f^\alpha_{nn'} \underset{R_\alpha\to 0}{\sim} 0, \tag{18}$$

$${}^{(r)}f^\alpha_{nn'} \underset{R_\alpha\to\infty}{\sim} \begin{cases} k_n^{-\frac{1}{2}}[\delta_{nn'}\sin(k_nR_\alpha - \ell_n\pi/2) + \\ \quad + {}^0K_{nn'}\Delta^\alpha_{nn'}\cos(k_nR_\alpha - \ell_n\pi/2)] & k_n^2>0; \\ (2|k_n|)^{-\frac{1}{2}}\{\delta_{nn'}\exp[|k_n|(R_\alpha - R_f)] + \\ \quad + {}^0K_{nn'}\Delta^\alpha_{nn'}\exp[-|k_n|(R_\alpha - R_f)]\}, & k_n^2<0, \end{cases} \tag{19}$$

where k_n is the wave vector for channel n, and R_f is a numerical parameter chosen to avoid overflow or underflow problems. In practice, we replace the sines and cosines in (19) by Riccati–Bessel functions; these behave the same as $R_\alpha \to \infty$, but they allow the boundary condition to be applied at a smaller value of R_α. The half–integrated Green's functions solve the inhomogeneous equations

$$[d^2/dR_\alpha^2 - \ell_n(\ell_n + 1)/R_\alpha^2 + k_n^2]\grave{g}^N_{n\beta}(R_\alpha) + \sum_{n'} U^{D\alpha}_{nn'}(R_\alpha)\grave{g}^N_{n'\beta}(R_\alpha)$$
$$= -\delta_{nn_\beta} t^\alpha_{m_\beta n_\beta}(R_\alpha) \qquad (20)$$

subject to the boundary conditions

$$\grave{g}^N_{n\beta} \underset{R_\alpha \to 0}{\sim} 0, \qquad (21)$$

$$\grave{g}^N_{n\beta} \underset{R_\alpha \to \infty}{\sim} d^\alpha_{\beta n} \begin{cases} k_n^{-\frac{1}{2}} \cos(k_n R_\alpha - \ell_n \pi/2) & k_n^2 > 0 \\ (2|k_n|)^{-\frac{1}{2}} \exp[-|k_n|(R_\alpha - R_f)], & k_n^2 < 0, \end{cases} \qquad (22)$$

where the matrix element in the boundary condition matrix is defined by

$$d^\alpha_{\beta n} = \Delta^\alpha_{nn_\beta} \int dR_\alpha \, {}^{(r)}f^\alpha_{n_\beta n}(R_\alpha) t^\alpha_{m_\beta n_\beta}(R_\alpha). \qquad (23)$$

We solve equations (17) and (20) using the finite difference boundary value method (FDBVM)[14,48,49] using a 9–point representation of the second derivative operator, with lower order approximations near the large–R_α edge of the grid in order to conveniently[14] impose the nonhomogeneous boundary condition. This numerical method is not the most efficient method to solve such coupled ODEs, but in the present context it offers many advantages over other methods. First of all we are concerned in this step with calculating the radial functions at all distances for use in integrals rather then just determining the asymptotic form of the radial functions as is done in standard close coupling calculations; the FDBVM yields the solutions at all R_α in a stable manner without requiring complicated reorthogonalization[50] procedures. A second advantage is that there is no restriction as to the uniformity of the stepsizes used for the finite difference grid. We take advantage of this flexibility by evaluating the integrals over radial functions with efficient quadrature rules which do not have evenly spaced nodes while avoiding the interpolation of the radial functions by including the quadrature points used in the radial integrals in the finite difference grid. In particular we use repeated Gauss–Legendre quadrature to evaluate the R_α and R_{α_0} integrals in Eqs. (8)–(13). This is important because in practice this allows the number, N^{QRS}_α, of radial quadrature points in the integration over R_α to be much less than the number, N^F_α, of points $\{R^F_{\alpha,j}\}$ in the finite difference grid in arrangement α. This saves both in storage, because it is only necessary to store the radial functions which will be used later at the quadrature points (rather than saving all the values for interpolation), and it saves in the time required for the radial integrals, especially the inter–arrangement integrals which involve two R_α–type integrations, as shown in Eqs. (11)–(13). (The time savings result from avoiding the time for interpolation and by keeping the number of points in the quadrature down because the order is high.)

One aspect of the storage of the radial functions not yet addressed concerns the order in which they will be accessed during the integration steps. In the FDBVM, a given radial function is calculated at all grid points simultaneously before any other radial function is calculated, thus it may seem most advantageous to store them in the same manner. However when performing the integrations, it is more convenient to access all radial functions together at a single integration point. Since the inter–arrangement integrals require accessing the same radial function at a given quadrature point many times, it is more efficient to originally store the radial functions so that they are sequentially accessed in these later steps. This is especially important if the radial functions are stored on disk rather than in memory. When the radial functions are stored on disk, we accomplish this ordering by storing the radial functions using standard FORTRAN–77 direct access files with a fixed record length equal to the square of the maximum number of channels per distortion potential block. Then record 1 contains the function ${}^{(r)}f^{1}_{nn'}$ for the first distortion potential block at the first quadrature point, records 2 through N(G) + 1 contains the functions $\grave{g}^{N}_{n\beta}$ for the first distortion potential block at the first quadrature point, where N(G) is the number of translational basis functions, record N(G) + 2 contains the function ${}^{(r)}f^{1}_{nn'}$ for the second distortion potential block at the first quadrature point, etc. When the radial functions are stored in memory, the same ordering is used except now a word rather than record addressing scheme is used to avoid wasting any space in memory caused by differing numbers of channels in the various distortion potential blocks.

The final advantage of the FDBVM is that once the distorted wave radial functions ${}^{(r)}f^{\alpha}_{nn'}$ have been calculated, the half–integrated Green's functions can be calculated relatively inexpensively. This is because the FDBVM equations take the form

$$\mathbf{\Lambda X} = \grave{\boldsymbol{\beta}}, \tag{24}$$

where $\mathbf{\Lambda}$ is a banded matrix of half band width $4N_\delta$, where N_δ is the number of channels in the distortion potential block, and $\mathbf{X}$ is the vector of radial functions determined by the boundary condition/inhomogenity vector $\grave{\boldsymbol{\beta}}$. It should be noted that the matrix $\mathbf{\Lambda}$ contains only the term on the left hand sides of Eqs. (17) and (20) and is therefore the same whether calculating the ${}^{(r)}f^{\alpha}_{nn'}$ or the $\grave{g}^{N}_{n\beta}$. The computational work required to solve Eq. (24) can be broken down into the two steps of factoring the nonsymmetric matrix $\mathbf{\Lambda}$ into its LU form followed by back substitution and forward elimination to obtain the solution.[51] The time to form the LU decomposition is much larger than the time for the forward substitution and backward elimination steps and so many different solution vectors can be generated with relatively little extra effort once the LU decomposition is known. A detailed analysis of the operation count require to solve Eq. (24) is complicated by the possibility of pivoting; however, if we ignore this complication then the relative work of factoring and solving can be assessed. From Ref. 51 we see that the work to factor will be proportional to $16N_{\alpha}^{F}N(D)^3$ for large N_{α}^{F} and $N(D)$, while the substitution and elimination time is proportional to $8N_{\alpha}^{F}N(D)^2$ to obtain a single solution or $8N_{\alpha}^{F}N(D)^3$ to obtain the entire ${}^{(r)}f_{nn'}^{\alpha}$ matrix for this distortion potential block. The time to solve for the $\grave{g}_{n\beta}^{N}$ for a single translational basis function will be $8N_{\alpha}^{F}N(D)^3$ also, so one can obtain the half-integrated Green's functions for three basis functions for about the same cost as is required to just calculate the ${}^{(r)}f_{nn'}^{\alpha}$. In addition, if the time to evaluate the potential coupling matrix $U_{nn'}^{D\alpha}$, is significant, then the calculation of the $\grave{g}_{n\beta}^{N}$ will be relatively even less expensive because the potential coupling matrix only appears in Λ. An additional point is that there exist efficient vectorized routines for the solution of standard forms of equation like (24). We use the LINPACK[52] routines SGBFA to factor Λ and SGBSL to solve for the radial functions.

We now address some aspects of the calculation of the Λ matrix. First consider the finite difference coefficients which must be generated for unevenly spaced grid points. We generate these coefficients by requiring that

$$d^2F/dR^2\big|_{R=R_j} = \sum_{i=1}^{N} c_i^j F(R_{j+i-(N-1)/2-1}), \tag{25}$$

where F is a polynomial of order N–1, R_j is the jth grid point, and the c_i^j are the finite difference coefficients. In particular, we write

$$F(R_k) = F(R_j) + \sum_{i=1}^{N-1} F^{(i)}(R_j)[(R_k - R_j)^i/i!],$$
$$k = j-(N-1)/2,\dots,j+(N+1)/2, \tag{26}$$

where $F^{(i)} = d^iF/dR^i$. We can write this in the form $\mathbf{Ac} = \mathbf{F}$, where $\mathbf{A}$ is the matrix of $[(R_k - R_j)^i/i!]$, $\mathbf{c}$ is the unknown vector of values of the derivatives $F^{(i)}$ and $\mathbf{F}$ is the vector of functions at the neighboring grid points. Multiplying through by $\mathbf{A}^{-1}$, we then see that

$$F^{(i)}(R_j) = \sum_{k=1}^{N} (A^{-1})_{i+1,k} F(R_{j-(N-1)/2+k-1}), \tag{27}$$

thus the elements $(A^{-1})_{3k}$ are the coefficients we seek.

Now consider the evaluation of the intra–arrangement potential $U^{\alpha}_{nn'}$. This is evaluated by expanding the interaction potential as

$$V^{int,\alpha} = \sum_{\lambda=0}^{\lambda_{max}} \nu_{\lambda}^{int,\alpha}(R_{\alpha},r_{\alpha})P_{\lambda}(\cos\gamma_{\alpha}), \tag{28}$$

where $\nu_{\lambda}^{int,\alpha}$ is an expansion coefficient, P_{λ} is a Legendre polynomial, and λ_{max} is a convergence parameter we take equal to the minimum of $N^{QA}_{\alpha\alpha}$ and $2j_{max} + 1$, where $N^{QA}_{\alpha\alpha}$ is the number of points in the Gauss–Legendre quadrature used for non–exchange angular integrals, and j_{max} is the maximum rotational quantum number. The expansion coefficients are determined by projection of the Legendre polynomials; the numerical quadrature involved in that process is written as a matrix multiply and evaluated using a fast matrix multiplication routine. Note that the potential expansion coefficients are independcent of channel indices, although they do depend on the arrangement index. The matrix elements are then assembled using Percival–Seaton coefficients,[53]

$$f^{J\lambda\alpha}_{nn'} = \int d\hat{R}_{\alpha}\, d\hat{r}_{\alpha}\, Y^{JM*}_{j_n\ell_n}(\hat{r}_{\alpha},\hat{R}_{\alpha})P_{\lambda}(\cos\gamma_{\alpha})Y^{JM}_{j_n\ell_n}(\hat{r}_{\alpha},\hat{R}_{\alpha}), \tag{29}$$

for the angular integrals and optimized vibrational quadrature for the r_{α} integral.

Finally we consider the imposition of the boundary conditions of Eq. (22). One way to impose these conditions is to calculate the $d_{\beta n}$ matrix elements and include them times the proper matching functions in the vector $\grave{\beta}$ so that radial functions with the proper boundary conditions are automatically produced. Another way is to use arbitrary boundary conditions in $\grave{\beta}$ to produce the function $\tilde{g}^{N}_{n\beta}$ which is related to the desired functions by

$$\tilde{g}^{N}_{n\beta} = \grave{g}^{N}_{n\beta}(R_{\alpha}) + \sum_{n'} {}^{(r)}f^{\alpha}_{nn'}(R_{\alpha})X_{n'\beta}, \tag{30}$$

where the matrix element $X_{n\beta}$ can be determined numerically from the asymptotic form of $\tilde{g}^{N}_{n\beta}$. This equation can also be used when changing from real to complex boundary conditions as described in Sect. 5. In our calculations we use the $d_{\beta n}$ matrix elements and the $\grave{\beta}$ vector to directly generate functions satisfying the proper boundary conditions.

2.2.3. Angular exchange integrals. We next consider the evaluation of the inter–arrangement integrals, $\mathscr{C}^{\alpha\alpha_0}_{nn_0}$ and $\mathscr{B}^{\alpha\alpha_0}_{nn_0}$. These take the form

$$\begin{aligned}\mathscr{C}^{\alpha\alpha_0}_{nn_0} &= 2\pi(\mathscr{M}^{\alpha\alpha_0})^3 R_\alpha R_{\alpha_0}[(2\ell_n+1)(2\ell_{n_0}+1)]^{\frac{1}{2}}(2J+1)^{-1} \times \\ &\times \sum_{\Omega}\sum_{\Omega'} (j_n\ell_n\Omega 0|j_n\ell_n J\Omega)(j_{n_0}\ell_{n_0}\Omega' 0|j_{n_0}\ell_{n_0}J\Omega') \\ &\times \int d\cos\Delta_{\alpha\alpha_0}\frac{1}{r_\alpha}\chi_{\alpha v_n j_n}(r_\alpha)\, Y_{j_n\Omega}(\gamma_\alpha,0) \times \\ &\times \left[\frac{-2\mu}{\hbar^2} V^{int,\alpha_0}(\mathbf{R}_{\alpha_0},\mathbf{x}_{\alpha_0})\right] \\ &\times \frac{1}{r_{\alpha_0}}\chi_{\alpha_0 v_{n_0} j_{n_0}}(r_{\alpha_0}) Y_{j_{n_0}\Omega'}(\gamma_{\alpha_0},0)\, d^J_{\Omega\Omega'}(\Delta_{\alpha\alpha_0}),\end{aligned} \tag{31}$$

where $\mathscr{M}^{\alpha\alpha_0}$ is a mass factor, J is the total angular momentum, $(\cdots|\cdots)$ is a Clebsch-Gordan coefficient, $\Delta_{\alpha\alpha_0}$ is the angle which rotates the coordinate system of one arrangement to the other, i.e., $\cos\Delta_{\alpha\alpha_0} = \hat{R}_\alpha\cdot\hat{R}_{\alpha_0}$, $Y_{j\Omega}$ is a spherical harmonic, and $d^J_{\Omega\Omega'}$ is a reduced rotation matrix element. The equation for $\mathscr{B}^{\alpha\alpha_0}_{nn_0}$ is the same as Eq. (31) except the $-2\mu V^{int,\alpha_0}/\hbar^2$ factor is replaced by unity. The coordinates in the integrand are functions of R_α, R_{α_0}, and $\Delta_{\alpha\alpha_0}$, i.e.,

$$r_\alpha = \mathscr{M}^{\alpha\alpha_0}[R^2_{\alpha_0} + (\overline{\mathscr{M}}^{\alpha\alpha_0}R_\alpha)^2 + 2\overline{\mathscr{M}}^{\alpha\alpha_0}R_\alpha R_{\alpha_0}\cos\Delta_{\alpha\alpha_0}]^{\frac{1}{2}} \tag{32}$$

$$r_{\alpha_0} = \mathscr{M}^{\alpha\alpha_0}[R^2_{\alpha} + (\overline{\mathscr{M}}^{\alpha\alpha_0}R_{\alpha_0})^2 + 2\overline{\mathscr{M}}^{\alpha\alpha_0}R_\alpha R_{\alpha_0}\cos\Delta_{\alpha\alpha_0}]^{\frac{1}{2}}, \tag{33}$$

$$\cos\gamma_\alpha = (-1)^{P_{\alpha\alpha_0}}\mathscr{M}^{\alpha\alpha_0}(\overline{\mathscr{M}}^{\alpha\alpha_0}R_\alpha + R_{\alpha_0}\cos\Delta_{\alpha\alpha_0})/r_\alpha, \tag{34}$$

$$\cos\gamma_{\alpha_0} = (-1)^{P_{\alpha_0\alpha}}\mathscr{M}^{\alpha\alpha_0}(\overline{\mathscr{M}}^{\alpha\alpha_0}R_{\alpha_0} + R_\alpha\cos\Delta_{\alpha\alpha_0})/r_{\alpha_0}, \tag{35}$$

where $\overline{\mathscr{M}}^{\alpha\alpha_0}$ is another mass factor and $P_{\alpha\alpha_0}$ is the parity of the permutation from (12) to $(\alpha\alpha_0)$. Both $\mathscr{M}^{\alpha\alpha_0}$ and $\overline{\mathscr{M}}^{\alpha\alpha_0}$ are symmetric with respect to α and α_0.

We first simplify Eq. (31) by writing it in terms of the body–frame matrix elements which are labeled by the quantum numbers, v,j,Ω,J,P rather than the laboratory–frame quantum numbers v,j,ℓ,J. Here Ω is the projection of the total angular momentum on the diatom bond axis, and P is the parity, defined as $P = (-1)^{j+\ell}$. Equation (31) then becomes

$$\mathscr{C}^{\alpha\alpha_0}_{nn_0} = \sum_{\Omega\geq 0} \sum_{\Omega'\geq 0} T^{Jj_n}_{\Omega\ell_n} T^{Jj_{n_0}}_{\Omega'\ell_{n_0}} \mathscr{C}^{\alpha\alpha_0 JP}_{v_n j_n \Omega v_{n_0} j_{n_0} \Omega'}, \tag{36}$$

where the body–frame–to–space–frame transformation is

$$T^{Jj}_{\Omega\ell} = \left[\frac{2}{1 + \delta_{\Omega 0}}\right]^{\frac{1}{2}} \left[\frac{2\ell + 1}{2J + 1}\right]^{\frac{1}{2}} (j\Omega\ell 0 | j\ell J\Omega). \tag{37}$$

We note that if $P(-1)^J = -1$, then the Ω and $\Omega' = 0$ terms are not present in Eq. (36), and the sum runs over $\Omega \geq 1$ and $\Omega' \geq 1$. Also it should be noted that this transformation is independent of the vibrational quantum numbers so that if channel pairs which only differ by v_n and v_{n_0} are grouped together, they can be transformed using vector instructions.

The body–frame matrix elements are written as

$$\begin{aligned}\mathscr{C}^{\alpha\alpha_0 JP}_{vj\Omega v'j'\Omega'} &= \pi(\mathscr{M}^{\alpha\alpha_0})^3 R_\alpha R_{\alpha_0} \int d\cos\Delta_{\alpha\alpha_0} \frac{1}{r_\alpha} \chi_{\alpha vj}(r_\alpha) Y_{j\Omega}(\gamma_\alpha, 0) \times \\ &\times \left[\frac{-2\mu}{\hbar^2} V^{int,\alpha_0}(R_{\alpha_0}, x_{\alpha_0})\right] \frac{1}{r_{\alpha_0}} \\ &\times \chi_{\alpha_0 v'j'}(r_{\alpha_0}) Y_{j'\Omega'}(\gamma_{\alpha_0}, 0) \tilde{d}^{JP}_{\Omega\Omega'}(\Delta_{\alpha\alpha_0}),\end{aligned} \tag{38}$$

and the modified rotation matrix elements are given by

$$\begin{aligned}\tilde{d}^{JP}_{\Omega\Omega'} &= [(1 + \delta_{\Omega 0})(1 + \delta_{\Omega' 0})]^{-\frac{1}{2}} [d^J_{\Omega\Omega'} + (-1)^{\Omega+\Omega'} d^J_{-\Omega-\Omega'} + \\ &+ P(-1)^{J+\Omega} d^J_{-\Omega\Omega'} + P(-1)^{J+\Omega'} d^J_{\Omega-\Omega'}].\end{aligned} \tag{39}$$

The body–frame matrix is then evaluated using Gauss–Legendre quadrature. In particular, we define the quantities

$$b_{\alpha v j\Omega i} = \frac{1}{r_\alpha(i)} \chi_{\alpha v j}[r_\alpha(i)] Y_{j\Omega}[\gamma_\alpha(i),0], \tag{40}$$

where $r_\alpha(i)$ is r_α evaluated at the ith quadrature point, etc.,

$$\mathscr{V}_i^{JP\Omega\Omega'} = \frac{-2\mu}{\hbar^2} V^{int,\alpha_0}[R_{\alpha_0}, x_{\alpha_0}(i)] R_\alpha R_{\alpha_0} \pi(\mathscr{M}^{\alpha\alpha_0})^3 w_i \mathfrak{d}^{JP}_{\Omega\Omega'}[\Delta_{\alpha\alpha_0}(i)], \tag{41}$$

where w_i is the weight for the ith quadrature point, and

$$\mathfrak{b}^{JP\Omega\Omega'}_{i\alpha v j} = b_{\alpha v j\Omega' i} \mathscr{V}_i^{JP\Omega\Omega'}, \tag{42}$$

so that the final result is a matrix multiply:

$$\mathscr{C}^{\alpha\alpha_0}_{vj\Omega v'j'\Omega'} = \sum_i b_{\alpha v j\Omega i} \mathfrak{b}^{JP\Omega\Omega'}_{i\alpha_0 v'j'}. \tag{43}$$

All of the quantities in Eqs. (39)–(42) are evaluated using vector instructions and the matrix multiply is evaluated using a fast matrix multiply routine. It should be noted that the coordinates specified by $\mathbf{x}_\alpha(i)$ and $\mathbf{x}_{\alpha_0}(i)$ all depend on $R_{\alpha'}$, R_{α_0}, and $\Delta_{\alpha\alpha_0}$ so that $b_{\alpha v j\Omega i}$ does also; however the $\mathfrak{d}^{JP}_{\Omega\Omega'}$ are independent of R_α and R_{α_0}.

An additional subtlety exists concerning the evaluation of the rotation matrix elements. This arises because the equations relating the $\mathbf{x}_\alpha$ to the $R_{\alpha'}$, R_{α_0} and $\Delta_{\alpha\alpha_0}$ only involve $\cos\Delta_{\alpha\alpha_0}$ and so we only store the cosine rather than the angle [see eqs. (32)–(35)]. The sign of the angle then is not specified and so it is necessary to keep track of the sense of the rotation based upon α and α_0. In particular, in Eq. (38), $\Delta_{\alpha\alpha_0}$ will be a positive rotation if $\alpha_0 = \alpha + 1$ (modulo 3) and a negative rotation if $\alpha_0 = \alpha - 1$ (modulo 3). The rotation matrix elements at negative angles are obtained by exchanging the subscripts, i.e. $\mathfrak{d}^{JP}_{\Omega\Omega'}$ is replaced by $\mathfrak{d}^{JP}_{\Omega'\Omega}$.

The matrix elements $\mathscr{B}^{\alpha\alpha_0}_{nn_0}$ are calculating in an analogous manner except that the $-2\mu V^{int,\alpha_0}/\hbar^2$ factor is not included in Eq. (41).

We note in passing that the matrix elements $W^{\alpha\alpha_0}_{nn_0}$, $\mathscr{C}^{\alpha\alpha_0}_{nn_0}$, and $\mathscr{B}^{\alpha\alpha_0}_{nn_0}$ are independent of energy, and the matrix elements $b_{\alpha vj\Omega i}$ used to calculate them are independent of J.

An alternative to the above procedure of evaluating the body–frame matrix elements and then transforming them into the space frame is to note that the entire calculation can be carried out in the body frame. This can lead to a sizeable reduction in operations because the transformation of Eq. (36), which needs to be performed inside of the integral over R_α and R_{α_0}, is eliminated if the body frame is used throughout. To perform the calculations in the body frame, it is necessary to apply the transformation $T^{Jj}_{\Omega\ell}$ to all quantities labeled by the ℓ quantum number, namely the radial functions ${}^{(r)}f^{\alpha}_{nn'}$ and $\grave{g}^{N}_{n\beta}$, the distortion potential reactance matrix ${}^{o}K_{nn'}$, and the intra–arrangement potential matrix elements $V^{\alpha}_{nn'}$ and $U^{D\alpha}_{nn'}$. With these changes, one proceeds as before, except Eq. (36) is now eliminated and it is necessary to back transform the final reactance or scattering matrix before calculating physical observables.

The only limitation of using the body frame throughout the entire calculation concerns the choice of the distortion potential. It is necessary that a given distortion potential block include the channels with all values of ℓ allowed by total angular momentum and parity constraints for a given v,j state. If this were not the case, the transformation $T^{Jj}_{\Omega\ell}$ will mix different distortion potential blocks and complicate the various computational steps. Two kinds of distortion potentials which do not allow the use of the body frame throughout are a single channel distortion potential and a distortion potential based on a centrifugal sudden decoupling index.[54]

2.2.4. Radial integrals. We next consider the radial integrals. For computational purposes, it is convenient to rewrite these equations by defining the quantities

$$\mathcal{F}_{nn_0} = \begin{cases} \sum_{n'} \Delta^{\alpha}_{nn'}\, {}^{(r)}f^{\alpha_0}_{n'n}(R_{\alpha_0}) V^{\alpha_0}_{n'n_0}(R_{\alpha_0}), & \alpha_n = \alpha_0; \\ \int dR_\alpha \sum_{n'} \Delta^{\alpha}_{nn'}\, {}^{(r)}f^{\alpha}_{n'n}(R_\alpha) W^{\alpha\alpha_0}_{n'n_0}(R_\alpha, R_{\alpha_0}), & \text{otherwise,} \end{cases} \tag{44}$$

$$\mathcal{G}_{\beta n_0} = \begin{cases} \sum_{n'} \Delta^{\alpha}_{n_\beta n'}\, \grave{g}^{N}_{n'\beta}(R_{\alpha_0}) V^{\alpha}_{n'n_0}(R_{\alpha_0}) & \alpha_\beta = \alpha_0; \\ \int dR_{\alpha_\beta} \sum_{n'} \Delta^{\alpha_\beta}_{n_\beta n'}\, \grave{g}^{N}_{n'\beta}(R_\alpha) W^{\alpha_\beta\alpha_0}_{n'n_0}(R_{\alpha_\beta}, R_{\alpha_0}), & \text{otherwise,} \end{cases} \tag{45}$$

and

$$\mathscr{K}_{\beta n_0} = \begin{cases} \Delta^{\alpha_0}_{n_\beta n_0} \grave{g}^{N}_{n'\beta}(R_{\alpha_0}), & \alpha_\beta = \alpha_0; \\ \int dR_{\alpha_\beta} \sum_{n'} \Delta^{\alpha_\beta}_{n_\beta n'} \grave{g}^{N}_{n'\beta}(R_\alpha) \mathscr{B}^{\alpha_\beta \alpha_0}_{n' n_0}(R_{\alpha_\beta}, R_{\alpha_0}), & \text{otherwise.} \end{cases} \tag{46}$$

With these definitions, Eqs. (8)–(13) become

$$\mathscr{K}^{B}_{nn_0} = \int dR_{\alpha_0} \sum_{n'} \Delta^{\alpha_0}_{n'n_0} \mathscr{F}_{nn'}(R_{\alpha_0}) \, {}^{(r)}f^{\alpha_0}_{n'n_0}(R_{\alpha_0}), \tag{47}$$

$$B_{\beta n_0} = \int dR_{\alpha_0} \sum_{n'} \Delta^{\alpha_0}_{n'n_0} \mathscr{G}_{\beta n'}(R_{\alpha_0}) \, {}^{(r)}f^{\alpha_0}_{n'n_0}(R_{\alpha_0}), \tag{48}$$

and

$$\begin{aligned} C_{\beta\beta'} = {} & \int dR_{\alpha_{\beta'}} \mathscr{K}_{\beta n_{\beta'}}(R_{\alpha_{\beta'}}) t^{\alpha_{\beta'}}_{m_{\beta'} n_{\beta'}}(R_{\alpha_{\beta'}}) - \\ & - \int dR_{\alpha_{\beta'}} \sum_{n'} \Delta^{\alpha_{\beta'}}_{n' n_{\beta'}} \mathscr{G}_{\beta n'}(R_{\alpha_{\beta'}}) \grave{g}^{N}_{n'\beta'}(R_{\alpha_{\beta'}}). \end{aligned} \tag{49}$$

Equations (47)–(49) take the form of a matrix multiply followed by integration, and we carry out the matrix multiplications using a fast matrix multiply routine.

In order to efficiently evaluate the sums over n and n′ inside of the R_α integral in Eqs. (44)–(46), we proceed as follows. The basic operation to be performed is the multiplication of a full matrix (with elements $W^{\alpha\alpha_0}_{n'n_0}$ or $\mathscr{B}^{\alpha\alpha_0}_{n'n_0}$) and a block diagonal matrix (${}^{(r)}f^{\alpha}_{n'n}$ or $\grave{g}^{N}_{n'\beta}$), and it is most efficient to organize these steps so that multiplication from the right is by the block diagonal matrix. This maximizes the vector lengths involved in the multiplication process. This ordering is easily accomplished by transposing the matrix equations. For illustration, we consider an example where there are two distortion potential blocks, then the product we wish to form is

$$\begin{bmatrix} A & B \\ C & D \end{bmatrix} \begin{bmatrix} E & \mathbf{0} \\ \mathbf{0} & F \end{bmatrix} = \begin{bmatrix} G & H \\ I & J \end{bmatrix}, \tag{50}$$

where $\mathbf{0}$ is the null matrix, and this is evaluated as

$$\begin{bmatrix} A \\ C \end{bmatrix} \begin{bmatrix} E \end{bmatrix} = \begin{bmatrix} G \\ I \end{bmatrix}, \tag{51}$$

and

$$\begin{bmatrix} B \\ D \end{bmatrix} \begin{bmatrix} F \end{bmatrix} = \begin{bmatrix} H \\ J \end{bmatrix}. \tag{52}$$

Equations (51)–(52) are each evaluated by a call to a fast matrix multiply routine.

The amount of work involved in the double radial integrals is substantial, so it is desirable to minimize it. We do this by two means. First of all, we use efficient Gaussian quadrature rules to evalute the integrals; this minimizes the number of terms in the quadrature sums. Secondly we note that many values of R_α, R_{α_0}, and $\Delta_{\alpha\alpha_0}$ lead to geometries which have no physical importance so that the $\mathscr{C}^{\alpha\alpha_0}_{nn_0}$ and $\mathscr{B}^{\alpha\alpha_0}_{nn_0}$ matrices will be negligible and not make any significant contributions to the quadrature sums. By detecting these geometries prior to the calculation of the $\mathscr{C}^{\alpha\alpha_0}_{nn_0}$ and $\mathscr{B}^{\alpha\alpha_0}_{nn_0}$ matrices, we can save a considerable amount of work. We do this by noting that the primary cause of the small values of these matrices are the non–classical values of the bond lengths r_α and r_{α_0} where the vibrational wavefunctions $\chi_{\alpha vj}$ are exponentially decaying to zero. Thus we specify limits on the bond lengths and if a particular combination of R_α, R_{α_0}, and $\Delta_{\alpha\alpha_0}$ produces values of r_α or r_{α_0} which lie outside of these limits, this point is omitted from the inner quadrature over $\cos\Delta_{\alpha\alpha_0}$. In practice, the limits are determined by comparing the vibrational wavefunctions to an input parameter and determining the distances where the absolute values of the $\chi_{\alpha vj}$ are less than this parameter. We always check that the results are converged with respect to decreasing this parameter.

The choice of which integration is carried out first, i.e., R_α or R_{α_0}, is important in determining the efficiency of the calculation of matrix elements between different arrangements. This is because Eq. (16) involves three parts, two of which depend explicitly on the two arrangements and one which depends only on the α_0 arrangement. Thus by performing the R_α integral first we avoid the inefficiency of repeatedly calculating the $U^\alpha_{nn'}$, or the storage problem of having to save them. Instead each matrix element is calculated as it is needed and then discarded. This is discussed in more detail in Sect. 4.

2.2.5. Final steps. Once all of the integrals have been evaluated, we solve for the reactance matrix by means of Eqs. (3), (3a), and (7). Then the scattering matrix is determined from

$$S = (1 - iK^{oo})^{-1}(1 + iK^{oo}), \tag{53}$$

where 1 is the unit matrix, and $i = \sqrt{-1}$. For computational purposes, this is rewritten as

$$\mathbf{S} = [1 + (\mathbf{K}^{oo})^2]^{-1}[1 - (\mathbf{K}^{oo})^2 + 2i\mathbf{K}^{oo}] \tag{54}$$

so that the real and imaginary parts of the scattering matrix are each determined using only real arithmetic.

3. Symmetry Decoupling

For systems with identical atoms, some numerical simplifications of the above equations are possible. First of all, some quantities, such as the vibrational wavefunctions and the radial functions, are common to more than one arrangement. Thus it is advantageous to calculate only the unique quantities. This applies to the integrals needed to construct the various matrices as well; because of the symmetry of the system, many matrix elements are either equal or related by a phase factor. Thus one can take advantage of this and only calculate the matrix elements that are unique by symmetry.

A second type of simplification present in systems having symmetry is that the diatom in at least one arrangement will be homonuclear. This can be exploited in several places. First of all, the work in the quadratures to determine the potential expansion coefficients of Eq. (28) can be cut in half because the interaction potential is an even function. This also means that only even values of λ occur in the Legendre expansion. Thus even and odd rotational quantum numbers are not coupled by the distortion potential and so the size of the distortion potential blocks decreases as does also the time to solve for the radial functions and the resources required to save them.

The final use of symmetry concerns the solution of Eq. (7) for the correction to the reactance matrix. So far in the discussion we have emphasized aspects of the calculation prior to this step, but when the basis set is large enough, most of the run time for a calculation will be spent in forming $\overline{\mathbf{B}}$. We can use symmetry here also to save on computational resources. Thus we seek to exploit the symmetry of the $\mathcal{K}$, $\mathcal{K}^B$, $\mathbf{B}$, and $\mathbf{C}$ matrices by introducing simple unitary transformations $\mathbf{U}_K$ and $\mathbf{U}_C$ into Eq. (7), i.e.

$$\mathbf{U}_K^\dagger \mathcal{K} \mathbf{U}_K = \mathbf{U}_K^\dagger \mathcal{K}^B \mathbf{U}_K + \mathbf{U}_K^\dagger \mathbf{B}^\dagger \mathbf{U}_C (\mathbf{U}_C^\dagger \mathbf{C} \mathbf{U}_C)^{-1} \mathbf{U}_C^\dagger \mathbf{B} \mathbf{U}_K, \tag{55}$$

where $\dagger$ denotes Hermitian conjugate and $\mathbf{U}_K$ and $\mathbf{U}_C$ are chosen such that $\mathcal{K}$, $\mathcal{K}^B$, $\mathbf{B}$, and $\mathbf{C}$ are block diagonalized. (In $\mathbf{B}$ the blocks are rectangular.) Then each block can be solved separately at much reduced cost.

To discuss the transformations in detail we consider a generic matrix $\mathbf{M}$, which may be $\mathcal{K}^B$, $\mathbf{B}$, or $\mathbf{C}$. Let the dimensions of $\mathbf{M}$ be $N_r \times N_c$. When $\mathbf{M}$ represents $\mathcal{K}^B$, it has elements $<\alpha n E|\hat{M}|\alpha' n' E>$ so $N_r = N_c = N$, the number of channels. When $\mathbf{M}$ represents $\mathbf{B}$, it has elements $<\beta|\hat{M}|\alpha n E>$ so $N_c = N$ and $N_r = M$, the number of basis functions. When $\mathbf{M}$ represents $\mathbf{C}$, it has elements

$<\beta|M|\beta'>$ so $N_r = N_c = M$. The order of the rows in **B** and of the rows and columns in **C** is such that all basis functions for a given distortion block occur consecutively.

In order to specify the transformations we now have to discuss in detail the form of the matrices for specific cases. We first consider the case where two atoms are the same and arrangement 1 is $A + B_2$; arrangements 2 and 3 both correspond to AB + B in this case, but differ in which B atom is bound. In arrangement 1, the channels are ordered such that the N^{even} channels with even j appear before the N^{odd} channels having odd j. Arrangements 2 and 3 are indistinguishable and have N^{other} channels each, thus the total number of channels is $N = N^{even} + N^{odd} + 2N^{other}$. A generic matrix will have the form

$$\mathbf{M} = \begin{bmatrix} \mathbf{A}^e & \mathbf{0} & \mathbf{C}^e & \mathbf{C}^e\mathbf{P}' \\ \mathbf{0} & \mathbf{A}^o & \mathbf{C}^o & -\mathbf{C}^o\mathbf{P}' \\ \mathbf{B}^e & \mathbf{B}^o & \mathbf{D} & \mathbf{E}\mathbf{P}' \\ \mathbf{P}\mathbf{B}^e & -\mathbf{P}\mathbf{B}^o & \mathbf{P}\mathbf{E} & \mathbf{P}\mathbf{D}\mathbf{P}' \end{bmatrix}, \tag{56}$$

where $\mathbf{A}^e$ is the submatrix containing elements between channels of arrangement 1 with even j, $\mathbf{B}^e$ is the submatrix containing elements between channels of arrangements 2 and channels of arrangement 1 with even j, etc. The phase matrices **P** (of dimension $N_r \times N_r$) and **P′** (of dimension $N_c \times N_c$) are diagonal with elements equal to $(-1)^{j_n}$, where j_n is the rotational quantum number for the appropriate channel of arrangement 3. If the matrix **M** is symmetric, e.g., $\mathcal{K}^B$ or **C**, then **E** is symmetric also, and $\mathbf{C}^e$ and $\mathbf{C}^o$ are transposes of $\mathbf{B}^e$ and $\mathbf{B}^o$, respectively.

We now discuss the reasons for this structure. First consider the matrix elements which are diagonal in arrangement quantum number. In arrangement 1, the diatom is homonuclear and the potential is an even function of the $\cos\gamma_1$, thus there will be nonzero coupling only between channels which are either both even or both odd functions of $\cos\gamma_1$. As a result channels with even j will not be coupled to channels with odd j. This is the reason for the null matrices in Eq. (56). Now turn to arrangements 2 and 3. Even though they are physically indistinguishable, they do not have identical radial functions, as we have defined them. This is because the angles γ_2 and γ_3, defined below Eq. (1), are related such that a geometry of arrangement 2 which is superimposable with a geometry of arrangement 3 will have

$$\gamma_2 = \pi - \gamma_3. \tag{57}$$

This causes a change in sign of the terms in the Legendre expansion of the potential which correspond to odd angular functions [see Eq. (28)]. This is equivalent to a phase change of $(-1)^{j_n+j_{n'}}$ in the radial functions because the Percival–Seaton coefficients are zero if $(-1)^{j_n+j_{n'}+\lambda} \neq 1$. Thus the lower right submatrix in Eq. (56) is $\mathbf{PDP}'$ rather than $\mathbf{D}$.

Now consider the matrices connecting arrangement 3 and arrangement 1. These are obtained from the <arrangement 2|arrangement 1> matrices as follows. First the radial functions of arrangement 2 are multiplied by $(-1)^{j_n+j_{n'}}$, where n and n′ are arrangement 2 channel labels, to produce arrangement 3 radial functions. The next change occurs in Eqs. (34) and (35), where the sign on the angles changes because the parity of the $(\alpha\alpha') = (13)$ [or (31)] permutation is opposite of the (12) [or (21)] permutation—this is equivalent to a phase change of $(-1)^{j_{n'}+j_{n''}}$, where n′ is an arrangement 2 channel label, and n" is an arrangement 1 channel label. The two $(-1)^{j_{n'}}$ phases cancel to give $\mathbf{PB}^e$ and $-\mathbf{PB}^o$. By the same argument, we see that the relation between <arrangement 1| arrangement 3> matrix elements and <arrangement 1|arrangement 2> matrix elements is similar, except now $\mathbf{P}'$ appears on the right because the arrangement 2 radial functions multiply from the right.

Finally consider <arrangement 3|arrangement 2> and <arrangement 2| arrangement 3> matrix elements. For concreteness we consider elements of $\mathbf{B}_{\beta n'}$, but the argument is the same for the other matrices. These matrix elements are written symbolically as

$$\mathbf{B}_{32} = \iint \mathbf{g}_3^T \mathbf{W}^{32} \mathbf{f}_2, \tag{58}$$

and

$$\mathbf{B}_{23} = \iint \mathbf{g}_2^T \mathbf{W}^{23} \mathbf{f}_3. \tag{59}$$

Introducing $\mathbf{g}_3^T = \mathbf{P}\mathbf{g}_2^T\mathbf{P}'$ and $\mathbf{f}_3 = \mathbf{P}'\mathbf{f}_2\mathbf{P}'$ due to Eq. (57) and $\mathbf{W}^{23} = \mathbf{P}'\mathbf{W}^{32}\mathbf{P}'$ from Eqs. (34) and (35), we have

$$\mathbf{B}_{32} = \iint \mathbf{P}\mathbf{g}_2^T \mathbf{P}'\mathbf{W}^{32} \mathbf{f}_2, \tag{60}$$

and

$$\mathbf{B}_{23} = \iint \mathbf{g}_2^T \mathbf{P}'\mathbf{W}^{32} \mathbf{f}_2\mathbf{P}', \tag{61}$$

or

$$\mathbf{B}_{23} = \mathbf{P}\mathbf{B}_{32}\mathbf{P}'. \tag{62}$$

Thus $\mathbf{PE}$ becomes $\mathbf{EP}'$.

The final statement about Eq. (56) to be verified is that $\mathbf{E}$ is symmetric when $\mathbf{M}$ is the C matrix or the $\mathscr{K}^B$ matrix. Consider the Born matrix for which Eq. (60) becomes

$$\mathcal{K}^{B}_{32} = \int\int \mathbf{P} f_2^T \mathbf{P} W^{32} f_2, \tag{63}$$

thus we must show that $\mathbf{PW}^{32}$ is "symmetric", i.e.

$$(-1)^{j_n} W^{32}_{nn'}(R_\alpha, R_{\alpha_0}) = W^{32}_{n'n}(R_{\alpha_0}, R_\alpha)(-1)^{j_{n'}}. \tag{64}$$

This follows from the sign change of the parity of the permutations in Eqs. (34) and (35). The same argument applies to the C matrix.

We are now ready to introduce the transformation U which we write as (see Ref. 4)

$$\mathbf{U} = \begin{bmatrix} \mathbf{1} & \mathbf{0} & \mathbf{0} & \mathbf{0} \\ \mathbf{0} & \mathbf{0} & \mathbf{1} & \mathbf{0} \\ \mathbf{0} & \mathbf{fP} & \mathbf{0} & \mathbf{fP} \\ \mathbf{0} & \mathbf{f1} & \mathbf{0} & -\mathbf{f1} \end{bmatrix}, \tag{65}$$

where f is $1/\sqrt{2}$. Then we have

$$\mathbf{U}^T\mathbf{MU} = \begin{bmatrix} \mathbf{A}^e & \mathbf{gC}^e\mathbf{P} & \mathbf{0} & \mathbf{0} \\ \mathbf{gPB}^e & \mathbf{PDP+PEP} & \mathbf{0} & \mathbf{0} \\ \mathbf{0} & \mathbf{0} & \mathbf{A}^o & \mathbf{gC}^o\mathbf{P} \\ \mathbf{0} & \mathbf{0} & \mathbf{gPB}^o & \mathbf{PDP-PEP} \end{bmatrix}, \tag{66}$$

where g is $\sqrt{2}$. Thus if the transformation of Eq. (65) is applied to C, it decouples Eq. (7b) into two independent matrix equations of order $N^{even} + N^{other}$ and $N^{odd} + N^{other}$ rather than one set of equations of order $N^{even} + N^{odd} + 2N_{other}$. If $N^{even} = N^{odd}$, then this results in a CPU time savings of $2^3/(1^3 + 1^3) = 4$ for large enough N.

Now consider the case when all atoms are the same, $A + A_2$. Here all arrangements are indistinguishable. The diatoms are all homonuclear, so again we will order the channels in a given arrangement so the N^{even} basis functions with j even occur before the N^{odd} basis vunctions with odd j. The total number of channels is then $N = 3N^{even} + 3N^{odd}$. Then, if we group the channels in a given arrangement with even j before those with odd j, the generic matrix is

$$\mathbf{M} = \begin{bmatrix} \mathbf{A}^{e} & \mathbf{0} & \mathbf{B}^{ee} & \mathbf{B}^{eo} & \mathbf{B}^{ee} & -\mathbf{B}^{eo} \\ \mathbf{0} & \mathbf{A}^{o} & -\mathbf{B}^{oe} & \mathbf{B}^{oo} & \mathbf{B}^{oe} & \mathbf{B}^{oo} \\ \mathbf{B}^{ee} & -\mathbf{B}^{eo} & \mathbf{A}^{e} & \mathbf{0} & \mathbf{B}^{ee} & \mathbf{B}^{eo} \\ \mathbf{B}^{oe} & \mathbf{B}^{oo} & \mathbf{0} & \mathbf{A}^{o} & -\mathbf{B}^{oe} & \mathbf{B}^{oo} \\ \mathbf{B}^{ee} & \mathbf{B}^{eo} & \mathbf{B}^{ee} & -\mathbf{B}^{eo} & \mathbf{A}^{e} & \mathbf{0} \\ -\mathbf{B}^{oe} & \mathbf{B}^{oo} & \mathbf{B}^{oe} & \mathbf{B}^{oo} & \mathbf{0} & \mathbf{A}^{o} \end{bmatrix}, \tag{67}$$

where **A** represents intra–arrangement coupling and **B** inter–arrangement coupling. When **M** is symmetric, then $\mathbf{B}^{ee}$ and $\mathbf{B}^{oo}$ are also symmetric and $\mathbf{B}^{eo}$ and $\mathbf{B}^{oe}$ are transposes. The signs in Eq. (67) follow from the discussion above, except here the radial functions are the same for all arrangements since the potential is an even function of $\cos\gamma_\alpha$. However, the pattern of **P** matrices remains because the **P** which arises from Eqs. (34)–(35) commutes with the radial functions. In writing Eq. (67) we have explicitly replaced the **P** matrix with ± signs.

In this case, the transformation matrix is complex, albeit unitary. In particular we have (see Ref. 4)

$$\mathbf{U} = \begin{bmatrix} \mathbf{1} & \mathbf{0} & \mathbf{1} & \mathbf{0} & \mathbf{1} & \mathbf{0} \\ \mathbf{0} & \mathbf{1} & \mathbf{0} & \mathrm{i}\mathbf{1} & \mathbf{0} & -\mathrm{i}\mathbf{1} \\ \mathbf{1} & \mathbf{0} & \epsilon^*\mathbf{1} & \mathbf{0} & \epsilon\mathbf{1} & \mathbf{0} \\ \mathbf{0} & \mathbf{1} & \mathbf{0} & \mathrm{i}\epsilon^*\mathbf{1} & \mathbf{0} & -\mathrm{i}\epsilon\mathbf{1} \\ \mathbf{1} & \mathbf{0} & \epsilon\mathbf{1} & \mathbf{0} & \epsilon^*\mathbf{1} & \mathbf{0} \\ \mathbf{0} & \mathbf{1} & \mathbf{0} & \mathrm{i}\epsilon\mathbf{1} & \mathbf{0} & -\mathrm{i}\epsilon^*\mathbf{1} \end{bmatrix}, \tag{68}$$

where i is $\sqrt{-1}$, ϵ is $\exp 2i\pi/3$, and * denotes complex conjugation. Applying this transformation we have

$$\mathbf{U}^{\dagger}\mathbf{M}\mathbf{U} = \begin{bmatrix} \mathbf{A}^{e}+2\mathbf{B}^{ee} & 0 & 0 & 0 & 0 & 0 \\ 0 & \mathbf{A}^{o}+2\mathbf{B}^{oo} & 0 & 0 & 0 & 0 \\ 0 & 0 & \mathbf{A}^{e}-\mathbf{B}^{ee} & \mathrm{h}\mathbf{B}^{eo} & 0 & 0 \\ 0 & 0 & \mathrm{h}\mathbf{B}^{oe} & \mathbf{A}^{o}-\mathbf{B}^{oo} & 0 & 0 \\ 0 & 0 & 0 & 0 & \mathbf{A}^{e}-\mathbf{B}^{ee} & \mathrm{h}\mathbf{B}^{eo} \\ 0 & 0 & 0 & 0 & \mathrm{h}\mathbf{B}^{oe} & \mathbf{A}^{o}-\mathbf{B}^{oo} \end{bmatrix}, \tag{69}$$

where h is $\sqrt{3}$. This transformation block diagonalizes the matrices into four blocks, two of which are the same. Thus, applying the transformation to **C** in (7b), rather than solving one set of matrix equations of order $N = 3N^{even} + 3N^{odd}$, we can solve three smaller sets of matrix equations of order N^{even}, N^{odd}, and $N^{even} + N^{odd}$. If $N^{even} = N^{odd}$, then the CPU time savings for large enough N will be

$6^3/(1^3 + 1^3 + 2^3) = 21.6$.

Further discussion of symmetry decoupling will be provided in a later publication.[41]

In addition to the CPU time savings discussed above, these block diagonalization procedures reduce the storage requirements for the calculations since the null blocks need not be stored.

4. Storage Manangement Strategies

4.1. GENERAL CONSIDERATIONS

It is a general principle of conventional programming practice that recalculation of quantities appearing more than once in an algorithm should be minimized. However this goal often comes into conflict with the physical limitations of the available computational resources. One is then faced with the choice of using more CPU resources and calculating the same quantity more than once or saving every quantity of potential later use and limiting the size of possible calculations because of finite memory or disk or slow input/output. Our strategy in the newest version of our code is to minimize the storage requirements subject to the constraint that excessive amounts of CPU resources will not be spent in recalculating expensive steps. We now discuss the choices we have made and give examples of the penalties incurred by not storing all quantities used two or more times. It should be emphasized that we are designing the present methodology for very large–scale calculations and to the extent that various parts of the calculation scale differently with the size of the problem, our methods will not be optimum for simple problems like the reaction of H with H_2 at low total energy and angular momentum.

One reusable set of quantities in our calculations are the intra–arrangement, intra–distortion–block potential matrix elements $U^{D\alpha}_{nn'}$, defined by Eq. (14a). The intra–arrangement, inter–distortion–block matrix elements $V^{\alpha}_{nn'}$, defined by Eq. (14) are required only in the nonreactive radial integrals of Eqs. (44) and (45), and hence they may be calculated, used once, and discarded without penalty, but the matrix elements $U^{D\alpha}_{nn'}$ appear at two steps of the calculations and require further consideration. The $U^{D\alpha}_{nn'}$ matrix elements are needed at every finite difference point to construct the FDBVM matrix in order to calculate the radial functions [Eqs. (17) and (20)], and they are used in the construction of the inter–arrangement matrix $W^{\alpha\alpha_0}_{nn_0}$ [Eq. (16)]. Thus it would seem advantageous to store them. However they occupy considerable space and so we consider recalculating them.

There are three considerations which make the recalculation of the nonreactive potential matrix elements feasible. The first is that in the limit of a large number N of channels, the time required for the calculation of this matrix will scale as the square of the number of channels while the solution of Eq. (7b) will scale as the cube of the number of channels. Thus comparatively little time will be spent calculating the matrix elements when N is large (see, however, Sect. 5). The second consideration concerns the calculation of the potential matrix itself — it is possible to calculate the matrix at a single distance independently of all other distances and still do it using vector instructions. The final point concerns the strategic way in which we have organized our calculations. On average each $U^{D\alpha}_{nn'}$ matrix element is used at most about 4/3 times. Most of this factor of 4/3 comes from Eqs. (17) and (20) where each matrix element is required once at every *finite difference* grid point. The remaining part comes from the integrations of Eqs. (44) and (45) [see also Eq. (16)]. By performing the integration over R_α as indicated in the inter–arrangement parts of Eqs. (44) and (45), each matrix element is required once at each R_α *quadrature* point and by using repeated Gauss–Legendre quadrature, there are typically only about a third as many quadrature points as finite difference grid points. In addition, these quadratures are only calculated provided the $\mathscr{C}^{\alpha\alpha_0}_{nn_0}$ and/or $\mathscr{B}^{\alpha\alpha_0}_{nn_0}$ matrices are non–negligible. Thus recalculating the matrix elements will not incur a large increase in computation time. We see then that an important consideration making the recalculation of the potential matrix elements feasible is our use of an efficient quadrature scheme. (It also should be noted that the matrix elements $V^{\alpha}_{nn'}$ coupling different distortion potential blocks of the same arrangement are also required only at the quadrature points rather than at all of the finite difference grid points.)

The analysis of timings for calculations we have made using simple potentials indicates that only a very small fraction of the overall run time is spent calculating the intra–arrangement potential coupling matrix elements. However, calculations[31–33] using complicated potentials such as the accurate H_3 DMBE[55] potential spend a nontrivial amount of time on this task. In this case one can consider the slightly different strategy of recalculating the $U^{\alpha}_{nn'}$, but saving the potential expansion coefficients $\nu^{int,\alpha}_{\lambda}$. These potential expansion coefficients are independent of the number of channels, so that eventually their evaluation will become negligible; however this is complicated by the fact that we need to retrieve or re–calculate each expansion coefficient for each distortion block, and as the number of channels increases, the number of distortion blocks usually increases also.

We now turn to the question of performing calculations at several energies. The quantities $U^{D\alpha}_{nn'}$ and $W^{\alpha\alpha_0}_{nn_0}$ are independent of energy while the radial functions are not and must be calculated at each new energy. Thus it might seem desirable to save the $U^{\alpha}_{nn'}$ and $W^{\alpha\alpha_0}_{nn_0}$. However these matrices would require considerable storage, especially $W^{\alpha\alpha_0}_{nn_0}$, which is a function of two radial distances. On the other hand, timing analyses indicate that only a small fraction of the overall run time is spent evaluating these matrices. Thus in our formulation of reactive scattering, there is no significant advantage in saving energy–independent quantities.

4.2. ARRAY REQUIREMENTS

The program is organized into eleven links performing the various steps of the calculation. When arrays are needed in only one link, they are stored in blank COMMON, which thereby becomes a scratch space overwritten by each link. The size of blank COMMON is therefore determined by the link with the largest temporary storage needs; this is link 10 which forms the C and B matrices. Arrays needed by more than one link are stored in labeled COMMON blocks.

The major array storage requirements for the GNVP calculations are listed below along with their sizes. The arrays are listed in the order they occur in the calculation, as discussed in Sect. 2.2. Only arrays in blank COMMON in link 10 or in labeled COMMON affect the memory requirements, and only these are considered. For simplicity the list has been prepared for a case with no symmetry; we will discuss the savings possible when symmetry is used at the end of this section. In addition, we assume for simplicity in tabulation that all the following parameters are independent of arrangement:

N_{α}^{HO}	the number of haramonic oscillator functions used to expand the diatomic wave functions for arrangement α.
N_{α}	the number of channels in arrangement α.
$N_{\alpha\alpha}^{QV}$	the number of nodes in the optimized vibrational quadratures for computing $U_{nn'}^{\alpha}$.
$N_{\alpha\alpha}^{QA}$	the number of nodes in the angular quadratures for computing U_{nn}^{α}
N_{α}^{QRS}	is defined below Eq. (23).
$\delta_{max,\alpha}$	number of distortion potential blocks in arrangement α.
$N_{rot,\delta}$	number of channels per distortion potential block.
N_{α}^{F}	number of finite difference grid points in arrangement α.
m_{δ}	number of translational basis functions per distortion block.

The largest arrays requiring storage and the leading terms in their dimensions are given in Table I. The quantity f^{NZ} is a user input parameter, which equals the fraction of Percival–Seaton coefficients that are nonzero. (Since this is not known, we use an estimate which must be greater than or equal to the true fraction. Even a cautious estimate though can save a significant amount of storage.)

For typical runs the last three arrays in the Table I account for about 90% of the storage. Thus the memory requirement scales approximately as the second power of the number of channels.

When symmetry is included, the storage needed for these N_{α}^{2} arrays is

TABLE I
Storage requirements for largest arrays (case of no symmetry)

array	defined in Sect.	storage required
$B_{\gamma}^{\alpha vj}$	2.2.1	$3N_{\alpha}N_{\alpha}^{HO}$
$w_{\alpha v'j'vj}$	2.2.1	$3N_{\alpha}^{2}N_{\alpha\alpha}^{QV}$
$f_{nn'}^{J\lambda\alpha}$ and associated packed indices	2.2.2	$6\,N_{\alpha\alpha'}^{QA},N_{\alpha}^{2}f^{NZ}$
$t_{mn}(R_{\alpha,j}^{R})$	2.1,2.2.2	$3N_{\alpha}^{F}m_{\delta}\delta_{max,\alpha}$
${}^{(r)}f_{nn'}^{\alpha}(R_{\alpha,j}^{F})$	2.2.2	$3N_{\alpha}^{QRS}N_{rot,\delta}N_{\alpha}$
$\grave{g}_{m\beta}^{N}(R_{\alpha,j}^{F})$	2.2.2	$3N_{\alpha}^{QRS}N_{rot,\delta}N_{\alpha}m_{\delta}$
$\mathscr{F}_{nn_0}(R_{\alpha,j}^{F})$	2.2.4	N_{α}^{2}
$\mathscr{G}_{\beta n_0}(R_{\alpha,j}^{F})$	2.2.4	$m_{\delta}N_{\alpha}^{2}$
$\mathscr{H}_{\beta n_0}(R_{\alpha,j}^{F})$	2.2.4	$m_{\delta}N_{\alpha}^{2}$
$\mathscr{K}_{nn_0}^{B}$	2.1	$9N_{\alpha}^{2}$
$B_{\beta n}$	2.1	$9N_{\alpha}^{2}m_{\delta}$
$C_{\beta\beta'}$	2.1	$9N_{\alpha}^{2}m_{\delta}^{2}$

decreased by a factor of 1/2 for the $A + B_2$ case and by a factor of 1/6 for the A_3 case.

5. Scattering Matrix Formulation

5.1. THEORY

We now discuss using complex rather than real boundary conditions to directly calculate the scattering matrix. This can have several advantages. First of all, the use of complex boundary conditions eliminates the possibility of spurious singularities in the final inversion step. Second, if iterative methods are used to solve the complex analog of Eq. (7b), it is possible to solve for a column of the scattering matrix corresponding to a particular initial state without having to solve for the other initial states; this can convert the final N^3 step to an N^2 one. An additional possible advantage is that fewer channels and gaussians may be required to converge an individual column of the transition matrix than to converge the whole reactance matrix.

A re–derivation of the GNVP equations in terms of scattering matrix boundary conditions shows that few of the equations presented in Sect. 2 are modified. The ones that change are Eq. (3), which becomes

$$S_{nn_0} = \delta_{\alpha\alpha_0}\, {}^{0}S_{nn_0} + \mathscr{S}_{nn_0}, \tag{70}$$

and Eqs. (4) and (7), which become

$$\mathscr{S} = \mathscr{S}^{B} + \tilde{B}^{T}\tilde{C}^{-1}\tilde{B}, \tag{71}$$

where ${}^{0}S_{nn_0}$ is the scattering matrix due to the distortion potential, and the quantities $\mathscr{S}^{B}$, $\tilde{B}$, and $\tilde{C}$ are given by the same equations as those for $\mathscr{K}^{B}$, B, and C in Sect. 2 except that the radial functions ${}^{(r)}f^{\alpha}_{nn_0}$ and $g^{N}_{n\beta}$, with $\alpha_n = \alpha_\beta = \alpha$, are replaced by the complex functions ${}^{(r)}\tilde{f}^{\alpha}_{nn_0}$ and $\tilde{g}^{N}_{n\beta}$ which satisfy the large–R_α boundary conditions

$$\tilde{f}^{\alpha}_{nn'} \underset{R_\alpha \to \infty}{\sim} \begin{cases} (\frac{1}{2i})^{\frac{1}{2}} k_n^{-\frac{1}{2}} \{ \delta_{nn'} \exp[-i(k_n R_\alpha - \ell_n \pi/2)] - \\ \quad - {}^{o}S_{nn'}\, \Delta^{\alpha}_{nn'} \exp[i(k_n R_\alpha - \ell_n \pi/2)]\}, & k_n^2 > 0; \\ i(2|k_n|)^{-\frac{1}{2}} \{ \delta_{nn'} \exp[|k_n|(R_\alpha - R_f)] - \\ \quad - {}^{o}S_{nn'}\Delta^{\alpha}_{nn'} \exp[-|k_n|(R_\alpha - R_f)]\}, & k_n^2 < 0, \end{cases} \tag{72}$$

and

$$\tilde{g}^{N}_{n\beta} \underset{R_\alpha \to \infty}{\sim} \tilde{d}^{\alpha}_{\beta n} \begin{cases} (\frac{1}{2i})^{\frac{1}{2}}\, k_n^{-\frac{1}{2}} \exp[i(k_n R_\alpha - \ell_n \pi/2)], & k_n^2 > 0 \\ i(2|k_n|)^{-\frac{1}{2}} \exp[-k_n|(R_\alpha - R_f)], & k_n^2 < 0, \end{cases} \tag{73}$$

where the new boundary condition matrix $\tilde{d}^{\alpha}_{\beta n}$ is given by Eq. (23) with the complex radial function ${}^{(r)}\tilde{f}^{\alpha}_{n_\beta n}$ substituted for the real one. It should be noted that the form of the matching functions in Eqs. (72) and (73) is required for the simple form of Eqs. (70) and (71) to be valid.

The matrices required for Eq. (71) can be calculated in two ways. First of all one can modify the boundary conditions used when solving the FDBVM equations to directly yield the complex radial functions and then proceed as with the real case. The second way of calculating the required matrices is to note that they can be constructed from the corresponding real matrices. This should be the more efficient method since the increased work of complex arithmetic is avoided until the end. In addition, the storage requirements should be less since real quantities require half the storage of complex ones. Thus we adopted this strategy.

In order to give the relations between the real and complex quantities, it is desirable to partition matrices into parts corresponding to open (i.e., energetically allowed) and closed (i.e., classically forbidden) channels. This is because of the different boundary conditions for these cases. Thus distortion potential block δ has a reactance matrix which we write, similarly to Eq. (3), as

$$ {}^{o}\mathbf{K}_{\delta} = \begin{bmatrix} {}^{o}\mathbf{K}_{\delta}^{oo} & {}^{o}\mathbf{K}_{\delta}^{oc} \\ {}^{o}\mathbf{K}_{\delta}^{co} & {}^{o}\mathbf{K}_{\delta}^{cc} \end{bmatrix}, \tag{74} $$

where o again stands for open and c stands for closed. Then the distortion potential scattering matrices ${}^{o}\mathbf{S}_{\delta}$ are given by

$$ {}^{o}\mathbf{S}_{\delta} = (1 - i\,{}^{o}\mathbf{K}_{\delta}^{oo})^{-1}(1 + i\,{}^{o}\mathbf{K}_{\delta}^{oo}). \tag{75} $$

We will transform the distorted wave radial functions for distortion blocked δ by

$$ {}^{(r)}\tilde{\mathbf{f}}_{\delta}^{\alpha} = {}^{(r)}\mathbf{f}_{\delta}^{\alpha}\mathbf{A}_{\delta}, \tag{76} $$

where the transformation matrix $\mathbf{A}_{\delta}$ is given by

$$ \mathbf{A}_{\delta} = \begin{bmatrix} -(2i)^{\frac{1}{2}}(1 - i\,{}^{o}\mathbf{K}_{\delta}^{oo})^{-1} & -(1 - i\,{}^{o}\mathbf{K}_{\delta}^{oo})^{-1}\,{}^{o}\mathbf{K}_{\delta}^{oc} \\ \mathbf{0} & i\,\mathbf{1} \end{bmatrix}. \tag{77} $$

This also means that the boundary condition matrix for distortion block δ is given by

$$ \tilde{\mathbf{d}}_{\delta}^{T} = \mathbf{d}_{\delta}^{T}\mathbf{A}_{\delta}. \tag{78} $$

Finally we use Eq. (30) to generate the half integrated Green's functions which satisfy Eq. (73), and for distortion block δ, the transformation matrix $\mathbf{X}_{\delta}$ is given by

$$\mathbf{X}_\delta = \begin{bmatrix} -\frac{1}{2}(1+i)\mathfrak{d}_\delta^{oo\,T} & -\frac{1}{2}(1+i)\mathfrak{d}_\delta^{oc\,T} \\ \mathbf{0} & \mathbf{0} \end{bmatrix}. \tag{79}$$

For the following equations, we define **A** and **X** to be the block diagonal matrices with diagonal blocks equal to $\mathbf{A}_\delta$ and $\mathbf{X}_\delta$, respectively.

With these definitions, the matrices we require are

$$\mathscr{I}^B = \mathbf{A}^{ooT}\mathscr{K}^{Boo}\mathbf{A}^{oo}, \tag{80}$$

$$\tilde{\mathbf{B}} = (\mathbf{B} + \mathbf{X}^T\mathscr{K}^B)\mathbf{A}, \tag{81}$$

and

$$\tilde{\mathbf{C}} = \mathbf{C} + \mathbf{X}^T\mathscr{D} - \mathscr{B}\mathbf{X} - \mathbf{X}^T\tilde{\mathscr{B}} - \mathbf{X}^T\mathscr{K}^B\mathbf{X}, \tag{82}$$

where the elements of $\mathscr{D}$ are given by

$$\mathscr{D}_{n\beta_0} = \int dR_{\alpha_0}\mathscr{F}_{nn_{\beta_0}}(R_{\alpha_0})t^{\alpha_0}_{m_{\beta_0}n_{\beta_0}}(R_{\alpha_0}) \tag{83}$$

with

$$\mathscr{F}_{nn_0} = \begin{cases} \Delta^{\alpha_0}_{nn_0}\;{}^{(r)}f^{\alpha_0}_{n_0n}(R_{\alpha_0}), & \alpha_n = \alpha_0; \\ \int dR_{\alpha_n}\sum_{n'}\Delta^{\alpha_n}_{nn'}\;{}^{(r)}f^{\alpha_n}_{n'n}(R_\alpha)\mathscr{B}^{\alpha_n\alpha_0}_{n'n_0}(R_{\alpha_n},R_{\alpha_0}), & \text{otherwise}, \end{cases} \tag{84}$$

and

$$\tilde{\mathscr{B}}_{n\beta_0} = \int dR_{\alpha_0}\sum_{n'}\Delta^{\alpha_0}_{n'n}\mathscr{F}_{nn'}(R_{\alpha_0})\grave{g}^{N}_{n'\beta_0}(R_{\alpha_0}). \tag{85}$$

Thus it is necessary to compute the additional two matrices $\tilde{\mathscr{B}}$ and $\mathscr{D}$ when complex boundary conditions are used. However, by virtue of their smaller dimensions, the extra work required will be small compared to the work required to generate the C matrix.

The symmetry decoupling procedures of the previous section apply here as well without change.

Another treatment of complex boundary conditions is provided in Ref. 40.

Equation (71) may be solved iteratively by extending iterative techniques developed already[56] for solving eq. (7). With iterative methods the computational effort for these steps scales with the number of channels as M^2, as compared to M^3 for direct methods. Thus it becomes more imperative to optimize the other M^2 steps in the calculation, as discussed in detail in Sect. 2.

5.2. EXAMPLES

As an example of the convergence properties for runs with employing symmetry decoupling and real and complex boundary conditions we consider the reaction D + $H_2 \rightarrow$ HD + H on the DMBE[55] potential energy surface. The A + B_2 symmetry was used in the same way for both sets of boundary conditions. We considered a total energy E of 0.98337 eV. For both of the arrangements, D + $H_2(\alpha = 1)$ and H + HD($\alpha = 2$), the number of harmonic functions used to expand the vibrational eigenstates is 78. The number of nodes in the optimized vibrational quadratures[47] employed for $U^{\alpha}_{nn'}$, is 14. The number of nodes $N^{QA}_{\alpha\alpha}$ in the angular quadratures for non-exchange integrals is 30. The finite difference grid extends from $R_\alpha = 0.80\ a_0$ to $20.0\ a_0$ for D + H_2 and from $0.40\ a_0$ to $15.0\ a_0$ for HD + H; in both cases the number N^F_α of grid points is 596.

For the exchange integrals the number $N^{QA}_{\alpha\alpha'}$ of angular quadrature points is 60, and the radial integrals are carried out with 16 repetitions of 12–point Gauss–Legendre quadrature ($N^{QGL}_\alpha = 16$, $N^{QR}_\alpha = 12$, $N^{QRS}_\alpha = N^{QGL}_\alpha N^{QR}_\alpha$). The maximum and minimum bond distances for which any vibrational wave

TABLE II
Parameters of the gaussian basis sets

	m=3	m=5	m=7	m=16
$\alpha = 1$ (D + H_2)				
$R^G_{\alpha,1}$ (a_0)	2.50	2.20	2.20	2.20
$\Delta(G)$ (a_0)	0.50	0.40	0.31	0.30
$R^G_{\alpha,m}$ (a_0)	3.50	3.80	4.06	6.70
$\alpha = 2,3$ (HD + H)				
$R^G_{\alpha,1}$ (a_0)	2.50	2.20	2.20	2.20
$\Delta(G)$ (a_0)	0.50	0.40	0.33	0.30
$R^G_{\alpha,m}$ (a_0)	3.50	3.80	4.18	6.70

function exceeds 10^{-9} were determined, and angles corresponding to distances outside this range were excluded from the inner (angular) quadrature loop in all exchange integrals.

We will compare four gaussian basis sets with complex boundary conditions and three with real boundary conditions. In all cases the gaussian width parameter[57] c, which determines the overlap of neighboring gaussians, was set equal to 1.4. The other parameters of the runs are given in Table II. The notation is as follows: m is the number of gaussians per channel, $R^G_{\alpha,1}$ is the center of the innermost gaussian, $\Delta(G)$ is the spacing between gaussians, and $R^G_{\alpha,m}$ is the center of the outermost gaussian.

Five vibrational levels with maximum rotational quantum numbers of 16, 15, 14, 13, and 12 were used in each arrangement. Thus the total number of channels for J = 0 is 225, and 55 of these are open. For J = 1, this vibrational–rotational basis yields 210 channels with 49 open for parity P = 1, and it yields 435 channels with 104 open for parity P = –1.

Reaction probabilities for $D + H_2(v,j,\ell) \rightarrow H + HD$ are defined by

$$P_{vj\ell} = \sum_{\alpha'=2}^{3} \sum_{v'} \sum_{j'} \sum_{\ell'} P_{1vj\ell\rightarrow\alpha'v'j'\ell'} . \tag{86}$$

where

$$P_{\alpha_n v_n j_n \ell_n \rightarrow \alpha_{n'} v_{n'} j_{n'} \ell_{n'}} = |S_{nn'}|^2, \quad n \neq n'. \tag{87}$$

In separate runs we determined that the values $P_{vj\ell}$ in the run with complex boundary conditions and m = 16 are well converged with respect to changing all numerical and basis parameters. (In fact, some of the parameters are overly cautious. For example, except for the very small probabilities, very good convergence is achieved for both real and complex boundary conditions by m = 11.) Thus we define the percentage deviation for other runs by

$$\text{per. dev.} = \frac{|P_{vj\ell}(m,\ R \text{ or } C) - P_{vj\ell}(m{=}16,\ C)|}{P_{vj\ell}(m{=}16,\ C)} \times 100\% \tag{88}$$

where R and C denote real and complex boundary conditions, respectively. Selected results for J = 0 are given in Table III, and Tables IV and V give selected results for the two parity blocks of J = 1. We see that real and complex boundary conditions converge to the same reaction probabilities. Furthermore, we see that many reaction probabilities are converged to better than 1% with m = 3, and almost all the reaction probabilities are converged to better than 1% with m = 7. For the reaction probabilities shown, the average deviation of the m = 7, C run from the m = 16, C run is 0.2%.

A few of the results for m = 5 show surprisingly large deviations. Additional calculations with all m in the range m = 3–11 and with both real and complex boundary conditions showed such instabilities only for m = 5 and—to a lesser

TABLE III
Convergence checks for D + H_2 → HD + H, J = 0

v j ℓ	$P_{vj\ell}$ m=16,C	percentage deviation m=3,R	m=5,R	m=7,R	m=3,C	m=5,C	m=7,C
0 0 0	6.14(−1)	0.30	1.0	0.034	1.1	0.29	0.0073
0 1 1	9.05(−1)	1.3	0.71	0.21	1.3	0.34	0.0053
0 2 2	5.57(−1)	0.19	0.18	0.081	1.3	0.0067	0.093
0 3 3	6.77(−1)	0.56	0.44	0.0015	0.67	0.29	0.043
0 4 4	7.08(−1)	0.073	0.16	0.036	0.66	0.30	0.014
0 5 5	4.62(−1)	0.90	1.1	0.11	0.85	0.85	0.12
0 6 6	2.21(−1)	0.79	0.12	0.22	1.2	0.44	0.22
0 7 7	5.39(−2)	3.9	0.51	0.077	1.5	0.10	0.059
1 0 0	2.96(−1)	7.8	22.	0.64	1.3	1.6	1.0
1 1 1	4.08(−1)	7.4	0.26	0.42	2.2	1.0	0.89
1 2 2	1.20(−1)	17.	36.	0.0095	5.8	0.89	0.40
1 3 3	8.65(−3)	13.	5.8	1.0	14.	6.4	1.4

TABLE IV
Convergence checks for D + H_2 → HD + H, J = 1, P = +1

v j ℓ	$P_{vj\ell}$ m=16,C	percentage deviation m=3,R	m=5,R	m=7,R	m=3,C	m=5,C	m=7,C
0 1 1	4.70(−1)	0.43	1.8	0.17	0.14	0.023	0.11
0 2 2	8.61(−1)	0.26	0.18	0.089	0.085	0.076	0.051
0 3 3	6.23(−1)	0.17	2.9	0.039	1.1	0.0096	0.062
0 4 4	5.79(−1)	0.16	0.16	0.0065	0.22	0.24	0.021
0 5 5	6.06(−1)	0.022	0.61	0.040	0.040	0.31	0.046
0 6 6	3.24(−1)	1.4	0.12	0.22	1.3	0.23	0.21
0 7 7	6.54(−2)	3.3	12.	0.045	2.7	0.27	0.047
1 1 1	5.50(−3)	16.	3.4	0.089	7.4	0.73	0.11
1 2 2	6.70(−3)	25.	3.2	0.41	7.8	0.86	0.29
1 3 3	1.49(−3)	19.	592.	0.23	11.	2.2	0.93

TABLE V
Convergence checks for D + H_2, HD + H, J = 1, P = –1

v j ℓ	$P_{vj\ell}$ m=16,C	percentage deviation m=3,R	m=5,R	m=7,R	m=3,C	m=5,C	m=7,C
0 0 1	6.14(–1)	2.3	0.15	0.15	0.99	0.29	0.046
0 1 2	7.62(–1)	0.024	0.016	0.086	0.85	0.24	0.0063
0 2 3	6.74(–1)	1.2	0.31	0.16	0.60	0.082	0.0000088
0 3 4	6.50(–1)	0.083	0.24	0.017	0.45	0.15	0.050
0 4 5	6.40(–1)	0.48	0.20	0.10	0.27	0.33	0.031
0 5 6	4.89(–1)	0.25	0.41	0.044	0.45	0.59	0.061
0 6 7	2.25(–1)	1.2	0.22	0.17	1.2	0.32	0.18
0 7 8	4.33(–2)	3.1	0.19	0.047	2.2	0.15	0.031
0 9 8	4.56(–5)	1.5	1.5	0.54	2.4	1.4	0.43
1 1 0	1.55(–1)	5.2	0.52	0.26	3.1	0.99	0.66
1 2 1	6.27(–2)	1.1	1.1	0.16	6.0	0.67	0.41
1 3 2	6.07(–3)	15.	4.7	1.2	14.	5.0	0.42

extent—for m = 10 and only for the real boundary conditions. No such instabilities have been observed in calculations with complex boundary conditions.

6. Concluding Remarks

With sufficient attention to algorithmic efficiency, it has become possible to solve very large–scale problems in quantum mechanical reaction dynamics. Section 5.2 discusses calculations on the reaction D + H_2(v=0,1) → HD + H both for J = 0 (see also Refs. 13, 14, and 36) and J ≠ 0, and, in previous work using the techniques described here, as well as earlier nonvariational and variational algorithms for $\mathscr{L}^2$ expansions of the reactive amplitude density, we and coworkers have performed converged calculations of several other reactive processes, including, for J = 0, H + H_2(v=0–2) → H_2(v′=0–2) + H,[19,20,31–33,39] O + H_2(v=0,1) → OH(v′=0,1) + H,[15,34,36] H + HBr(v=0) → H_2(v′=0,1,2) + Br,[35,39] and Cl + H_2(v=0) → HCl(v′=0) + H.[58] We have also obtained accurate results for H + H_2(v=0,1) → H_2(v′=0,1) + H for J = 0–20,[33] and accurate results have been obtained for O + HD(v=0) → OH(v′=0) + D and OD(v′=0) + H for J = 0–2.[39] We anticipate further improvements in several areas, e.g., more efficient strategies for the Green's functions, better basis sets, and physical or decoupling approximations as initial guesses for iterative techniques. These improvements should allow even more progress in the future.

7. Acknowledgments

We are grateful to several colleagues for collaboration on related aspects of this research program. This work was supported in part by the National Science Foundation and the Minnesota Supercomputer Institute.

8. References

1. H. Feshbach, Ann. Phys. (NY) **19**, 287 (1962).
2. K. M. Watson and J. Nuttall, *Topics in Several Particle Dynamics* (Holden–Day, San Francisco, 1967).
3. P. G. Burke and K. Smith, Rev. Mod. Phys. **34**, 458 (1962).
4. W. H. Miller, J. Chem. Phys. **50**, 407 (1969).
5. D. G. Truhlar, J. Abdallah, Jr., and R. L. Smith, Adv. Chem. Phys. **25**, 211 (1974).
6. I. Shavitt, in *Methods of Electronic Structure Theory*, Vol. 3, edited by H. F. Schaefer III (Plenum, New York, 1977), p. 189.
7. M. J. Seaton, Phil. Trans. A **245**, 469 (1953).
8. P. G. Burke, Comments At. Molec. Phys. **3**, 31 (1971).
9. R. K. Nesbet, Adv. Quantum Chem. **9**, 215 (1975).
10. W. M. Huo, T. L. Gibson, M. A. P. Lima, and V. McKoy, Phys. Rev. A. **36**, 1632 (1987).
11. Y. C. Tang, in *Few–Body Systems and Multiparticle Dynamics*, edited by D. A. Micha (American Institute of Physics Conference Proceedings No. 162, New York, 1987), p. 174.
12. D. J. Kouri, in *Theory of Chemical Reaction Dynamics*, edited by M. Baer (CRC Press, Boca Raton, Florida, 1985), Vol. I, p. 163.
13. K. Haug, D. W. Schwenke, Y. Shima, D. G. Truhlar, J. Z. H. Zhang, and D. J. Kouri, J. Phys. Chem. **86**, 6757 (1986).
14. J. Z. H. Zhang, D. J. Kouri, K. Haug, D. W. Schwenke, Y. Shima, and D. G. Truhlar, J. Chem. Phys. **88**, 2492 (1988).
15. J. Z. H. Zhang, Y. Zhang, D. J. Kouri, B. C. Garrett, K. Haug, D. W. Schwenke, and D. G. Truhlar, Faraday Discussions Chem. Soc. **84**, in press.
16. D. W. Schwenke, D. G. Truhlar, and D. J. Kouri, J. Chem. Phys. **86**, 2772 (1987).
17. W. H. Miller and B. M. D. D. Jansen op de Haar, J. Chem. Phys. **86**, 6213 (1987).
18. J. Z. H. Zhang, S.–I. Chu, and W. H. Miller, J. Chem. Phys. **88**, 6233 (1988).
19. D. W. Schwenke, K. Haug, D. G. Truhlar, Y. Sun, J. Z. H. Zhang, and D. J. Kouri, J. Phys. Chem. **91**, 6080 (1987).
20. D. W. Schwenke, K. Haug, M. Zhao, D. G. Truhlar, Y. Sun, J. Z. H. Zhang, and D. J. Kouri, J. Phys. Chem. **92**, 3202 (1988).
21. R. G. Newton, *Scattering Theory of Particles and Waves* (McGraw–Hill, New York, 1966), Sec. 11.3.
22. A. Messiah, *Quantum Mechanics* (John Wiley & Sons, New York, 1966), chapter XIX.
23. N. F. Mott and H. S. W. Massey, *The Theory of Atomic Collisions*, 3rd edition (Clarendon Press, Oxford, 1965), p. 349.
24. R. D. Levine, *Quantum Mechanics of Molecular Rate Processes* (Oxford University Press, London, 1969), Section 3.2.3.

25. D. G. Truhlar and J. Abdallah, Jr., Phys. Rev. A **9**, 297 (1974).
26. G. R. Satchler, *Direct Nuclear Reactions* (Oxford University Press, New York, 1983), pp. 264ff.
27. L. M. Hubbard, S. Shi, and W. H. Miller, J. Chem. Phys. **78**, 2381 (1983).
28. G. C. Schatz, L. M. Hubbard, P. S. Dardi, and W. H. Miller, J. Chem. Phys. **81**, 231 (1984).
29. B. H. Choi, R. T. Poe, and K. T. Tang, J. Chem. Phys. **81**, 4979 (1984).
30. P. S. Dardi, S. Shi, and W. H. Miller, J. Chem. Phys. **83**, 575 (1985).
31. M. Mladenovic, M. Zhao, D. G. Truhlar, D. W. Schwenke, Y. Sun, and D. J. Kouri, Chem. Phys. Lett. **146**, 358 (1988).
32. M. Zhao, M. Mladenovic, D. G. Truhlar, D. W. Schwenke, Y. Sun, D. J. Kouri, and N. C. Blais, J. Amer. Chem. Soc., in press.
33. M. Mladenovic, M. Zhao, H. Dong, D. G. Truhlar, D. W. Schwenke, Y. Sun, and D. J. Kouri, unpublished calculations.
34. K. Haug, D. W. Schwenke, D. G. Truhlar, Y. Zhang, J. Z. H. Zhang, and D. J. Kouri, J. Chem. Phys. **87**, 1892 (1987).
35. Y. C. Zhang, J. Z. H. Zhang, D. J. Kouri, K. Haug, D. W. Schwenke, and D. G. Truhlar, Phys. Rev. Lett. **60**, 2367 (1988).
36. K. Haug, D. W. Schwenke, Y. Shima, D. G. Truhlar, J. Z. H. Zhang, Y. Zhang, Y. Sun, D. J. Kouri, and B. C. Garrett, in *Science and Engineering on Cray Supercomputers* (Cray Research, Minneapolis, 1987), p. 427.
37. D. W. Schwenke, K. Haug, D. G. Truhlar, R. H. Schweitzer, J. Z. H. Zhang, Y. Sun, and D. J. Kouri, Theoret. Chim. Acta **72**, 237 (1987).
38. D. J. Kouri, Y. Sun, R. C. Mowrey, J. Z. H. Zhang, D. G. Truhlar, K. Haug, and D. W. Schwenke, in *Mathematical Frontiers in Computational Chemical Physics Physics*, edited by D. G. Truhlar (Springer–Verlag, New York, 1988), in press.
39. P. Halvick, D. J. Kouri, M. Mladenovic, D. W. Schwenke, O. Sharafeddin, Y. Sun, D. G. Truhlar, C.–H. Yu, and M. Zhao, unpublished calculations.
40. Y. Sun, C.–H. Yu, D. J. Kouri, D. W. Schwenke, P. Halvick, M. Mladenovic, and D. G. Truhlar, to be published.
41. Y. Sun, D. J. Kouri, D. W. Schwenke, M. Zhao, P. Halvick, and D. G. Truhlar, to be published.
42. J. M. Blatt and L. C. Biedenharn, Rev. Mod. Phys. **24**, 258 (1952).
43. M. A. Brandt, D. G. Truhlar, and R. L. Smith, Computer Phys. Commun. 5, 456 (1973).
44. D. A. Micha, Arkiv Fysik **30**, 411 (1965).
45. D. H. Bailey, SIAM J. Sci. Stat. Comput. **9**, 603 (1988).
46. V. Strassen, Numer. Math. **13**, 354 (1969).
47. D. W. Schwenke and D. G. Truhlar, Computer Phys. Commun. **34**, 57 (1984).
48. D. G. Truhlar and A. Kuppermann, J. Amer. Chem. Soc. **93**, 1840 (1971).
49. K. Haug, Ph.D. thesis, University of Minnesota, Minneapolis, 1987 (unpublished).
50. M. E. Riley and A. Kuppermann, Chem. Phys. Lett. **1**, 537 (1968).
51. G. H. Golub and C. F. Van Loan, *Matrix Computations* (Johns Hopkins University Press, 1983), page 92.
52. J. J. Dongarra, C. B. Moler, J. R. Bunch, and G. W. Stewart, *LINPACK User Guide* (Society for Industrial and Applied Mathematics Philadelphia, Pennsylvania, 1979).

53. I. C. Percival and M. J. Seaton, Proc. Camb. Phil. Soc. **53**, 654 (1957).
54. K. Onda and D. G. Truhlar, J. Chem. Phys. **71**, 5097 (1979).
55. A. J. C. Varandas, F. B. Brown, C. A. Mead, D. G. Truhlar, and N. C. Blais, J. Chem. Phys. **86**, 6258 (1987).
56. C. Duneczky, R. E. Wyatt, D. Chatfield, K. Haug, D. W. Schwenke, D. G. Truhlar, Y. Sun, and D. J. Kouri, Computer Phys. Commun., to be published.
57. I. P. Hamilton and J. C. Light, J. Chem. Phys. **84**, 306 (1986).
58. Y. C. Zhang, J. Z. H. Zhang, D. J. Kouri, K. Haug, D. W. Schwenke, and D. G. Truhlar, unpublished calculations.

HOW VARIATIONAL METHODS IN SCATTERING THEORY WORK

B. RAMACHANDRAN and ROBERT E. WYATT
Department of Chemistry and The Institute for Theoretical Chemistry
University of Texas, Austin, TX 78712
U.S.A.

ABSTRACT. We explore the factors responsible for the rapid convergence of the Schwinger and Newton variational principles in scattering theory. We find that, contrary to conventional wisdom, these variational methods yield high accuracy not because the error associated with the computed quantity is second order in the error in the wavefunction, but because variational methods find wavefunctions that are far more accurate in relevent regions of the potential, compared to nonvariational methods.

1. Introduction

Variational methods are at present used extensively in the study of inelastic and reactive scattering involving atoms and diatomic molecules[1-5]. Three of the most commonly used variational methods are due to Kohn (the KVP)[6], Schwinger (the SVP)[7] and Newton (the NVP)[8]. In the applications of these methods, the wavefunction is typically expanded in a set of basis functions, parametrized by the expansion coefficients. These linear variational parameters are then determined so as to render the variational functional stationary. Unlike the variational methods in bound state calculations, the variational principles of scattering theory do not provide an upper or lower bound to the quantity of interest, except in certain special cases.[9] Neverthless, variational methods are useful because, the minimum basis size with which an acceptable level of accuracy can be achieved using a variational method is often much smaller than those required if nonvariational methods are used. The reason for this is generally explained by showing (as we do below) that an error of δf in the scattering wavefunction $f(r)$ is turned into an error of the order of $(\delta f)^2$ in the stationary quantity–K, S or T matrix elements–defined by the variational functional, where as nonvariational methods convey an error of order δf to the quantity of interest.

This suggests that the ability of a certain method, variational or nonvariational, to achieve a certain accuracy for a given basis size is related to the error $\delta f(r)$ in the computed wavefunction. However, this says nothing about the *nature* of the error. Should it be uniformly distributed over the entire wavefunction? Or should it be confined almost entirely to certain regions of the potential? Do a particular method work better when the error is of a certain type, as opposed to another method? This line of thought leads to more unexplored territory. The three variational methods mentioned above, and the nonvariational methods that have been used in scattering calculations, involve solving linear algebraic systems of the form **Ac** = **b**, for the unknown expansion coefficients **c** of the wavefunction. Since each method solves a different set of equations, it is natural to expect that the coefficients **c** are different in each case, except possibly in the limit where all methods have reached a

A. Laganà (ed.), Supercomputer Algorithms for Reactivity, Dynamics and Kinetics of Small Molecules, 169–185.

reasonable level of convergence. Naturally, the different solutions can also be expected to suffer different magnitudes of error $\delta f(r)$ which would, presumably, be distributed over the range of r in quite different manners. Apart from pure academic curiosity in the behavior of $\delta f(r)$ produced by the different methods, we feel that an understanding of the behavior of $\delta f(r)$ and its relationship to the error in quantities such as K, T or S matrix elements computed from the different methods could lead to a deeper understanding of the way in which variational and nonvariational methods work. Moreover, such understanding can lead to practical advantages as well, such as being able to devise strategies that reduce basis sizes by concentrating the effort to converge a calculation to only certain regions of the wavefunction, and deliberately leaving certain other regions incompletely represented.

Such considerations have, in large measure, motivated the present enquiry. We examine the behavior of the error $\delta f(r)$ in the wavefunctions computed by different methods for different basis sizes, for potential scattering from three one dimensional interaction potentials. We also compute the K matrix element using the different methods, and examine the behavior of the error in the K matrix elements as compared to the behavior of the error in $f(r)$. To supplement this enquiry, we examine a class of methods refered to as "variationally correct, but nonstationary" methods,[10] obtained by inserting the wavefunction from one method -variational or nonvariational- into the functional of another variational method, to study how the *same* error in the wavefunction is reflected in the results from the different methods.

We begin, in the following section, by discussing the relationship between the three variational methods mentioned above, from the point of view of the Kato identity and the Lippmann-Schwinger equation. For reasons presented there, we confine attention to the variational methods based on the integral equation formulation, viz., the SVP and the NVP. As a representative of nonvariational methods, we also consider a method which makes use of the integral equation approach to scattering, viz., the method of moments for the amplitude density (the MMAD).[10] In this Section, we also outline the calculations we propose to carry out, in order to study the behavior of the error in the wavefunctions generated by the different methods. In Section 3, we present our results and discuss them. Section 4 is a brief summary of the work presented here, followed by a few concluding remarks.

2. Theoretical and Computational Considerations

2.1. THEORETICAL ASPECTS

The Kato identity,[11] which forms the basis for the three most commonly used variational methods in scattering theory, the KVP, the SVP and the NVP, is given as follows, for the K matrix element:

$$-k\mathrm{K}_{\mathrm{ex}} = -k\mathrm{K}^{(0)} + \langle f_1|(\mathrm{H}-\mathrm{E})|f_2\rangle \qquad (1)$$

where $\mathrm{K}^{(0)}$ is a zero-order result (see below), which is refined by the term containing the functions $f_1(r)$ and $f_2(r)$. The quantity K_{ex} is of course, the exact K matrix element, and $k = (2\mu E/\hbar^2)^{1/2}$, where μ is the reduced mass and E, the scattering energy. Throughout the course of this work, we use units in which $\hbar^2/2\mu = 1$, so that $k^2 = E$.

If, in (1), we set $f_1 = f_2 = f_t$ where f_t is a trial wavefunction, we get the KVP. The K matrix versions of the KVP have been somewhat unpopular in the past, due to the so called

Kohn anomalies.[9] A recent S matrix version of KVP[2] avoids this problem, and has been successfully applied to three dimensional reactive scattering problems.[1(c),2]

In order to establish the connection between Eq.(1) and the SVP[12], we consider the following: in (1), we set

$$-k\mathrm{K}^{(0)} = \langle f_2|(\mathrm{H-E})|s\rangle + \langle s|(\mathrm{H-E})|f_2\rangle, \quad (2)$$

where $|s\rangle = \sin(kr)$. It is easy to see that as $f_2(r)$ approaches the exact wavefunction of H in the limit of convergence, the second terms in (1) and (2) vanish, leaving us with the first term in (2), which by definition, is the K matrix element. Next, we recall that the Lippmann-Schwinger equation,

$$f(r) = s(r) + \mathrm{G}^0\mathrm{V}f(r), \quad (3)$$

which is used in the Born series, to generate successively higher order approximations to the exact wavefunction as follows:

$$f^{(n)} = s + \mathrm{G}^0\mathrm{V}f^{(n-1)}, \quad (4)$$

where G^0 is the principal value Green function, whose kernel, $\mathrm{G}^0(r,r')$ is given by

$$\mathrm{G}^0(r,r') = -(1/k)\sin(kr_<)\cos(kr_>),$$

with $(r_<,r_>)$ being the lesser and greater respectively, of (r,r'). If, in (1) we use $f_1 = f^{(0)}$ and $f_2 = f^{(1)}$, as defined by (4), we can show that the result is

$$-k\mathrm{K} = 2\langle s|\mathrm{V}|f^{(0)}\rangle - \langle f^{(0)}|\mathrm{V}-\mathrm{VG}^0\mathrm{V}|f^{(0)}\rangle, \quad (5)$$

which is the SVP functional. The NVP functional given below is the result of using $f_1 = f_2 = f^{(1)}$ in (1).[12]

$$-k\mathrm{K} = \mathrm{K}_1^{\mathrm{B}} + 2\langle s|\mathrm{VG}^0\mathrm{V}|f^{(0)}\rangle - \langle f^{(0)}|\mathrm{V}(\mathrm{G}^0-\mathrm{G}^0\mathrm{VG}^0)\mathrm{V}|f^{(0)}\rangle, \quad (6)$$

where $\mathrm{K}_1^{\mathrm{B}} = \langle s|\mathrm{V}|s\rangle$ is the first Born term. Thus we see that these three variational methods are intimately connected to one another.

Let us now examine the consequences of an error of $\delta f^{(0)}$ in the wavefunction $f^{(0)}$, for the K matrix element. Let $f^{(0)} = f_{ex} + \delta f^{(0)}$. Inserting this into, say, the SVP functional of (5), and recalling from (3) that $(1-\mathrm{G}^0\mathrm{V})f_{ex} = |s\rangle$, we get the relationship

$$-k\mathrm{K} = -k\mathrm{K}_{ex} + \langle \delta f^{(0)}|\mathrm{V}-\mathrm{VG}^0\mathrm{V}|\delta f^{(0)}\rangle. \quad (7)$$

Thus we see that the error in the computed K matrix element is one order higher than the error in the corresponding wavefunction. From the Kato identity point of view, we see that the error term in the functional would be $\langle \delta f^{(0)}|\mathrm{V}|\delta f^{(0)}\rangle$ for the KVP, $\langle \delta f^{(0)}|\mathrm{V}|\delta f^{(1)}\rangle$ for the SVP and $\langle \delta f^{(1)}|\mathrm{V}|\delta f^{(1)}\rangle$ for the NVP. Since we expect $f^{(1)}$ to be a more accurate wavefunction than $f^{(0)}$, the above relationships show that a given wavefunction $f^{(0)}$ is transformed to a better wavefunction $f^{(1)}$, simply by virtue of being inserted into the NVP functional.

In the present study, we have chosen to exclude the KVP, for the following reasons: the trial wavefunction $f^{(0)}$ used in the KVP is required to satisfy the boundary conditions of the problem. This could be accomplished by including two non-L^2 terms that represent the asymptotic part of the wavefunction in the expansion of $f^{(0)}$. This would mean of course, that in order for our comparisons to the SVP, the NVP and nonvariational methods to be objective and meaningful, we would have to use the same expansion of $f^{(0)}$ in the other methods as well. However, in the applications of the SVP and the NVP, the tendency has been to use a purely L^2 basis for the expansion of $f^{(0)}$ since the Green function enforces the boundary conditions. Further, most of the nonvariational methods used in scattering theory[10] are also based on the integral equation formulation, where again the trial function is freed from the responsibility of satisfying boundary conditions, due to the presence of the Green function. In order to make our investigations and their results useful and relevant to a broader spectrum of methods, we confine our investigation of variational methods to those based on the Lippmann-Schwinger integral equation, viz., the SVP and the NVP, in which we use a purely L^2 basis for expanding $f^{(0)}$, in essentially the same manner as in existing applications.

A large number of nonvariational methods for scattering calculations are known. Of these, we examine the method of moments for the amplitude density (MMAD), as an example of a nonvariational method. This method is attractive because of its simplicity. It also represents a large class of nonvariational methods that can be derived from it, or formulated as extensions of it.[10] Moreover, the studies of Staszweska and Truhlar indicate that the method of moments for the amplitude density (the MMAD) is capable of generating a more accurate wavefunction than the analogous method for the wavefunction.[10]

The MMAD is implemented as follows: let the amplitude density $\zeta^{(0)}(r) = V(r)f^{(0)}(r)$ be expanded in a set of basis functions as follows:

$$\zeta^{(0)}(r) = \sum_{n=1}^{N} \alpha_n |n> \tag{8}$$

where the $|n>$ are L^2 basis functions. Substituting (8) into (3), we get

$$f^{(0)} = |s> + \sum_{n=1}^{N} \alpha_n G^0 |n>.$$

Multiplying both sides by V, and collecting terms containing $\zeta^{(0)}$, we get

$$\sum_{n=1}^{N} \alpha_n (1-VG^0)|n> = V|s>.$$

The amplitude density is now required to satisfy the equations

$$\sum_{n=1}^{N} <m|1-VG^0|n>\alpha_n = <m|V|s>; \qquad m = 1,......,N, \tag{9}$$

which are then solved for the expansion coefficients $\{\alpha_n\}$. The wavefunction $f^{(0)}(r)$ is generated from these coefficients by simply dividing (8) by $V(r)$.

In our applications of the SVP and the NVP, we expand the trial wavefunction $f^{(0)}(r)$ in terms of L^2 basis functions:

$$f^{(0)}(r) = \sum_{n=1}^{N} c_n |n>.$$

Substitution of this expansion into the functionals of (5) and (6), followed by extremization of the respective functionals with respect to the coefficients $\{c_n\}$, lead, respectively, to the following sets of equations:

$$\sum_{n=1}^{N} \langle m|V\text{-}VG^0V|n\rangle c_n = \langle m|V|s\rangle; \qquad m = 1,\ldots\ldots,N, \tag{10}$$

for the SVP, and

$$\sum_{n=1}^{N} \langle m|V(G^0\text{-}G^0VG^0)V|n\rangle c_n = \langle m|VG^0V|s\rangle; \qquad m = 1,\ldots\ldots,N, \tag{11}$$

for the NVP. These equations, when solved, yield the expansion coefficients $\{c_n\}$, with which we then construct the wavefunction $f^{(0)}$.

Before going any further, let us establish an unambiguous system of notation for refering to the various quantities that we propose to compute and examine. We shall denote the wavefunctions from the different methods by a right subscript. So, for example, $f^{(0)}_{MMAD}$ refers to the wavefunction obtained by dividing (8) by V(r). We adopt the following convention for the K matrix elements: a K matrix element is denoted by $[K_m]_f$, where the inner subscript 'm' refers to the method by which the wavefunction (or amplitude density, in the case of MMAD) was computed, and the outer subscript 'f' refers to the functional from which the K matrix element was evaluated. Thus, $[K_{MMAD}]_{SVP}$ would be a K matrix element obtained by inserting the wavefunction $f^{(0)}_{MMAD}$ into the SVP functional of (5). We refer to the error in the wavefunction computed from the different methods also by the same convention used for $f^{(0)}(r)$, namely, a right subscript to $\delta f^{(0)}(r)$. Another quantity we shall be interested in, is the summed residual of the amplitude density for method 'm', which is obtained as

$$\Xi^2{}_m = \sum_{i=1}^{M} [\zeta_{ex}(r_i)-\zeta^{(0)}{}_m(r_i)]^2. \tag{12}$$

The quantity $\Xi^2{}_m$ will be particularly useful in obtaining a measure of the error in the wavefunction, in the range of r where the potential is far from zero.

From each of our methods, we extract four types of K matrix elements $[K_m]_f$, which are:
1) the direct calculation,

$$[K_m]_D = \langle s|V|f^{(0)}{}_m\rangle, \tag{13}$$

2) corresponding to $f^{(0)}{}_m$ once iterated using the Lippmann-Schwinger equation, given by

$$[K_m]_{LS} = K_1^B + \langle s|VG^0V|f^{(0)}{}_m\rangle, \tag{14}$$

3) $[K_m]_{SVP}$, and 4) $[K_m]_{NVP}$.

2.2. COMPUTATIONAL ASPECTS

We base our studies on potential scattering from three one-dimensional potentials V(r), which are

	Potential A:	$V(r) = 10[1-e^{-(3/2)(r-2)}]^2 - 10,$
	Potential B:	$V(r) = -10e^{-r},$
and	Potential C:	$V(r) = 10e^{-r},$

at a scattering energy of $E = k^2 = 4$. These potentials have been used by us before[13,14] in order to compare the convergence characteristics of the KVP, the SVP and the NVP, and the exact K matrix elements at this energy were computed to high accuracy in that work,[14] by converged finite-difference calculations. Since in the present investigation, we are predominately concerned with the error in the wavefunctions, we wish to avoid errors from sources other than those inherent in the methods, such as numerical integrations etc., as far as possible. Therefore, we choose for the basis functions {|n>}, a set of normalized one-dimensional box energy eigenfunctions: $(2/a)^{1/2}\sin(n\pi r/a)$ where the length of the box, a, is chosen to be 10. With this basis and the Green function $G^0(r,r')$ of Section 2.1, we can evaluate all the integrals of the types $\langle n|V|s\rangle$, $\langle m|V|n\rangle$, $G^0|n\rangle$ and $G^0V|n\rangle$ analytically, for the potentials listed above. Matrix elements where G^0 appears twice, such as $\langle m|VG^0VG^0V|n\rangle$ in the NVP functional of (11) were evaluated by the half-integrated Green function method of Schwenke et. al..[3,13,14] In these evaluations, the outermost integral was evaluated numerically, using the trapezoidal rule. We have verified that our integrations are converged to at least eight decimal places with respect to step-size in all cases.

The exact wavefunction $f_{ex}(r)$ was computed by pushing the MMAD to convergence with respect to basis size, and the wavefunction was evaluated to one order higher in the Born series, by using the relationship

$$f_{ex}(r) = |s\rangle + G^0\zeta^{(0)}$$

and the corresponding K matrix element evaluated using (14), but with $|\zeta^{(0)}\rangle$ substituted in the place of $V|f^{(0)}{}_m\rangle$. For each potential, error in the K matrix element from this wavefunction with respect to the previously computed exact values[14] is less than 10^{-5}%.

The error in the wavefunctions $\delta f^{(0)}{}_m$ is computed as $\delta f^{(0)}{}_m(r) = f_{ex}(r) - f^{(0)}{}_m(r)$, for different basis sizes N. We have found it most illumnating to present these as "error-scapes" or three-dimensional perspective plots in which $\delta f^{(0)}{}_m(r)$ is plotted as a function of the coordinate r, and the number of basis functions N. The error in the K matrix elements are presented as fractional errors, given by $(K_{ex}-[K_m]_f)/K_{ex}$, which are tabulated for each $f^{(0)}{}_m$.

3. Results and Discussion

3.1. BEHAVIOR OF THE ERRORS IN $f^{(0)}{}_m$

We present, in Fig.1, the error-scape for $f^{(0)}{}_{MMAD}$ for potential A. The value of the coordinate r increases from 0 on the right to 10 at the left of this figure. The number of basis functions N in the expansion of the wavefunction increases from 5 at the top of the figure, to 35 at the bottom. The extreme values of the error $\delta f^{(0)}{}_{MMAD}$ have been artificially set to ±10, which explains the "cropped" look of the features in the large r region, so that the vertical scale of the figure would not render the smaller features invisible. The fractional error in the K matrix elements computed from these wavefunctions with N varying from 25 to 30 are presented in Table I, as well as the quantity $\Xi^2{}_{MMAD}$.

Let us first examine Fig. 1. The most striking aspect of this figure is that the large errors in the intermediate-to-large values of r at the top of the figure (small N) get progressively pushed to larger values of r as N increases. However, we see that the internal region is not

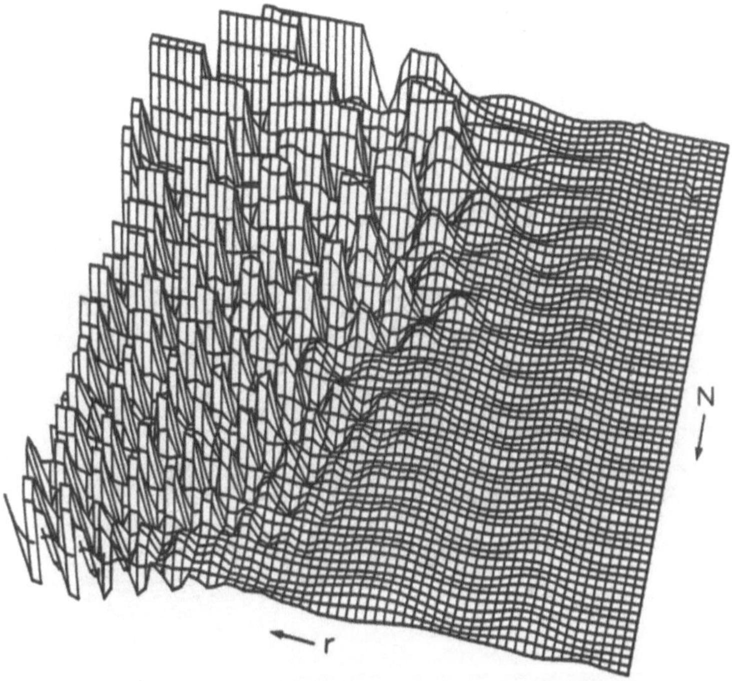

Fig.1. The error-scape for the MMAD. The values of N run from 5 to 35 in steps of 1. The coordinate r increases from 0 at the right extreme to 10 at the left.

entirely flat even in the foreground of the figure, where N is large, indicating that $f^{(0)}_{MMAD}$ still has not converged completely to f_{ex} in this region. A careful examination of the small r region also reveals that the gentle oscillations we see in the foreground extend all the way to fairly small values of N. The implication of this is that the the wavefunction in the internal region converges to a certain extent very quickly as N increases, but the small, oscillatory error in this region is extremely slow to disappear.

In Fig.2, we present the error-scape for $f^{(0)}_{SVP}$ for potential A. The ranges of r and N in this figure are the same as that in Fig.1. We have also cropped the large features at ± 10, as in Fig.1. The difference between Fig.1 and Fig.2 is immediately apparent. Once again, the small N region has large errors at almost all values of r. However, the large errors are pushed to the large r region with remarkable rapidity as N increases. Fig.2 creates the impression that the smaller errors in the internal region at small N are larger for $f^{(0)}_{SVP}$ than for $f^{(0)}_{MMAD}$. This impression is created by a different vertical scaling in this figure, which was necessary to make these features clearly visible. (We opted against using the same vertical scale in Fig.1 for purely aesthetic reasons. The exaggerated vertical scale renders the extremely rough terrain on the left half of Fig.1 all but impossible for the plotter to handle.) The foreground of the figure, in spite of the vertical exaggeration, is remarkably flat, indicating that the wavefunction has achieved a high degree of convergence. This immediately suggests that variational methods such as the SVP achieve higher accuracy in

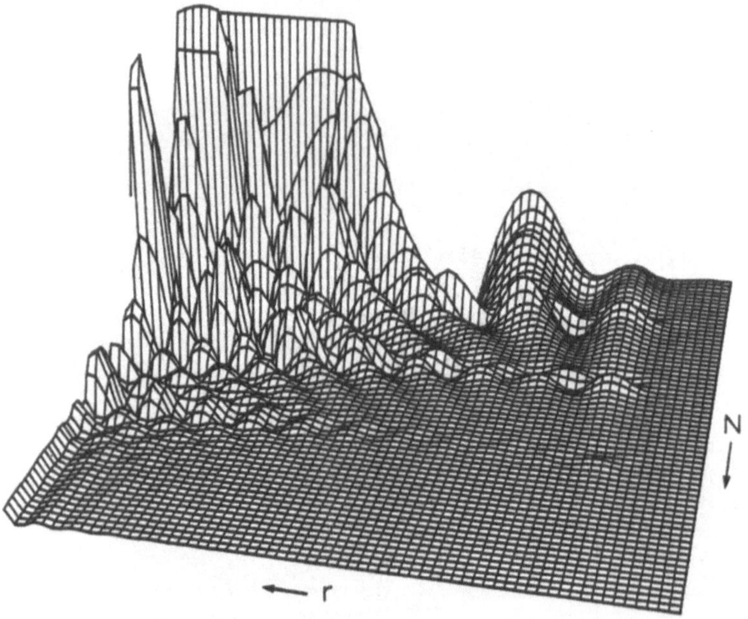

Fig.2. The error-scape for the SVP. The vertical scale is different from that of Fig.1. The ranges of N and r are the same as that of Fig.1.

computing K (and S or T) matrix elements, not only because the error in the computed quantity is second order in $\delta f^{(0)}$, but also because, compared to nonvariational methods such as the MMAD, they achieve far higher accuracy in the wavefunction itself, for a given basis size N. This conclusion is reinforced by an examination of the fractional errors in $[K_{SVP}]_f$ and the quantity $\Xi^2{}_{SVP}$ which we present in Table II, and discuss below.

The error in $f^{(0)}{}_{NVP}$ is presented in Fig.3. The vertical scale of this figure is the same as Fig.2. Once again, the larger features in the error-scape has been cropped at ±10. As in the cases of the MMAD and SVP wavefunctions, the error in the internal region is large when N is small. And, as in the case of the other two methods, the large errors are pushed into the asymptotic regions very quickly as N increases. However, in striking contrast to the SVP, the large errors in the asymptotic regions persist even to the largest values of N. Yet another difference between this figure and Fig.2 is that in the case of the NVP, the errors in the intermediate r region disappear rapidly as N increases, and, in spite of the vertical exaggeration, the small-intermediate r region becomes almost completely flat by the time N reaches a value of approaximately 10-12, whereas undulations in the same regions of the SVP error-scape persist to much higher values of N. The fractional errors in the K matrix elements computed using the NVP wavefunctions, as well as $\Xi^2{}_{NVP}$, are tabulated in Table III.

Even a casual comparison of Fig.1 to Figs. 2 and 3 bring out the dramatic difference in the behavior of the wavefunctions generated from nonvariational and variational methods. It is clear that all three methods attempt to converge the wavefunction in the internal region

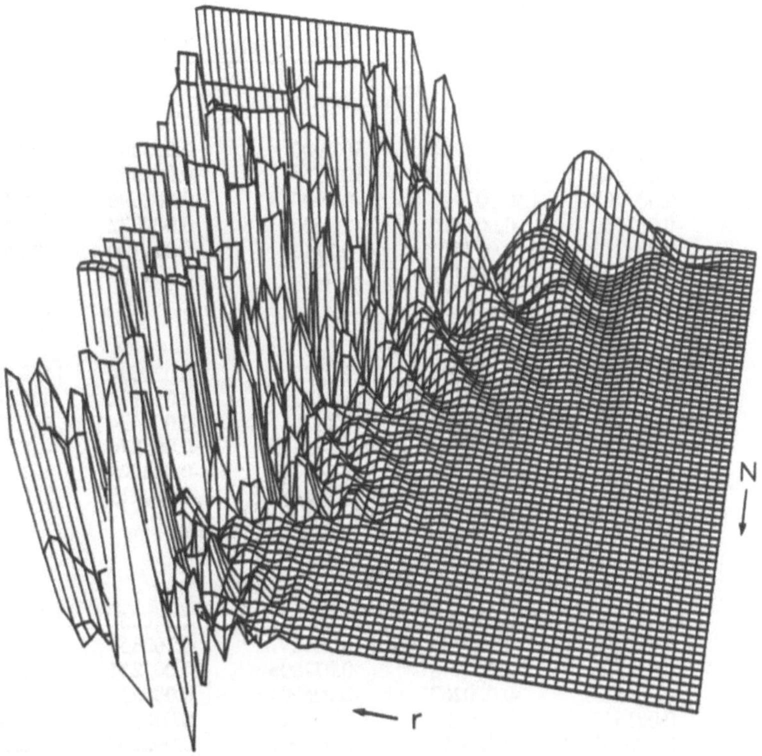

Fig.3. The error-scape for the NVP. The ranges of N and r are the same as Figs.1 and 2. The vertical scale is the same as that of Fig.2.

first. The difference between them lies mostly in the extent to which each of them succeed. In the case of the MMAD, the large errors in the wavefunctions constructed from a small basis are damped rather slowly as N increases. Although the largest features on the surface are pushed to the asymptotic region as N increases, the range of r over which the errors are large is still considerable even at N = 35, when compared to the NVP surface of Fig.3 and especially the SVP surface of Fig.2. Moreover, even for large basis sizes, the internal region of Fig.1 is not free of error. The gentle oscillations in this region extend in the N direction, all the way back to extremely small values of N.

Compared to the MMAD, both variational methods are highly successful in converging the small r region of the wavefunction. However, the NVP is the more successful of the two in achieving this goal. It appears as if the NVP *knows*(!) that the internal region is the part of the wavefunction where a high degree of accuracy is required, and chooses to ignore the asymptotic region until the internal region is well-converged. The SVP seems to posses less of such practical wisdom, but we shall see in the next subsection that the SVP functional is incapable of high accuracy in the K matrix elements unless the wavefunction is converged over a much larger range of r than required by the NVP functional.

TABLE I. Fractional errors in the various $[K_{MMAD}]_f$'s

N	$[K_{MMAD}]_D$	$[K_{MMAD}]_{LS}$	$[K_{MMAD}]_{SVP}$	$[K_{MMAD}]_{NVP}$	Ξ^2_{MMAD}
			Potential A		
25	0.017161	0.002904	50.7900	0.005805	99.6355
26	0.013246	0.003850	12.2518	0.002532	39.7616
27	0.000703	0.001922	0.290049	0.002096	26.1862
28	-0.002310	0.001526	2.36068	0.002072	27.4869
29	0.008717	0.002566	6.51187	0.001700	26.5707
30	-0.006202	0.000191	7.63698	0.001105	20.3153
			Potential B		
15	-0.027579	-0.014694	-16.3826	-0.026694	1459.97
16	-0.014942	-0.024682	-11.1153	-0.015722	978.142
17	-0.010103	-0.002892	-7.11386	-0.009494	661.098
18	-0.005383	-0.010934	-4.865123	-0.005876	457.539
19	-0.004106	0.000141	-3.244704	-0.003724	321.739
20	-0.002109	-0.005443	-2.264872	-0.002415	230.756
			Potential C		
5	1.64088	-2.420802	-529.862	-1.24805	14654.0
6	0.412664	0.214734	-10.4642	0.162359	684.585
7	-0.028381	0.007448	-0.972806	0.002251	5.75765
8	-0.001351	-0.000262	-0.006095	0.000230	2.697574
9	0.001557	0.000624	-0.010310	0.000204	2.607406
10	0.000583	0.000319	-0.002000	0.000200	2.573637

The behavior of the error in $f^{(0)}$ for the different methods as revealed in Figs.1-3 for potential A, appears to be typical of the methods, at least for potential scattering, and similar patterns are observed for potentials B and C also. So, the information presented and discussed above can be reliably extended to these cases also. Instead of presenting the error-scapes for these cases, we now turn attention to a matter that, in the final count, is more important than the convergence of the wavefunctions themselves, viz., the convergence of the K matrix elements. Here we present the results for all three potentials, and discuss the three cases.

3.2. CONVERGENCE OF $[K_m]_f$

In Table I, we summarize the calculations carried out using the wavefunction $f^{(0)}_{MMAD}$, by listing the fractional errors in the various $[K_{MMAD}]_f$, as well as the quantity Ξ^2_{MMAD}, which provides a reliable measure of the error in the wavefunction in the range of r where $V(r)$ is large. Much of these results come as no surprise. The errors in general decrease as Ξ^2_{MMAD} decreases, for all the three potentials. The errors for $[K_{MMAD}]_{LS}$ are less than those for $[K_{MMAD}]_D$, which is not surprising, since the insertion into the

Lippmann-Schwinger equation improves the wavefunction by raising its order in the Born series from $f^{(0)}$ to $f^{(1)}$. From our discussion in Section 2, concerning the relationships between the three variational methods, the small error in $[K_{MMAD}]_{NVP}$ is also not surprising. However, $[K_{MMAD}]_{SVP}$ has large errors, and these errors are very large in most cases compared to the other $[K_{MMAD}]_f$'s. The explanation for this is found in the behavior of the errors in the terms that constitute the SVP functional of (5). Among the terms on the right hand side of of (5), the slowest to converge is the term $\langle f^{(0)}|V|f^{(0)}\rangle$. In terms of the error in $f^{(0)}$, this term would contribute $\langle \delta f^{(0)}|V|\delta f^{(0)}\rangle$ to the overall error given in (7). From Fig.1, we see that for the values of N listed in Table I for potential A, $\delta f^{(0)}_{MMAD}$ is large in the region of $r \geq 6$. The absolute value of these errors frequently get as large as 10, and as the "cropped" features of Fig.1 indicate, they exceed 10 at several points in the interval $7 \leq r \leq 10$. For potential A, the value of $V(r)$ is approximately $-2.5\mathrm{x}10^{-3}$ at $r = 8$. If we view the evaluation of the integral $\langle \delta f^{(0)}|V|\delta f^{(0)}\rangle$ in terms of some numerical quadrature, this implies that several terms in the sum will be of the order of 0.25 or more, before being multiplied by the appropriate weighting factor. If the term $\langle \delta f^{(0)}|VG^0V|\delta f^{(0)}\rangle$ is approximately equal to $\langle \delta f^{(0)}|V|\delta f^{(0)}\rangle$, and had the same sign, it is obvious that the errors in (7) would nearly cancel, resulting in a fairly accurate K matrix element. However, this does not happen for two reasons: first, the term $\langle \delta f^{(0)}|VG^0V|\delta f^{(0)}\rangle$ is second order in V, which means that even if the Green function G^0 was not present, the unweighted terms in the quadrature in the same range of r mentioned above, would only be of the order of $6.0\mathrm{x}10^{-4}$. The second reason is that the Green function G^0, acts in many ways, as a 'filter' to the error, or noise, in the region, and therefore, helps decrease the magnitude of the term even further. Thus it appears, from this analysis, that it is is the non-uniform convergence of the terms in the SVP functional that leads to the large errors in $[K_{MMAD}]_{SVP}$ that we see in Table I.

This leads to another question: why are the K matrix elements $[K_{MMAD}]_D$ and $[K_{MMAD}]_{LS}$, which are first order in the error $\delta f^{(0)}_{MMAD}$, so accurate, if the error in the wavefunction is so large in the interval $7 \leq r \leq 10$? The error associated with $[K_{MMAD}]_D$ is of course, $\langle s|V|\delta f^{(0)}\rangle$. At the values of r where, as we pointed out above, the unweighted terms contributing to the integral $\langle \delta f^{(0)}|V|\delta f^{(0)}\rangle$ would be 0.25 or larger, the corresponding terms in $\langle s|V|\delta f^{(0)}\rangle$ would be at least an order of magnitude smaller, since $|s\rangle$ cannot exceed the value of 1. The quantity $[K_{MMAD}]_{LS}$ of course, will benefit from being second order in V, and from the presence of the Green function, as explained above in connection with the term $\langle \delta f^{(0)}|VG^0V|\delta f^{(0)}\rangle$.

The analysis presented above also explains the reasons for the excellent convergence of $[K_{MMAD}]_{NVP}$ for all three potentials. One is that *every* term in the NVP functional contains V to at least a power of two. Second, every term in the NVP functional also contains the Green function G^0 at least once, which as mentioned above, helps to smooth out the errors. If the errors in each of the terms of the NVP functional are individually small, it is not surprising that the K matrix element contains only a small error. From the Kato identity point of view, of course, we also see that *any* wavefunction $f^{(n)}$, gains one order in the Born series simply by virtue of being inserted into the Newton functional.

TABLE II. Fractional errors in the various $[K_{SVP}]_f$'s

N	$[K_{SVP}]_D$	$[K_{SVP}]_{LS}$	$[K_{SVP}]_{SVP}$	$[K_{SVP}]_{NVP}$	Ξ^2_{SVP}
			Potential A		
25	0.007421	0.007329	0.007421	0.046065	265.2041
26	0.001637	0.001637	0.001637	0.001387	14.7710
27	0.000601	0.000568	0.000601	0.000244	4.42994
28	0.000176	0.000163	0.000176	0.000119	1.94425
29	0.000074	0.000051	0.000074	0.000057	0.930695
30	0.000027	0.0000011	0.000027	0.000012	0.306902
			Potential B		
15	-0.001737	-0.001711	-0.001737	-0.000196	15.1296
16	-0.000943	-0.000957	-0.00943	-0.000097	8.30403
17	-0.000532	-0.000520	-0.000532	-0.000049	4.72950
18	-0.000307	-0.000313	-0.000307	-0.000026	2.76978
19	-0.000184	0.000178	-0.000184	-0.000014	1.67735
20	-0.000112	-0.000114	-0.000112	-0.000008	1.03904
			Potential C		
5	-0.836800	-0.110821	-0.836800	0.981230	2527.24
6	-0.091517	-0.057092	-0.091517	0.032125	115.659
7	-0.001466	-0.000377	-0.001466	0.000064	0.230633
8	-0.000674	-0.000256	-0.000674	0.000050	0.368651
9	-0.000347	-0.000133	-0.000347	0.000010	0.116524
10	-0.000294	-0.000131	-0.000294	0.000018	0.197714

Finally, before leaving Table I, we also note that the quantity Ξ^2_{MMAD}, which provides a measure of the error in the wavefunction in the range of r where the influence of the potential is mainly felt, is fairly large for all three potentials, and diminishes in magnitude rather slowly as the basis size is increased, except in the case of potential C. This is something we have anticipated from the slow convergence of the oscillatory error in the internal region, that we noticed in Fig.1. It is also noticeable that, in the case of potential C, as Ξ^2_{MMAD} decreases to about 2.6 for N = 9-10, the fractional errors in even $[K_{MMAD}]_{SVP}$, the most inaccurate of the different computed K matrix elements, reach fairly small values. This suggests that the explanation offered above for the poor performance of SVP as a nonstationary method–the slow convergence of $\langle f^{(0)}_{MMAD}|V|f^{(0)}_{MMAD}\rangle$ due to large, persistent errors in $f^{(0)}_{MMAD}$ in the internal region–is correct.

Turning to Table II, where we summarize the results of using $f^{(0)}_{SVP}$ to compute the various $[K_{SVP}]_f$'s, we see many differences from those of Table I. First of all, each column of Table II contain numbers smaller than the corresponding columns of Table I. As is clear from Fig.2 for potential A, and from the small values of Ξ^2_{SVP} for the other potentials, the wavefunction $f^{(0)}_{SVP}$ has converged much better than $f^{(0)}_{MMAD}$ for the same values of N.

TABLE III. Fractional errors in the various $[K_{NVP}]_f$'s

N	$[K_{NVP}]_D$	$[K_{NVP}]_{LS}$	$[K_{NVP}]_{SVP}$	$[K_{NVP}]_{NVP}$	Ξ^2_{NVP}
			Potential A		
25	-0.006848	0.000330	1.49208	0.000321	6.29618
26	0.011353	0.000115	3.71506	0.000096	3.52791
27	0.033177	0.000149	32.5653	0.000314	15.4963
28	-0.002527	0.000019	0.225792	0.000018	0.524584
29	0.000523	0.000010	0.005990	0.000010	0.247302
30	0.000413	0.000010	0.003191	0.000010	0.245215
			Potential B		
15	-0.000921	-0.000086	-0.027417	-0.000086	6.52062
16	0.000490	-0.000043	-0.016379	-0.000043	3.70187
17	-0.000380	-0.000022	-0.009981	-0.000022	2.17436
18	0.000228	-0.000012	-0.006165	-0.000012	1.31690
19	-0.000177	-0.000006	-0.003901	-0.000006	0.820496
20	0.000115	-0.000003	-0.002500	-0.000003	0.524340
			Potential C		
5	0.260683	-0.035980	-14.6788	-0.035980	106.864
6	0.060902	0.002405	-0.156809	0.002405	7.19032
7	0.001736	0.000012	-0.005350	0.000012	0.177103
8	0.009795	-0.000002	-0.060446	0.000002	0.250808
9	0.000262	0.000009	-0.000443	0.000009	0.114927
10	0.001163	0.000008	-0.000963	0.000008	0.093800

As a consequence, all K matrix elements, including $[K_{SVP}]_{SVP}$, are computed to a high accuracy. It is interesting, however, that the fractional error in $[K_{SVP}]_D$ and $[K_{SVP}]_{SVP}$ are identical for all cases, indicating that the error arising from the SVP functional, $<\delta f^{(0)}_{SVP}|V-VG^0V|\delta f^{(0)}_{SVP}>$, is about the same order of magnitude as that from the direct calculation, $<s|V|\delta f^{(0)}_{SVP}>$, at least to six decimal places. Improving $f^{(0)}_{SVP}$ by insertion into the Lippmann-Schwinger equation does improve the results in most cases, as does insertion into the NVP functional. The latter strategy in fact, achieves an order of magnitude or more in accuracy, for potentials B and C.

Table III summarizes the results of similar calculations using $f^{(0)}_{NVP}$. The errors in $[K_{NVP}]_{SVP}$ are larger than those in $[K_{SVP}]_{SVP}$. Once again, we can use arguments similar to those presented in connection with $[K_{MMAD}]_{SVP}$, to explain the errors in $[K_{NVP}]_{SVP}$. The persistent large errors in the near-asymptotic regions of the wavefunction (see Fig.3) would cause the term $<\delta f^{(0)}_{NVP}|V|\delta f^{(0)}_{NVP}>$ to be fairly large, and to decrease slower than the term $<\delta f^{(0)}_{NVP}|VG^0V|\delta f^{(0)}_{NVP}>$. Comparing Table III to Table I, however, we see that except for one case (N=27, potential A) the error in $[K_{MMAD}]_{SVP}$ is much larger than $[K_{NVP}]_{SVP}$

for all potentials. The explanation for this is readily given by the fact that the larger errors in $f^{(0)}_{NVP}$ occur at larger values of r compared to those in $f^{(0)}_{MMAD}$, and the potential at this larger value of r is much smaller. Table III also shows that out of all the calculations of K matrix elements in the three tables, $[K_{NVP}]_{NVP}$ is the most accurate. We can see from Fig.1 for potential A, and deduce from the values of Ξ^2_{NVP}, that $f^{(0)}_{NVP}$ is the best converged of the wavefunctions generated by the three methods, in the internal region. This accuracy in the wavefunction combined with the factors that help make the NVP functional relatively insensitive to the errors in $f^{(0)}_{NVP}$ in the asymptotic regions, such as each term being second order in V, and containing the Green function G^0, are responsible for the high accuracy of $[K_{NVP}]_{NVP}$. From comparing $[K_{NVP}]_{LS}$ and $[K_{NVP}]_{NVP}$, we also see that $\langle s|VG^0V|\delta f^{(0)}_{NVP}\rangle$, which is first order in the error in the wavefunction, is approximately equal to the error term associated with the NVP functional, which is second order in the error $\delta f^{(0)}_{NVP}$: $\langle \delta f^{(0)}_{NVP}|V(G^0-G^0VG^0)V|\delta f^{(0)}_{NVP}\rangle$.

3.3. WHY ARE VARIATIONAL METHODS SUPERIOR?

The great popularity of variational methods in scattering theory has always been attributed to the fact that they achieve a higher level of accuracy for a given basis size N. Their superiority over nonvariational methods have been, as already mentioned, explained by the fact that the variational functionals for quantities such as K, S or T matrix elements are second order in the error $\delta f^{(0)}$ in the computed wavefunction, while in nonvariational methods, the error in the quantity of interest remains first order in $\delta f^{(0)}$. Implicit in this explanation is a tacit assumption that the quantity $\langle \delta f^{(0)}|A|\delta f^{(0)}\rangle \ll \langle s|B|\delta f^{(0)}\rangle$, where explicit definitions of the operators A and B depend on the variational method used. We have seen above, that this expectation is not always met. In the case of the SVP, we have seen that $\langle \delta f^{(0)}|A|\delta f^{(0)}\rangle = \langle s|B|\delta f^{(0)}\rangle$, whereas for the NVP, $\langle \delta f^{(0)}|A'|\delta f^{(0)}\rangle \cong \langle s|B'|\delta f^{(0)}\rangle$, where $A = (V-VG^0V)$, $B = V$, $A' = V(G^0-G^0VG^0)V$, and $B' = VG^0V$.

Moreover, the explanation above does not prohibit the same analysis being extended to variationally correct, but nonstationary methods, in which a wavefunction from a nonvariational method is inserted into a variational functional. This renders the functional nonstationary, but the error in the quantity computed by the functional is still, in principle, a smaller quantity. However, from Table I, we see that in practice, this is not always the case. Inserting $f^{(0)}_{MMAD}$ into the SVP functional is nothing short of disastrous, and while the NVP functional does reduce the error, it is not a very dramatic improvement compared to $[K_{MMAD}]_D$, and is, in most cases, comparable to $[K_{MMAD}]_{LS}$, both of which are first order in the error in the wavefunction.

This suggests, it seems to us, that the traditional explanation for the superiority of variational methods must be incorrect. From the previous two subsections, we see that the main difference between the two variational methods, the SVP and the NVP, and the nonvariational method MMAD, is that for a given basis size, the variational methods find a far more accurate wavefunction than the nonvariational method. The variational methods also differ from the nonvariational ones in that the former are characterized by a stationary principle. Let us now examine the consequences of this aspect of variational methods, taking the SVP functional as an example.

The requirement that the SVP functional (5) be stationary with respect to a small change in the wavefunction, results in the following relationship:

$$\partial[K_{SVP}]_{SVP} = 2\langle s|V|\partial f^{(0)}\rangle - 2\langle \partial f^{(0)}|V(1-G^0V)|f^{(0)}\rangle = 0.$$

We can see that this condition will be satisfied if, *and only if* the wavefunction $f^{(0)}$ satisfies the Lippmann-Schwinger equation, so that $(1-G^0V)|f^{(0)}\rangle = |s\rangle$. This, we feel, is where the power of a variational method over a nonvariational one originates. The requirement that the functional (5) be stationary with respect to small changes in $f^{(0)}$ sets a level of accuracy with which the equation $(1-G^0V)|f^{(0)}\rangle = |s\rangle$ must be satisfied, which in turn forces an accurate solution for $f^{(0)}$. A comparison of Fig.1 to Figs. 2 and 3 supports this argument, and shows that the level of accuracy in $f^{(0)}$ achieved thus is far greater than that in the absence of a stationary principle. Variational methods also gain accuracy by concentrating the effort to converge a solution only to those regions of the potential where high accuracy is necessary. The reason for this can be readily found in the form of the functionals of each method. Examining the functionals for the SVP and the NVP, we see that the equations to be satisfied are 'weighted' in different ways in each case, i.e., in the case of the SVP, the equations to be satisfied are

$$\langle m|V(1-G^0V)|f^{(0)}\rangle = \langle m|V|s\rangle; \qquad m = 1,....,N,$$

while for the NVP, they are

$$\langle m|VG^0V(1-G^0V)|f^{(0)}\rangle = \langle m|VG^0V|s\rangle; \qquad m = 1,....,N.$$

The weighting factors in each case play decisive roles in determining which regions of the wavefunction converge first, and to what extent. They also determine how sensitive the computed K matrix element would be, to large errors in the unconverged regions of the wavefunction.

A comparison of Fig.2 to Fig.3 illustrates this last point. We mentioned above, that terms such as $\langle f^{(0)}|V|f^{(0)}\rangle$ in the SVP functional are very sensitive to errors in the wavefunction for intermediate values of *r*. From Fig.2, we see that the SVP wavefunction converges rapidly, over a large range of *r* as the basis size is increased. From the comments made above, we can relate this behavior to the nature of the weighting in the SVP functional. On the other hand, terms that contain the Green function G^0 at least once and the potential V(*r*) twice are less sensitive to errors in the wavefunction at intermediate-large values of *r*. These terms are, however, much more sensitive to errors in the small *r* region of the wavefunction, where the potential is very large. So, it comes as no surprise that the NVP wavefunction in Fig.3 converges very quickly, and very well in the internal region. This can also be verified by comparing Ξ^2_{SVP} in Table II and Ξ^2_{NVP} in Table III. The errors in the wavefunction in the near-asymptotic regions do not influence the NVP functional much, and hence, the convergence of these regions are poor.

A bit of practical wisdom that comes out of this discussion is that if one were attempting to converge an SVP calculation using, say, the very popular distributed Gaussian basis,[15] it would be advisable to spread the Gaussians over a region large enough that V(*r*) is very small at the upper limit of the integration. On the other hand, if one were trying to converge an NVP calculation, the same number of Gaussians, or presumably a smaller number of them, can be put to maximum use by concentrating most of them over a much smaller range.

4. Summary and Conclusion

We have examined the nature of the error in the wavefunctions for potential scattering in three one dimensional potentials, computed by one nonvariational method, the MMAD, and two commonly used variational methods, the SVP and the NVP. We have found that the nature of the error is dramatically different in each case. All three methods generate wavefunctions that are more accurate in the internal region than in the asymptotic region. However, the MMAD wavefunctions have errors in the internal region that tend to disappear rather slowly, whereas the variational methods banish the error to the asymptotic regions extremely rapidly, as the basis size is increased.

Having computed the wavefunctions, we used them to compute K matrix elements by different methods. We found that, as expected, the variational methods yielded the more accurate K matrix elements, the NVP results being more accurate than those from the SVP. Variationally correct, but nonstationary methods, obtained by inserting the wavefunction from a nonvariational method into a variational functional provide greater accuracy only in certain cases. In particular, large errors in the wavefunction at intermediate values of r results in large errors from the SVP functional. We also found that the bilinear NVP functional generally provides better results when used as a nonstationary method, since *any* wavefunction gains an order of improvement in the Born series, simply by virtue of being inserted into the Newton functional.

The traditional explanation for the superiority of variational methods over nonvariational ones has been that an error of order $\delta f^{(0)}$ in the wavefunction is transformed into an error of order $(\delta f^{(0)})^2$ in the computed stationary quantity. Implicit in this explanation is the expectation that the error term of order $(\delta f^{(0)})^2$ is much smaller than the error term that results from a direct calculation, which is first order in $\delta f^{(0)}$. Our comparison of the K matrix elements computed by different methods from the SVP and NVP wavefunctions has led to the discovery that the results from the SVP functional using the SVP wavefunction, $[K_{SVP}]_{SVP}$, is only as accurate as $[K_{SVP}]_D = -(1/k)\langle s|V|f^{(0)}{}_{SVP}\rangle$, while $[K_{NVP}]_{NVP}$ is only as accurate as $[K_{NVP}]_{LS} = -(1/k)\langle s|V|f^{(1)}{}_{NVP}\rangle$, where the superscripts to the f's refers to the order of the wavefunction in the Born series. This suggests that the traditional explanation is incorrect, since the expectation implicit in it, and mentioned above, is not met. The reason for the rapid convergence of variational methods, as suggested by the results of this investigation, is due to the requirement that the variational functional be stationary with respect to small changes in the wavefunction. This requirement forces a much more accurate solution for the wavefunction. Also, due to the nature of the linear algebraic equations solved in each case, the convergence of certain regions of the wavefunction is accelerated at the expense of certain other regions. The variational functionals are relatively insensitive to large errors in these latter regions. This finding leads to the practical possibility of designing basis sets so that they are ideally suited for the particular variational method that we wish to use.

Acknowledgements

This work was supported in part by grants from the Robert A. Welch Foundation and the National Science Foundation. Grants of CPU time on the CRAY X-MP/24 at the University of Texas Center for High Performance Computing is gratefully acknowledged.

References

1. (a) W.H. Miller, and B.M.D.D. Jansen op de Haar, J.Chem.Phys. **86**,6213 (1987); (b) J.Z.H. Zhang,D.J. Kouri, K. Haug, D.W. Schwenke, Y. Shima and D.G.Truhlar, J.Chem.Phys. **88**, 2492 (1988); (c) J.Z.H. Zhang and W.H.Miller, Chem.Phys.Lett. **140**, 329 (1987).
2. J.Z.H. Zhang, S.-I. Chu, and W.H. Miller, J.Chem.Phys. **88**, 6233 (1988).
3. D.W. Schwenke, K. Haug, M. Zhao, D.G. Truhlar, Y. Sun, J.Z.H. Zhang, and D.J.Kouri, J.Phys. Chem. **92**, 3202 (1988).
4. (a) K. Haug, D.W. Schwenke, D.G. Truhlar, Y. Zhang, J.Z.H. Zhang and D.J. Kouri, J.Chem.Phys. **87**, 1892 (1987); (b) J.Z.H. Zhang, and W.H. Miller, J.Chem. Phys. **88**, 4549 (1988).
5. (a) A.C. Peet and W.H. Miller, Chem.Phys.Lett. (to be published); (b) D.E. Manolopoulos and R.E.Wyatt, Chem.Phys.Lett. (to be published).
6. W. Kohn, Phys. Rev. **74**, 1763 (1948).
7. J. Schwinger, Phys. Rev. **72**, 742 (1947).
8. R.G. Newton, *Scattering Theory of Particles and Waves*, 2nd. Ed. (Springer, New York, 1982).
9. R. K. Nesbet, *Variational Methods in Electron-Atom Scattering Theory* (Plenum, New York, 1980).
10. G. Staszewska and D.G. Truhlar, J.Chem.Phys. **86**, 2739 (1987).
11. T. Kato, Phys.Rev. **80**, 475 (1950); Prog.Theor.Phys. **6**,295,394 (1951).
12. K. Takatsuka, R. Luccheese and V. McKoy, Phys.Rev. **A24**, 1812 (1981).
13. B. Ramachandran, T.-G. Wei and R.E. Wyatt, J.Chem.Phys. (to be published).
14. B. Ramachandran, T.-G. Wei and R.E. Wyatt, Chem.Phys.Lett. (to be published).
15. I.P. Hamilton and J.C. Light, J.Chem.Phys. **84**, 3061 (1986).

QUANTUM DYNAMICS OF SMALL SYSTEMS USING DISCRETE VARIABLE REPRESENTATIONS*

J. C. LIGHT, R. M. WHITNELL, T. J. PARK, and S. E. CHOI
James Franck Institute and Department of Chemistry
University of Chicago, Chicago, Illinois 60637, U.S.A.

ABSTRACT. The discrete variable representation (DVR) on Gaussian quadrature points for orthogonal polynomials is defined. It is shown that the Hamiltonian for multidimensional systems is easy to evaluate and sparse in the DVR. Methods of solution of the time dependent Schrödinger equation in the DVR are presented. A highly efficient method of solution of the time independent Schrödinger equation for many eigenvalues and eigenvectors is presented which is based (in the DVR) on sequential adiabatic-sudden partitioning diagonalization, truncation, and recoupling of lower dimensional solutions. Brief summaries are presented of the application of these techniques to predissociation of NaI, vibrational states of H_3^+, and the thermal rate constant of the $H+H_2$ reaction.

1. Introduction

Over the past twenty years or so there have been unparalleled advances in our understanding of elementary dynamical processes in chemical physics.[1,2] These advances have been powered on the experimental side primarily by the laser which now provides unprecedented power, time resolution, and frequency resolution. Thus processes such as intramolecular relaxation, molecular rotation in liquids, photodissociation, absorption and emission by transient species, and molecular collisions can be studied in great detail. On the theoretical side the orders of magnitude increases in speed and memory of computers over the last two decades have powered an equally rapid advance. These computational advances have permitted the development and application of classical simulation techniques to many processes in chemical physics ranging from bimolecular collisions in the gas phase,[3] to clusters and surface interactions,[4] and even to condensed systems with interfaces.[5]

The exact dynamics of microscopic systems is, of course, governed by quantum mechanics. Although the quantum equations are linear partial differential equations, as opposed to the non-linear coupled classical equations resulting from Hamilton's equations, the solution of the Schrödinger equation for complex dynamical processes remains a formidable problem. The fundamental reason for this discrepancy is the "delta function" nature of classical mechanics versus the "distributed" nature of quantum mechanics. Thus in quantum mechanics each degree of freedom for each particle must be represented by many coupled equations, in general, rather than by two equations per degree of freedom for classical systems. Hence, exact quantum dynamics calculations have been limited to (very) small systems. Quantum dynamics of complex systems by Monte Carlo path integral techniques may soon be a viable option, however.[6]

A. Laganà (ed.), Supercomputer Algorithms for Reactivity, Dynamics and Kinetics of Small Molecules, 187–213.

Since many processses demonstrate substantial quantum effects of tunneling, wave packet break-up and interference, and, obviously, discrete energy spectra, symmetry induced selection rules, etc., it is clearly desirable to develop methods by which more complex dynamical problems can be solved quantum mechanically both accurately and efficiently. There is a reciprocity between the number of particles which can be treated quantum mechanically and the number of states of importance. Thus the ground states of many electron systems can be determined as can the bound state (and continuum) dynamics of diatomic molecules. Our focus in this manuscript will be on nuclear dynamics of few particle systems which are not restricted to small amplitude motion. This can encompass vibrational states and isomerizations of triatomic molecules, photodissociation and exchange reactions of triatomic systems, some atom-surface collisions, etc.

In this manuscript we will focus first on the representations one can use for the quantum dynamical processes of interest. In particular we believe that the advent of large scale computing has changed the preferred approaches to many problems in theoretical chemical physics, i.e. larger computers should not just be used to solve larger problems with standard methods, but the methods should be tailored to large problems insoluble by standard approaches. The simplest example is vibrations of triatomic molecules: the standard normal mode analysis,[7] while yielding the lower vibrational states simply and accurately, cannot be extended to treat large amplitude vibrations and isomerization. Thus a more general approach which can be implemented on computers, and can give both lower eigenstates and large amplitude motion states is desired. Localized representations,[8-10] in particular discrete variable (DVR) representations,[11] permit a facile representation of the Hamiltonian for a variety of problems.

The second focus of the manuscript must be on the solution of problems once the representation is established. Again the DVR facilitates solution of quantum dynamical processes because of the "structured sparsity" of the multi-dimensional Hamiltonian in the DVR, $\mathbf{H}^{DVR}$. For time dependent problems this permits efficient solution by second order differencing[12] and by Lanczos propagation[13] since the matrix vector multiply operation, $\mathbf{H}^{DVR}\mathbf{c}$, is very fast. It also permits split exponential propagation since the kinetic energy coupling is "one dimensional". This has been applied to photodissociation of NaI.[14] For time independent (Green's function) approaches or for eigenvalue problems, the DVR permits sequential diagonalization (by coordinate), truncation, and <u>exact</u> re-coupling of the adiabatic solutions in the higher dimensions.[10,15] This has permitted for the first time the accurate evaluation of high vibrational states of floppy molecules[9,10,15] and the exact evaluation of the thermal rate coefficient for reactive scattering via flux-flux autocorrelation functions.[16]

The DVR is related to, but distinct from pseudo-spectral and collocation methods[17] of solving differential equations. For the DVR there is an <u>orthogonal</u> transformation which defines the relation of the DVR to the finite basis representation (FBR).[11,18] Thus, for example, the Hermitian character of operators remains obvious in the DVR. Both pseudo-spectral and collocation methods, however, use a "mixed" representation of operators and, as such, do not display the Hermitian character of operators such as **H**. Thus the advantages of the DVR are that the accuracy is that of a Gaussian quadrature and it is a true representation, while the collocation methods permit more freedom in the choice of points, a distinct advantage in some multidimensional problems.[19]

In the next section we define multi-dimensional DVR's and list some caveats in their application. In Section 3 we look at approaches to time dependent problems, and summarize some of our results for the NaI photodissociation compared with experiments of the Zewail group[20] in Section 4. In Section 3 we, also, summarize the sequential diagonalization-truncation approach to the solution of multi-dimensional problems and in Section 4 we, also, summarize results on triatomic vibrational states of H_3^+ and present some very recent results on the first exact evaluation of 3D rate constants via the quantum flux-flux autocorrelation function approach of Miller.[21-23] We then give a brief summary, conclusion, and prognosis.

2. Discrete Variable Representations

Although pointwise "representations" (such as finite differences) of numerical problems have been used for many years,[24] and the more powerful collocation or pseudospectral methods for more than a decade,[17] the DVR is relatively new. Harris et al.[25] generated the DVR for a harmonic oscillator basis (the Gauss-Hermite points) and used the DVR-FBR transformation matrix to evaluate the potential matrix in the FBR. Dickinson and Certain[26] then established that the points were the Gaussian quadrature points. Only recently, however, was the DVR[11,27] (or the equivalent Discrete Ordinate Method[28]) recognized as a desirable representation in which many quantum problems could be formulated and solved more easily. In particular the advantages of multi-dimensional DVR's are just now being recognized.[15]

We first look at the definition of a DVR in one dimension. Let an operator, $\mathcal{H}^o$, have a set of orthonormal (L^2) eigenfunctions $\{|\varphi_i\rangle\}$ and associated eigenvalues $\{\varepsilon_i^o\}$ on a specified range of the variable, x. Thus we have, in this representation (the FBR),

$$(H^o)_{ij} \equiv \langle\varphi_i|\mathcal{H}^o|\varphi_j\rangle = \varepsilon_i^o\,\delta_{ij} \tag{2.1a}$$

$$\langle x|\varphi_i\rangle = \varphi_i(x) \tag{2.1b}$$

$$\langle\varphi_i|\varphi_j\rangle = \int \varphi_i^*(x)\,\varphi_j(x)\,dx = \delta_{ij}. \tag{2.1c}$$

In this representation the $\mathcal{H}^o$ operator, containing the differential operators, is diagonal, but the coordinate, x, is not. If we form the matrix representation of x and diagonalize it, we generate the orthogonal (or unitary) DVR-FBR transformation:

$$(X_{ij}) \equiv \langle\varphi_i|x|\varphi_j\rangle \tag{2.2a}$$

$$T_x^+ X\, T_x = \mathsf{x}(\text{diag}). \tag{2.2b}$$

Since x is an Hermitian operator, T_x is unitary. The FBR of any operator, in particular the Hamiltonian, can now be transformed to the DVR. If H is the Hamiltonian in the FBR

$$H \equiv H^o + V \tag{2.3}$$

we can then find the equivalent "exact" DVR:

$$\tilde{H} = \tilde{H}^o + \tilde{\tilde{V}} \qquad (2.4a)$$

$$\tilde{H}^o \equiv T^+ H^o T = T^+ \varepsilon^o T \qquad (2.4b)$$

$$\tilde{\tilde{V}} \equiv T^+ \tilde{V} T \qquad (2.4c)$$

We now define an approximation to the "exact" DVR which __is__ the normal DVR by replacing the transformed potential matrix, $\tilde{\tilde{V}}$, by

$$(\tilde{V})_{ij} \equiv V(x_j)\, \delta_{ij} \approx (\tilde{\tilde{V}})_{ij}. \qquad (2.5)$$

Thus we do not evaluate the potential matrix in the DVR by the transformation 2.4c, but rather __approximate__ it by 2.5.

For basis sets corresponding to classical orthogonal polynomials (harmonic oscillator functions, Chebyshev polynomials, Legendre and associated Legendre polynomials, etc.) the approximation 2.5 is equivalent to the approximation of V by Gaussian quadrature. For N Legendre polynomials, P_n, for example, with $x = \cos\theta$, we have

$$\delta_{nm} = \int P_n(x) P_m(x) dx \qquad (2.6a)$$

$$(T)_{nj} = \omega_j^{1/2} P_n(x_j) \qquad (2.6b)$$

where x_j and ω_j are the points and weights of the N point Gauss-Legendre quadrature. The points, χ_i, are obtained automatically as eigenvalues of x from eq. 2.2b.[26] We can approximate V in the FBR by the Gauss-Legendre quadrature and find:

$$(V)_{nm} = \int_{-1}^{1} P_n(x) V(x) P_m(x) dx \approx \sum_{i=1}^{N} P_n(x_i)\omega_i^{1/2}\, V(x_i) P_m(x_i)\omega_i^{1/2} = (T\, \tilde{V}\, T^\dagger)_{mn}$$

where $\tilde{V}$ is defined in 2.5. Thus the DVR approximation made in 2.5 becomes of very high accuracy as N, the dimension of the FBR and DVR increases.

We also note that the DVR is quite easy to find and to use. For an orthogonal polynomial basis, the matrix X in 2.2a is tri-diagonal and easily obtained from the recurrence relations for the polynomials.[29] Finally we note that the elements of the eigenvectors of the DVR Hamiltonian, H, are __not__ necessarily the amplitudes of the eigenfunctions at the DVR points although they are proportional to them. Thus if $\tilde{c}_j$ is the jth eigenvector of $\tilde{H}$, then

$$\psi_j(x_i) = (\tilde{c}_j)_i\, \omega_i^{-1/2}. \qquad (2.7)$$

Although the 1D DVR's are useful, the use of direct product DVR's for multidimensional problems is much more highly advantageous. There are three reasons for this. First the Hamiltonian matrix in the multi-dimensional DVR is easy to construct. Second, for a DVR in an orthonormal coordinate system, the Hamiltonian is sparse. Third, the "low connectivity" of the Hamiltonian matrix in the multi-dimensional DVR permits efficient solution of both time dependent and time independent problems.

We use a three dimensional Hamiltonian as an example:

$$\mathcal{H} = \sum_{\alpha=1}^{3} f_\alpha(\{q^\beta\}')h_\alpha{}^o(q^\alpha) + V(\{q\}) \tag{2.8}$$

where $h_\alpha{}^o(q^\alpha)$ contains all the differential operators in coordinate q^α, $f_\alpha(\{q^\beta\}')$ is a constant or a function of the other coordinates, $q^\beta \neq q^\alpha$, and each $h_\alpha{}^o$ defines a basis:

$$h_\alpha{}^o\, \varphi_i{}^\alpha\,(q^\alpha) = \varepsilon_i{}^\alpha\, \varphi_i{}^\alpha\,(q^\alpha). \tag{2.9}$$

We determine the DVR for each basis by diagonalization of the appropriate coordinate matrix as in eqn. 2.2. Thus for each coordinate we have a basis, say with N_α functions and the corresponding quantities for the DVR as defined in equations 2.1-5:

$$\{|\varphi_i{}^\alpha>\}_{N_\alpha},\ T_\alpha,\ \{q_i{}^\alpha\}_{N_\alpha},\ \varepsilon^\alpha(\text{diag}),\ \tilde{\varepsilon}^\alpha \equiv T_\alpha{}^\dagger\, \varepsilon^\alpha\, T_\alpha. \tag{2.10}$$

The quantities listed in 2.10 can now be used to write, by inspection, the DVR of $\mathcal{H}$ in 3D:

$$\begin{aligned}(\tilde{H}_{i,j,k}{}^{i',j',k'})' &= f_\alpha(q_j{}^\beta, q_k{}^\gamma)(\tilde{\varepsilon}^\alpha)_{i,i'}\, \delta_{jj'}\, \delta_{kk'} \\ &+ f_\beta(q_i{}^\alpha, q_k{}^\gamma)\, (\tilde{\varepsilon}^\beta)_{jj'}\, \delta_{ii'}\, \delta_{kk'} \\ &+ f_\gamma(q_i{}^\alpha, q_j{}^\beta)\, (\tilde{\varepsilon}^\gamma)_{jk'}\, \delta_{ii'}\, \delta_{jj'} \\ &+ \tilde{V}(q_i{}^\alpha, q_j{}^\beta, q_k{}^\gamma)\, \delta_{ii'}\, \delta_{jj'}\, \delta_{kk'}\end{aligned} \tag{2.11}$$

where i,i' run over the N_α points of the DVR in α, etc. The proliferation of Kroneker deltas in the above expression clearly shows the sparseness of $\tilde{H}$. Of the $(N_\alpha N_\beta N_\gamma)^2$ elements of $\tilde{H}$, only $(N_\alpha + N_\beta + N_\gamma - 2)N_\alpha N_\beta N_\gamma$ elements are non-zero.

Before looking at approaches to solutions using the DVR we want to make several comments and caveats. First, because DVRs correspond in accuracy to quadratures, one should use several more points in each dimension than would be required for the basis functions alone. In other words, a DVR with too few points will lead to quadrature errors which may make $\tilde{H}$ have eigenvalues which are not variational. Second, ∇^2 in many coordinate systems (spherical polar, hyperspherical, etc.) contains singular effective potentials. It is important that the h^o's include these as much as possible since quadratures near singularities are of uncertain reliability. This problem is relieved somewhat by the fact that we need retain in the DVR only points at which the desired solutions will be non-zero.

Thus, for low energy solutions, points in very high regions of the potential energy may usually be truncated. This local truncation is also very useful in reducing the size of the DVR basis.

3. Solutions Using the DVR

Because of the sparseness and the structure of the Hamiltonian matrix in a direct product DVR, solutions of both time dependent and time independent (eigenvalue) problems are made much more efficient compared with standard approaches. The two features exploited in the time dependent problems are the sparseness of $\mathbf{H}$ and the fact that the kinetic energy operators couple only one dimension at a time. This latter feature is exploited in the solution of time independent problems by sequential diagonalization and truncation in which "adiabatic" eigenvectors in lower dimensions are recoupled (exactly, within the basis) in the higher dimensions after truncation. We turn first to the time dependent problems.

3.1. TIME DEPENDENT SOLUTIONS OF THE SCHRÖDINGER EQUATION

The solution of the time dependent Schrödinger equation is naturally considered when the Hamiltonian itself is time dependent and for such scattering events like photodissociation, which require solutions corresponding to relatively few initial states. Solutions of the time dependent Schrödinger equation using the fast Fourier transform (FFT), pioneered, for chemical physics problems, by Feit and Fleck[30] and Kosloff et al.,[31] have been used by a number of scientists for a variety of problems.[32] The computer time for the FFT approach scales as $(d \ln n)n^d$ for n points in each of d dimensions whereas the DVR will scale as $(d\, n)n^d$, since these quantities are proportional to the number of multiples required for multiplication of a solution vector by the Hamiltonian, H. Although on the basis of this scaling alone the FFT would seem to be preferred, we believe that for some problems the flexibility in basis afforded by the DVR will make it highly advantageous. We therefore present below, very briefly, how the DVR can be used in three approaches to the solution of the time dependent Schrödinger equation:

$$\partial/\partial t\, \psi(\{q\},t) = -i\mathcal{H}/\hbar\, \psi(\{q\},t). \tag{3.1}$$

3.1.1. *Second Order Differencing (SOD).* When $\mathcal{H}$ is time dependent, the standard second order differencing approach[12,31] to the time evolution of the wave vector representation c(t) (Eq. 2.7) on the DVR points is probably required and is straightforward:

$$\mathbf{c}(t_{n+1}) = \mathbf{c}(t_{n-1}) - 2i\, \Delta t\, \mathbf{H}\, \mathbf{c}\,(t_n)/\hbar. \tag{3.2}$$

Because the representation is finite (L^2), one must be careful that (a) the basis is adequate to "support" the dynamics of the true wave packet under the operation of the true Hamiltonian operator, H, i.e. contains basis functions of high enough energy to represent the wave packet; and (b) that the range of the basis and treatment of "edges" is adequate to avoid spurious "edge" effects, i.e. reflections of the wave packet from the edge of the basis, etc. These problems, common to all L^2 methods including the FFT, can be handled by a sufficiently dense basis and imaginary (optical) potentials near the edges of the basis to avoid spurious reflections.[33] For time steps below a critical size, Δt_{crit}, the above method is stable and accurate. This method is reliable but often takes substantial computer time for large n^d and long times.

3.1.2. *Split Exponential Propagator*. Since the true time evolution due to 3.1 is unitary (for real $\mathcal{H}$), and the exponential propagator is unitary and accurate over short times,[34] we can approximate the exact unitary propagator by

$$U(t,t_o) \approx \exp\{-i/\hbar \int_{t_o}^{t} \mathcal{H}\, dt + O(t-t_o)^3\}. \tag{3.3}$$

A solution can clearly be constructed as the sequence

$$\tilde{c}(t_n) = \tilde{U}(t_n, t_{n-1})\tilde{c}(t_{n-1}) \tag{3.4}$$

accurate to $1/2 \int_{t_{n-1}}^{t_n} dt \int_{t_{n-1}}^{t} dt' \, [\mathcal{H}(t), \mathcal{H}(t')]$ in each interval or to $O(\Delta t^2)$ overall.

For constant or slowly varying Hamiltonians, therefore, the unitary exponential propagator seems excellent. A major problem, however, is that the representation of the exponential of an operator may be difficult. Tal-Ezer and Kosloff,[35] for example, used a Chebyshev polynomial expansion of the exponential.

An alternative, useful for both the DVR and the FFT approaches, is to use a short time split exponential operator [30] where the exponential kinetic energy and potential energy contributions are evaluated sequentially:

$$\tilde{U}(\Delta t) \approx \exp(-i\tilde{V}\Delta t/2\hbar)\, \exp(-i\tilde{\mathcal{H}}^o\Delta t/\hbar)\, \exp(-i\tilde{V}\Delta t/2\hbar) \tag{3.5}$$

where the symmetric decomposition makes the error of order $(\Delta t)^3$ per step due to the fact that $\tilde{\mathcal{H}}^o$ and $\tilde{V}$ do not commute. In the DVR the matrix of $\exp(-iH^o\Delta t/\hbar)$ is evaluated by transformation

$$\exp(-i\tilde{H}^o\Delta t/\hbar) = T^\dagger \exp(-i\varepsilon^o\Delta t/\hbar)T. \tag{3.6}$$

For higher dimensional DVR's, we note that $\exp(-iH^o\Delta t/\hbar)$ can be represented as the product of one sparse matrix per dimension. For two dimensions, for example, we have in the DVR:

$$[\exp(-i\tilde{H}_\alpha{}^o\Delta t/\hbar)]_{ij}{}^{i'j'} = (T_\alpha{}^\dagger \exp(-i\,\varepsilon_\alpha{}^o\Delta t/\hbar)T_\alpha)_{ii'}\,\delta_{jj'} \tag{3.7}$$

and <u>usually</u>, the kinetic energy operators commute:

$$\exp(-i(\mathcal{H}_\alpha{}^o+\mathcal{H}_\beta{}^o)\Delta t/\hbar) = \exp(-i\mathcal{H}_\alpha{}^o\Delta t/\hbar)\, \exp(-i\mathcal{H}_\beta{}^o\Delta t/\hbar). \tag{3.8}$$

(If the kinetic energy operators do not commute, we use a symmetric decomposition such as 3.5 for them also.)

Combining equations 3.5-8, we see that in the DVR the operation of the split exponential operator on a vector, $\tilde{c}(t_n)$, is essentially a sequence of ~(d+1) sparse matrix vector multiples. For time independent Hamiltonians, this is usually highly advantageous over the SOD method because a much larger time step (Δt) can be used with good accuracy.[14,30,32]

3.1.3. *Lanczos Propagation.*[36] Finally we mention here, but do not describe in detail, a method applicable to time independent Hamiltonians which will be advantageous when the system accesses a relatively small subspace in its time evolution. The Lanczos method generates, from the initial vector, $c(t_o) = c_o$, a set $\{c\}_p$ of p orthogonal vectors containing c_o. This set of vectors is in a Krylov space generated by operating on c_o with H (p-1) times. In this reduced space the Hamiltonian, H_p, is tri-diagonal and easily diagonalized. Thus the unitary time evolution operator can be evaluated exactly within the p-dimensional subspace. When amplitude reaches the "edge" of the subspace, say at time t', the vector c(t') is used to generate a new Lanczos reduced representation for another time period. Although this procedure was found to be accurate for wave packet propagation,[36] extensive testing and comparisons have not been performed.

In summary, because of the structured sparseness of the DVR Hamiltonian, time evolution techniques relying on operations by the Hamiltonian (or a function of $\mathcal{H}$) on a vector can be implemented quite efficiently. Although the theoretical efficiency of multidimensional DVR's is not as high as that for the analogous FFT approach, the flexibility and ability to truncate the DVR may make it advantageous for time dependent problems.

3.2. TIME INDEPENDENT SOLUTIONS OF THE SCHRÖDINGER EQUATION: SEQUENTIAL DIAGONALIZATION AND TRUNCATION (OR SEQUENTIAL ADIABATIC-SUDDEN DECOMPOSITIONS)

The DVR for multi-dimensional problems (at least to d=3 which includes large amplitude vibration-rotation states of triatomic molecules) appears to be the representation of choice because

(a) $\tilde{\mathcal{H}}$ is sparse and simple to evaluate;

(b) The basis can be truncated by removing points lying in uninteresting regions, usually where $V(\{q\}) > V_{MAX}$ (a parameter);

(c) Solving the set of 1D Hamiltonian matrix equations for coordinate α, yields eigenvectors $\tilde{c}_n^{jk}$ and 1D eigenvalues $\tilde{\varepsilon}_n^{jk}$ where n denotes the nth 1D eigensolution along q^α at $q_j^\beta\, q_k^\gamma$. This is then used as a <u>local</u> basis for subsequent truncation and solution (as shown below). This procedure permits the construction of a very efficient basis for the higher dimensional problem.[15,16] The procedure is illustrated in Figure 1 where the non-zero elements of the Hamiltonian are shown at each stage (for coordinates χ, θ, ρ).

Here we outline the sequential diagonalization, truncation, recoupling procedure for the solution of a 3D problem in the DVR. If we choose to solve in the order γ first, then β, then α, we would first set up the matrix equation in γ for each value of q^α and q^β; say q_i^α, q_j^β

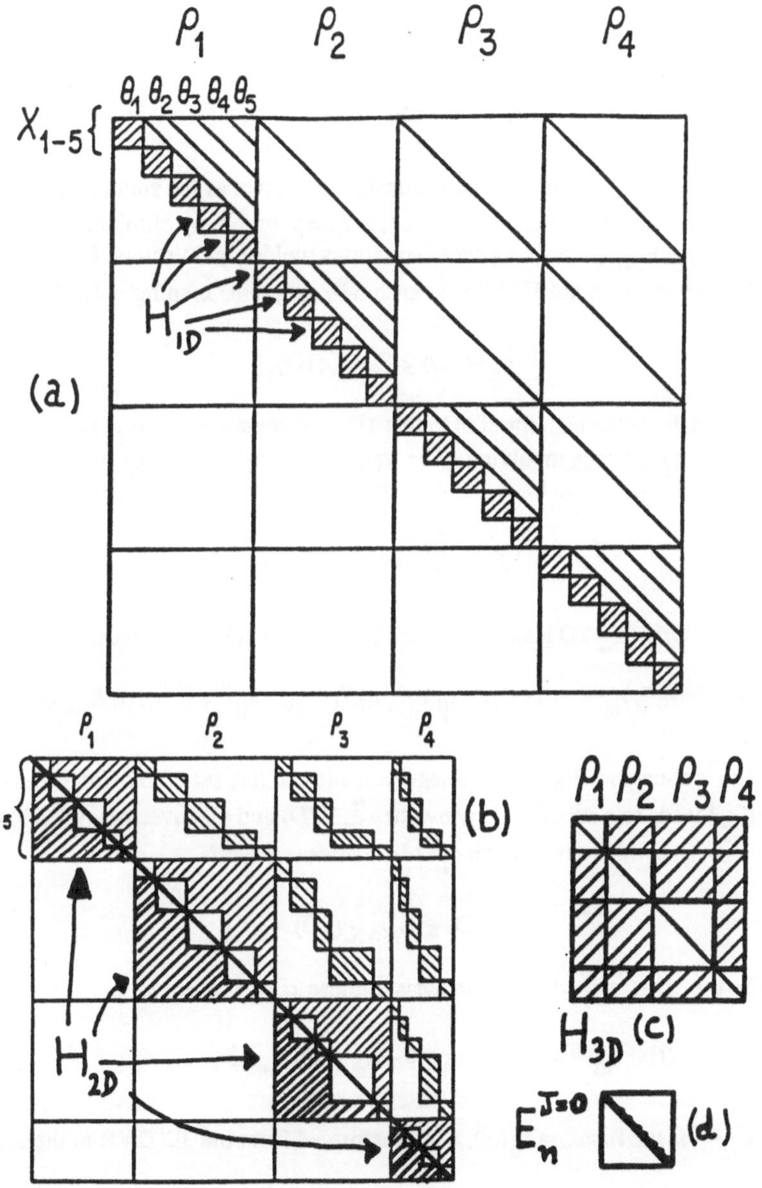

Figure 1. Solution of 3D DVR Hamiltonian by successive diagonalization-truncation. (a) 3D H in the DVR; (b) H in the truncated 1D basis, 2D DVR; (c) H in the truncated 2D basis, 1D DVR; (d) H in the truncated 3D eigenvector basis.

$$(\tilde{h}_{kk'}{}^{ij}) \equiv f_\gamma(q_i^\alpha, q_j^\beta)(\tilde{\varepsilon}^\gamma)_{kk'} + V(q_i^\alpha, q_j^\beta, q_k^\gamma)\delta_{kk'} \tag{3.9}$$

and solve for the 1D eigenvalues and eigenvectors

$$\tilde{h}^{ij}\, C^{ij} = \tilde{\varepsilon}^{ij}(1D)\, C^{ij}. \tag{3.10}$$

There are $N_\alpha N_\beta$ such small eigenvalue problems, shown as the small diagonal blocks of Figure 1a, each of which can be truncated, before solution, by eliminating points with $V(q_i^\alpha q_j^\beta q_k^\gamma) > V_{MAX}$. Solving these equations yields eigenvectors, C^{ij}, on each $q_i^\alpha q_j^\beta$ line. We now truncate the 1D basis at each q_i^α, q_j^β point, keeping only eigenvectors with

$$\tilde{\varepsilon}_m{}^{ij}(1D) \le E_{MAX}(1D). \tag{3.11}$$

We now transform the Hamiltonian to this 1D representation. This actually requires only that remaining functions involving q^γ be transformed to the 1D eigenvector basis. In the basis of q_i^α, q_j^β |n; ij> we have, for example, a function of q_i^α, q_j^β, and q_k^γ transformed as

$$f_{nm}(q_i^\alpha q_j^\beta) \equiv \sum_k C_{kn}{}^{ij}\, f(q_i^\alpha q_j^\beta q_k^\gamma) C_{km}{}^{ij}. \tag{3.12}$$

We now can write N_α 2D Hamiltonian matrices in the q_j^β, |n; ij> basis:

$$\left(\tilde{H}^{2D}(q_i^\alpha)\right)_{nj}{}^{mj'} = \sum_k C_{nk}{}^{ij}\, f_\beta(q_i^\alpha, q_k^\gamma)\, C_{mk}{}^{ij'}\, \varepsilon_{jj'}{}^\beta + \tilde{\varepsilon}_n{}^{ij}(1d)\, \delta_{nm}\, \delta_{jj'}. \tag{3.13}$$

This stage is shown as Figure 1b. Diagonalization of this set of 2D Hamiltonians at each value of q^α leads to a set of 2D eigenvalues $\tilde{\varepsilon}_m{}^i(2D)$ and eigenvectors, $C^i(2D)$ at each q_i^α. We now truncate this basis at each q_i^α to eigenvectors with

$$\tilde{\varepsilon}_m{}^i\,(2D) \le E_{MAX}\,(2D) \tag{3.14}$$

and write the 3D Hamiltonian in the basis of these 2D eigenvectors as

$$(\tilde{H}^{3D})_{ni}{}^{mi'} \equiv \Gamma_{ni}{}^{mi'}\, (\tilde{\varepsilon}^\alpha)_{ii'} + \tilde{\varepsilon}_m{}^i(2D)\, \delta_{nm}\, \delta_{ii'} \tag{3.15}$$

where $\Gamma_{ni}{}^{mi'}$ is the function $f_\alpha(q^\beta, q^\gamma)$ transformed from the 3D DVR to the 2D basis at q_i^α and $q_{i'}^\alpha$:

$$\Gamma_{ni}{}^{mi'} \equiv \sum_{p,q,j} C_{np}{}^{ij}(2D) \left[\sum_k C_{pk}{}^{ij}\, f_\alpha(q_j^\beta, q_k^\gamma)\, C_{qk}{}^{i'j}\right] C_{mq}{}^{i'j}(2D). \tag{3.16}$$

Note that these transformations are done sequentially at each dimension. Although the transformation is of 4th order in the number of points per dimension say ℓ, and Γ itself is "4th order" in ℓ, it is rarely invoked fully since the $f_\alpha(q_j^\beta, q_k^\gamma)$ prefactors are often unity or a function of one variable only (e.g. q^γ). In any case the sums over p and q are over only the truncated basis sets in 1D, and n and m only range over the truncated 2D eigenvector bases. The result is a 3D Hamiltonian (3.15) in a basis of good 2D eigenfunctions with exact coupling provided by $\Gamma\varepsilon^\alpha$. This means that a large fraction (.1 to .25) of the eigenvalues of 3.15 are usually very accurate (errors in 5th significant figure). Typical truncations are to retain only 0.1 to 0.5 of the eigenfunctions to be re-coupled in the next higher dimension. Thus a 20 x 30 x 40 = 24,000 point basis is reduced to a 10^2-10^3 basis 3D problem, yielding, perhaps, up to 120 very accurate eigenvalues.

We note that this procedure is equivalent to a sequence of adiabatic-sudden decompositions, with solutions of all the adiabatic Hamiltonians, but then with exact re-coupling of the truncated basis via the $\tilde{h}^o$ operators for the other degrees of freedom (e.g. $\mathcal{E}^\alpha$, etc. of eqn. 2.11). This permits good physical sense to be made out of the intermediate "adiabatic" results in some cases.[10] We note, however, that no adiabatic approximation is made since the exact re-coupling is straightforward.

We also note that the use of a DVR in one or two dimensions and a different basis (e.g. distributed Gaussians) for the remaining dimensions permit most of the advantages of the full DVR approach in terms of sequential diagonalization, truncation, and recoupling.[9-11] However the transformations (like 3.16) are less efficient since the function f_α will not be diagonal in a basis representation as it is in the DVR (e.g. the inner summation is over an additional index). We now turn to the applications of these techniques to several demanding problems in quantum dynamics.

4. Some Numerical Results

Since this article is focused on the DVR methodology rather than on a specific scientific problem *per se*, in this section we will present several examples of the applications of the DVR, but only samples of results for each. We will, however, try to give sufficient details of the implementation to be of use to the prospective new "DVR user".

4.1. PHOTODISSOCIATION OF NaI

Recent experiments by Zewail's group[20] have shown that it is possible to prepare a wave packet on the upper potential energy surface in a time short with respect to the vibrations on that surface, and to see the pulses of product produced each time the wave packet crosses the avoided crossing region and transfers to the dissociative surface. Schematic potential energy curves and initial wave packet are shown in Figure 2, together with schematic optical potentials used to absorb flux on the lower surface and the coupling potential, V_{12}. The DVR was that for Chebyshev polynomials, i.e. equally spaced and weighted points between R_i (R'_i on the lower surface) and R_f. Specifically the DVR transformation for N functions is

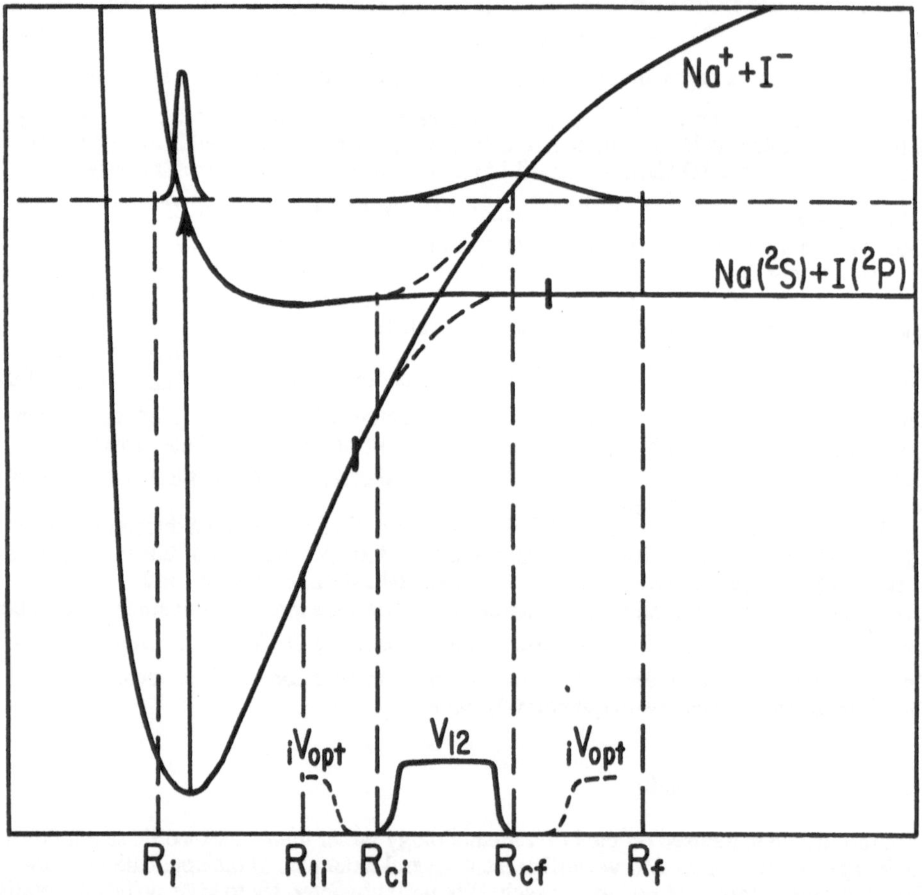

Figure 2. Schematic potential energy surfaces for NaI. V_{12} is the potential coupling the two surfaces, and iV_{opt} is the optical potential.

$$T_{\ell\ \alpha} = \omega^{1/2} \mathcal{F}_\ell(R_\alpha) \tag{4.1}$$

$$\Delta R = R_f - R_i$$

$$\omega = \Delta R/N$$

$$\mathcal{F}_\ell(R_\alpha) \equiv A_\ell \cos[\ell\pi(R_\alpha - R_i)/\Delta R]$$

$$A_\ell = [(2 - \delta_{\ell,o})/\Delta R]^{1/2}.$$

The maximum kinetic energy "supported" by this grid is

$$K_{MAX} = \hbar^2/2\mu\, \omega^2 = \hbar^2 N^2/2\mu(\Delta R)^2. \tag{4.2}$$

Because, for NaI, the reduced mass is large and the kinetic energy over the well is also large, a very dense grid would be required to do the exact quantum dynamics on the lower surface over the well. Since the system is largely adiabatic, we assume that amplitude on the lower surface over the well will make one oscillation and then follow an adiabatic passage to dissociation. We therefore absorb it with the optical potential and count it as dissociated together with the "dissociative" amplitude absorbed by the optical potential (on the lower covalent surface) near R_f. With N ~ 600, $R_f - R_i \sim 16$Å, we followed the time evolution of the initial wave packet using the split exponential propagator (eq. 3.5) for times up to ~5 picoseconds, corresponding to five passages over the coupling region, and resulting in dissociation of about 80% of the initial wave packet. In Figure 3 we plot the autocorrelation function,

$$C(t) \equiv |\langle \psi(o)\, \psi(t) \rangle| \tag{4.3}$$

which clearly shows the decaying oscillations of the system. The Fourier transform of C(t) yields the vibrationally resolved absorption spectrum of the adiabatic upper state. Comparison with experiment yielded a value of the coupling constant ($V_{12} = 310$ cm^{-1}).

Because this model system is one dimensional and dissociative, making a uniform grid appropriate, it is likely that the FFT split exponential propagator[30,32] would be advantageous. In multi-dimensional systems the DVR should be competitive.

4.2. VIBRATIONAL STATES OF H_3^+ (J=0)[15]

Because H_3^+ is the simplest polyatomic molecule, substantial effort has gone into the measurement of its vibrational spectrum,[37-39] the calculation of the ground state potential energy surface,[40,41] and the calculation of the vibrational quantum states.[8,41,42] The vibrational states are difficult to calculate because the normal mode description breaks down for states about 12 000 cm^{-1} above the minimum of the potential energy surface. Although the equilibrium is the equilateral triangle configuration, above about 12 000 cm^{-1}, the energy exceeds the barrier to "isomerization", i.e. the collinear configuration. Thus diffuse states with very large amplitude motion begin in this energy range, and a description appropriate for large amplitude motion is required.

We[15] recently completed an accurate calculation of the vibrational states of J=0 H_3^+ in the energy range up to about 23 000 cm^{-1} above the potential energy minimum. The potential is an accurate fit to the most accurate ab initio surface[40] and was recently used by both Meyer et al[41] and Miller and Tennyson[42] to calculate the lower J=0 vibrational states accurately. Since this full 3D DVR illustrates many of the procedures, problems, and advantages of the multi-dimensional DVR, we review our calculation in some detail.

Since H_3^+ has D_{3h} symmetry, it is advantageous to use an orthogonal coordinate system in which the symmetry operations are easily carried out. The hyperspherical coordinate system recently presented by Pack[43] permits the symmetry decomposition of the problem into the A'_1, A'_2, and E' irreducible representations to be carried out easily.[15] After wave

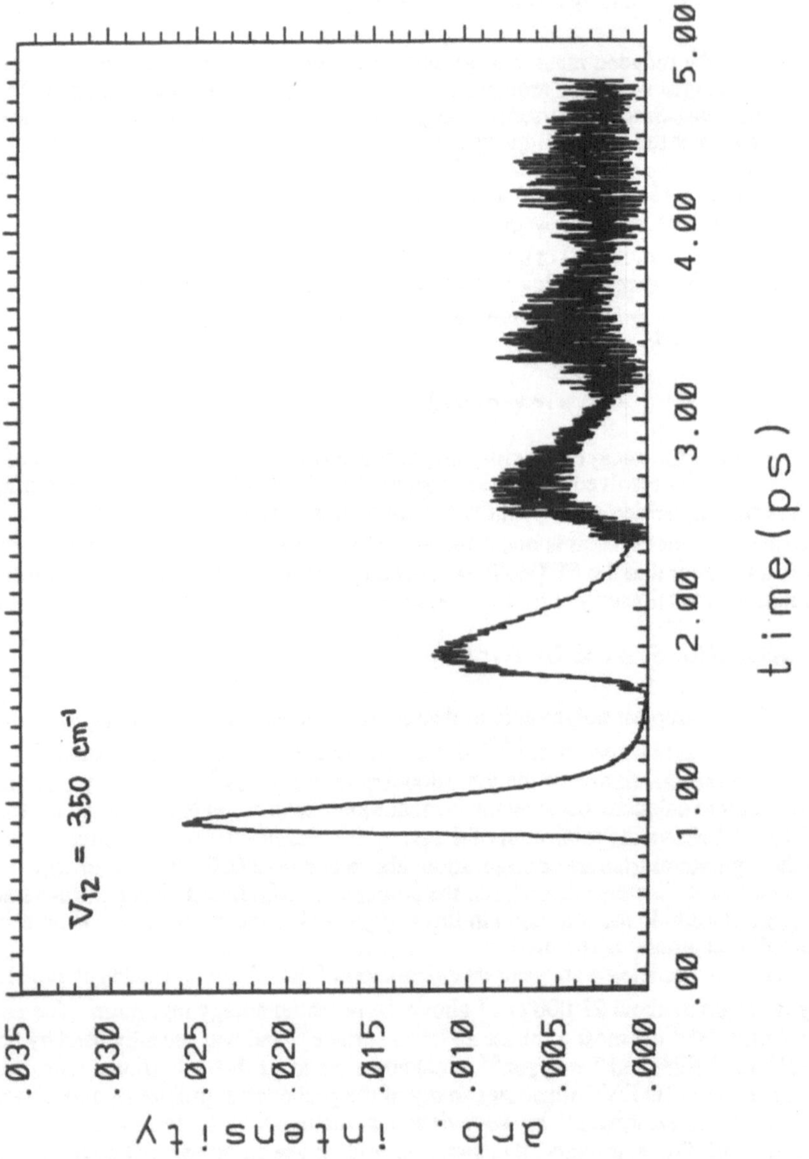

Figure 3. |C(t)| vs t for NaI.

function factorization, the Hamiltonian in these coordinates becomes

$$H = -\hbar^2/2\mu \left(\partial^2/\partial\rho^2 + 16/\rho^2\sin 2\theta \ \partial/\partial\theta \ \sin 2\theta \ \partial/\partial\theta\right.$$

$$\left. + 1/\rho^2\sin^2\theta \ \partial^2/\partial\chi^2 - 15/4\rho^2\right) + V(\rho,\theta,\chi) \tag{4.4}$$

where the coordinate ranges, volume element, and reduced mass are

$$0 \le \rho \le \infty$$

$$0 \le \theta \le \pi$$

$$0 \le \chi \le 2\pi$$

$$d\tau = 1/32 \ d\rho \ \sin\theta \ d\theta \ d\chi$$

$$\mu = m/\sqrt{3}. \tag{4.5}$$

Physically, $\theta = 0$ corresponds to the equilateral triangle and $\theta = \pi$ corresponds to the collinear configuration. As χ ranges from 0 to 2π, the internal configuration space is spanned twice, being related by reflection through the triatom plane.

The bases chosen are triginometric functions in χ;

$$\cos (n-1)\chi, \ \sin n\chi \qquad n = 1,2, \ldots \tag{4.6}$$

Jacobi[29] polynomials in $\cos\theta$;

$$P_{\ell}^{(m,o)}(\cos\theta) \qquad \ell = 0, 1, \ldots \quad m = \text{parameter} \tag{4.7}$$

(actually associated Legendre functions)

and Chebyshev polynomials in x

$$nx = n\pi\rho/(\rho_{max} - \rho_{min})$$

$$T_n \ \alpha \ \cos nx. \tag{4.8}$$

The basis in χ is then decomposed to a basis for each irreducible representation of the D_{3h} symmetry and the DVR in $\cos 6\chi$ corresponding to each representation is determined. Thus the actual DVR points in χ range only over the physically unique portion of the potential energy surface, 0 to $\pi/6$.

The choice of the basis in θ, the Jacobi polynomials, illustrates one of the problems inherent in the direct product basis required (with our current understanding) for a multi-dimensional DVR. The eigenfunctions of the hyperspherical angular operator as a whole (both θ and χ terms) are hyperspherical harmonics (similar to spherical harmonics). These form a direct sum basis, not a direct product basis, and thus we cannot generate a corresponding DVR. When we use the direct product basis, above, for χ and θ, we find that the effective potential in θ becomes $n_\chi^2/\sin^2\theta$ due to the $1/\sin^2\theta\ d^2/d\chi^2$ term in H. It is singular at $\theta = 0,\pi$ for all but the lowest (n=1) basis function in χ. Because both the equilateral triangular ($\theta = 0$) and collinear ($\theta = \pi$) configurations are energetically accessible, the basis in θ must be chosen to handle the singularity. For each n_χ this can be done by choosing the parameter, m, of the Jacobi polynomial to be $m = n_\chi/2$. Since this does not generate a direct product basis, however, it cannot be used for the direct product DVR.

The problem is adequately solved, at some cost in efficiency of convergence, by choosing a single m for each symmetry representation which yields the proper behavior for the lowest energy state of that symmetry. Specifically we use[15]

$$m_{A'_1} = 0; \quad m_{A'_2} = 3; \quad m_E = 1$$

and then use the direct product basis of the Jacobi polynomials with the symmetry adapted functions of χ. Thus a typical basis function for the A'_2 representation would be

$$\phi_{n,\ell,k}^{A_2} = C_{n\ell k} \cos nx(\rho)\, P_\ell^{(3,0)}(\cos\theta) \sin(6(k+1)\chi) \tag{4.9}$$

where $C_{n\ell k}$ is a normalization constant, and the last factor can also be written in terms of Jacobi polynomials as

$$\sin(6(k+1)\chi) \;\alpha\; \sin 6\chi\;\; P_k^{(1/2,1/2)}(\cos 6\chi). \tag{4.10}$$

For each symmetry, then, the DVR basis is the direct product of equally spaced points in ρ, Gauss associated Legendre points in θ, and the eigenvalues of cos 6χ in the appropriate symmetry basis of polynomials in cos 6χ.[15] For each symmetry (e.g. A_2) the DVR-FBR transformation matrices are the direct product of the appropriate 1D FBR-DVR transformation matrices:

$$T^{DVR} \equiv T_\rho^{DVR} \otimes T_\theta^{DVR,m=3} \otimes T_\chi^{DVR,A_2}. \tag{4.11}$$

The Hamiltonian is then constructed in the DVR (eq. 2.11) and solved by the sequential diagonalization-truncation procedure. For all symmetries ρ is the final coordinate to be coupled. For A'_1 the order is (θ,χ,ρ) while for A'_2 and E' the order is (χ,θ,ρ). (This permits the singularity of effective potential to be treated more adequately.) For simplicity we will review here only the parameters of the A'_1 calculation and a sample of the results. More details and full results are given elsewhere.[15]

Once the basis is defined, the corresponding DVR is determined by the numbers of basis functions in each degree of freedom n_ρ, n_θ, n_χ. These determine the quadrature error and convergence possible within the basis. The sequential diagonalization-truncation procedure defined in the last section depends on three cut-off parameters, V_{MAX}, $E_{MAX}(1D)$, and $E_{MAX}(2D)$. These affect the variational error and efficiency. We illustrate these two aspects in Table I for the 1D eigenvalues in θ for fixed ρ and χ. We find that eigenvalues < $V_{MAX}/2$ are not much affected by truncation, but, in this case at least, it is preferable not to truncate the basis points.

TABLE I. Convergence of 1D eigenvalues with respect to n_θ and V_{MAX} for $\rho = 2.25\ a_0$ and $\chi = 0.4581$. Energies (eV) are shown for the most exact calculation, and $\Delta E_i = E_i - E_i^{ex}$ are shown for less exact calculations

State	V_{MAX}(eV) ∞	∞	∞	3.0
	$n_\theta =$ 24	12	16	24(22)
1	.296766	.000005	.000001	.000000
3	1.41305	.00001	.00000	.000000
5	2.45616	.00003	.00000	.00574
7	3.31087	.00008	.00000	.20369

The 2D (χ,θ) Hamiltonians at each value of ρ are represented in the 1D eigenvector basis (χ) at each value of θ, with the 2D Hamiltonian being given by 3.13. Only 1D eigenvectors with eigenvalues less than $E_{MAX}(1D)$ are retained. Thus the size of the 2D basis will be different at each value of ρ and will also depend on $E_{MAX}(1D)$. In Table II we show the convergence of the 2D eigenvalues at $\rho = 2.25$, again for the A'_1 states. The convergence, while reasonable, is not as rapid as we often see in 2D vibrational problems using the DVR. Presumably this is due to the singular effective potentials. The final 3D basis, consisting of 2D eigenvectors at each ρ is again governed by $E_{MAX}(2D)$, with all 2D eigenvectors having eigenvalues below this retained. We also retain a minimum of two 2D eigenvectors at each ρ if they exist. Convergence of the 3D A'_1 eigenvectors with

TABLE II. Convergence of 2D A'_1 eigenvalues at $\rho = 2.25$ as a function of E_{MAX}(1D) and n_θ:[a] $\Delta E = E(n)-E(141)$ are shown (eV).

State	E_{MAX}(1D)(eV)	3.0	4.0	5.0	6.0[b]
	n_{2D}[c]	67	88	108	141 E(141)
1		.00001	.00001	.00001	.328885
4		.00178	.00051	.00006	1.66132
7		.06334	.00338	.00031	2.61633
9		.434	.0200	.00258	3.00141

[a] $n_\theta = 22$ except for $E_{MAX} = 6.0$.
[b] $n_\theta = 24$.
[c] n_{2D} = size of 2D basis.

E_{MAX}(2D) are shown in Table III. In Table IV we compare with results of Miller and Tennyson[42] and Meyer et al[41] for (we believe) the same Hamiltonian operator, with

TABLE III. Convergence of 3D A'_1 eigenvalues with respect to E_{MAX}(2D). ($\Delta E = E(n)-E(374)$ (eV) are shown.)

State	E_{MAX}(2D)	3.0	3.4	3.8
	n_{3D}	217	292	374
1		.000000	.000000	.541068
14		.00099	.00005	2.07424
29		.01326	.00099	2.61316
41		.05371	.00477	2.92259

eigenvalues converted to cm^{-1}. We note that Meyer et al[42] used a Watson Hamiltonian in which some coupling terms are neglected. In this calculation we believe all states up to ~22 000 cm^{-1}, the limit of validity of the potential, were converged to about 3 cm^{-1}. This included 36 states of A'_1 symmetry, 18 states of A'_2, and 29 states of E' (to 19,696 cm^{-1} only).

Despite the fact that this calculation provides many accurate eigenvalues for H_3^+ on this potential energy surface to much higher energies than possible before, the calculation is

TABLE IV. Comparison of 3D A'_1 eigenvalues with Miller and Tennyson,[42] and Meyer, Botschwina, and Burton.[41] Eigenvalues in cm^{-1}.

State	Our work[a,15]	M&T	MB&B[b]
1	4363.97	4363.5	4363.5
2	7542.65	7541.9	7541.85
3	9141.45	9140.9	9140.8
4	10626.6	10625.9	10626.4
5	11647.2	11651.0	11659.9
6	12134.0		12132.5
10	14957.3		
20	19031*		
30	21482*		

[a] $n_\theta = 22$, $n_\chi = 18$, $n_\rho = 30$, $V_{MAX} = 7.2$ eV, $E_{MAX}(1D) = 5.7$ eV, $E_{MAX}(2D) = 4.0$ eV, $n_{3D} = 452$.
[b] A (slightly different) Watson Hamiltonian was used.
*1-3 cm^{-1} convergence.

less "efficient" than earlier applications of the sequential diagonalization-truncation techniques using DVR's or DVR's combined with distributed Gaussian bases (DGB's).[9-11] There appear to be two reasons for this which may be taken as cautions about indiscriminate DVR approaches. First, the hyperspherical coordinate system leads to very sharp valleys in the 2D potential energy contour plots for some values of ρ. Consequently, a large density of DVR points in θ,χ is required to represent the wave functions adequately in these valleys. A different coordinate system might be better although the hyperspherical coordinates permit simple utilization of the D_{3h} symmetry. Second, the singular effective potentials, although ultimately treated adequately in the DVR, require a higher density of points for adequate quadrature. Development of DVR's corresponding to hyperspherical harmonics would clearly be advantageous, although currently non-direct product multi-dimensional DVR's are unknown.

Despite these somewhat negative aspects of the H_3^+ J=0 calculations, we emphasize that it is a very powerful approach, permitting the calculation of all physically meaningful vibrational states on the potential energy surface.

4.3. EXACT QUANTUM RATE CONSTANTS FOR H+H_2 REACTION VIA FLUX-FLUX AUTOCORRELATION FUNCTION[16]

As a final sample of the application of multidimensional DVR's to quantum dynamical problems, we present a few of Park's[16] recent results on the H+H_2 reaction. This is the first full "exact" quantum calculation of a rate constant, all others containing approximations for the J≠0 contributions. The success of these calculations is based on the following:

(a) Use of the flux-flux autocorrelation[21,22] function formulations of the exact quantum rate constant;

(b) The simplicity of representations of the Hamiltonian and flux operators in the DVR using Pack's[43] hyperspherical coordinate system;

(c) Use of parity, symmetry, and total angular momentum decompositons of the problems;

(d) The fact that the $H+H_2$ system is highly quantum mechanical with low mass; and that the reaction is direct; and

(e) The use of the DVR and sequential diagonalization-truncation to generate an adequate L^2 basis of eigenvectors of the J=0 Hamiltonian in a transition state region.

In this brief section we can only summarize the approach taken and define the parameters of the calculations. We then give a selected sample of the results and information obtained.

Some time ago Miller[21] derived an exact quantum formulation of the thermal rate constant, and more recently Miller et al.[22] re-cast it in terms of the time integral of the flux-flux autocorrelation function. Although, as might be expected for an exact formulation, the time integral is to $t \rightarrow \infty$, the rapid decay of the autocorrelation functions in the transition state region of $H+H_2$ permits essentially exact results to be obtained on a short ($< 30 \times 10^{-15}$ sec) time scale. Thus we may solve the problem in an L^2 representation which will represent the dynamics accurately in the transition state region for this time scale.

We use the defintions of the thermal rate constant derived by Miller et al.[22] Let a surface, S=0, divide reactants and products. Then we have the rate constant given by

$$k(T) = Z_R^{-1} \int_0^\infty dt \; C_f(t,\beta) \tag{4.12}$$

$$C_f(t,\beta) \equiv tr\left[F\, e^{-\gamma H}\, F\, e^{-\gamma^* H}\right]$$

$$\gamma = \beta/2 - i\, t/\hbar$$

$$F \equiv 1/2\left[\delta(s)\,(p_s/m) + (p_s/m)\,\delta(s)\right]$$

where $\beta = 1/kT$, Z_R is the reactant partitions function, H is the full Hamiltonian, and F is the flux operator, the expectation value of which is the flux across the surface, S=0. The quantity $C_f(t,\beta)$ is the trace of the flux-flux autocorrelation operator. This formulation has been solved several times by various techniques for model 1D and collinear reaction problems, and was applied approximately to the 3D $H+H_2$ reaction by Yamashita and Miller,[23] but has not been evaluated exactly for a 3D reactive system.

We use Pack's[43] hyperspherical coordinates with the surface defining reaction from the reactant channel to the two possible (labeled) product channels given by $\chi = \pi/6$ and $7\pi/6$ for reaction $\alpha \rightarrow \beta$ and $\chi = 5\pi/6$ and $11\pi/6$ for reactions $\alpha \rightarrow \gamma$ (α,β,γ represent the three labeled chemical channels). The flux operator for $\alpha \rightarrow \beta$ is defined explicitly as

$$F^{\alpha\to\beta} = \hbar/i\mu\rho^2 \sin^2\theta\{F_\chi(\pi/6) + F_\chi(7\pi/6)\}$$

$$F_\chi(y) = 1/2\{(\chi-y)\,\partial/\partial\chi + \partial/\partial\chi\,\delta(\chi-y)\} \qquad (4.13)$$

Because the reactant channel is singled out, the flux operator has C_{2v} symmetry. Thus the Hamiltonian was also decomposd in C_{2v} symmetry as well as by total angular momentum and parity.

If the eigenfunctions of the Hamiltonian are known for a particular symmetry decomposition, then the contribution of that symmetry block to the flux-flux correlation function is

$$C_f^{sym}(t) = \sum_{i,j(sym)} \exp\{-\beta/2(E_i+E_j)\} \cos\{(E_i-E_j)t/\hbar\} \; |(T^TFT)_{ij}{}^{sym}|^2 \qquad (4.14)$$

where F is the matrix of the flux operator in the DVR, T is the transformation matrix from the DVR to the eigenvectors of the Hamiltonian (of the symmetry under consideration). Equation 4.14 is instructive in that it shows the role of the Boltzmann factor in making the sum converge, and the role of the flux operators. Since F represents a very local operator at the surface, S=0, only eigenvectors of the Hamiltonian with significant projections on this surface will contribut to the T^TFT term. The time integral of 4.14 in an L^2 basis is truncated after $C_f(t)$ falls to about 10^{-3} of its initial value. The behavior of $C_f(t)$ at 700 K is shown in Figure 4.[15]

We used a 3D DVR for this calculation in Pack's hyperspherical coordinate system. For the final calculations for J=0 we used a Chebyshev basis in ρ, $N_\rho = 16$, with ρ_{min} = 2.5 a_o, $\rho_{max} = 7.0$ a_o, a symmetry (C_{2v}) adapted trigonometric basis in χ, $\cos n\chi$, $\sin(n+1)\chi$ ($n = 0, 1, ...$) with $n_\chi = 36$, and a Legendre polynomial basis in θ with $n_\theta = 24$, for a total basis size of 13824. Since the potential for the $H+H_2$ system is very large at the equilateral triangular configuration where the effective potential also has a singularity, the wave functions are all excluded from this troublesome region and no convergence difficulties (like those for H_3^+) were encountered.

The Hamiltonian was diagonalized by the sequential diagonalization-truncation procedure leading to a 3D basis of $N_{3D} < 500$. Diagonalization of this yielded 190 (A_1) and 179 (B_1) 3D eigenvectors with $E_n < 1.2$ eV. These yield fully converged autocorrelation functions for J=0 and T up to 1500 K, and J=0 thermal rate constants which agree with the exact rate constant from close coupling calculations[44] to about 3-10%, as shown in Table V.

For J≠0, the J=0 eigenvectors were used as a basis. The diagonal blocks of the Hamiltonian for each J, K_J, where K_J is the projection of the total angular momentum on the body fixed axis were evaluated in the J=0 basis and diagonalized leading to a set of approximate eigenvalues and eigenvectors for each, J, K_J. The <u>energies</u> of these states were then corrected by second order perturbation theory for the K, K±2 coupling, ignoring the Coriolis coupling. The projection of the flux operators on these states and the evaluation of the flux autocorrelation function via 4.14 using the <u>corrected</u> eigenvalues then permitted the evaluation of the rate constants for J≠0.

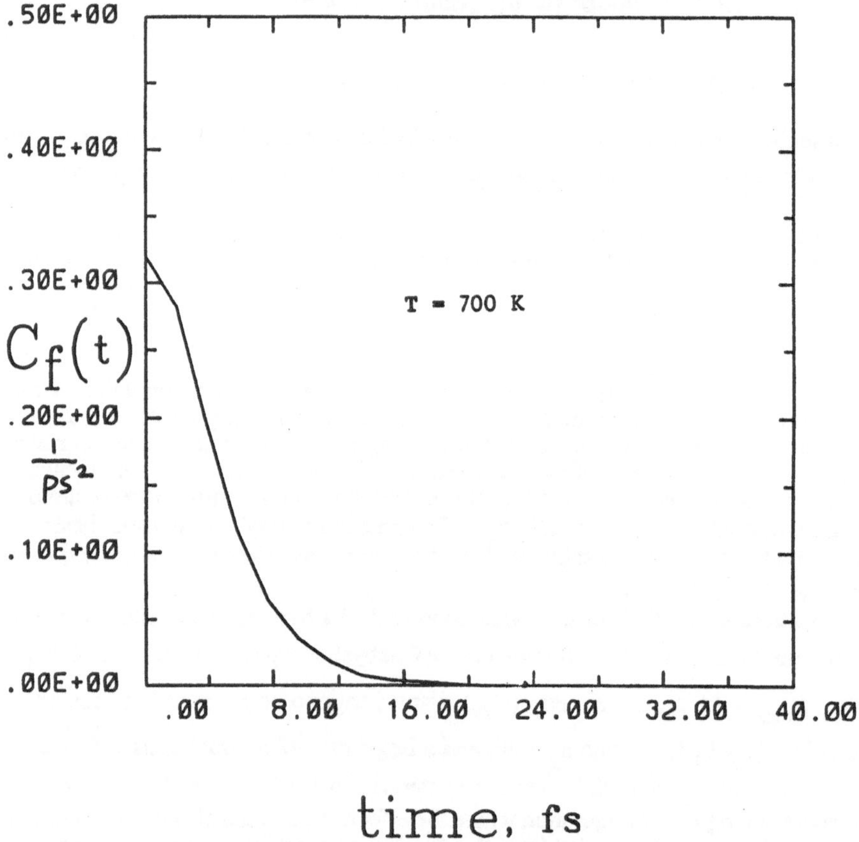

Figure 4. $H+H_2$ flux-flux autocorrelation function vs t at T = 700 K.

TABLE V. J=0 thermal rate constants (cm^3 $molecule^{-1}$ sec^{-1}).

T (K)	k^o (T)[a]	k^o (T)[b]
500	9.48 (-16)	10.7 (-16)
700	6.08 (-15)	6.42 (-15)
900	1.62 (-14)	1.66 (-14)
1100	2.92 (-14)	2.97 (-14)
1300	4.25 (-14)	4.36 (-14)
1500	5.46 (-14)	5.71 (-14)

[a]T. J. Park[16]
[b]F. Webster and J. C. Light[44]

The overall rate constant is then given by

$$k(T) = \sum_{J,K_J,sym} g_{sym}\, k^{J,K_J,sym}(T)\,(2J+1) \tag{4.15}$$

where g_{sym} is the proper nuclear spin weighting factor (e.g. 2 for A_1 and 6 for B_1 for even parity).[16] In Figure 5 we plot the ratio of rate constants $k^J(T)/k^o(T)$ at T = 700 K. (The factor 2J+1 is omitted.) Although the rate constant drops rapidly with J, including the degeneracy factor makes the total contribution a maximum at about J=5 with J values up to ~15 required for convergence with J.

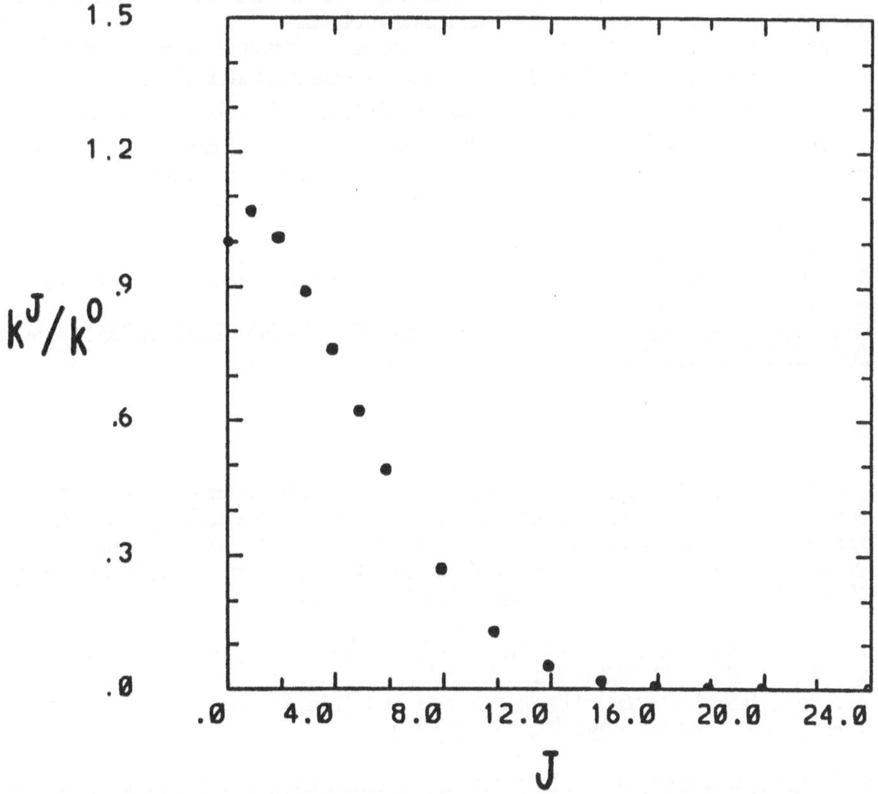

Figure 5. Variation of rate constant ratio with J at 700 K. (2J+1) degeneracy factor is omitted.

This calculation demonstrates that exact thermal rate constants for elementary reactions can be evaluated in a computationally efficient manner using the DVR to construct a zero order eigenvector basis which is then used for the J≠0 contributions. Extension of these results to include Coriolis coupling and to asymmetric systems is relatively straightforward.

5. Summary

In this paper we have presented a general approach to the exact quantum solution of few particle problems with large amplitude motions. Based on a rigorous pointwise representation, the DVR, the Hamiltonian for multidimensional systems was shown to be very easy to construct and very sparse. Two methods of solution which exploit this sparseness were presented. The first was based on the evolution of a wave vector in this representation via either Lanczos propagation or time evolution via a split time exponential propagator. In the second approach a set of accurate eigenvectors and eigenvalues of the Hamiltonian is sought. The sparseness and structure of the Hamiltonian is now exploited by a sequential diagonalization-truncation procedure in which lower dimensional problems are solved, the eigenvector basis is truncated, and then the remaining solutions are re-coupled to provide a good representation for the next higher dimensional problem.

The approaches were illustrated by three examples--the photodissociaiton of NaI via a two surface wave packet calculation; the J=0 vibrational states of all symmetries for H_3^+; and the exact quantum thermal rate constant for the H+H_2 reaction via the 3D DVR formulation of the flux-flux autocorrelation function. The split time exponential propagator was used for the first problem, and the sequential diagonalization-truncation approach for the latter two problems. Each of these cases illustrates the substantial advances represented by the DVR, permitting the efficient solution of problems on a much larger scale than heretofore possible.

We expect the DVR to become the approach of choice for a number of problems in both spectroscopy and dynamics.

This research was supported by NSF CHE-8505001, by NSF CHE-8806514, and by DOE DE-FG02-87ER13679.

6. References

1. R. D. Levine and R. B. Bernstein, *Molecular Reaction Dynamics and Chemical Reactivity*, 2nd ed. (Clarendon, Oxford, 1987); R. B. Bernstein, Ed., *Atom-Molecule Collision Theory* (Plenum, New York, 1979); W. H. Miller, Ed., *Dynamics of Molecular Collisions* (Plenum, New York, 1976); D. C. Clary, Ed., *The Theory of Chemical Reaction Dynamics* (Reidel, New York, 1986).

2. D. H. Levy, Ann. Rev. Phys. Chem. **31**, 197 (1980); J. A. Beswick and J. Jortner, Adv. Chem. Phys. **47**, 363 (1981); G. Scoles, Ed., *Atomic and Molecular Beam Methods* (Oxford University Press, London, 1987); M. Dantus, M. J. Rosker, and A. H. Zewail, J. Chem. Phys. **87**, 2395 (1987).

3. M. Karplus, R. Porter, and R. D. Sharma, J. Chem. Phys. **49**, 3259 (1965); R. N. Porter and L. M. Raff, in *Dynamics of Molecular Collisions, Part A*, W. H. Miller, Ed. (Plenum, New York, 1976).

4. J. C. Tully, Acc. Chem. Res. **14**, 188 (1981).

5. C. H. Mak, H. C. Andersen, and S. M. George, J. Chem. Phys. **88**, 4052 (1988).

6. R. E. Cline, Jr., and P. G. Wolynes, J. Chem. Phys. **88**, 4334 (1988); R. D. Coalson, D. L. Freeman, and J. D. Doll, J. Chem. Phys. **85**, 4567 (1986).

7. E. B. Wilson, Jr., J. C. Decius, and P. C. Cross, *Molecular Vibrations* (McGraw-Hill, New York, 1955).

8. I. P. Hamilton and J. C. Light, J. Chem. Phys. **84**, 306 (1986); I. P. Hamilton, J. Chem. Phys. **87**, 774 (1987).

9. Z. Bacic and J. C. Light, J. Chem. Phys. **85**, 4594 (1986); ibid. **86**, 3065 (1987); Z. Bacic, D. Watt, and J. C. Light, J. Chem. Phys. **89**, 947 (1988).

10. J. C. Light and Z. Bacic, J. Chem. Phys. **87**, 4008 (1987).

11. J. V. Lill, G. A. Parker, and J. C. Light, Chem. Phys. Lett. **89**, 483 (1982); J. C. Light, I. P. Hamilton, and J. V. Lill, J. Chem. Phys. **82**, 1400 (1985); Z. Bacic, R. M. Whitnell, D. Brown, and J. C. Light, Comp. Phys. Commun. (in press).

12. A. Askar and A. Cakmak, J. Chem. Phys. **68**, 2794 (1978).

13. T. J. Park and J. C. Light, J. Chem. Phys. **85**, 5870 (1986).

14. S. E. Choi and J. C. Light, private communication.

15. R. M. Whitnell, Ph.D. Thesis, Univ. of Chicago, 1988 (submitted); R. M. Whitnell and J. C. Light, J. Chem. Phys. (in press).

16. T. J. Park, Ph.D. Thesis, Univ. of Chicago, 1988 (submitted); T. J. Park and J. C. Light, private communication.

17. D. Gottlieb and S. Orzag, *Numerical Analysis of Spectral Methods, Theory and Applications* (SIAM, Philadelphia, PA, 1977); L. Fox and I. B. Parker, *Chebyshev Polynomials in Numerical Analysis* (Oxford University Press, London, 1968); for recent applications in chemical physics, see R. Friesner, J. Chem. Phys. **85**, 1462 (1986); ibid. **86**, 3522 (1987).

18. L. Fox, in *Methods of Numerical Approximation*, D. C. Hanscomb, Ed. (Pergamon, New York, 1966).

19. R. A. Friesner, J. Phys. Chem. **92**, 3091 (1988).

20. T. S. Rose, M. J. Rosker, and A. H. Zewail, J. Chem. Phys. **88**, 6672 (1988); M. J. Rosker, T. S. Rose, and A. H. Zewail, Chem. Phys. Lett. **146**, 175 (1988).

21. W. H. Miller, J. Chem. Phys. **61**, 1823 (1974).

22. W. H. Miller, S. C. Schwartz, and J. W. Tromp, J. Chem. Phys. **79**, 4889 (1983).

23. K. Yamashita and W. H. Miller, J. Chem. Phys. **82**, 5475 (1985).

24. See, for example, C. Lanczos, *Applied Analysis* (Prentice-Hall, Englewood Cliffs, New Jersey, 1956).

25. D. O. Harris, G. G. Engerholm, and W. D. Gwinn, J. Chem. Phys. **43**, 1515 (1965); P. F. Endres, J. Chem. Phys. **47**, 798 (1967).

26. A. S. Dickenson and P. R. Certain, J. Chem. Phys. **49**, 4209 (1968) (see, also, ref. 18).

27. R. W. Heather and J. C. Light, J. Chem. Phys. **79**, 147 (1983); R. M. Whitnell and J. C. Light, J. Chem. Phys. **86**, 2007 (1987); J. V. Lill, G. A. Parker, and J. C. Light, J. Chem. Phys. **85**, 900 (1986).

28. B. Shizgal and R. Blackmore, J. Comput. Phys. **55**, 313 (1984); R. Blackmore and B. Shizgal, Phys. Rev. A**31**, 1855 (1985).

29. M. Abramowitz and I. A. Stegun, *Handbook of Mathematical Functions*, Nat. Bur. Std. Applied Mathematics Series 55 (U. S. Govt. Printing Office, Washington, D.C., 1964); P. Dennery and A. Krzywicki, *Mathematics for Physicists* (Harper & Row, New York, 1967).

30. M. D. Feit and J. A. Fleck, J. Chem. Phys. **78**, 301 (1982); **80**, 2578 (1984).

31. R. Kosloff and R. Kosloff, J. Chem. Phys. **79**, 1823 (1983); D. Kosloff and R. Kosloff, J. Comput. Phys. **52**, 35 (1983); R. Kosloff and C. Cerjan, J. Chem. Phys. **81**, 3722 (1984).

32. D. Kouri and R. C. Mowrey, J. Chem. Phys. **86**, 2087 (1987); **84**, 6466 (1986); G. Wahnstrom and H. Metiu, Chem. Phys. Lett. **134**, 531 (1987); B. Jackson and H. Metiu, J. Chem. Phys. **86**, 1026 (1987); R. Heather and H. Metiu, J. Chem. Phys. **87**, 5497 (1987).

33. R. Heather and H. Metiu, J. Chem. Phys. **86**, 5009 (1987); G. Jolicard, C. Le Forestier, and E. J. Austin, J. Chem. Phys. **88**, 1026 (1988).

34. P. Pechukas and J. C. Light, J. Chem. Phys. **44**, 3897 (1966).

35. H. Tal-Ezar and R. Kosloff, J. Chem. Phys. **81**, 3967 (1984).

36. T. J. Park and J. C. Light, J. Chem. Phys. **85**, 5870 (1986).

37. T. Oka, Phys. Rev. Lett. **45**, 531 (1980).

38. J. K. G. Watson, S. C. Foster, A. R. W. McKellar, P. Bernath, T. Amano, F. S. Pan, M. W. Crofton, R. S. Altman, and T. Oka, Can. J. Phys. **62**, 1875 (1984); W. A. Majewski, M. D. Marshall, A. R. W. McKellar, J. C. W. Johns, and J. K. G. Watson, J. Mol. Spec. **122**, 341 (1987).

39. B. Rehfuss, M. Bawendi, and T. Oka, private communication.

40. C. E. Dykstra and W. C. Swope, J. Chem. Phys. **70**, 1 (1979).

41. W. Meyer, P. Botschwina, and P. G. Burton, J. Chem. Phys. **84**, 891 (1986).

42. S. Miller and J. Tennyson, J. Mol. Spec. **128**, 132, 530 (1988); G. D. Carney and R. N. Porter, J. Chem. Phys. **65**, 3547 (1976).

43. R. T. Pack, Chem. Phys. Lett. **108**, 333 (1984); R. T. Pack and G. A. Parker, J. Chem. Phys. **87**, 3888 (1987).

44. F. Webster and J. C. Light, J. Chem. Phys. **85**, 4744 (1986); F. Webster, Ph.D. Thesis, Univ. of Chicago, 1988.

FINITE ELEMENT CALCULATIONS OF SCATTERING MATRICES FOR ATOM-DIATOM REACTIVE COLLISIONS. EXPERIENCES ON AN ALLIANT FX/8

J. LINDERBERG
Department of Chemistry
Aarhus University
DK-8000 Aarhus C, Denmark

ABSTRACT. The determination of hyperangular eigenstates from a Finite Element Method formulation of the quantum mechanical three particle problem and their subsequent use in FEM propagation of the Wigner R-matrix is discussed in relation to program execution on vectorprocessors and parallel architecture computers.

1. Introduction

Vector operations dominate in many quantum mechanical calculations and the advent of computers with a design that greatly enhances the rate at which they can be performed has made it feasible to consider also some very demanding problems, such as the determination of state-to-state cross sections in reactive scattering theory. My report here concerns some aspects of a three-dimensional treatment of atom-diatom collisions based on the Finite Element Method employed in a hyperspherical coordinate formulation.

The general formulation of the problem is given elsewhere [1] and is available in a partial form in a review [2]. Thus only a statement of the relevant equations is given in the next section. A generalized eigenvalue problem is the rate determining step in the present formulation and we offer timings and profiles for some cases in the main part of the paper. A discussion of the conclusions drawn from the data finishes the paper.

2. Statement of Problem

A general quantum mechanical problem involves finding a solution to Schrödinger's equation with appropriate boundary conditions. Reactive atom-diatom scattering provides rather simple boundary conditions since there are often no interactions to worry about when the particles are well separated. We can accordingly restrict ourselves to finding approximate solutions to the Schrödinger equation in a limited domain of configuration space. The convenient and algebraically straightforward procedure we use is the variational R-matrix theory introduced by Wigner and Eisenbud [3].

We define a set V in configuration space and its boundary S as being adequate for the characterization of the scattering interactions, that is the reaction domain. A set of functions, $\{u_j(x)\}$, is introduced in the Sobolev space $H^1(V)$:

A. Laganà (ed.), Supercomputer Algorithms for Reactivity, Dynamics and Kinetics of Small Molecules, 215–221.

$u_j(x) \in H^1(V),\ X \in V.$

They define matrix elements

$$A_{jk} = \int_V dx\{[E-W(x)]u_j^*(x)\, u_k(x) - \nabla u_j^* \cdot \nabla u_k\} \tag{1}$$

where ∇u_k denotes the gradient of the appropriate dimension and W(x) is the potential function for the system in units where Planck's constant and the masses are absorbed. Wigner's R-matrix has the form of an integral kernel on the boundary set S and is calculated from the expression [1,2]

$$R(x,x') = \Delta(x,x')/\Delta, \tag{2}$$

with

$$\Delta(x,x') = \begin{vmatrix} A_{11} & A_{12} & --- & u_1^*(x') \\ A_{21} & A_{22} & --- & u_2^*(x') \\ | & & & | \\ | & & & | \\ u_1(x) & u_2(x) & --- & 0 \end{vmatrix}, \tag{3}$$

and

$$\Delta = \begin{vmatrix} A_{11} & A_{12} & --- \\ A_{21} & A_{22} & --- \\ --- & --- & \end{vmatrix}. \tag{4}$$

Equations (2) and (3) apply when x and x' are in the boundary set S. It is clear that the determinant Δ vanishes for certain eigenvalues, $E = \varepsilon_v$, and that the kernel has a spectral form

$$R(x,x') = \Sigma_v\, \varphi_v(x)(\varepsilon_v - E)^{-1}\, \varphi_v^*(x'). \tag{5}$$

A complete spectral resolution is generally not feasible since the dimension of the matrix becomes very large, that is presently several hundred thousands.

A convenient basis for the calculations is provided by the Finite Element Method [4]. The set V is then subdivided into disjoint subsets, each of which supports an elementary basis. A finite element space can then be constructed such that it is a subspace of $H^1(V)$ and has "small" support. The scattering problem seems to lend itself to a physically natural way [5] for defining subsets of V. We consider a splitting such that

$$V = \bigcup_j \{x \mid q_j > |x| > q_{j+1}\} \tag{6}$$

and a set of hyperradii $\{q_j\}$ defines boundary sets:

$$S_j = \{x \mid |x| = q_j\}. \tag{7}$$

The finite element space will, for the present purpose, similarly be separated into subspaces with bases:

$$\mathbf{U}_j(x) = (u_{j1}(x), u_{j2}(x),...). \tag{8}$$

An element $u_{j\nu}(x)$ has local support in the sense that

$$u_{j\nu}(x) = 0 \text{ for } (|x| > q_{j\nu}) \vee (|x| < q_{j+1}), \tag{9}$$

and this results in a block form of the matrix **A**.

The determinant form (2) is now interpreted as consisting of blocks $\mathbf{A}_{jk}$ and row vectors $\mathbf{U}_j$ as well as columns $\mathbf{U}_j^\dagger$. Only blocks on the diagonal and the first sub- and superdiagonal are different from zero. We construct the R-matrix kernel by means of the following algorithm constructed by means of blockwise Gaussian elimination:

$$(\mathbf{A}_{jj} + \mathbf{A}_{jj-1}\,\mathbf{B}_{j-1})\mathbf{B}_j + \mathbf{A}_{jj+1} = 0, \tag{10}$$

$$\mathbf{B}_o = 0, \tag{11}$$

$$(\mathbf{A}_{jj} + \mathbf{A}_{jj-1}\,\mathbf{B}_{j-1})\mathbf{r}_j\,(x') + \mathbf{V}_j^\dagger\,(x') = 0, \tag{12}$$

$$\mathbf{V}_{j+1}(x) = \mathbf{V}_j(x)\mathbf{B}_j, \tag{13}$$

$$\mathbf{V}_1(x) = \mathbf{U}_1\,(x). \tag{14}$$

It should be recognized that the boundary set S_1 is identified with the boundary set S and thus there are no amplitudes in the form (2) from $j>1$. The kernel is then given by the expression

$$R(x,x') = \Sigma\, \mathbf{V}_j(x)\, \mathbf{r}_j(x')\ . \tag{15}$$

The purpose of using this form, rather than one obtained by successive inversion with decreasing j, is the expectation that the series (15) terminates in a way that admits a study of the convergence as the 'propagation' brings in smaller hyperspherical radii.

The equation systems (10) and (12) may have large dimensions and they need to be solved for many values of the energy parameter E. Thus we are looking for a spectral form for the relevant matrices.

It is worth noticing that the algorithm can be started with a different initial condition than (13) in order to get scattering amplitudes [1].

3. Eigenpairs

The matrix blocks $\mathbf{A}_{jk}$ are themselves sparse matrices in the finite element space we have under consideration. Their dimension has reached 4000 in some of our applications but, since no more than 21 elements can be different from zero in the particular choice of space we have made, storage requirements are modest. We expect that the matrix block $\mathbf{A}_{jj}$ varies slowly with the value q_j and that $\mathbf{B}_j$ also may be represented in a form where only modest changes occur. This is the motivation for a search for a diagonal form for $\mathbf{A}_{jj}$. We define the metric matrix $\mathbf{M}_{jj}$ as

$$\mathbf{M}_{jj} = d\mathbf{A}_{jj}/dE, \tag{16}$$

and search for eigenpairs of the problem

$$\mathbf{A}_{jj}\, \mathbf{C}_{jv} = \mathbf{M}_{jj}\, \mathbf{C}_{jv}\, a_v, \tag{17}$$

with

$$\mathbf{C}^{\dagger}_{j\mu}\, \mathbf{M}_{jj}\, \mathbf{C}_{jv} = \delta_{\mu v}\,. \tag{18}$$

The eigenvalues a_v are linear functions of E while the eigenvectors are independent. We expect to represent $\mathbf{B}_j$ and $\mathbf{r}_j(x')$ with sufficient accuracy through truncated expansions,

$$\Sigma_\gamma\, \mathbf{C}_{jv}\, \mathbf{B}_{jv} \sim \mathbf{B}_j \tag{19}$$

and

$$\Sigma_\gamma\, \mathbf{C}_{jv}\, \mathbf{r}_{jv}\,(x) \sim \mathbf{r}_j(x). \tag{20}$$

The truncation is governed by a condition

$$|a_v| \leq \Delta E \tag{21}$$

and applies in an interval (E–ε, E+ε) with ε < ΔE.

Ericsson and Ruhe [6] have implemented and analyzed an algorithm which performs the task of identifying a subset of eigenpairs for the problem posed by Eqs. (18) and (19). It performs well for the problems we have considered and the program was easily installed on the computers at our disposal.

Ericsson's and Ruhe's STLM package uses, in the standard version, a profile storage for the two matrices and it appears that the cost of obtaining eigenpairs increases almost proportional to the number of matrix elements required in storage. The finite element method gives in our implementation that the cost per eigenvector components increases only slowly with the matrix dimension. Results for several computers are given in Table I.

TABLE I. Cost for each eigenvector component of the generalized eigenvalue problem as obtained with the STLM package. Matrix dimensions from 300 to 4005.

VAX11/780	20-35	CPUms
VAX 8650	3~6	CPUms
Alliant FX/8	2.2~4.2	CPUms
Amdahl VP1100	.2~.24	CPUms

No thorough optimization has been undertaken in order to utilize available system software in conjunction with STLM. The vector optimization option of the FORTRAN compiler was invoked on the Alliant FX/8, but it has proven to be advantageous for the overall utilization of the capacity to refrain from concurrency operation.

There are utilities for profiling the program operation on the Alliant that are very useful for the optimization of the possible usage of vector facilities. We show two examples in Table II, that demonstrate, in a modest way, the shifting balance between different subprograms for numerical problems of varying size. The scalar product routine scpr is called more than 6.5 million times in the larger problem and dominates the picture. It is not ideal for vectorization and an improvement of the algorithm requires a reconsideration of its role.

TABLE II. Detailed specification of CPU-time, in seconds, spent in the different subprograms for two runs.

Matrix dimension	528	4005
Subprogram		
scpr (STLM)	67.45	1334.74
subv (STLM)	23.85	323.38
other (STLM)	32.44	182.78
main	62.38	572.63

The significance of proper vectorprogramming is emphasized by a comparison of straight forwardly compiled code and vectoroptimized code. Table III details the time for various operations in an example where the matrix dimension was 630. A total speedup of a factor 5 is achieved by the utilization of the tools for adjusting the program to improve its vector possibilities.

TABLE III. Times, in seconds, for various matrix operations in the STLM package for a sparse matrix problem of dimension 630.

Alliant FX/8	FORTRAN CODE		OPTIMIZED CODE[a)]	
	case 1	case 1	case 1	case 2
Total time (s)	270	380	54	57
time/eigenpair(s)	6.7	10	1.4	1.5
Decomposition ($LDL^{\dagger}$)	5.1	5.2	1.6	1.5
Solution ($LDL^{\dagger}X=b$)	.38	.37	.058	.058
Matrix times vector	.37	.36	.058	.053

a) with the option –Ogv

4. Conclusion

Vector operations are the dominant feature of calculations with the finite element method. Our examples from atom-diatom reactive scattering theory demonstrate that the capabilities of vector processors greatly enhance the possibilities of considering also relatively demanding problems. The transfer of programs to the Alliant FX/8 has been smooth and the tools provided with the system have been easy to use. The compiler indicates where possible vectorization and/or concurrency is inhibited and the profiling facility identifies the subprograms where gains are particularly significant. There are disadvantages connected with the use of concurrency for the simultaneous use of the three computing elements in the cluster available to us. A computing element which idles in a cluster process is not available to other users while the cluster process is active.

The Alliant FX/8 at UNI.C, the national computer center in Denmark, is available to the theoretical chemistry group through grants from the Danish Natural Sciences Research Council and the College of Natural Sciences at Aarhus University.

References

1. J. Linderberg, Y. Öhrn, S.B. Padkjær, B. Vessal, to be published.
2. J. Linderberg, *Comp. Phys. Rep.* **6**, 209 (1987).
3. E.P. Wigner and L. Eisenbud, *Phys. Rev.* **72**, 29 (1947).
4. P.G. Ciarlet, *'The Finite Element Method for Elliptic Problems'*, (North-Holland, Amsterdam, 1978).
5. U. Fano, *Phys. Rev.* **A24**, 2402 (1981).
6. T. Ericsson and A. Ruhe, *Math. of Comp.* **35**, 1251 (1980), STLM, A Software Package for the Spectral Transform Lanczos Method (Umeå University, Umeå, Sweden, 1984).

INVESTIGATIONS WITH THE FINITE ELEMENT METHOD.
The collinear $H + H_2, F + H_2$ and $Ne + H_2^+$ reactions

Ralph Jaquet
Theoretische Chemie
Universität Siegen
POB 101240
D5900-Siegen, West-Germany

Abstract. We are investigating systematic the application of the finite element method (FEM) for solving the Schrödinger equation. The subsequent work is devoted to the calculation of vibrational transition probabilities for the collinear reactive system $A + BC$ (i.e. $H + H_2$ and their isotopes, $F + H_2$ and $Ne + H_2^+$). We have performed an extensive analysis of FEM on the vector-computer Cyber 205 and have developed a vector code for the efficient use in two mathematical dimensions. The implementation of a three dimensional program is now in progress. The details of our FEM calculations are the following: The integration area is discretized into triangles where quadratic polynomials for the local wavefunctions are defined. With this simple ansatz convergent results can be reached for most reactions with ≈ 10000 grid points. In case of 3D calculations a much larger number of grid points is necessary. Different 1D-3D eigenvalue problems with higher degree polynomials are investigated to reduce the number of grid points and to optimize the geometry of the finite elements. Applying this experience to reactive problems, much less grid points will be needed for comparable accuracy. The main computing time results from solving linear equations and generalized eigenvalue equations for very large, but sparse matrices. Direct methods (e.g. Cholesky or Jacobi) are needed for starting, and if good starting vectors are available, iterative methods (e.g. conjugated gradients or coordinate overrelaxation) can be used.

1. Introduction

The accurate treatment of the equation of motion, e.g. for the reaction (atom + molecule)

$$A + BC(v, j) \rightarrow AB(v', j') + C$$

(v, v': vibrational states; j, j': rotational states) is conventionally accomplished by 'close-coupling'-type methods. Because the number of coupled equations is too large, one tries to find an accurate solution of the Schrödinger–equation for the 2 or 3 most important degrees of freedom. The remaining degrees of freedom are approximated by wavefunction expansion. Because of the special features of present supercomputers, i.e. well designed to matrix algebra, FEM seems to be a very promising method for the 'exact' solution in few dimensions.

Our aim is to use FEM for the investigation of dynamical problems, especially the reactive scattering dynamics between atoms and diatomic molecules. There has been significant progress in treating molecular collisions theoretically [1], but up to now the $H + H_2$ system is the only one where accurate 3D calculations [2-10] have been performed.

Substantial advantage of FEM is the flexibility of choosing irregular meshes, treatment of special boundary conditions and optimizing the local wavefunctions inside each element [11-13].

A. Laganà (ed.), Supercomputer Algorithms for Reactivity, Dynamics and Kinetics of Small Molecules, 223–233.

Askar, Cakmak and Rabitz [14-16] first showed how to use FEM for reactive scattering ($H+H_2$) and how to use FEM for different two-dimensional eigenvalue problems [16]. We have investigated in addition different symmetric and asymmetric configurations for $H + H_2$ (and their isotopes) and $F + H_2$ [17-19]. FEM is also partly used for solving the 3D reactive scattering dynamics of $H + H_2$ and $F + H_2$ by propagating the eigenfunctions (calculated by FEM) in the two hyperspherical angle coordinates [5,9,10]. Askar and Rabitz [20] have performed the first full 3D rotational inelastic scattering for $H + H_2$. Further works with FEM in the last fifteen years are summarized in refs. [17,19].

We have used FEM for the calculation of vibrational transition probabilities. In this paper we will summarize the important aspects of our work. In addition new preliminary results for the system $Ne + H_2^+$ will be shown where a potential minimum for NeH_2^+ in the reactive region has to be accounted for.

2. Theory

2.1 General formalism of FEM

In this chapter a short summary will be given, for further details see [17]. The time independent Schrödinger equation

$$(H - E)\varphi = 0 \tag{1}$$

on domain G with Dirichlet boundary conditions ($\delta\varphi = 0$ on the boundary) is equivalent to the following variational problem:

$$\delta L = \int_G \left[\frac{\hbar^2}{2\mu}\nabla\delta\varphi\nabla\varphi + \delta\varphi(V - E)\varphi\right] dG = 0 \quad . \tag{2}$$

The solution of this equation is a stationary wavefunction for a given boundary condition. In order to solve the scattering problem one needs a linear set of independent wave solutions corresponding to the number of open channels.

The general idea of FEM is now the following [11]:

a) The domain G (see the v-shape in fig. 1) is discretized into many subdomains called elements (in case of collinear reactions we use triangles). This can be done very flexibly, so that the form of the elements can optimally fitted to the problem.
b) On each element the wavefunction φ is approximated by a local function u. The simplest choice are polynomials of different degress

$$u(x, y) = \sum_{i,j} c_{ij} x^i y^j \quad . \tag{3}$$

c) On each element a certain number of grid points, so called nodes, are chosen and the function u on the element e is expanded as

$$u^{(e)}(x, y) = \sum_{i=1} u_i^{(e)} N_i^{(e)}(x, y) \quad ; \quad u_i^{(e)} = u(x_i, y_i) \quad , \tag{4}$$

where the formfunctions $N_i^{(e)}$ (i.e. the 'basis') are defined to have the following interpolating properties:

$$N_i^{(e)}(x_j, y_j) = \begin{cases} 1 & j = i \\ 0 & j \neq i \end{cases} \quad . \tag{5}$$

The unknown variables to be determined are the $u_i^{(e)}$.

d) The integral in eq. 2 has to be taken over the whole domain G and is a sum over all elements

$$\delta L = \sum_e \delta L_e \quad , \tag{6}$$

where δL_e is the integral over the element e with the area G_e.

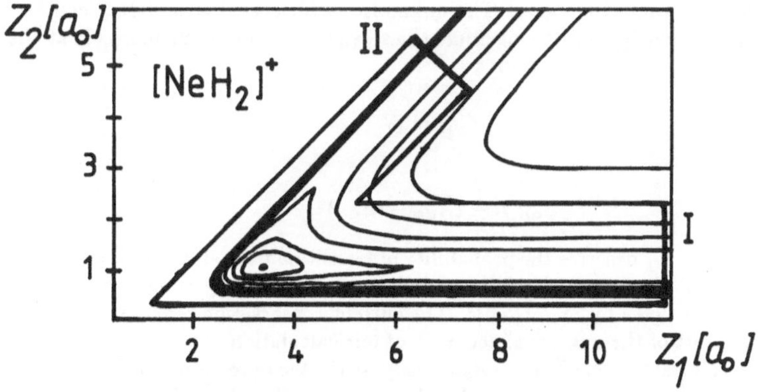

Fig. 1. $Ne + H_2^+$ potential surface: contour plot in mass-weighted coordinates . The contour lines are -.018, -.009, 0.0, 0.018, 0.036,...(eV). The integration area was determined so that $r_{H_2^+}$ and r_{NeH^+} had equal length. I: $Ne + H_2^+$ channel and II: $NeH^+ + H$ channel.

One of the important properties of FEM is that the coordinates of the domain G can be transformed to "unit triangle coordinates" so that the integrals over the elements can be solved simply and analytically. For eigenvalue problems we use polynomials up to the 8th degree in one to three dimensions. For these high degrees it is easier to do the integration numerically (Gauß integration).

The contribution δL_e of the element e can be expressed in terms of matrix elements over the formfunctions N_i. In the case of quadratic 2D polynomials (6 nodes) one has 6 x 6 matrices for the hamiltonian H and metric(overlap) M:

$$\delta L_e = < \delta \underline{\varphi}^{(e)} | \underline{\underline{H}}^{(e)} - E\, \underline{\underline{M}}^{(e)} | \underline{\varphi}^{(e)} > \quad . \tag{7}$$

Summing up these matrices (eq. 6) a matrix of the size of the total number of grid points N is build up. The final N x N matrix equation

$$\underline{\underline{H}}\, \underline{\varphi} = E\, \underline{\underline{M}}\, \underline{\varphi} \tag{8}$$

is to be solved, where $\underline{\varphi}$ is the wavefunction at the grid points. In the case of scattering problems with asymptotic boundary conditions, one has a system of linear equations for each scattering energy E

$$(\underline{\underline{H}}' - E\underline{\underline{M}}')\, \underline{\varphi}' = \underline{F} \quad . \tag{9}$$

The prime and the vector F result from including the wavefunction boundary conditions into eq. 8.

2.2 The calculation of the transition probability P

For the calculation of the transition probabilities P the scattering problem has to be solved in two stages. The stationary wavefunctions in eq. 2 which one obtains for different boundary conditions have to be matched to the exact asymptotic solution of the Schrödinger equation (see [15]).

The calculated stationary wavefunction φ_α with special boundary conditions α can be asymptotically written as

$$\varphi_\alpha = \sum_{n=1}^{M} (A_{\alpha n} \cos(k_n \xi) + B_{\alpha n} \sin(k_n \xi))\ \chi_n(\eta) \tag{10}$$

and has to be matched to the exact asymptotic solution. M is the sum of all open states on both asymptotic sides. The internal coordinates ξ and η describe the translational and vibrational coordinates in the asymptotic region. The unknown coefficients $A_{\alpha n}$ and $B_{\alpha n}$ can be determined by projection of φ_α and $\frac{\partial \varphi_\alpha}{\partial \xi}$ on the eigenfunctions $\{\chi_n\}$ of the diatomic molecule at the boundary.

The complex S-matrix results from

$$\underline{\underline{S}} = \underline{\underline{Z}}^{*^{-1}}\ \underline{\underline{Z}} \tag{11}$$

with

$$Z_{\alpha n} = \frac{1}{2}\ (A_{\alpha n} - iB_{\alpha n}) k_n{}^{\frac{1}{2}} \tag{12}$$

The matrix elements S_{ij} describe the probability of transition from state i to state j. For a complete solution of the S-matrix one needs M linear independent boundary conditions α. The transition probability is simply given by $P_{ln} = |S_{ln}|^2$. The current conservation requires $\sum_n |S_{ln}|^2 = 1$, which can be used as a test of the numerical accuracy of our calculation.

In order to calculate the coefficients $A_{\alpha n}$ easily we use the pure vibrational states of the molecule (for example $\chi_{v=0}, \chi_{v=1}$ or $-\chi_{v=0}, -\chi_{v=1}$) as boundary conditions in the asymptotic region.

3. The structure of the FEM code for the Cyber 205

The FEM code is specially adapted to the vector computer Cyber 205. For the chosen physical problems high accuracy and accordingly many grid points are needed. This may result in long vector lengths, allowing optimal vector coding and keeping the computing time short. The Cyber 205 works efficiently for large vector lengths ($N \geq 200$). We have developed a program for different kinds of mathematically 2D problems in inelastic and reactive scattering and bound state systems.

FEM needs simple matrix algebra. But because of the short bandwidth (i.e. 100 - 250) of the large matrices, i.e. N = 6000–30000 (see tab. 1 in [17]), it is not easy to vectorize the matrix operations optimally. For an improved vectorisation it is advantageous to have a large core memory of at least 1.5 Mwords. In this case a Mflop rate between 20–50 has been reached which is roughly one third of the maximum rate.

The calculation of integrals and the building up of the matrix $\underline{\underline{A}} = (\underline{\underline{H'}} - E\underline{\underline{M'}})$ (eq. 9) do not need much time compared to solving the linear equation with several boundary conditions. The analysis of the stationary states for calculating the transition probability is again negligible in time. The solution of the linear equation is achieved with two different methods: a) the direct Cholesky method [11] and b) the iterative conjugate gradient method [11]. In the Cholesky method a decomposition of the matrix $\underline{\underline{A}}$ in two triangular matrices $\underline{\underline{A}} = \underline{\underline{L}} \cdot \underline{\underline{L}}^T$ is used. This decomposition is the time consuming step, so that for different boundary conditions $\underline{F}$ (see eq. 9) the stationary solution $\underline{\varphi'}$ in $\underline{\underline{A}}\underline{\varphi'} = \underline{F}$ can be solved efficiently. Having enough information about the energy dependence of the wavefunction one can use an iterative method with starting vectors.

Special details of the numerical procedures are given in ref. [17,19]. The Cholesky code is not optimal, because the algorithm needs the values of the matrix recursively. This complicates vectorisation. The effective vector length is equal to the semi–bandwidth of the matrix $\underline{\underline{A}}$ and is greater than 200 only for large dimensions of $\underline{\underline{A}}$. The iterative procedure can be vectorized much easier. The conjugate gradient method is used and vectorized as described in [21]. The maximum vector length can be as large as the dimension of $\underline{\underline{A}}$.

For all scattering energies the calculation of $\underline{\underline{H}}$ and $\underline{\underline{M}}$ is needed only once. Then in case of direct solution for each scattering energy the decomposition of the matrix $(\underline{\underline{H'}} - E\underline{\underline{M'}})$ is solved once and for each boundary condition a stationary solution has to be calculated. The triangular decomposition of the matrix takes a large amount of computer time, the time for the different boundary conditions is negligible. Representative CPU-times are given in tab. 1 of refs. [17,19]. For Cholesky the decomposition costs for N=6201 to 31951 grid points between 6.7 and 105.7 sec CPU-time. Then $\underline{\varphi'}$ is calculated for each boundary condition in 0.6 to 6.2 sec CPU-time.

4. Application

4.1 Introduction

With FEM we have already investigated the collinear $H + H_2$ reaction (and different isotopes) [17,19], $F + H_2$ [18,19] and some simple L-shape potential energy surfaces (PES) with potential barriers or minima [19]. In addition calculations have been performed for the system $Ne + H_2^+$.

We have used FEM for different characteristic types of reactions: a) symmetric and asymmetric reactions, b) small skew angle and large skew angle (i.e. HLH or LHL systems) and c) including many open channels. The light (L) and heavy (H) atoms are H, D, T, Mu, F, Ne. Our next aim is the calculation of specific problems in 3D. So the first step is to solve the 2D problem in comparatively short times.

For the treatment of the reaction A + BC → AB + C mass-weighted coordinates are used. Usually the kinetic energy is expressed by the new coordinates Z_1 and Z_2, related to the interparticle distances r_{AB} and r_{AC}:

$$Z_1 = r_{AB} + [m_C/m_B + m_C]\, r_{AC} \quad , \quad Z_2 = \sqrt{\frac{m_{BC}}{m}} \cdot r_{AC}$$
$$m = m_A(m_B + m_C)/(m_A + m_B + m_C) \quad , \quad m_{BC} = m_B m_C/(m_B + m_C) \quad .$$

4.2 $H + H_2$ and isotopes

The $H + H_2$ reaction was investigated [17] by using the Porter-Karplus surface. The boundary values for the asymptotes are the eigenfunctions of the diatomic Morse oscillator. We have shown that the reactive probabilities for $H + H_2(v) \rightarrow H_2(v') + H$ with up to 3 open vibrational channels exactly agree with the literature values [22-25]. In that case calculations were done for $K = 5000$ finite elements and $N = 10251$ grid points (this number of elements has been proven to give convergent results as reported in [17]). Convergence was tested at several energies for a different number of grid points. For E=0.8 eV the convergence in the transition probability P is reached very early ($N_{grid} \leq 1700$), but $(\frac{\partial \varphi}{\partial \xi})$ is still locally not correct. Some other systems have shown much slower convergence for P (e.g. $H + MuH$). Using many open channels, one naturally has to include more points to describe the oscillations of the eigenfunctions accurately. $H + MuH$ and $F + H_2$ need an increased number of points in translational direction. For $H + H_2$ itself one finds no complications.

The results for the isotope reactions $H + DH$, $D + HD$, $H + MuH$ (symmetric reactions) and $Mu + D_2$, $H + D_2$, $D + H_2$ (asymmetric reactions) [17] can be summarized as follows: These systems are investigated for the first time with FEM and there is exellent agreement with the known 'exact' literature values. In the case of $H + MuH$ the results are more sensitive to discretization, more than 32 000 grid points had to be used. This number of grid points seems to be very large.

Using polynomials of higher degree a considerable reduction of points can be expected.

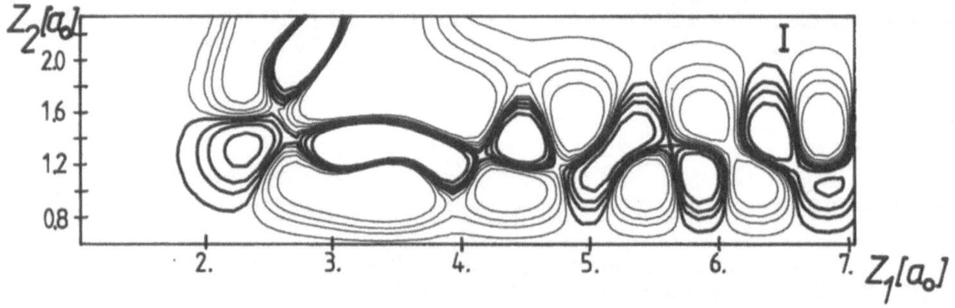

Fig. 2. $H + H_2$: One special stationary wavefunction φ_α for boundary conditions α. The contour lines are $\pm$(2.,1.,0.5,0.1). The symmetric wavefunction at the boundaries of the integration area (like in fig. 1) is set to zero except at the asymptotic boundaries I and II (χ are the vibrational eigenfunctions of H_2, six different boundary conditions are necessary at $E_{tot} = 1.305eV$): I : $\chi_{H_2(v=0)}$ and II : $\chi_{H_2(v=0)}$.

We have analysed in detail for $H + H_2$ [19] how φ_α changes near the resonance energy of $\approx$ 1.3 eV. Rapid changes of φ_α can be seen along the translational direction. Because the asymptotic part of the potential energy surface for $H + H_2$ is divided by a potential barrier, φ_α (see fig. 2) shows in the reactive region less nodal structure than in the asymptotic region; in principle less finer discretization can be used in the inner region.

4.3 $F + H_2$

In the collinear exchange reaction $F + H_2(v) \rightarrow FH(v') + H$, at least 3 vibrational channels of FH are open. This means that in the simplest calculation 4 different boundary conditions have to be accounted for. In our previous work [18] we have shown that FEM is a real alternative to other methods [26-31] and that for the energy range of E_{trans} = 0-0.4eV $\approx$10000 grid points are sufficient to obtain convergent results typically better than 1%.

We have analysed different sizes of the whole integration area. Because of the size of the chosen integration area about 240 grid points are needed in the translational direction with $\approx$20-30 nodes in φ. As a demonstration, one of the five stationary wavefunctions φ_α for E_{trans}=0.3 eV is plotted in fig. 3.

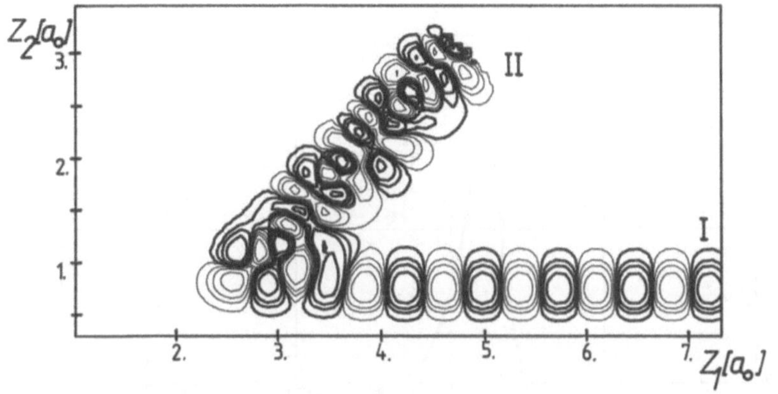

Fig. 3. $F + H_2$: One special stationary wavefunction φ_α for boundary conditions α. The wavefunction at the boundaries of the integration area (like in fig. 1) is set to zero except at the asymptotic boundaries I and II (χ are the vibrational eigenfunctions of the diatomic molecules H_2 and FH, five different boundary conditions are necessary at $E_{Trans} = 0.3eV$): I : $\chi_{H_2(v=0)}$ and II : $\chi_{FH(v=0)}$.

4.4 $Ne + H_2^+$

In the case of $[Ne + H_2]^+$ a 3D PES for the $^2A'$ ground state ($^2\Sigma^+$ symmetry in collinear configuration) was calculated at the CEPA-PNO-level [32]. This surface describes the reaction $Ne + H_2^+ \rightarrow NeH^+ + H$, which is endoergic by 0.55 eV and has been studied experimentally by Herman et al. [33] at 0.7-3.5 eV collision energy. The reaction dynamics has been investigated so far only by quasiclassical trajectories [34] based on a SCF- and DIM-surface [35]. Our CEPA calculations [36] yield a minimum at collinear $[Ne..H..H]^+$ geometry $[R_{NeH} = 2.29\ a_0,\ R_{H_2} = 2.08\ a_0]$ with a well depth of 0.47 eV with respect to $Ne + H_2^+$. An analytical fit of the collinear part of the surface (fig. 1) has been used for FEM calculations of the reaction probabilities. Because in our previous FEM work only potential surfaces with a barrier in the inner region were used we tested for the L-shape potential (with potential minimum) [37] if a finer grid size must be used (see fig. 4 and [19]). In tab. 1 and fig. 5 preliminary results of the collinear reaction $Ne + H_2^+$ for the total reaction probabilities P are given. In the first calculations, with $\approx$20000 grid points and for the energy range 0.7-1.4 eV (4-9 open channels), the P's were obtained within a few percent accuracy. In fig. 6 one stationary wavefunction is plotted. It shows (in comparison to $H + H_2$ and $F + H_2$) much larger nodal structure in the inner region of the PES. So a finer grid size has been used to reach convergent results. A detailed analysis of the FEM calculations is given in ref. 36.

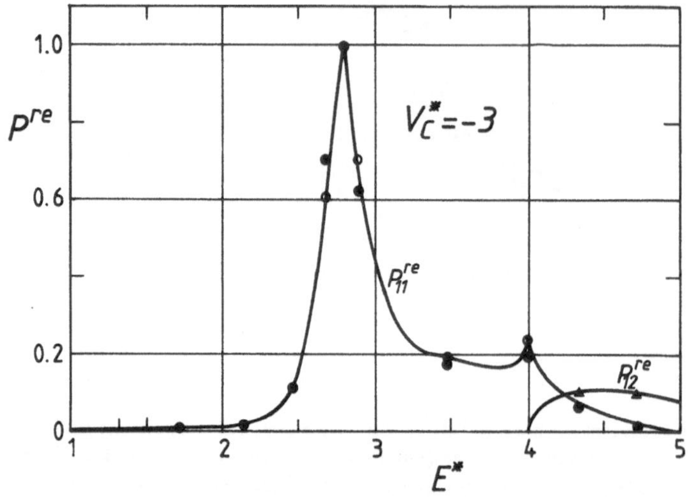

Fig. 4. Reaction probabilities P_{11}^{re} and P_{12}^{re} for the L-shape potential ([37]: $V_C^* = -3$ and $V_I^* = V_T^* = 0$; V_C^*= potential minimum in the inner region; reduced energy units). Analytical exact values of [37]:—; FEM results (• = 6601 and ∘ = 10251 grid points).

Table 1. Total reaction probabilities P_v^{re} a) for $Ne + H_2^+(v = 0, 1, 2, 3, 4) \rightarrow NeH^+(v') + H$

E_{tot} (eV)	P_0^{re}	P_1^{re}	P_2^{re}	P_3^{re}	P_4^{re}
0.70	5.E-5	8.E-5	—	—	—
0.75	0.426	0.490	0.065	—	—
0.80	0.397	0.523	0.081	—	—
0.85	0.397	0.510	0.091	—	—
0.90	0.388	0.533	0.080	—	—
0.95	0.416	0.513	0.070	—	—
1.00	0.419	0.479	0.101	0.002	—
1.10	0.606	0.520	0.663	0.214	0.002
1.20	0.593	0.563	0.649	0.188	0.003

a) Results calculated with 71 points in vibrational and 284 points in translational direction. Size of the integration area: see fig. 1.

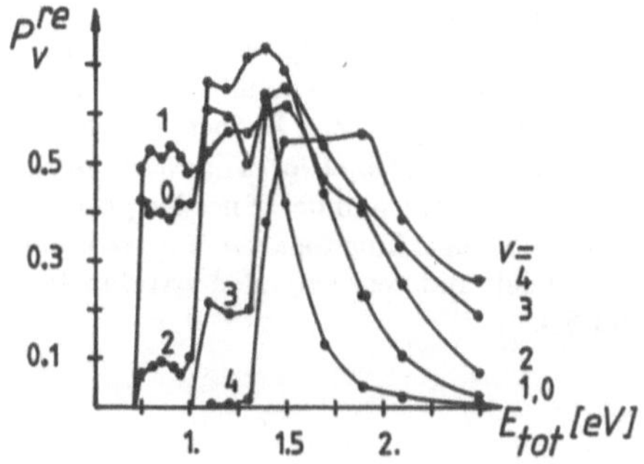

Fig. 5. Total reaction probabilities P for $Ne + H_2^+(v = 0, 1, 2, 3, 4) \rightarrow NeH^+ + H$: P_v^{re} as a function of total energy $E = 0.7 - 2.5eV$.

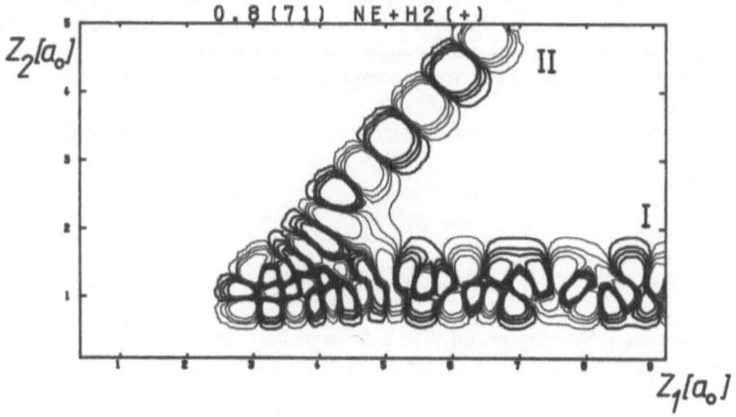

Fig. 6. One special stationary wavefunction φ_α for boundary condition α. The wavefunction at the boundaries of the integration area (like in fig. 1) is set to zero except at the asymptotic boundaries I and II (χ are the vibrational eigenfunctions of the diatomic molecules H_2^+ and NeH^+, four different boundary conditions are necessary at $E_{tot} = 0.8eV$): I : $\chi_{H_2^+(v=0)}$ and II : $\chi_{NeH^+(v=0)}$.

4.5 Eigenvalue problems in 1D-3D

For 3D reactive scattering a much larger number of grid points (≈100000) seems to be necessary for a quadratic function. In order to reduce the number of nodes we are investigating in 1D-3D different eigenvalue problems (e.g. Morse-,double gaussian-, Henon-Heiles (2D,3D)-potentials) [19,38] with higher degree polynomials (up to the 8th degree) and try to optimize the geometry of the finite elements. In 1D we could show that for optimized variable length of the elements (i.e. the element size depends on the potential [39]) the accuracy can be increased by orders of magnitude in comparison to equidistant elements, which are optimal for a 'particle in a box' problem. In addition by increasing the degree of the polynomial the number of needed grid points can be reduced considerably to get even higher eigenvalues and eigenfunctions much more accurate. This we need for reactive scattering with many open channels and very extended wavefunctions. The experience in 1D has now been extended to 2D and 3D.

5. Conclusion

As has been shown in the last years [5,9,10,17-19] FEM is a very useful and accurate tool for the solution of reactive scattering problems. In the collinear case we can calculate the whole scattering wavefunction with FEM in one step and the FEM procedure allows to obtain realistic estimates of the accuracy and convergence properties [17].

Our aim is the systematic investigation of the usefulness of FEM for dynamical problems and for electronic structure calculations, too. In the present work we have shown that mathematical 2D scattering problems can be solved in reasonable time. In the scattering code the wavefunction has been locally approximated by quadratic polynomials. The inclusion of optimal elements and of polynomials of higher degree will cut down the number of necessary grid points considerably. We have shown for eigenvalue problems [19] that this results in a large decrease of computing time. For the collinear case (i.e. 2D) we think that roughly 1000 grid points or less will be enough. These further improvements will make FEM more competitive in comparison to conventional basis set expansion methods.

The general complexity will increase for higher dimensions. We have started to develop a FEM code for 3D, and scattering calculations in 3D seem to be a feasible project on the basis of the present experience.

Acknowledgement

The author thanks V.Staemmler and W.H.E.Schwarz for fruitful discussions. The computations have been carried out at the computer center of the Ruhr-Universität (Bochum).

References

1. a) D.C.Clary, *The theory of chemical reaction dynamics*, Reidel, Dordrecht, 1986
   b) R.B.Bernstein, *Atom-molecule collision theory*, Plenum Press, New York, 1979
   c) M.Baer, *Theory of chemical reaction dynamics*, **I-IV**, CRC Press, Boca Raton, 1985
2. a) A.Kuppermann and G.C.Schatz, *J.Chem.Phys.*, **62** (1975) 2502
   b) G.C.Schatz and A.Kuppermann, *J.Chem.Phys.*, **65** (1976) 4642, 4668
   c) G.C.Schatz and A.Kuppermann, *Phys.Rev.Lett.*, **35** (1975) 1266
3. a) A.B.Elkowitz and R.E.Wyatt, *J.Chem.Phys.*, **62** (1975) 2504; **63** (1975) 702
   b) M.J.Redmon and R.E.Wyatt, *Chem.Phys.Let.*, **63** (1979) 209
4. a) R.B.Walker, E.B.Stechel and J.C.Light, *J.Chem.Phys.*, **69** (1978) 2922
   b) F.Webster and J.C.Light, *J.Chem.Phys.*, **85** (1986) 4744

5. a) A.Kuppermann and P.G.Hipes, *J.Chem.Phys.*, **84** (1986) 5962
   b) P.G.Hipes and A.Kuppermann, *Chem.Phys.Let.*, **133** (1987) 1
6. a) K.Haug, D.W.Schwenke, Y.Shima, D.G.Truhlar, J.Zhang, and D.J.Kouri, *J.Phys.Chem.*, **90** (1986) 6757
   b) J.Z.H.Zhang, D.J.Kouri, K.Haug, D.W.Schwenke, Y.Shima, and D.G.Truhlar, *J.Chem.Phys.*, **88** (1988) 2492
7. M.Baer, *J.Phys.Chem.*, **91** (1987) 5846
8. a) J.Z.H.Zhang and W.H.Miller, *Chem.Phys.Let.*, **140** (1987) 329
   b) J.Z.H.Zhang, S.-I.Chu, and W.H.Miller, *J.Chem.Phys.*, **88** (1988) 6233
9. a) R.T.Pack and G.A.Parker, *J.Chem.Phys.*, **87** (1987) 3888
   b) C.A.Parker, R.T.Pack, B.J.Archer, and R.B.Walker, *Chem.Phys.Let.*, **137** (1987) 564
10. J.Linderberg, B.Vessal, *Int.J.Quant.Chem.*, **31** (1987) 65
11. H.R.Schwarz, *Methode der finiten Elemente*, Teubner, Stuttgart, 1980
12. K.J.Bathe, E.L.Wilson, *Numerical methods in finite element analysis*, Prentice Hall, Englewood Cliffs, New Jersey, 1977
13. R.Gruber, *Finite elements in physics*, Graduate Summer course on Comp.Physics, 1-10 Sept. 1986, EPF Lausanne
14. H.Rabitz, A.Askar and A.Cakmak, *Chem.Phys.*, **29** (1978) 61
15. A.Askar, A.Cakmak and H.Rabitz, *Chem.Phys.*, **33** (1978) 267
16. M.Duff, H.Rabitz, A.Askar, A.Cakmak and M.Ablowitz, *J.Chem.Phys.*, **72** (1980) 1543
17. R.Jaquet, *Theor.Chim.Acta*, **71** (1987) 425
18. R.Jaquet, *Chem.Phys.*, **118** (1987) 17
19. R.Jaquet, *Habilitationsschrift*, Univ. Siegen (1987)
20. A.Askar, H.Rabitz, *J.Chem.Phys.*, **80** (1984) 3586
21. R.Diekkämper, *Proceedings of the 1983 Conferences on Cyber 200 in Bochum*, ed. by H.Ehlich, K.-H.Schloßer (1983)
22. A.Kuppermann, J.A.Kaye and J.P.Dwyer, *Chem.Phys.Let.*, **74** (1980) 257
23. J.Römelt, *Chem.Phys.Let.*, **74** (1980) 263
24. J.Römelt, *Chem.Phys.*, **79** (1983) 197
25. D.K.Bondi and J.N.L.Connor, *J.Chem.Phys.*, **82** (1985) 4383
26. J.C.Light, R.B.Walker, *J.Chem.Phys.*, **65** (1976) 4274
27. J.Römelt, *Chem.Phys.Let.*, **87** (1982) 259
28. G.C.Schatz, J.M.Bowman and A.Kuppermann, *J.Chem.Phys.*, **58** (1973) 4023
29. J.T.Adams, R.L.Smith and E.F.Hayes, *J.Chem.Phys.*, **61** (1974) 2193
30. J.N.L.Connor, W.Jakubetz and J.Manz, *Mol.Phys.*, **35** (1978) 1301; **39** (1980) 799
31. a) B.Lepetit, J.M.Launay, M.LeDourneuf, *J.Phys.B:At.Mol.Phys.*, **19** (1986) 2779
    b) J.M.Launay and M.LeDourneuf, *J.Phys.B*, **15** (1982) L455
32. V.Staemmler and R.Jaquet, *Theor.Chim.Acta*, **59** (1981) 487
33. Z.Herman and I.Koyano, *J.Chem.Soc.,Farad.Trans 2*, **83** (1987) 127
34. C.Stroud and L.M.Raff, *Chem.Phys.*, **46** (1980) 313
35. E.F.Hayes, A.K.Q.Siu,F.M.Chapman Jr., and R.L.Matcha, *J.Chem.Phys.*, **65** (1976) 1901
36. V.Staemmler, J.Urban, and R.Jaquet, to be published
37. J.O.Hirschfelder and K.T.Tang, *J.Chem.Phys.*, **64** (1976) 760
38. R.Jaquet, eigenvalue problems in ref. [19] to be published
39. I.P.Hamilton and J.C.Light, *J.Chem.Phys.*, **84** (1986) 306

Calculation of Multichannel Eigenvalues and Resonances

Roger W. Anderson
Department of Chemistry
University of California
Santa Cruz, California 95064

ABSTRACT. The calculation of bound eigenvalues and resonances (position and width) in multichannel systems is discussed. Methods to efficiently determine multichannel eigenvalues with perturbatively corrected trigonometric propagators are reviewed. The propagators can also be generalized the calculate multichannel resonances though complex scaling. For many multichannel problems the location of eigenvalues and resonances is best approached by first finding them for the Born-Oppenheimer or Adiabatic approximation to the coupled channel problem. Trigonometric propagators can be used to determine the eigenvalues for the resulting single channel problems. A new implementation of the variable phase method is effective in determining the location and width of resonances. The method directly calculates the first and second derivative of the phase shift with respect of the collision wave number. Some zeros of the second derivative have energies close to resonances, and the precise location and width of the resonances can be found from the energy dependence of the phase shift and its derivatives in the region of the zeros. The derivatives allow accurate estimation of zero energy phase shifts to determine the number of bound states in a potential. The methods have been applied to collinear reactive scattering and other examples. New resonances at threshold for reactive scattering of vibrationally unexcited HHH, HDH, and HTH are reported.

1. Introduction

There has been much interest for since the beginnings of chemical physics in the experimental and theoretical characterization of bound states and resonances. Spectroscopy provides precise values for the differences between the energy levels in molecules, and the energy and widths of resonances that correspond to predissociating states. Molecular scattering can also provide information about the existence of resonances, but the information is less precise than that provided by spectroscopy. If the internal motion of a molecule can be well separated into motion on one-dimensional potentials, it is a simple matter to calculate the energies of bound states and the position and width of resonances. However for many systems including diatomic molecules, the states must be calculated by mixing states corresponding to several "separable" states. Perturbations and most predissociations must be treated in this way, and many states of van der Waals molecules cannot be described as motion on a one-dimensional potential.

This paper reports some recent results on techniques to find to give bound states and resonances. First the integration of coupled radial Schrödinger equations is reviewed to give real energies corresponding to bound states and the complex energies corresponding to resonances. The bulk of the paper describes a new implementation of the variable phase method. This latter topic is very useful for characterizing resonances. Since the method was primarily developed in Italy, it is a fitting topic for this workshop.

A. Laganà (ed.), Supercomputer Algorithms for Reactivity, Dynamics and Kinetics of Small Molecules, 235–249.

The work is directed toward the problem of adjusting potential parameters to fit the spectroscopy of molecules. A potential fitting program should simultaneously fit the experimental bound (negative energy) and the predissociating (positive energy) states. It should also fit the experimental intrinsic line intensities. The program should make transparent calculation of perturbations, coriolis coupling, lambda doubling,.... Here only partial progress toward this goal is made. Supercomputers are indicated for the effective implementation of potential fitting programs. The programs involve large amounts of calculations, and much of the calculations can be vectorized for efficient execution.

2. Integration of Coupled Radial Equations

Many problems in spectroscopy and collision dynamics are treated by solving a system of coupled radial Schrödinger equations:

$$((d^2/dr^2 + k^2)\mathbf{I} - \mathbf{U}(r))\Psi = 0, \tag{1}$$

where **I** is the unit matrix, k^2 is the collision energy multiplied by twice the reduced mass, and U(r) is the matrix of the potential expanded in N basis functions that treat the internal energy states of the interacting particles. Ψ is a matrix wavefunction, and after integration its elements at large r yield scattering matrices and the existence of bound and resonance states. In this paper atomic units are used and all matrices are $N \times N$.

There has been much work in the past 20 years on methods of solving Eqn (1). The effective methods make extensive use of vector and matrix computations, and they are well suited for implementation on supercomputers and parallel processor computers. At Santa Cruz we [1,2,3,4] have concentrated on approximate potential methods where the radial coordinate range is divided into a sufficient number of intervals (usually of equal length, $h = r_2 - r_1$), and the potential matrix within each interval is approximated as low order polynomial. For a linear method, the potential matrix is given by its average, $\langle \mathbf{U} \rangle$, and its first derivative[5], $\mathbf{U}'$. $\langle \mathbf{U} \rangle$ and $\mathbf{U}'$ are found from the values of the potential at the zeros of the second degree Legendre polynomial that spans the interval. The average potential is then diagonalized to give N local eigenvalues, $\mathbf{a}$, and the transformation that diagonalizes $\langle \mathbf{U} \rangle$ is also used to transform $\mathbf{U}'$ to give the first derivative of the potential, β, in the local basis. Since the energy eigenvalues, $\mathbf{a}$, correspond to separation of the radial and internal coordinates we call them the Born-Oppenheimer approximation to the energy curves in this paper. Propagators, C1, C2, C3 and C4, can be constructed from k^2, $\mathbf{a}$ and β to give the wavefunction, Ψ, and its radial derivative, Ψ', at r_2 in terms of their values at r_1. We have $\Psi_2 = (\mathbf{C}1)\Psi_1 + (\mathbf{C}2)\Psi'_1$, and $\Psi'_2 = (\mathbf{C}3)\Psi_1 + (\mathbf{C}4)\Psi'_1$. To minimize numerical instabilities found with nonclassical channels, log derivatives of Ψ at r_2 is calculated as $\Psi'_2\Psi_2^{-1}$. This is used as the derivative at the end of the interval, and **I** is used for the wavefunction. If the propagators (considered as a supermatrix of dimension $2N$) are unitary, then the log derivatives at the end of the interval will be real symmetric for real potential and kinetic energies. The Ψ and Ψ' are then transformed into the local basis of the next interval, to continue the propagation. Details of the method have been described [1,2,3,4].

3. Calculations of Multichannel Eigenvalues and Resonances

The bound states of molecular systems can be formulated as a multichannel eigenvalue problem. Such a treatment is especially simple if the Born-Oppenheimer approximation is valid for the singlet states of a diatomic molecule. In this case, the vibrational and rotational states are accurately given by the eigenvalues and eigenfunctions of a single Schrödinger equation. However, to accurately treat spectral perturbations or the bound states of a molecule like NaAr, a multichannel eigenvalue problem must be solved. Similarly predissociations that involve two or more electronic states are examples of multichannel resonances.

3.1 EIGENVALUES

The calculations of multichannel eigenvalues requires proper boundary conditions and evaluation of the continuity of the wavefunction and its radial derivative. The basic idea has been known for many years [5,6,7]. For a radial coordinate that extends from 0 to large positive values, suitable boundary conditions are **0** for Ψ at $r=0$ and at $r=r_r$. The large radius end point, r_r, is positioned to assure nonclassical behavior for all eigenstates of interest. Ψ' is assumed to be **I** and **-I** at the two ends of the integration range. A matching point, r_m is assumed to be the boundary between the $(i-1)^{\text{th}}$ and the ith intervals. Two propagations are performed. The first from $r=0$ to r_m which produces in the local basis at the end of the $(i-1)^{\text{th}}$ interval the log derivative Ψ'_l, and the second from r_r to r_m (negative h) which produces in the local basis at the beginning of the ith interval the log derivative Ψ'_r. The outward log derivative, Ψ'_l, is then transformed into the local basis for the ith interval. The requirement for a multichannel solution is then that Eqn (2) be satisfied:

$$(\Psi'_r - \Psi'_l)\Psi = 0. \tag{2}$$

This equation will be satisfied if the wavefunction and its derivative are continuous at r_m. The requirement, that $|\Psi'_r - \Psi'_l| = 0$, has often been used in the past to signal the location of an eigenvalue. We have found it very useful, however, to note that $(\Psi'_r - \Psi'_l)$ is a real symmetric matrix, and hence its eigenvalues and eigenvectors are easily found. At a bound state, at least one of the eigenvalues will become zero. The corresponding eigenfunction will give the expansion of the eigenfunction in the Born-Oppenheimer basis for the ith interval. The components of the expansion indicate whether the eigenfunction is a mixed (perturbed) state or if the Born-Oppenheimer approximation is valid for the description of the bound state. The eigenvalues and the eigenfunctions of the difference in log derivatives is also of great value in limiting the eigenvalue search to bound state associated with a particular local basis state or potential energy curve. The desired energy eigenstate is easily found because the appropriate eigenvalue of the difference in log derivatives changes sign as it passes through zero. The root bracketing that is important for many root finding algorithms is easily accomplished since the eigenvalues of the log derivative difference increase as k^2 is increased.

The location of the matching radius, r_m, is also important. It is well known that it should not be located near a node of the desired eigenfunction. It is convenient to diagonalize Ψ'_r to display its eigenvalues. One prescription for the location of r_m is to follow these eigenvalues during an integration inward from r_r. The matching radius can be assigned to the interval nearest the first zero of the eigenvalue corresponding to the desired composition. At this point the magnitude of the eigenvalue of Ψ' will be much less than the corresponding q_i. To assure efficient root finding, it is common to find that the location of the matching radius must be changed for different energy regions and for bound states that correspond to different local basis functions.

The identification of an eigenstate as a vibrational state with a certain number of nodes can be accomplished by counting the zeros of the eigenvalues of the log derivative during an integration at the energy corresponding to a bound state. However, it appears to be a better procedure to initially identify the state and its approximate energy before refining the root with multichannel integrations. The Born-Oppenheimer energies, **a**, are useful for the initial root search. The WKB phase can be easily evaluated for radial motion on one of the Born-Oppenheimer potential curves. The approximate (often very accurate!) location of the multichannel bound states is given by the values of k^2 where the phase has a value corresponding to a bound state. The Born-Oppenheimer curves may be improved by using the diagonal elements of the nonadiabatic radial coupling to generate Adiabatic potentials. The WKB phases are also easily calculated on these potentials.

We have used this method to calculate the vibrational and rotational levels for the three lowest excited electronic states of NaAr [8].

3.2 RESONANCES

The calculation of resonances is an interesting and important research area. The simplest method is to calculate scattering matrices for a fine grid of energies and look for characteristic structure in particular elements of the S matrix. The magnitude of these matrix elements can display peaks, valleys, and combinations when they are plotted against energy. Argand plots of the imaginary and real parts of S matrix elements as a function of system energy also display clearly rapid evolution of phase in the vicinity of a resonance. The particular energy dependence of either the magnitudes or the phase plots depends on the presence of overlapping resonances and the evolution of the background or nonresonant phase. One method to separate the components is to assume Breit-Wigner energy behavior for a resonance and to fit energy dependence of the magnitude and phase of a calculated S matrix element with least square analysis. The important information that must be extracted from a resonance is its position, E_r, and width, Γ.

In a time-dependent picture a resonance can be considered as a state that decays in time. This is consistent with the assignment of a complex energy to the resonance, $E = E_r - i\Gamma/2$. The imaginary part of E determines the lifetime, τ, of the decaying state as $\tau = 1/\Gamma$. The scattering procedure described above can give these real and imaginary parts. The complex energies that correspond to resonances can also be directly calculated. The Schrödinger equation is solved for bound states with complex energies. The only difficulty is that even with complex energies, wavefunctions that decay exponentially at large radius are only possible if the radius is assumed to be complex with a positive imaginary component. The use of a complex radial coordinate is called complex scaling, complex rotation, and other names. Most work [9]in complex scaling has expanded both the internal and radial degrees of freedom in a large basis set. The computer codes used for electronic structure can be used to diagonalize the resulting Hamiltonian matrix to determine the complex energies.

The complex scaling technique can also be applied to the integration of coupled radial equations[10]. Brändas and Rittby [11,12,13,14] have done such calculations for both the single and multichannel problems. With the adoption of "exterior scaling," the calculations are easily described and implemented with the propagators described earlier in this paper. The coupled equations are integrated with complex values for k^2, corresponding to a guess for $E_r - i\Gamma/2$. This results in complex values for the q_i. The boundary conditions for $r = 0$ are simply $\Psi = 0$ and $\Psi' = \mathbf{I}$. The boundary conditions at large radius are more interesting. To calculate the resonance as a bound state with an imaginary energy, the wavefunction must decay in magnitude for large r. Inspection of the wavefunction for real r shows that even complex q_i do not satisfy this boundary condition. When the radius has large magnitude, it must also have a sufficiently large positive imaginary component to meet this boundary condition. What is interesting is that the entire integration need not be done with complex radius. The radius need only be complex scaled in the exterior region where the potential for the system can be expressed as an expansion in inverse powers of the radius. The radial coordinate is chosen to be r for $|r| < R_0$, and $R_0 + (r - R_0)e^{i\theta}$ for $|r| \geq R_0$ (where r and R_0 are real numbers). Hence the potential and h are real for $r < R_0$, and only the q_i are complex. For $r > R_0$, the potential, h, and the q_i assume complex values. The angle θ must be large enough to assure that the imaginary part of the product, q_i times the radius, is positive at r_r. The calculation proceeds as before for bound states with the complication of complex entries for the matrices involved. The eigenvalues of the difference in the log derivatives are now complex, and the zeros are found for $E = E_r - i\Gamma/2$. The IMSL routine ZANLYT which uses the Muller method is useful to refine the complex roots. The value for θ can be chosen to limit the search for resonances with large or small values for Γ.

4. Variable Phase Treatment of Resonances and Bound States

One characteristic of resonance is that the phase shift will more or less abruptly increase by π as the system energy is increased through a resonance. The well-known difficulty with methods that determine the phase shift by matching the numerical solution of Eqn (1) asymptotically to Riccati-Bessel functions is that the phase shift is only determined modulo π. This can confuse a search for increases by π in the phase shift. Truhlar [15] has developed a procedure to keep track of the absolute phase shift while using propagator methods to solve coupled radial equations. However, here we take another look at the variable phase approach [16] to scattering. For a single channel problem with $l = 0$, the variable phase method gives the following equation for the absolute phase shift:

$$\delta' = -U(r)\sin^2(kr + \delta)/k \tag{3}$$

where δ' is the derivative of δ with respect to r and δ is zero for $r = 0$. This equation is integrated to large r to give the asymptotic phase shift. The variable phase method has been used in the past to detect resonances and to determine the number of bound states of a potential by Levinson's theorem. For both of these purposes it is convenient to have the first and second derivatives of δ with respect to k. If Δ is defined as $\delta + kr$, then Eqns (4) are easily derived from Eqn. (3).

$$\begin{aligned} \Delta' &= -U(r)\sin^2(\Delta)/k + k, \\ d\Delta'/dk &= 2 - (\Delta + U(r)\sin(2\Delta)(d\Delta/dk))/k, \\ d^2\Delta'/dk^2 &= (2 - U(r)(\cos(2\Delta)(d\Delta/dk)^2 + \sin(2\Delta)(d^2\Delta/dk^2)) - d\Delta'/dk)/k. \end{aligned} \tag{4}$$

These equations may be simultaneously integrated to give the phase δ_1 and its derivative at large r. In terms of these phases and derivatives, $\delta = \Delta - kr$, $d\delta/dk = d\Delta/dk - r$, and $d^2\delta/dk^2 = d^2\Delta/dk^2$.

Knowledge of the phases and their derivatives is of great value for the two principle uses of the variable phase method. The quantity, δ/π, at zero energy indicates the number of bound states in the potential. Workers [17] in the past have calculated δ for a very small energy to evaluate $\delta(0)$. However, for small k the appearance of this quantity in the denominator of Eqn. (3) will cause δ considered as a function of r to make very sharp changes at particular r. This can cause numerical problems. If the derivatives of the phase shift are known for small k, then the zero energy phase shift is simply given:

$$\delta(0) = \delta(k) - k(d\delta(k)/dk) + k^2(d^2\delta(k)/dk^2). \tag{5}$$

The derivatives also can assist in characterizing resonances. The phase shift will increase by π in the vicinity of a resonance, and such a characterization is unambiguous for a sharp resonance. The increase by π becomes much less obvious in the case of a broad resonance. Then the special features of the first and especially the second derivative become useful. The first derivative with respect to k has a maximum near a resonance. The second derivative will have a zero where the derivative passes from positive to negative values with increasing k near a resonance. The zeros of the second derivative can be easily found by searching for sign changes. For simplicity the derivatives of the phase shift are taken with respect to k. The corresponding energy derivatives can be easily generated, and must be taken for the multichannel case (see below).

The position and width of the resonance can be easily related to the derivatives with respect to k. Assuming the Breit-Wigner expression for the resonant part of the phase shift, the following equations result:

$$\begin{aligned}
\delta &= \delta_b + \delta_r = \delta_b - \tan^{-1}(m\Gamma/(k^2 - k_r^2)),\\
d\delta/dk &= d\delta_b/dk + d\delta_r/dk = d\delta_b/dk + 2mk\Gamma/(m^2\Gamma^2 + (k^2 - k_r^2)^2),\\
d^2\delta/dk^2 &= d^2\delta_b/dk^2 + d^2\delta_r/dk^2\\
&= d^2\delta_b/dk^2 + 2m\Gamma/(m^2\Gamma^2 + (k^2 - k_r^2)^2)\\
&\quad - 8mk^2\Gamma(k^2 - k_r^2)/(m^2\Gamma^2 + (k^2 - k_r^2)^2)^2,
\end{aligned} \tag{6}$$

where δ_b is the background phase shift, and m is the reduced mass of the system. The energies, E_r and E, are given as $k_r^2/2m$ and $k^2/2m$. The background phase is normally assumed to slowly vary with energy, and if this is true, the first and second derivatives with respect to k should be small. The second of Eqns. (6) shows that $\Gamma = (2k/m)/(d\delta/dk)$ for $k = k_r$, and the last shows that $d^2\delta_r/dk^2$ is greater than zero for $k = k_r$.

The analog of Eqns. (4) can also be written for nonzero angular momentum, l. In this case, we have for $\Delta = \delta_l + \gamma_l$:

$$\Delta' = -U(r)D_l^2 \sin^2(\Delta)/k + \gamma_l',$$

and the derivatives of Δ with respect to k are given by taking derivatives of the magnitude, D_l and the phase γ_l, of the Riccati-Bessel functions, j_l and n_l, and the total phase, Δ. The phase of the Riccati-Bessel functions is zero, but the magnitude is unbounded at $r = 0$. Since Δ increases slowly at small r, the integration may be safely started for r where j_l has a value greater than a given threshold. The r values corresponding to this threshold can be given by the small argument expansion of j_l.

Multichannel variants of the phase equation can also be defined, and this is where supercomputers would become attractive for implementing the method. Calogero defines a first order differential equation for the reactance matrix, **R**, and the scattering matrix, **S**. It is more enlightening to consider the radial evolution of **S**. **S** is complex, but it is bounded and S(r) shows directly the evolution of the transition probabilities as r is increased. This can display the radial regions where coupling is strong. The differential equation for S(r) is Eqn (7):

$$\mathbf{S}'(r) = i[\mathbf{S}(r)\mathbf{H}^{(1)}(r) - \mathbf{H}^{(2)}(r)]\mathbf{U}(r)[\mathbf{H}^{(1)}(r)\mathbf{S}(r) - \mathbf{H}^{(2)}(r)], \tag{7}$$

where $\mathbf{S}(0) = \mathbf{I}$, and the value of **S** for large radius is the asymptotic scattering matrix. The diagonal matrices $\mathbf{H}^{(1)}(r)$ and $\mathbf{H}^{(2)}(r)$ have elements given by the Riccati-Hankel functions.

We have:

$$H^{(1,2)}(r) = -N(r) \pm i J(r),$$

where the diagonal matrices J(r) and N(r) have elements, $j_{li}((2mE_i)^{½} r)/(2mE_i)^{¼}$ and $n_{li}((2mE_i)^{½} r)/(2mE_i)^{¼}$, where the j_{li} and n_{li} are Riccati-Bessel functions of order given by the orbital angular momentum for the ith channel. The E_i are the asymptotic values for the kinetic energy in the ith channel, $E - \varepsilon_i$. It is important to note that U(r) is defined differently in Eqn. (7) than in Eqn. (1). The diagonal elements of U in Eqn. (7) approach zero for large r, while they approach the channel energies, ε_i, in Eqn. (1). The diagonal elements of U in Eqn. (7) are the Eqn. (1) values $U_{ii} - \varepsilon_i$. In both cases the off diagonal elements approach zero for large r. Eqn. (7) is nominally a system of N^2 differential equations. However, the S matrix is a unitary matrix so only $N(N+1)/2$ equations need to be integrated, if the lower diagonal elements of S are chosen to satisfy the unitarity condition. The advantage of this procedure would be that the unitarity of S is assured.

The $H^{(1,2)}$ matrices in Eqn. (7) are expressed in terms of the system energy to direct the calculation of the energy derivatives. Derivatives with respect to energy are required in the multichannel case because an energy change will increase each channel energy by the same amount, but an increase in the wave number for one channel would not correspond to the same increase for the other channels. The differential equation for the energy derivatives of S can be written as follows:

$$\begin{aligned} dS'/dr &= i[S'H^{(1)} + SH^{(1)\prime} - H^{(2)\prime}]U[H^{(1)}S - H^{(2)}] \\ &\quad + i[SH^{(1)} - H^{(2)}]U[H^{(1)\prime}S + H^{(1)}S' - H^{(2)}], \\ dS''/dr &= iS''H^{(1)} + 2S'H^{(1)\prime} + SH^{(1)\prime\prime} - H^{(2)\prime\prime}]U[H^{(1)}S - H^{(2)}] \\ &\quad + i[SH^{(1)} - H^{(2)}]U[H^{(1)\prime\prime}S + 2H^{(1)\prime}S' + H^{(1)}S'' - H^{(2)\prime\prime}] \\ &\quad + 2i[S'H^{(1)} + SH^{(1)\prime} - H^{(2)\prime}]U[H^{(1)\prime}S + H^{(1)}S' - H^{(2)\prime}], \end{aligned} \tag{8}$$

where the single and double primes indicate first and second derivatives with respect to energy. Integration of the system of differential equations defined by Eqns. (7) and (8) will give the S matrix and its first and second energy derivatives in the asymptotic region.

There are nominally $3N^2$ equations to be integrated in Eqns. (7) and (8). It would be desirable to be able to integrate only the energy dependence of desired S_{ij} in addition to determining the S matrix elements. In this way a small fraction of Eqns. (8) need be integrated. Numerical experimentation needs to be made with Eqns. (8) to determine the effect of neglecting elements of S′ and S″. If these matrices can be completely ignored on the right sides of Eqns. (8), then a simple quadrature will give asymptotic S′ and S″. The calculations will be simpler if only a small number of S'_{ij} and S''_{ij} must be carried on the right side of Eqns. (8).

The results can now be applied to resonances in multichannel problems. For a particular S matrix element, S_{ij}, that described transitions between states i and j, we now have its energy derivatives. If we write this element as $S = \rho\exp(2i\delta)$, where ρ and δ are real, it is easy to show the following relationships:

$$d\delta/dE = (½)Im\{d\ln S/dE\},$$

$$d\ln\rho/dE = \mathrm{Re}\{d\ln S/dE\},$$

$$d^2\delta/dE^2 = (½)[\mathrm{Im}\{(1/S)(d^2S/dE^2) - (d\ln S/dE)(d\ln\rho/dE)\}$$

$$-2\mathrm{Re}\{(d\ln S/dE)(d\delta/dE)\}]$$

$$d^2\ln\rho/dE^2 = \mathrm{Re}\{(1/S)(d^2S/dE^2) - (d\ln S/dE)(d\ln\rho/dE)\}$$
$$+ 2\mathrm{Im}\{(d\ln S/dE)(d\delta/dE)\}.$$

The energy derivatives of δ are useful to characterize the position and width of resonances. They also allow construction of a time delay matrix for the collision process.

5. Applications

In this section some the variable phase method is applied to the calculation of resonances of a single channel model problem, and the resonances predicted by the DIVAH model for collinear reactive scattering on the Karplus-Porter potential surface.

5.1 RESONANCES IN A MODEL POTENTIAL

To initially investigate the use of the variable phase method in characterizing resonances, the simple potential in atomic units: $V(r) = 7.5r^2\exp(-r)$, was used with a reduced mass of 1. The potential has been used previously in investigations of resonances[9,18], and it has a maximum value of 4.06 at $r = 2$. Eqns. (4) have been integrated for this system, and the results are presented in Figures 1, 2, and 3. The phase shift is given for 1000 energy points in Figure 1. One resonance is obvious near and energy of 3.4 and a broad resonance is indicated near 4.8. The first and second derivatives with respect to k (Figures 2 and 3) clearly show the resonance features at the two energies. The arctangent function is used as a nonlinear mapping to allow display of the large derivatives corresponding to the sharp resonance and the much smaller derivatives for the broad resonance. The phase shift for an energy of 0.0001 is -0.1212, and Eqn. (5), predicts an extrapolated value of 0.000015 for zero energy. Clearly there are no bound states in the potential.

A program has been written to extract the position and width of resonances from the derivatives calculated from Eqn. (5). The second derivatives are calculated with a coarse energy grid over the energy region of interest. The approximate positions of the resonances are located where the second derivative changes sign from positive to negative values. A finer grid is then used in the vicinity of each appropriate zero crossing. If the latter grid indicates that the resonance is very sharp, then the position is located from the second derivative zero crossing by bisection. The first derivative at the zero crossing yields Γ. If the resonance is not very sharp, its position and width are found by fitting the last of Eqns. (6) to the second derivatives obtained with the second grid.

The root finding program finds the energies of the two resonances at 3.426 and 4.852, and the corresponding widths (FWHM), Γ, are 0.0255 and 2.199. The broad resonance has an energy above the barrier in the potential.

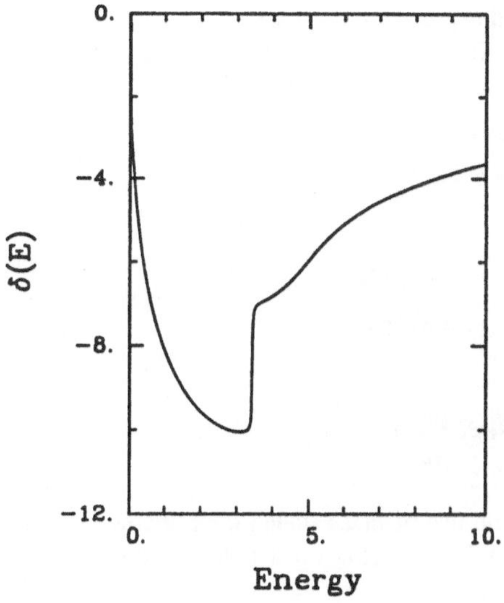

Figure 1. Phase shift in radians for the model potential. Energy in atomic units.

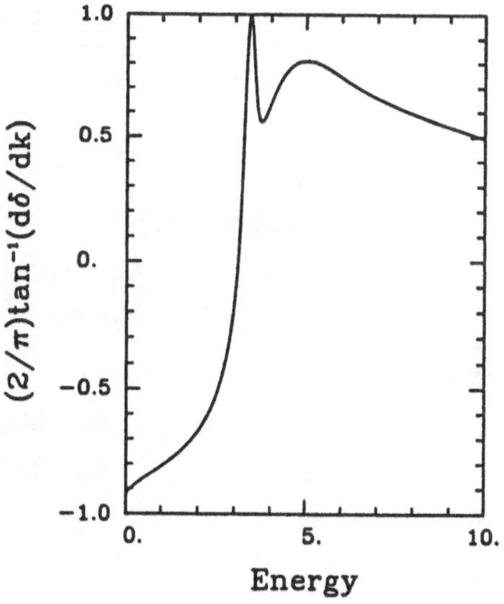

Figure 2. Scaled arctangent of the first derivative of the phase shift with respect to k for the model potential.

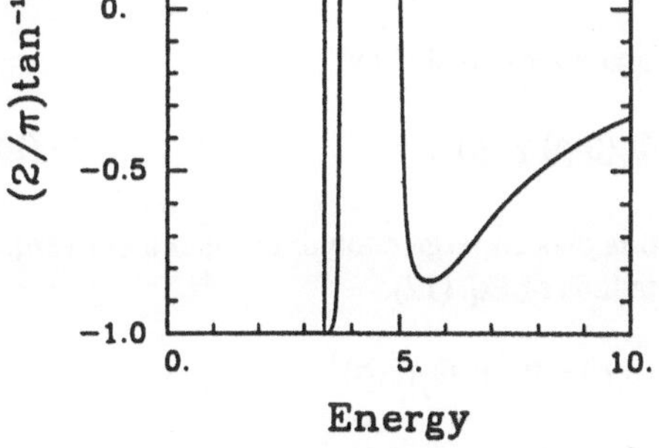

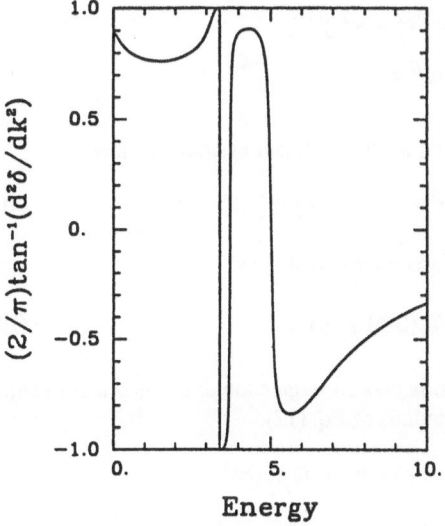

Figure 3. Scaled arctangent of the second derivative of the phase shift with respect to k for the model potential.

5.2 RESONANCES IN THE DIVAH MODEL FOR HLH REACTIONS

An atom-diatom collinear system can be described in terms of the diatom separations, r'_α and r'_γ, and the distances of the atoms from the diatom center of mass, R'_α and R'_γ. The use of Jacobi mass scaled coordinates results in a kinetic energy term with the single mass

$$\mu = \left[\frac{M_A M_B M_C}{M_A + M_B + M_C} \right]^{1/2} .$$

These coordinates are defined as

$$R_\alpha = a_\alpha R'_\alpha \qquad r_\alpha = a_\alpha^{-1} r'_\alpha$$
$$R_\gamma = \alpha_\gamma R'_\gamma \qquad r_\gamma = a_\gamma^{-1} r'_\gamma$$

where

$$a_\alpha = (\mu_{A,BC}/\mu_{BC})^{1/4} ,$$
$$a_\gamma = (\mu_{C,BA}/\mu_{BA})^{1/4} .$$

The subscripted μ terms are reduced masses. The use of mass scaled coordinates casts the 3 body collinear problem into that of a single particle moving on a potential energy surface skewed by the angle β,

$$\beta = \tan^{-1} \left[\left[\frac{M_B(M_A+M_B+M_C)}{M_A M_C} \right]^{1/2} \right] .$$

The angle β becomes smaller as the central atom mass, M_B, decreases relative to the masses of A and C. Collinear hyperspherical coordinates are defined in terms of the mass scaled coordinates as;

$$\rho = (r_\alpha^2 + R_\alpha^2)^{1/2} = (r_\gamma^2 + R_\gamma^2)^{1/2} ,$$
$$\tan(\theta) = r_\alpha / R_\alpha , \tag{9}$$

where the range of the angle, θ, is from 0 to β.

In polar coordinates the Schrödinger equation for the collinear system becomes:

$$\left[\frac{1}{\rho} \frac{\partial}{\partial \rho} (\rho \frac{\partial}{\partial \rho}) + \rho^{-2} \frac{\partial^2}{\partial \theta^2} + k^2 - 2\mu V(\rho,\theta) \right] \Psi(\rho,\theta) = 0 , \tag{10}$$

where $k^2 = 2\mu E$. The wavefunction is expanded and approximated as;

$$\Psi(\rho,\theta) = \rho^{-1/2} \sum_{j=1}^{n_c} \Phi_j(\theta,\rho) \chi_j(\rho) . \tag{11}$$

where the number of terms, n_c, in the summation is chosen large enough to yield a converged scattering calculation. If the $\Phi_j(\theta,\rho)$ are eigenfunctions of Eq. (12):

$$\left[\frac{-1}{2\mu\rho^2} \frac{\partial^2}{\partial \theta^2} + V(\rho,\theta) \right] \Phi_j(\theta,\rho) = W_j(\rho) \Phi_j(\theta,\rho) , \tag{12}$$

then Eq. (13) is found after multiplying by $\Phi_i^*(\theta,\rho)$ and integrating over θ.

$$\left[\frac{d^2}{d\rho^2} + \frac{1}{4\rho^2} + k^2 - 2\mu W_i(\rho) \right] \chi_i(\rho) = -\sum_j (2P_{ij} \frac{\partial}{\partial \rho} + Q_{ij}) \chi_j(\rho) , \tag{13}$$

where the elements of the **P** and **Q** matrices are given by,

$$P_{ij} = \int_0^\beta \Phi_i^*(\theta,\rho) \frac{\partial \Phi_j(\theta,\rho)}{\partial \rho} d\theta \ , \tag{14a}$$

$$Q_{ij} = \int_0^\beta \Phi_i^*(\theta,\rho) \frac{\partial^2 \Phi_j(\theta,\rho)}{\partial \rho^2} d\theta \ . \tag{14b}$$

Romelt [19,20] has presented an approximate method called the Diagonal corrected Vibrational Adiabatic Hyperspherical (DIVAH) model. In the DIVAH model all but the diagonal terms of **P** and **Q** are ignored. Now the coupled equations in Eq. (13) become a set of n_c uncoupled equations;

$$[\frac{d^2}{d\rho^2} - U_i(\rho) + k^2]\,\chi_i(\rho) = 0 \ ,$$

where the potential energy is given as ;

$$U_i(\rho) = 2\mu W_i(\rho) - \frac{1}{4\rho^2} - Q_{ii}(\rho) \ . \tag{15}$$

The Q_{ii} term is the DIVAH correction. Only Q_{ii} is needed because the P_{ii} are zero. The Q_{ii} are easily found in terms of the overlap of the angular eigenfunctions, Φ_i in different intervals. We have

$$Q_{ii}(\rho) = \frac{1}{2h^2}[\left\langle \Phi_i^{n+1} \mid \Phi_i^{n-1} \right\rangle - 1] \ .$$

Since the set of $\Phi_i(\theta,\rho)$ are orthonormal the value of Q_{ii} is always negative or zero. When this is used in the equation for $U_i(\rho)$, it introduces a repulsive term in the potential.

Each DIVAH potential (Eqn. (15)) is numbered with an integer, n, ($n = i$) with a lowest value of 0. Two types of resonances are characterized by the bound states and the resonances for each n. If a positive energy resonance is found for the nth potential, then a shape resonance is predicted for the full quantum system (Eqn. 13) for reactants with initial excitations corresponding to that n. Reactants corresponding to other n will exhibit Feshbach or closed channel resonances. If a bound (negative energy) state is found for a DIVAH potential, then no system resonance is predicted at that energy for approach on n. However, other reactant vibrational preparations will still exhibit Feshbach resonances.

The shape or open channel resonances are found from integrations of Eqn. (4) and the root finding program described above. This procedure gives both the position and the width of the resonances expected for the full quantum system. The bound state eigenvalues [21] of the DIVAH potentials give the positions of the Feshbach or bound state resonances. No information can be obtained about the width of these resonances from the DIVAH model. The phase shift for zero energy on a DIVAH potential gives the number of bound states in the potential. The results for the symmetric hydrogen isotope exchange reactions are given in Table I and Figures 4 and 5. Comparisons with the full quantum calculations of Mattson and Anderson [22] are also found in the Table and the Figures.

Table I. Resonance Positions[a] and Widths[a] for Collinear Reactions. DIVAH energies and widths determined by variable phase method.

System	Quantum[b] Position	Quantum[b] Width	DIVAH (This work) n[c]	E_r[d]	Γ[d]	$\delta(0)$[e]
HMuH	0.62354	0.00046	0	0.62335	0.00050	0
	1.5715	0.0006	1	1.569	bound	π
	1.704	0.012	1	1.7031	0.0088	
	2.4709	0.00015	2	2.468	bound	π
	2.5967	0.0006	2	2.5949	0.00044	
			3	no shape		2π
			4	no shape		2π
			5	no shape		2π
HHH				0.485	0.136	0
	0.8733	0.023	1	0.8725	0.0137	0
	1.307	0.03	2	1.2936	0.000033	0
	1.748	0.04	3	1.716	bound	π
			3	no shape		
			4	no shape		π
			5	no shape		π
HDH			0	0.497	0.167	0
	0.82	0.01	1	0.813	0.034	0
	1.18	0.06	2	1.1721	0.0026	0
			2	1.2317	0.0051	
	1.53	0.02	3	1.5320	0.00000045	0
			4	no shape		π
			5	no shape		π
HTH			0	0.502	0.178	0
	0.83	<0.01	1	0.793	0.043	0
	1.12	0.04	2	1.1265	0.0052	0
			3	1.4639	0.000024	0
			3	1.490	0.086	
			4	1.847	0.066	π
			5	no shape		π

a. Energies in eV.

b. Quantum calculations of T. Mattson and R. Anderson [22]

c. Numbering of potential curve. The first has $n = 0$.

d. Shape resonance on curve n have values for E_r and Γ. Resonances corresponding to bound states on curve n are labeled. The absence of shape resonance is indicated.

e. $\delta(0)/\pi$ is the number of bound states in curve n.

Figure 4. P_{00} for HMuH (bottom) and HHH (top). Short vertical arrows denote position of DIVAH resonances.

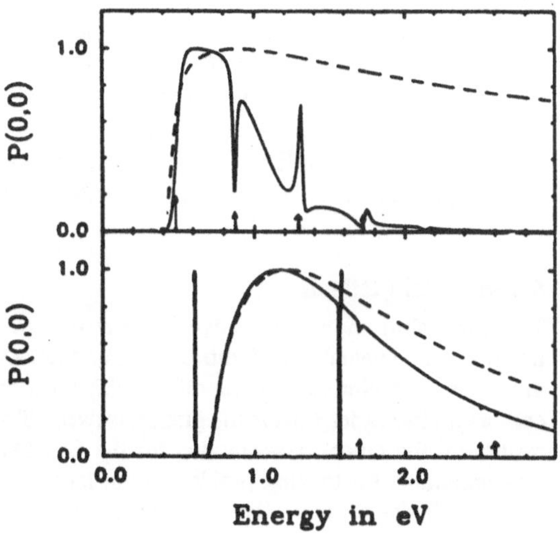

Figure 5. Solid lines are P_{00} for HHH (bottom), HDH (middle), HTH (top). Short arrows show Divah resonances for HDH and HTH.

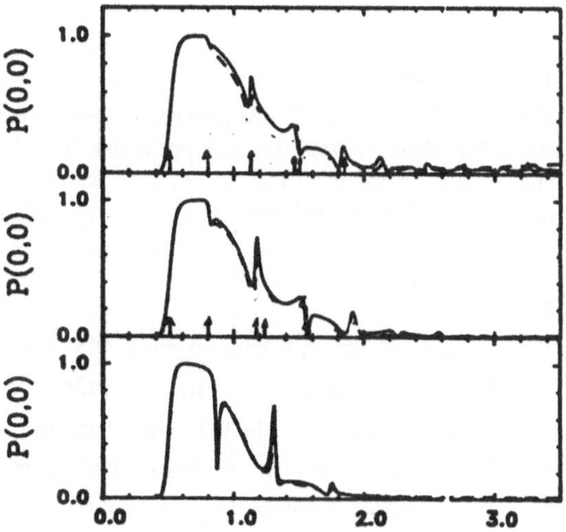

The agreement between the predictions of the DIVAH model and the full quantum calculations is very good as has been noted before. However three new broad resonances have been found with the variable phase method. These are at 0.485, 0.497, and 0.502 eV for HHH, HDH, and HTH respectively[23]. The low energy resonance of HMuH has been known for many years, and it is reasonable to expect $n = 0$ resonances for the other isotopic variants. The new resonances are broad, because they have energies greater than the barrier. The HMuH resonance has an energy below the barrier. Inspection of Figures 4, and 5 shows the evidence for the low energy resonances in the full quantum calculation. The reaction probability for the heavier isotopes has a definite curvature at threshold that is missing in the threshold probability for HMuH at energies just above the sharp resonance.

The usefulness of the extrapolation of phase shifts to zero energy is also found for this application. For example the phase shift for scattering at $E = 0.00307$ eV on the $n = 0$ HTH DIVAH potential is -2.733, and Eqn. (5) predicts a zero energy phase shift of 0.128.

6. Numerical Methods

The phase shift and its derivatives with respect to k for the model potential and the DIVAH potentials are evaluated with the Los Alamos DDRVB2 differential equation solver. This program has the option to treat the differential equations as either stiff or nonstiff. In the former case a variable order Adams integration is used. The Adams integrator was found to give good results for the model potential and for the DIVAH potentials. However, the Gear integrator was preferable for treating problems with large angular momentum. It should be possible to devise an integrator that will provide more efficient integration.

For the single channel problems in this paper, the variable phase methods takes more than ten times longer to calculate an absolute phase shift than an efficient propagator method which calculates a modulo π phase shift.

7. Conclusions

The methods described in this paper are effective for the determination of bound and resonance states for single and multichannel problems. The direct calculation of phase shift derivatives is very useful for the determination of zero energy phase shifts and the positions and widths of resonances. The methods are suitable for use on supercomputers.

8. References

1. R.W. Anderson, J. Chem. Phys. **77**, 4431 (1982).
2. R.W. Anderson, J. Chem. Phys. **77**, 5426 (1982).
3. T.G. Mattson, M.R. Mitchel, R.W. Anderson, Mol. Phys. **50**, 251 (1983).
4. T.G. Mattson and R.W. Anderson, Mol. Phys. **52**, 319 (1984).
5. R.G. Gordon, J. Chem. Phys. **51**, 14 (1969).
6. D. R. Hartree, Proc. Cambridge Phil. Soc. **24**, 105 (1928).
7. B.R. Johnson, J. Chem. Phys. **69**, 4678 (1978).
8. R.W. Anderson, unpublished.
9. N. Moiseyev, P.R. Certain, and F. Weinhold, Mol. Phys. **36**, 1613 (1978).
10. J.N.L. Connor and A.D. Smith, J. Chem. Phys. **78**, 6161 (1983).
11. M. Rittby, N. Elander, and E. Brändas, Phys. Rev. **A24**, 1636 (1981).
12. M. Rittby, N. Elander, and E. Brändas, Int. J. Quantum Chem. **23**, 865 (1983).

13. M. Rittby, N. Elander, and E. Brändas, Int. J. Quantum Chem. **22**, 445 (1982).

14. M. Rittby, N. Elander, and E. Brändas, Chem. Phys. **87**, 55 (1984).

15. D.G. Truhlar and D.W. Schwenke, Chem. Phys. Lett. **95**, 83 (1983).

16. F. Calogero, *Variable Phase Approach To Potential Scattering*, (Academic Press, New York and London, 1967).

17. J.B. Braga and J.N. Murrell, Mol. Phys. **53**, 295 (1984).

18. E. Engdahl, E. Brändas, M. Rittby, and N. Elander, Phys. Rev. **A37**, 3777 (1988).

19. J. Römelt, Chem. Phys. **79**, 197 (1983).

20. J. Römelt, in D. C. Clary (ed.), *The Theory of Chemical Reaction Dynamics*, (D. Reidel Publishing Company, Dordrecht, 1986), p. 77.

21. R.W. Anderson, unpublished.

22. T.G. Mattson and R.W. Anderson, in press J. Phys. Chem.

23. Workshop comment from D. Truhlar that G. Schatz had reported these resonances in his thesis.

ACCURATE DETERMINATION OF POLYATOMIC INFRARED SPECTRA

C.IUNG and C.LEFORESTIER
Laboratoire de Chimie Théorique (UA 506)
Université de Paris-Sud
91405 Orsay
France

ABSTRACT. We present an application of the Lanczos algorithm to the determination of infrared spectra of polyatomic molecules. This method presents two definite improvements on the conventional approach :i)the Lanczos algorithm allows to treat directly ultra-large basis sets (10^5 or more states in this calculation). ii) it is possible to tune selectively the calculation to specific components of the spectrum (e.g. a given combination band $n\nu_i + m\nu_j$), thus allowing an easy assignment of the lines. This algorithm has been implemented on a Cray 2 computer, using both vectorization and multitasking facilities. The method is illustrated on the CD_3H molecule.

1. INTRODUCTION

The absorption spectrum at zero temperature {1} $I_0(\omega)$ is given by

$$I_0(\omega) = \sum_\alpha |\langle\Psi_\alpha|\mu|\Psi_0\rangle|^2 \, \delta(\hbar\omega-\varepsilon_\alpha) \tag{1}$$

where $\{|\Psi_\alpha\rangle\}$ are the molecular eigenstates
$\{\varepsilon_\alpha\}$ the corresponding eigenvalues
and μ the dipole operator responsible for the transitions under consideration.

Starting from the potential energy surface, supposed already known, the calculation of the spectrum implies the determination of the molecular eigenstates and eigenvalues. The direct diagonalization of the hamiltonian matrix is limited to a small subspace (typically ~ 10^3) of the total basis set. Higher excited states must be computed through C.I type techniques {2}.

In this paper, we present an alternative way to compute the spectrum, based on the Lanczos method {3-4}. It should be noted that the Lanczos algorithm has already been applied to the calculation of different spectra {1,5-6}. The novel feature presented in this paper concerns the role devoted to the dipole operator μ entering eq.(1), which allows to selectively determine specific components of the

A. Laganà (ed.), Supercomputer Algorithms for Reactivity, Dynamics and Kinetics of Small Molecules, 251–259.

spectrum and makes a very easy spectroscopical assignment of the computed lines.

2. THE LANCZOS METHOD

One can see from eq.(1) that the amount of informations needed to compute the spectrum is in fact rather limited : only the eigenvalues ε_α and the projections $\langle\Psi_\alpha|\mu|\Psi_0\rangle$ of the ket $\mu|\Psi_0\rangle$ onto the eigenvectors are needed, and not all their components. These informations are provided straightforwardly through the Recursive Residue Generation (RRG) method of Wyatt and coworkers {7-8}, which is based on the Lanczos method.

Defining $\mu|\Psi_0\rangle=|u_0\rangle$ as the initial vector of the Lanczos recursion scheme,

$$\beta_{n+1}\ |u_{n+1}\rangle = (H-\alpha_n)\ |u_n\rangle - \ \beta_n\ |u_{n-1}\rangle \qquad (2)$$

where $\alpha_n = \langle u_n|H|u_n\rangle$, $u_{-1}=0$ and $\langle u_n|u_m\rangle = \delta_{nm}$

one first builds the tridiagonal representation $\underline{\underline{T}}^{(\bar{N})}$ of the hamiltonian matrix in the Lanczos basis set $\{|u_n\rangle,\ n=0,\bar{N}-1\}$,

$$\underline{\underline{T}}^{(\bar{N})} = \begin{bmatrix} \alpha_0 & \beta_1 & & & 0 \\ \beta_1 & \alpha_1 & \ddots & & \\ & \ddots & \ddots & \ddots & \\ & & \ddots & \ddots & \beta_{\bar{N}-1} \\ 0 & & & \beta_{\bar{N}-1} & \alpha_{\bar{N}-1} \end{bmatrix} \qquad (3)$$

where $\bar{N}-1$ is the number of recursions retained. The Scott-Wyatt modified QL algorithm {9} then provides the eigenvalues $\{\varepsilon_\alpha\}$ and the first line of the eigenvector matrix $\{ \langle u_0|\Psi_\alpha\rangle \equiv \langle\Psi_\alpha|\mu|\Psi_0\rangle\}$. It must be emphasized that the order $\bar{N}$ of the tridiagonal matrix can be chosen in general much smaller than the size N of the initial basis set. The reason is that the Lanczos scheme can only unveil the eigenvectors contained in the initial vector $|u_0\rangle$, and that the ket $\mu|\Psi_0\rangle(\equiv |u_0\rangle)$ can be tailored such that it displays non negligible components in a restricted subspace of the molecular basis set, as will be shown below.

In order to assign the calculated lines of the spectrum, we can use for μ any function : It will not change the line positions but only their intensities. Consider first as a simple example the case of a purely harmonic potential

$$V(\underline{q}) = \sum_j f_{jj}\ q_j^2$$

and a dipole function of the form $\mu(\underline{q}) = q_i$. One can see that the Lanczos algorithm (eq.(2)) using $\mu|\Psi_0\rangle$ as initial vector $|u_0\rangle$ will stop at the first recursion, $|u_0\rangle$ being already an eigenvector. As a consequence the tridiagonal T matrix will be of dimension 1, leading to only one eigenket, namely $|1\nu_i\rangle$. Similarly a dipole function of the form $\mu(q)=q_i+q_i^2$ will result in a tridiagonal T matrix of dimension 3, spanned by the eigenvectors $|\Psi_0\rangle$, $|1\nu_i\rangle$ and $|2\nu_i\rangle$.

In an actual case, anharmonicity and intermode coupling allow more eigenstates to contribute to the tridiagonal matrix, but the above basic trends will still be observed :

i) the tridiagonal matrix size $\bar{N}$ can be in general taken much smaller than the size of the molecular basis set, depending of the dipole function.

ii)an ad hoc dipole function will unveil well defined lines in the spectrum. Thus by choosing a dipole function characteristic of a given band $n\nu_i$ (or a combination band $n\nu_i + m\nu_j$), one can assign unambiguously the lines of the spectrum. Numerical examples will be given in section 4. Once all the line positions have been assigned, a final calculation using the exact dipole function μ will provide the desired spectrum.

3. IMPLEMENTATION ON A CRAY COMPUTER

3.1. Handling of the H matrix

For very large systems(10^5 states or more), the hamiltonian matrix $\{H_{ij}\}$ cannot be stored in core memory. Also, defining this matrix as an external file cannot be envisionned, due to its large access time. The way to represent H acting on a vector is guided by our knowledge of the system under study. For example, the vibrational hamiltonian of a molecule is often given using the normal (or local) modes $\{Q_i\}$ as a zero order description :

$$H = \sum_i \{ P_i^2 + 1/2\, V_{ii}\, Q_i^2 \} \qquad (= H^0)$$
$$+ \sum_{ijn} V_{ijn}\, Q_i\, Q_j\, Q_n \quad + \dots \qquad (= W).$$

In a similar way, the total hamiltonian H is split into a zero order hamiltonian H^0 and coupling terms W_i , $H = H^0 + W_1 + W_2 + \dots$, each coupling term being of the form :

$$W_i=\{ A^{(1)}(Q_j) + A^{(2)}(Q_{j'}) + \dots \}\{ B^{(1)}(Q_k) + B^{(2)}(Q_{k'}) + \dots \} \,..\, \{\dots\}$$

The above expansion is quite flexible and has been found very convenient in different types of molecular systems {10-11}.

The basis set used in the calculation is chosen as the eigenstates of H^0:

$$H^0\, |v_1 v_2 \dots v_M\rangle = E(v_1, v_2, \dots, v_M)\, |v_1 v_2 \dots v_M\rangle .$$

Acting H on any given vector

$$|u_n\rangle = \sum_{v_1 v_2 .. v_M} C_{v_1 v_2 .. v_M} |v_1 v_2 \dots v_M\rangle$$

can be obtained by successive applications of the single mode operators $A^{(k)}$ on the vector :

$$A^{(k)} |u_n\rangle = \sum_{v'_k v_k} \sum_{v_1..v_{k-1}} \sum_{v_{k+1}..v_M} A^{(k)}_{v'_k v_k} C_{v_1 v_2 .. v_M} |v_1 v_2 ... v_M\rangle \quad (4)$$

In fact, the A matrices being sparse, only their non zero elements are kept in core memory. For example, in a recent application, although the overall basis set size was 120000, the number of non zero matrix elements to be stored in core memory summed up to 10^5 only. As a result, the core memory needed is typically 7-8 times the size of the basis set used.

The above scheme can be efficiently vectorized by choosing as inner do-loop in eq.(4) the sum over either $v_1..v_{k-1}$ or $v_{k+1}..v_M$ depending on their length. This has been shown to lead to a vectorization ratio in the range 6-10.

3.2. Multitasking {12}

One can take advantage of the above definition of the hamiltonian in order to split the calculation between the different processors. The method used is to associate one task to each W_i term. We report below the CPU usage which has been obtained on a non dedicated machine using 3CPU's. This example corresponds to the HCN {11} molecule, where the potential has been expanded in 18 couplings terms, using 130 Lanczos recursions : The total number of tasks was 2340. The same job ran on a single processor in 185 seconds.

Multitasking CPU Breakdown

	1 CPU	:	15.76 s	=	15.76 s
	2 CPU	:	15.42 s	=	30.85 s
	3 CPU	:	36.75 s	=	110.26 s
	4 CPU	:	0.45 s	=	1.82 s
	-----		--------		------
Average CPU usage :	2.32	:	68.40 s	=	158.69 s

4. CD_3H SPECTRUM CALCULATION

We have used the Lanczos method as presented in section 2. This method proceeds in two steps :

i) The determination of the ground vibrational state $|\Psi_0\rangle$.

ii) The calculation of line positions ε_α and line intensities $I_0(\omega_\alpha)$ as given by eq.(1).

4.1. Molecular Basis Set

The CD_3H molecule possesses six vibrational modes, three of them being doubly degenerate due to the C_{3v} symmetry {13}. The zero-order description used in the calculation consists of the direct product of the single mode basis sets

$$\{|v_1,v_2,v_3,v_4,\ell_4,v_5,\ell_5,v_6,\ell_6\rangle\}$$

$$\equiv\{|v_1\rangle\}x\{|v_2\rangle\}x\{|v_3\rangle\}x\{|v_4,\ell_4\rangle\}x\{|v_5,\ell_5\rangle\}x\{|v_6,\ell_6\rangle\} \quad (5)$$

where the $|v_i>$ states (i=1,2,3) are Morse eigenstates and the $|v_i,\ell_i>$ (i=4,5,6) correspond to doubly degenerate harmonic oscillator eigenstates associated to the bending motions{14}.

In order to reduce the overall basis set size used in the calculation, we have defined three combination (merged) modes, namely $\nu_1+\nu_5$, $\nu_2+\nu_6$ and $\nu_3+\nu_4$, and kept only the combined states $|v_i,v_j,\ell_j>\equiv|v_i>x|v_j,\ell_j>$ located below a given energy threshold E_{ij}. As will be seen in section 4.3., the actual basis set used depends on the band under study : The E_{ij} threshold is taken larger when dealing with the $n\nu_i$ or $m\nu_j$ band. The effective basis set is then the direct product of the three combination basis sets:

$$\{|v_1,v_2,v_3,v_4,\ell_4,v_5,\ell_5,v_6,\ell_6>\}=\{|v_1,v_5,\ell_5>\}x\{|v_2,v_6,\ell_6>\}x\{|v_3,v_4,\ell_4>\}$$

each $\{|v_i,v_j,\ell_j>\}$ basis set being restricted as described above. The maximum basis set size used in the calculation has been 120000 states. This method represents a trade-off between two extreme cases:

i) the direct product case, as in eq.(5) where ultra large basis sets can be handled (see the method discussed in section 3), but displaying a lot of very high energy states.

ii) keeping only the direct product states satisfying a given energy criterion, which implies to build once for all the hamiltonian matrix, and thus drastically reducing the size of the basis set used.

It must be noted that a spectrum calculation requires high energy states to be present in the basis set anyway, the latter case corresponding more to the relaxation from an initial state.

4.2. Ground vibrational state calculation

Experience {11} has shown that high accuracy description of this state is needed in order to obtain the correct intensities $|<\Psi_\alpha|\mu|\Psi_0>|^2$.

The reason is that errors in its description (e.g. $|\Psi_0>=|\Psi_0^{exact}> +|\delta\Psi_0>$) will tend to pollute the computed intensities to the first order, as compared to the second order error on the ε_0 eigenvalue resulting from the variational principle. This error can have a dramatic effect when dealing whith low intensity lines associated to direct high overtone transitions such as $0\nu_1$ --> $n\nu_1$.

The Lanczos algorithm can also be used to solve this problem very efficiently. First we build the tridiagonal $\underline{\underline{T}}$ matrix resulting from using the <u>zero-order</u> ground vibrational state $|0,0,0,0,0,0,0,0,0>$ as initial vector $|u_0>$ of the recursion scheme (see eq.(2)). The lowest eigenvalue ε_0 of this matrix corresponds to the ground state energy. It must be noted that, even we are using the full molecular basis set, the number of Lanczos recursions needed to reach convergence is small (typically 50), as we are interested only in the lowest eigenvalue.

Once this eigenvalue obtained, the corresponding eigenvector $|\Psi_0>$ is then determined using the iterative conjuguate gradient method {15-16} to solve the linear system ($\underline{H}$ - ε_0) $|\Psi_0>$ = 0 . A high accuracy solution ($<\Psi_0|H - \varepsilon_0|\Psi_0> < 10^{-9}$ a.u.) is obtained in 15 iterations.

4.3. Spectrum Calculation

Before using the exact dipole function $\mu(\underline{q})$ in eq.(1), one must first assign unambiguously the energy levels associated to the potential energy surface. Even for a small polyatomic molecule like CD_3H, the number of states lying in any given region of the spectrum is so large that it makes this assignment very difficult : as an illustration, we present on figure 1 the density of vibrational states as a function of the energy.

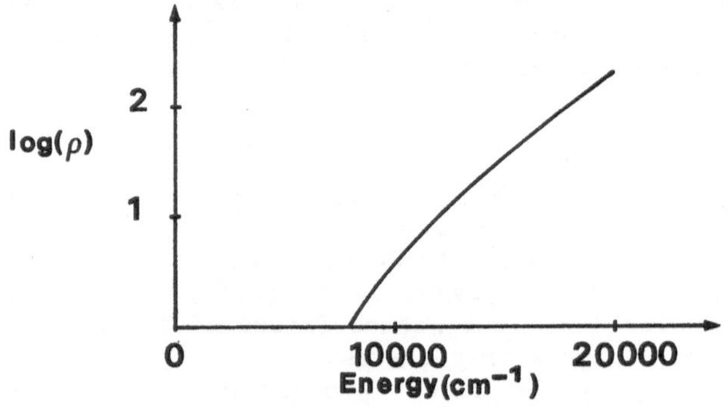

figure 1:Density of states per wavenumber

This difficulty can be circumvented by tailoring the dipole function as discussed in section 2. As an example, figure 2 displays the results obtained by using the dipole function $\mu(\underline{q})=(q_1+q_5)^3$, tailored to observe the $n\nu_1+m\nu_5$ combination band. In this calculation, we have used 400 Lanczos recursions, which leads to line positions converged within $0.1cm^{-1}$. One can see from the figure that the $n\nu_1+m\nu_5$ lines emerge in the spectrum, their associated intensities $\langle n\nu_1+m\nu_5|\mu|\Psi_0\rangle^2$ being at least two orders of magnitude larger than the background lines.

figure 2 :Relative intensities (in arb. units) observed for a dipole function of the form $\mu(\underline{q})=(q_1+q_5)^3$ in eq.(1).

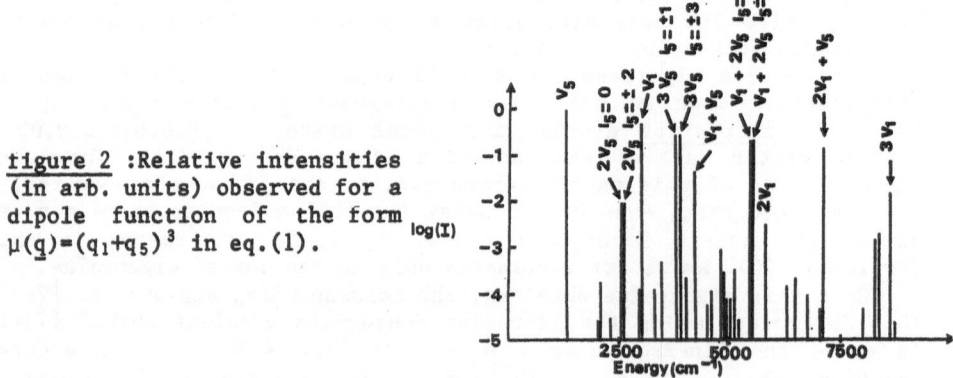

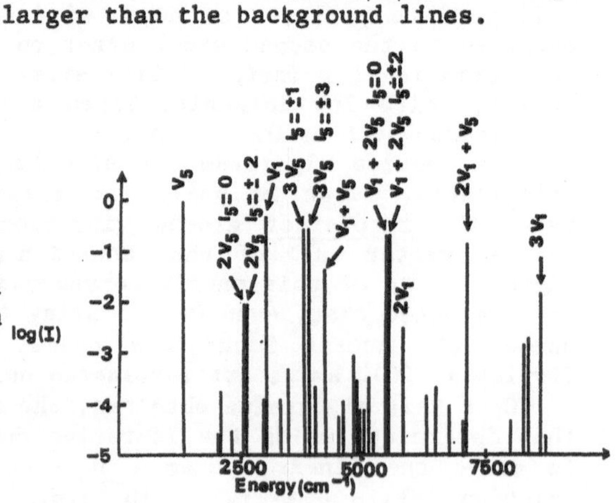

Special emphasis has been put on this mode as numerous experimental studies have been devoted to the C-H chromophore in CX_3H molecules {17-18}. The spectrum obtained with a $\mu(\underline{q}) = q_1$ dipole function is shown in figure 3. Convergence of the $n\nu_1$ levels has necessitated a basis set comprising also highly excited states of the ν_5 , which shows the crucial role played by the 2:1 Fermi resonance between the stretch and bend motions, namely the basis set comprised states up to 9 and 10 quanta in the ν_1 and ν_5 modes respectively. As a result of this Fermi resonance, combination states $n\nu_1 + 2m\nu_5$ appear in this spectrum. On the other hand, these results depend only slightly on the number of states used for the other modes. 1000 Lanczos recursions have been used in this calculation in order to converge the $6\nu_1$ line position.

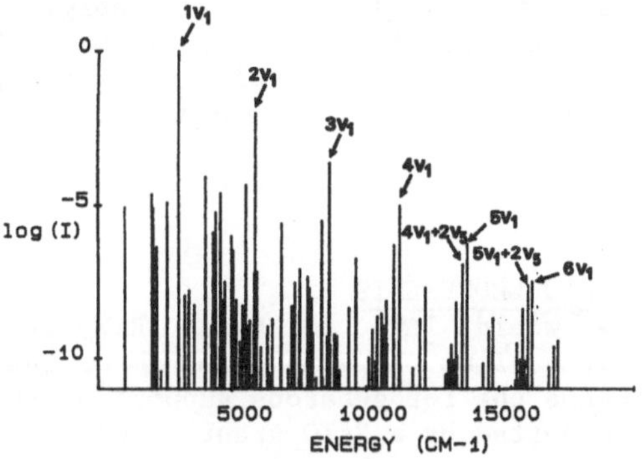

figure 3 :Relative intensities (in arb. units) observed for a dipole function of the form $\mu(\underline{q})=q_1$ in eq.(1).

5. DISCUSSION

In this paper, we have presented an application of the Lanczos method to the IR spectrum calculation of a polyatomic molecule. The power of this approach comes from the use of the Lanczos algorithm, in conjunction with tailored dipole functions, which allows to tune selectively the calculation to some specific components of the spectrum : it is equivalent to split the whole spectrum calculation into smaller, nearly independent parts. This has been shown explicitly in the case of the C-H stretch overtones, which have been converged up to c.a 16000 cm^{-1}, in a region where the density of states is about 50 per cm^{-1}. Contrary to a previous conclusion drawn by Sibert {19}, the Lanczos method is able to provide highly excited overtone line positions, by choosing an ad hoc dipole function in eq.(1), as discussed in sections 2 and 4.3.

The Lanzos algorithm appears ideally suited to handle ultra-large basis set. The reason is that, contrary to standard diagonalization methods, one never has to build explicitly the hamiltonian matrix $\underline{\underline{H}}$: only the effect of $\underline{\underline{H}}$ acting on a given vector is needed. This can be efficiently achieved in terms of single-mode (or merged modes) matrices as has been described in section 3. The method could also

greatly benefit from recent developpements aimed at designing better zero order basis sets such as :
i) the SCF vibrational method {20-21} which provides an optimized single mode basis set description.
ii)the adiabatic DVR method recently developped and applied by Light and coworkers {22-23} which allows to sample efficiently the configuration region of interest
iii)a canonical transformation of the hamiltonian operator, as recently proposed by Sibert {19}

An important application of the method presented here concerns the fitting of the potential energy surface. By comparing the experimental line positions with those arising from a spectrum calculation using a parametrized potential function, one can adjust the parameters until agreement has been reached {24}. Such a method has been used in order to fit a CD_3H potential energy surface, including all vibrational degrees of freedom {25}. We have thus been able to reproduce the x_{ii} and g_{ii} experimental spectroscopical parameters of Gray and Robiette {26} to within $0.5cm^{-1}$ for most of them. A similar fitting procedure could be used for the dipole function, once an accurate potential energy surface is known, by adjusting the lines intensities.

ACKNOWLEDGEMENTS

We would like to thank R.Wyatt and R.Friesner for very helpful discussions. We are grateful to the Centre de Calcul Vectoriel pour la Recherche for generous support of the Cray2 computer time. This work was supported by a NATO grant to Claude Leforestier.

REFERENCES

{1} H.Koeppel, W.Domcke and L.S.Cederbaum, Ad.Chem.Phys. 57, 59 (1984)
{2} See for example the paper by J.Tennyson, S.Miller and B.Sutcliffe in this issue
{3} C.Lanczos, J.Res.Natl.bul.Std. 45, 255 (1950)
{4} J.K.Cullum and R.A.Willoughby, Lanczos algorithms for large Symmetric Eigenvalue Computations, Vols.I,II, Birkhaüser,Boston,1985
{5} G.Moro and J.H. Freed, J.Chem.Phys. 74, 3757(1981)
{6} G.Moro and J.H. Freed, Large-Scale Eigenvalue Problems edited by J.K.Cullum and R.Willoughby, Mathematics Studies Series, Vol.127, North-Holland, Amsterdam,1986
{7} A.Nauts and R.E.Wyatt, Phys.Rev.Letters 81, 2238 (1983); A.Nauts and R.E.Wyatt, Phys.Rev. A30, 872 (1984)
{8} R.E.Wyatt, Ad.Chem.Physics, to be published
{9} R.E.Wyatt and D.S.Scott,op.cit. ref6
{10} R.A.Friesner, J.P.Brunet, R.E.Wyatt and C.Leforestier, Int.J.Supercomputer Applications 1, 9(1987); R.A.Friesner,R.E.Wyatt,C.Hempel and B.Crimer, J.Comp.Phys. 64, 220 (1986)

{11} J.P.Brunet,R.A.Friesner,R.E.Wyatt and C.leforestier to be published
{12} Cray-2 Multitasking Programmer's Manual SN-2026
{13} S.Califano, Vibrational States, Wiley, New York, 1976
{14} P.Barchewitz,Spectroscopie Atomique et Moléculaire, Masson, Paris,1971
{15} E.Strang, Introduction to applied Mathematics, pp419-422, Wellesley-Cambridge,Wellesley, 1986
{16} K.V.Vasavada,D.J.Schneider and J.H Freed, J.Chem.Phys. 86, 647(1987)
{17} J.S.Wong,W.H.Green,C.Cheng,C.B.Moore, J.Chem.Phys. 86, 5994(1987);
S.Peyerimhoff,M.Lewerenz and M.Quack, Chem.Phys.Letters 109, 563(1984);
J.Segall,R.N.Zare,H.R.Dübal,M.Lewerentz and M.Quack, J.Chem.Phys. 86, 634(1987);
M.Lewerenz and M.Quack, Chem.Phys.Letters 123, 197(1986);
H.R.Dübal and M.Quack, J.Chem.Phys. 81, 3779(1984);
H.R.Dübal and M.Quack, Chem.Phys.Letters 72, 342(1980);
A.Amrein,H.R.Dübal,M.Lewerenz and M.Quack, Chem.Phys.Letters 112, 387(1984);
J.E Bagott,M.Chuang,R.N.Zare,H.R.Dü bal and M.Quack, J.Chem.Phys. 82, 1186(1985);
H.R.Dübal and M.Quack, Molecular Physics 53, 257(1984);
A.Campargne and F.Stoeckel, J.Chem.Phys. 85, 1220(1986)
{18} J.W.Perry,D.J.Moll,A.Kuppermann and A.H.Zewail, J.Chem.Phys. 82, 1185(1985);
G.A.Voth,R.A.Marcus and A.H.Zewail, J.Chem.Phys. 81, 5494(1984)
{19} E.L.Sibert, J.Chem.Phys. 88, 4378(1988)
{20} F.L.Tobin and J.M.Bowman, Chem.Phys. 47, 151(1980);
M.A.Ratner and R.B.Gerber, J.Phys.Chem. 90, 20(1986)
{21} R.B.Gerber and M.A. Ratner, Ad.Chem.Phys. 70,97 (1988)
{22} J.C.Light,I.P.Hamilton and J.V.Lill, J.Chem.Phys. 82, 1400(1985);
{23} Z.Bacic and J.C.Light, J.Chem.Phys. 85, 4594(1986)
{24} J.N.Murrell,S.Carter,S.C.Farantos,P.Huxley and A.J.C.Varandas, Molecular potential energy function, J.Wiley,New York, 1984
{25} C.Iung and C.Leforestier ,J.Chem.Phys., submitted for publication
{26} D.L.Gray and A.G.Robiette, Molecular Physics 37, 1901(1979)

THE CALCULATION OF RO-VIBRATIONAL SPECTRA USING SUPERCOMPUTERS

Jonathan Tennyson and Steven Miller,
Department of Physics and Astronomy,
University College London,
London WC1E 6BT, U.K.

Brian T. Sutcliffe,
Chemistry Department,
University of York,
Heslington, York YO1 5DD, U.K.

ABSTRACT. Developments in the use of supercomputers for the generation of first principles, synthetic, ro-vibrational spectra for triatomic molecules are discussed. Refinements to a previously published two-step procedure for calculating rotationally excited states are given. The algorithm presented is shown to drive a single Cray-XMP processor at over 140 Mflops. A sample H_2D^+ pure rotational spectrum and line strengths for H_3^+ 'hot bands' are presented. The experimental implications of these results are briefly discussed.

1. INTRODUCTION

It is now 10 years since Carney, Sprandel and Kern (1) published their much cited review of variational ro-vibrational calculations on triatomic systems. It is therefore interesting to consider how the subject has progressed in the intervening period and in particular to focus on the new areas of theoretical spectroscopy that can now be explored with modern supercomputers. At the time of the review in 1978, it was taken for granted that the Eckart Hamiltonian was the one to choose for studying the nuclear motions of polyatomic systems. It was further widely assumed that the role of electronic structure calculations in the solution of the nuclear motion problem was to obtain force constants and rotational constants to be used in perturbation-theoretic analysis.

In pioneering work, Carney and Porter (2,3) showed that for small systems performing large amplitude motions, perturbation theory was inadequate. Their approach involved computing an extensive electronic potential energy surface for which bound states were found variationally using a basis set expansion.

In this article we consider current state-of-the-art ro-vibrational calculations. We focus on the H_3^+ molecule and its isotopomers, not only because this was the system studied by Carney and Porter, but also because

A. Laganà (ed.), Supercomputer Algorithms for Reactivity, Dynamics and Kinetics of Small Molecules, 261–270.

it provides a severe test of any approach. Not only is there highly accurate experimental data for comparison, but there remains as a challenge to theory the unassigned near-dissociation spectra of Carrington and Kennedy (4).

Supercomputers have played an important part in the calculation of potential energy surfaces of high accuracy for systems like H_3^+ (5). Here we will concentrate on the results of nuclear motion calculations on the corrected MP-7/87CGTO surface of Meyer et al. (6).

It is now accepted that the Eckart Hamiltonian is not reliable for large amplitude nuclear motions. Instead we have used Hamiltonians formulated in internal coordinates to obtain results of outstanding accuracy for triatomic systems.

Thus we have reproduced the vibrational fundamentals of H_3^+ and its isotopomers to better than 0.1% (7). Rotational constants, including distortion and Coriolis coupling constants, have been calculated to an accuracy competitive with experiment, and which can only be represented by going to sixth-order perturbation theory (7). Detailed predictions of the band origins and rotational structure of the doubly-excited states of H_3^+ (8) has led to the first experimental assignments involving these states (9,10).

The calculations have also been extended to states with very high total angular momentum ($J \sim 30$) covering the breakdown of vibration-rotation separation (11,12) and onto the point at which there are no true bound states $J \sim 50$ (13). Dipole transition moment calculations have been used to synthesise spectra (14,15). In particular, we have predicted the "forbidden" pure rotational spectrum of H_3^+ (14) refining previous estimates which were based on perturbation theory (16). These results are now being used in the search for H_3^+ in interstellar clouds.

In what follows we shall show how algorithms have been developed for supercomputers to yield these results. It will be clear that such calculations would not have been possible without their use.

2. THEORY

The programs (17,18) used to solve the ro-vibrational problems discussed here are based substantially on two theoretical developments: the use of a flexible body-fixed coordinate system (19) and a two-step variational method for rotationally-excited states (12). The problem has been formulated as a generalised body-fixed Hamiltonian which allows internal coordinates and the orientation of the axis system to be input parameters in the programs. This Hamiltonian has the form:

$$\hat{H} = \hat{K}_V^{(1)} + \hat{K}_V^{(2)} + \hat{K}_{VR}^{(1)} + \hat{K}_{VR}^{(2)} + V(r_1, r_2, \theta) \qquad (1)$$

where the subscript V denotes vibrational only and VR vibration-rotation. For systems with zero total angular momentum, J = 0, the $\hat{K}_{VR}$ operators vanish. In the special case of atom-diatom scattering coordinates, with which we will concern ourselves here, the operators $\hat{K}_V^{(2)}$ and $\hat{K}_{VR}^{(2)}$ vanish. The exact form of these operators can be found elsewhere (19).

In Eq.(1) $V(r_1,r_2,\theta)$ represents the electronic potential in internal coordinates. In the present method, the eigenfunctions which are solutions to Eq.(1) are expressed in terms of a set of basis functions which are products of angular, rotational and radial parts.

The angular and rotational parts of the wavefunction are represented by Condon and Shortley $\Theta_{jk}(\theta)$ functions and Wigner rotation functions $D^J_{M,k}(\alpha,\beta,\gamma)$, where J is the total angular momentum and k the projection of J on the body-fixed z-axis. The problem is split into two symmetry blocks by considering symmetric and antisymmetric combinations of functions; these blocks are labelled $p = 0$ and 1 respectively. In this representation, the total parity of the wavefunction is given by $(-1)^{J+p}$.

In our calculations, the radial parts of the wavefunction, $\phi_m(r_1)$ and $\phi_n(r_2)$ are represented by Morse-oscillator-like functions, with parameters which can be variationally optimised to minimise the expectation value of $\hat{H}$.

The total wavefunction representing the Lth solution of $\hat{H}$ can thus be written:

$$|J,p,L\rangle = \sum_{k=p}^{J} \sum_{j,m,n} d^{JpL}_{kjmn} |j,k,m,n\rangle \tag{2}$$

with eigenenergy $E^{J,p}_L$.

The two-step variational method consists of first obtaining a set of functions

$$|k,i\rangle = \sum_{j,m,n} c^{Jki}_{jmn} |j,k,m,n\rangle \tag{3}$$

which are eigenfunctions of the Hamiltonian

$$\hat{H}_k = \hat{K}^{(1)}_V + \hat{K}^{(2)}_V + \delta_{kk'} \hat{K}^{(1)}_{VR} + V(r_1, r_2,\theta) \tag{4}$$

with energy ε_{ki}. The assumption that k is a good quantum number is equivalent to the neglect of off-diagonal Coriolis interactions which couple functions involving k with $k' = k \pm 1$.

The $|k,i\rangle$ can then be used as a new basis set for the full problem, which now has eigenfunctions given by

$$|J,p,L\rangle = \sum_{k,i} b^{JpL}_{ki} |k,i\rangle . \tag{5}$$

It can easily be seen that

$$d^{JpL}_{kjmn} = \Sigma_i \, b^{JpL}_{ki} \, c^{Jki}_{jmn} . \tag{6}$$

3. COMPUTATIONAL CONSIDERATIONS

Several efficient algorithms have been suggested for obtaining variational solutions to vibrational Hamiltonians such as $\hat{H}_k$ (20-22). Generally for triatomic systems, the time taken to construct the secular matrix is dwarfed by that required to obtain the eigenvalues and eigenvectors of interest. As diagonalisation is inherently a matrix operation, such procedures are well suited to supercomputers.

The two-step procedure outlined above has the following advantages:

(i) Not all the eigenvectors of $\hat{H}_k$ are required to converge solutions for the full problem;

(ii) The secular matrix produced by $\hat{H}$ acting on $|k,i\rangle$ has a sparse-blocked structure which has great computational advantages.

In addition, if the no-Coriolis approximation is a good one and k is nearly a good quantum number - as is the case for most van der Waals complexes - the secular matrix is dominated by the ε_{ki} and can be rapidly diagonalised. When the Coriolis terms are significant the second variational step dominates the computational requirements for anything but very low J.

The second variational step calculation falls naturally into two parts: construction and diagonalisation of the secular matrix. From Eqs.(1),(4) and (5) it can be shown that

$$\langle k',i'|\hat{H}|k,i\rangle = \delta_{kk'}\,\delta_{ii'}\,\varepsilon_{ki} + \delta_{k\pm1,k'}\,\langle k',i'|K_{VR}^{(1)} + K_{VR}^{(2)}|k,i\rangle \tag{7}$$

$\underline{H}$ can thus be stored as a vector of diagonal elements containing the ε_{ki} and a series of matrices containing the off-diagonal elements. The matrix containing elements between $|k,i\rangle$ and $|k+1,i'\rangle$ will be denoted by $\underline{O}^k$. As the secular matrix is symmetric, it is not necessary to compute the matrix elements between $|k,i\rangle$ and $|k-1,i'\rangle$. There are J-p $\underline{O}^k$ blocks, and only storing these saves a factor of approximately J in storage compared to storing the whole secular matrix. As the $|k,i\rangle$ are usually chosen as basis functions on the criterion of energy ordering (23), the $\underline{O}^k$ are in general rectangular matrices.

The algorithm used for constructing and diagonalising the secular matrix may be summaried:

1. Read in all ε_{ki} and select N lowest.

2. Read integrals over vibrational basis functions.

3. For $k = 0$ to J
   a) Read ε_{ki} and c_{jmn}^{Jki}
   b) If $k > 0$: form $\underline{O}^{k-1}$ and write to disk
   c) Copy c_{jmn}^{Jki} to c_{jmn}^{Jk-1i}

4. For $p = 0$ and $p = 1$
   a) Read the $\underline{O}^k$ from disk
   b) Diagonalise the secular matrix to obtain E_L^{Jp} and b_{ki}^{JpL}
   c) Compute d_{kjmn}^{JpL} Eq.(6)
   d) Write E_L^{Jp} and d_{kjmn}^{JpL} to disk.

In this algorithm the integrals needed to form the off-diagonal matrix elements are computed in the first variational step. Step 4c is included to transform the eigenvectors back to coefficients of the original vibrational basis set. These coefficients can then be used in subsequent transition moment calculations.

This algorithm includes some recent developments. Writing O^k to a scratch file allows the secular matrix to overlap the vectors $\underline{c}^{Jki}$ and $\underline{c}^{Jk-1i}$ saving up to 40% of the fast store in some cases. As the p = 1 secular matrix corresponds to the p = 0 matrix with blocks involving k = 0 removed, the results for both parities can be obtained from a single matrix.

The rate limiting steps in these algorithms are the construction of the secular matrix, step 3b, and diagonalisation, step 4b. It is thus worth considering these in some detail.

If the calculation is being performed in scattering coordinates (r_1 is the diatom bondlength, r_2 the distance from the diatomic centre-of-mass to the third atom and θ the angle between $\underline{r}_1$ and $\underline{r}_2$) then the integrals linking the blocks with k - k' = 1 have a particularly simple form in terms of the original vibrational basis (19).

$$I^{k,k+1}_{jmn,j'm'n'} = \delta_{jj'}\,\delta_{mm'}\,C^{+}_{Jk}\,C^{+}_{jk}\,\langle n|\frac{1}{2\mu_2 r_2{}^2}|n'\rangle \tag{8}$$

if $\underline{r}_2$ is the body-fixed z-axis and

$$I^{k,k+1}_{jmn,j'm'n'} = \delta_{jj'}\,\delta_{nn'}\,C^{+}_{Jk}\,C^{+}_{jk}\,\langle m|\frac{1}{2\mu_1 r_1{}^2}|m'\rangle \tag{9}$$

with z embedded along $\underline{r}_1$. In these expressions

$$C^{+}_{Jk} = \left[J(J+1) - k(k+1)\right]^{1/2}; \tag{10}$$

μ_1 and μ_2 are the appropriate reduced masses (19); $|m\rangle$ is a radial basis function in the r_1 coordinate and $|n\rangle$ a function in the r_2 coordinate.

Transforming these integrals to matrix elements between basis functions of the second variational step, $|k,i\rangle$, gives

$$O^{k}_{i,i'} = \sum_{jmn}\sum_{j'm'n'} c^{Jki}_{jmn}\,c^{Jk+1i'}_{j'm'n'}\,I^{k,k+1}_{jmn,j'm'n'}\;. \tag{11}$$

Because of the structure of $\underline{I}$, the most efficient way of constructing O^k is to make the loops over i and i' the innermost ones which allows them to vectorise. The innermost operation can thus be written

$$O^{k}_{i,i'} = O^{k}_{i,i'} + p_i\,q_{i'} \tag{12}$$

where

$$p_i = c^{Jki}_{jmn}\,I^{kk+1}_{jmn,j'm'n'},\quad q_{i'} = c^{Jk+1i'}_{j'm'n'}. \tag{13}$$

This operation dominates the matrix construction time.

Table 1. Cray-XMP single processor times in seconds and Mflops rates for constructing and diagonalising a 2400 dimensional secular matrix for H_2D^+ with J = 7.

		Vectorisation	Matrix construction	Diagonalisation[a] p = 0	p = 1	Total Mflops
ROTLEV (17)		OFF	1848.8	1711.4	1319.4	5
		ON	236.9	131.1	100.0	59
plus FORTRAN	MXMB	OFF	1536.9	1698.7		6
		ON	110.3	134.7	108.6	32
CAL	MXMB	ON	98.6	56.1	43.2	141
same using	CFT77	ON	100.6	54.6	41.6	142

[a]Lowest 20 eigenvalues for each symmetry to a tolerance of approximately $0.001 cm^{-1}$.

Table 1 compares timings for (a) programming the loops as discussed above, (b) replacing Eq.(12) by a call to a vector-matrix multiply routine MXMB and (c) replacing MXMB by a CAL (Cray Assembler Language) equivalent routine (25). Algorithm (a) was used in the previously published version of this program, ROTLEV (17).

The blocked structure of the secular matrix lends itself naturally to iterative diagonalisation procedures. The key step in the iterative process (26,27) is the multiplication of the secular matrix by an image vector ($\underline{z}$):

$$w_r = \sum_{s=1}^{N} H_{rs} z_s . \qquad (14)$$

Employing the structure of $\underline{H}$ means that this can be written

$$w_{k,i} = \varepsilon_{ki} z_{k,i} + \sum_{i'=1}^{N_{k+1}} O^{k}_{i,i'} z_{k+1,i'} + \sum_{i'=1}^{N_{k-1}} O^{k-1}_{i',i} z_{k-1,i'} \qquad (15)$$

where

$$N = \sum_{k=0}^{J} N_k \qquad (16)$$

and N_k is the number of basis functions with a particular value of k selected from the solutions of $\hat{H}_k$.

Using this structure replaces a vector-matrix product of dimension N with a dot product of dimension N and 2(J-p) vector-matrix products of dimension N_k, where N_k is roughly N/(J+1). This is a considerable saving. Again the vector matrix multiplies can be performed using routine MXMB (25). Table 1 compares results for using FORTRAN and CAL versions of this routine.

Table 1 demonstrates a large speed-up, approximately a factor of 10, upon vectorisation. A further factor, of about 2.5, is obtained by judicious choice of algorithm and use of one assembler routine. The final program drives a single processor of the Cray-XMP at about 60% of the theoretical maximum value of 235 million floating point operations per second (Mflops). This program, ROTLEVD, is currently being prepared for publication as part of a program suite for calculation of triatomic ro-vibrational energy levels, transition moments and synthetic stick spectra (18).

4. SAMPLE RESULTS

To illustrate the results obtainable using current supercomputers, we discuss the calculated rotational spectrum of H_2D^+ and the "hot bands" of H_3^+.

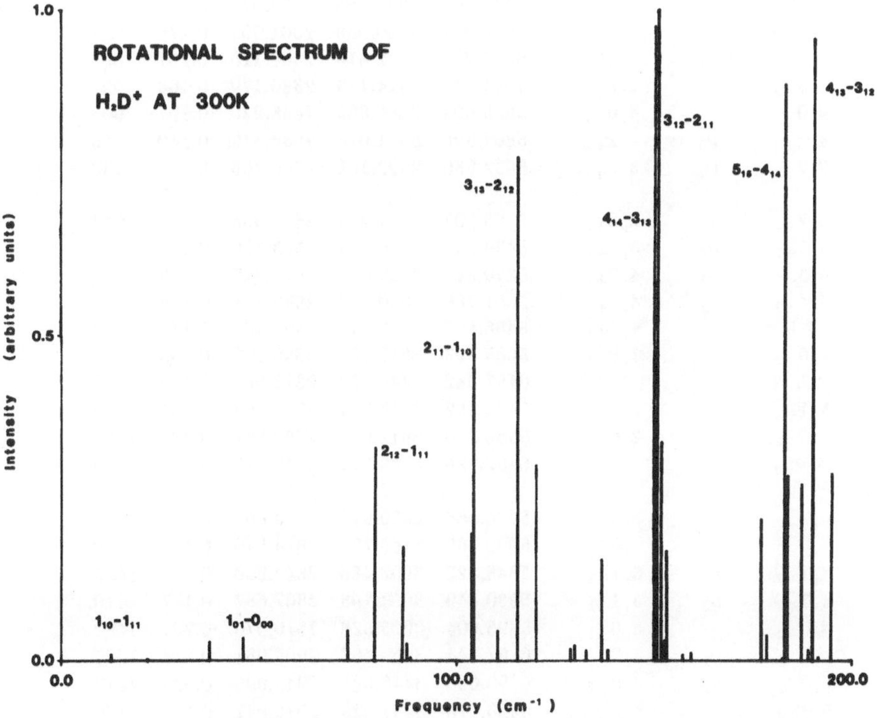

Figure 1 shows a synthetic stick spectrum of H_2D^+ at 300K. The frequency range covered, 0 - 200cm^{-1}, means that only pure rotational transitions are present. Although we have assignments for all these

TABLE II. Line strengths for "hot band" transitions in H_3^+.

Final state			Initial state	E'	E"	ω_{if}	S(f-i)	A_{if}
v'(l_2)	J',K'/G'$_U$	v"	J",K"/G"$_U$	(cm^{-1})	(cm^{-1})	(cm^{-1})	(D^2)	s^{-1}
$\nu_1+\nu_2$	$3,3_{+1}$	ν_1	$4,3$	5909.649	3820.625	2089.024	0.163	465
$\nu_1+\nu_2$	$3,4_{+1}$	ν_1	$4,4$	5764.679	3667.003	2097.676	0.216	627
$\nu_1+\nu_2$	$2,3_{+1}$	ν_1	$3,3$	5653.682	3485.273	2168.409	0.156	498
$2\nu_2(2)$	$3,2_{+2}$	ν_2	$4,3_{+1}$	5430.347	3233.027	2197.320	0.194	646
$2\nu_2(2)$	$3,4_{+2}$	ν_2	$4,4_{+1}$	5298.686	3069.003	2229.683	0.290	1010
$2\nu_2(2)$	$3,5_{+2}$	ν_2	$4,5_{+1}$	5104.913	2863.711	2241.202	0.393	1390
$2\nu_2(0)$	$3,1$	ν_2	$3,1_{-1}$	5280.799	3002.486	2278.313	0.165	614
$2\nu_2(0)$	$3,0$	ν_2	$3,0_{-1}$	5304.036	3025.528	2278.508	0.198	734
$2\nu_2(0)$	$4,2$	ν_2	$4,2_{-1}$	5542.504	3259.663	2282.840	0.197	734
$2\nu_2(0)$	$4,3$	ν_2	$4,3_{-1}$	5608.682	3325.518	2283.164	0.268	1000
$2\nu_2(2)$	$2,3_{+2}$	ν_2	$3,3_{+1}$	5180.662	2876.591	2304.071	0.184	705
$2\nu_2(2)$	$2,4_{+2}$	ν_2	$3,4_{+1}$	5031.976	2719.308	2312.668	0.277	1070
$\nu_1+\nu_2$	$4,2_{-1}$	ν_1	$4,2$	6275.437	3927.953	2347.485	0.184	745
$\nu_1+\nu_2$	$4,1_{+1}$	ν_1	$4,1$	6341.605	3991.586	2350.019	0.244	994
$\nu_1+\nu_2$	$3,1_{-1}$	ν_1	$3,1$	6022.981	3660.983	2361.998	0.175	722
$\nu_1+\nu_2$	$3,0_{-1}$	ν_1	$3,0$	6046.732	3682.614	2364.118	0.207	859
$2\nu_2(2)$	$1,3_{+2}$	ν_2	$2,3_{+1}$	4994.314	2614.135	2380.179	0.163	691
$2\nu_2(2)$	$4,0_{-2}$	ν_2	$4,0_{-1}$	5895.604	3446.685	2448.920	0.197	909
$2\nu_2(2)$	$4,2_{+2}$	ν_2	$4,2_{+1}$	5887.398	3351.023	2536.375	0.140	715
$2\nu_2(2)$	$3,2_{+2}$	ν_2	$3,2_{+1}$	5532.956	2992.161	2540.796	0.142	730
$2\nu_2(2)$	$2,2_{+2}$	ν_2	$2,2_{+1}$	5265.703	2723.765	2541.938	0.126	647
$2\nu_2(2)$	$3,3_{+2}$	ν_2	$3,3_{+1}$	5430.347	2876.591	2553.756	0.142	743
$2\nu_2(2)$	$4,3_{+2}$	ν_2	$4,3_{+1}$	5810.346	3233.027	2577.319	0.169	906
$2\nu_2(2)$	$4,4_{+2}$	ν_2	$4,4_{+1}$	5651.768	3069.003	2582.764	0.154	832
$2\nu_2(0)$	$5,5$	ν_2	$4,5_{+1}$	5458.853	2863.711	2595.142	0.148	809
$2\nu_2(2)$	$2,0_{-2}$	ν_2	$1,0_{-1}$	5285.893	2616.487	2669.406	0.153	916
$\nu_1+\nu_2$	$4,3_{-1}$	ν_1	$3,3$	6157.342	3485.273	2672.069	0.170	1020
$2\nu_2(2)$	$3,1_{-2}$	ν_2	$2,1_{-1}$	5485.152	2755.291	2729.861	0.204	1300
$2\nu_2(2)$	$3,0_{-2}$	ν_2	$2,0_{-1}$	5566.330	2812.642	2753.687	0.208	1360
$\nu_1+\nu_2$	$4,0_{-1}$	ν_1	$3,0$	6452.726	3682.614	2770.112	0.153	1020
$2\nu_2(2)$	$4,2_{-2}$	ν_2	$3,2_{-1}$	5714.864	2930.998	2783.866	0.240	1630
$2\nu_2(2)$	$5,3_{-2}$	ν_2	$4,3_{-1}$	5969.366	3144.790	2824.576	0.185	1310
$2\nu_2(2)$	$4,1_{-2}$	ν_2	$3,1_{-1}$	5845.422	3002.486	2842.936	0.199	1430
$2\nu_2(2)$	$4,1_{+2}$	ν_2	$3,1_{+1}$	5930.849	3063.193	2867.657	0.177	1310
$2\nu_2(2)$	$4,0_{-2}$	ν_2	$3,0_{-1}$	5895.604	3025.528	2870.076	0.201	1490
$2\nu_2(2)$	$5,2_{-2}$	ν_2	$4,2_{-1}$	6167.744	3259.663	2908.081	0.214	1650
$2\nu_2(2)$	$5,0_{-2}$	ν_2	$4,0_{-1}$	6390.680	3446.685	2943.996	0.313	2510
$2\nu_2(2)$	$5,2_{+2}$	ν_2	$4,2_{+1}$	6326.970	3351.023	2975.947	0.136	1120
$2\nu_2(2)$	$5,3_{+2}$	ν_2	$4,3_{+1}$	6212.584	3233.027	2979.557	0.138	1150
$2\nu_2(2)$	$5,3_{-2}$	ν_2	$4,3_{-1}$	6129.580	3144.790	2984.790	0.199	1660

transitions, for clarity only the most intensive have been labelled.

So far only 2 pure rotational transitions of H_2D^+ have been observed: $1_{10} - 1_{11}$ at 12.414cm^{-1} (28,29) and $2_{20} - 2_{21}$ at 5.200cm^{-1} (30). These frequencies compare with our estimates of 12.412cm^{-1} and 5.193cm^{-1} respectively. It is interesting to note that these transitions are very weak in comparison with many others at typical experimental temperatures. We are confident that our results, which will be presented in detail elsewhere (15), will enable further transitions to be assigned and the rotational spectrum of this highly asymmetric top ($\kappa \simeq -0.06$) to be unravelled.

Table 2 presents transition frequencies, line strengths and Einstein A coefficients for transitions involving the H_3^+ 'hot bands' $\nu_2 \rightarrow 2\nu_2$ and $\nu_1 \rightarrow \nu_1 + \nu_2$. For space reason only the 40 strongest transitions involving $J \lesssim 4$ are given. The J sublevels are assigned following Watson (31). In practice these assignments become increasingly difficult for higher J levels due to increasing overlap between vibrational states (12).

There are many unassigned lines in the hydrogen discharges which are used to produce laboratory H_3^+ (9,10). Analysis using our data has led to the first identification of transitions involving the 'hot band' $\nu_2 \rightarrow 2\nu_2$ (9) and the overtone $\nu_0 \rightarrow 2\nu_2$(10).

In conclusion, we have shown how the use of supercomputers has transformed the goal of first principles nuclear motion calculations. Detailed studies of novel spectroscopic regions are now possible allowing accurate predictions to be made as an aid to the experimentalist. Supercomputers are thus opening new doors in molecular spectroscopy.

ACKNOWLEDGEMENTS

We thank the staff of the Atlas Computing Centre at SERC's Rutherford Appleton Laboratories on whose Cray-XMP48 this work was carried out and Miss M.Morris for her careful typing of this article.

References

(1) G.D. Carney, L.L. Sprandel and C.W. Kern, Advan. Chem. Phys., 37, 305 (1978).

(2) G.D. Carney and R.N. Porter, J. Chem. Phys., 60, 4251 (1974).

(3) G.D. Carney and R.N. Porter, J. Chem. Phys., 65, 3547 (1976).

(4) A. Carrington and R.A. Kennedy, J. Chem. Phys., 81, 91 (1984).

(5) J. Tennyson and B.T. Sutcliffe, J. Chem. Soc. Faraday Trans. 2, 82, 1151 (1986).

(6) W. Meyer, P. Botschwina and P.G. Burton, J. Chem. Phys., 84, 891 (1986).

(7) S. Miller and J. Tennyson, J. Mol. Spectrosc., 128, 183 (1987).

(8) S. Miller and J. Tennyson, J. Mol. Spectrosc., 128, 530 (1988).

(9) T. Oka, private communication.

(10) J.K.G. Watson, private communication.

(11) J. Tennyson and B.T. Sutcliffe, Mol. Phys., 58, 1067 (1986).

(12) J. Tennyson, S. Miller and B.T. Sutcliffe, J. Chem. Soc. Faraday Trans. 2, (in press).

(13) S. Miller and J. Tennyson, Chem. Phys. Letts., 145, 117 (1988).

(14) S. Miller and J. Tennyson, Ap. J., (in press).

(15) S. Miller, J. Tennyson and B.T. Sutcliffe, Mol. Phys., (to be submitted).
(16) F.-S. Pan and T. Oka, Ap. J., 305, 518 (1986).
(17) J. Tennyson, Computer Phys. Comms., 42, 257 (1986).
(18) J. Tennyson and S. Miller, Computer Phys. Comms., (to be submitted).
(19) B.T. Sutcliffe and J. Tennyson, Mol. Phys., 58, 1053 (1986).
(20) J. Tennyson, Computer Phys. Rep., 4, 1 (1986).
(21) S. Carter and N.C. Handy, Mol. Phys., 57, 175 (1986).
(22) Z. Bacic and J.C. Light, J. Chem. Phys., 85, 4594 (1986) and 86, 3065 (1987).
(23) B.T. Sutcliffe, S. Miller and J. Tennyson, Computer Phys. Comms., (in press).
(24) J. Tennyson and B.T. Sutcliffe, J. Chem. Phys., 77, 4061 (1982) and 79, 43 (1983).
(25) V.R. Saunders, private communication.
(26) P.J. Nikolai, ACM Trans. Math. Software, 5, 188 (1979).
(27) F02FJF, NAG Fortran Library Manual, Mark 11, vol.4 (1983).
(28) M. Bogey, C. Demuynck and J.L. Destombes, Astron. Astrophys., 138, L11 (1984).
(29) H.E. Warner, W.T. Conner, R.H. Petrmichl and R.C. Woods, J. Chem. Phys., 81, 2514 (1984).
(30) S. Saito, K. Kawaguchi and E. Hirota, J. Chem. Phys., 82, 45 (1984).
(31) J.K.G. Watson, J. Mol. Spectrosc., 103, 350 (1984).

APPROXIMATE QUANTUM TECHNIQUES FOR ATOM DIATOM REACTIONS

A. Laganà
Dipartimento di Chimica
Università di Perugia
Perugia
Italy

E. Garcia
Departamento de Quimica Fisica
Universidad del Pais Vasco
Bilbao
Spain

O. Gervasi
Centro di Calcolo
Università di Perugia
Perugia
Italy

ABSTRACT. In this paper, the theoretical basis and the program architecture of some reduced dimensionality quantum reactive scattering computational procedures are illustrated. The aim is to evidence to what extent it is possible to take advantage of parallel and vector performances of modern supercomputers for carrying out extensive calculations of the reactive properties of atom diatom systems. Some efforts have been paid to indicate alternative ways of formulating both the theoretical approaches and the computational codes. Speed-up factors of the suggested solutions have been calculated for some test cases. Results of extensive calculations performed for two prototype reactive systems are presented.

1. Introduction

Accurate quantum numerical studies of reactive encounters, although affordable in principle on any type of mid-sized computer, need such a large amount of memory space and cpu time to make unpractical an extended calculation of detailed and global reactive properties even for the simple case of an atom diatom (A+BC) system. Exact three-dimensional quantum calculations have been performed mainly for light systems containing at least two hydrogen atoms.[1] Preliminary results for a fully asymmetric system have also been presented very recently.[2]

More popular than exact quantum calculations are those approximate techniques which

A. Laganà (ed.), Supercomputer Algorithms for Reactivity, Dynamics and Kinetics of Small Molecules, 271–294.

have been formulated by enforcing some reasonable constraints on the dynamics of the reactive system so as to reduce the dimensionality of the problem.[3] Several computer codes have been developed following recipes provided by these reduced dimensionality approaches. For some of them while the demand of computer resources has lowered to a level that makes it possible to use supercomputers for carrying out a systematic investigation of the reactive properties of several atom diatom reactions, the level of accuracy of the theoretical treatment is still quite high.

To make a program run fast on supercomputers those parts which are critical for time consumption need to be restructured. Such a process can be governed directly by the operating system of the computer which makes use of specific utilities. In this case, the code structure is reorganized without external interventions during the compilation step by introducing the appropriate concurrence among the available resources. This restructure, of course, cannot take advantage of the physical nature of the problem. It is driven only by the particular criteria built-in into the machine software for optimizing computer programs once that they have been coded. In other words, it can only make explicit those concurrencies which are already built-in into the program and optimize the overlap of their execution for the particular architecture of the machine. A deeper restructuring has to profit from the knowledge of the physical problem and single out the theoretical approach best suited for being coded using the architectural features of supercomputers. At this stage the restructuring has to be handled directly by the user. Two characteristics of the problem formulation are of particular importance for determining the success of a restructuring effort:

a. the number of independent tasks it is made of (parallelism);

b. the amount of data to be processed in the same way (vectorialism).

When solving a realistic physical problem it is difficult to be able to fully profit from both these aspects. For obtaining that, theoretical as well as computational foundations of the problem formulation have to be revisited and the computer program extensively modified. Because of the large amount of time involved in this process more often the restructuring is performed at an intermediate level. In this case the sections of the program critical for determining the consumed cpu time are singled out and, if possible, redesigned. For the remainder of the program an automatic vectorizing is invoked and, when needed, minor alterations are introduced to make it more efficient.

In the present paper we shall discuss the problems arisen when restructuring two computer programs based on different quantum approaches to the calculation of the reactive probability. These two types of approximate quantum treatments reduce the solution of the reactive scattering problem to the integration of a two mathematical dimension Schroedinger equation. One of them is based on the rotating model (RM). Such a model has been originally developed for the collinearly dominated reactions (RLM).[4] In its more elaborated version this type of approach includes a correction for the amount of energy confined into the bending motion of the triatom (BCRLM).[5] More recently it has been extended to systems having a bent transition state (BCNRM).[6] The other approach is of the Infinite Order Sudden (IOS)[7] type in the version adapted to the light heavy light systems.[8] Minor additional assumptions have been made to adapt this computational scheme to the investigated reaction.[9] Both approaches show features suitable for vector and parallel restructuring. Choices made to this purpose are discussed in the paper and resulting computational advantages illustrated.

The paper is organized as follows. In section (2) a sketch of the formalism and of the numerical procedures aimed at evidencing where restructuring can be profitably performed is given. In section (3) the structure of the computer programs developed following these approaches is illustrated. In section (4) outcomes of test runs performed using the two restructured programs are discussed. In section (5) results of extended calculations carried out for two prototype atom-diatom reactions are reported.

2. Theoretical guidelines

In an IOS approach the application of both the energy sudden[10] and the centrifugal sudden[11] decoupling scheme reduces the exact Schroedinger equation for the atom-diatom scattering into a set of fixed angle of approach (γ) two mathematical dimension ones. In the Body Fixed formulation these equations read as

$$\left[-\frac{\hbar^2}{2\mu}\left(\frac{1}{Q_\lambda}\frac{\partial^2}{\partial Q_\lambda^2}Q_\lambda+\frac{1}{q_\lambda}\frac{\partial^2}{\partial q_\lambda^2}q_\lambda-\frac{l_\lambda(l_\lambda+1)}{Q_\lambda^2}-\frac{j_\lambda(j_\lambda+1)}{q_\lambda^2}\right)+V(Q_\lambda,q_\lambda;\gamma_\lambda)-E\right]$$
$$*\Phi(Q_\lambda,q_\lambda;\gamma_\lambda)=0 \quad (1)$$

In equation (1) λ labels the type of atom-diatom arrangement being considered ($\lambda=\alpha$ indicates the reactant channel, $\lambda=\beta$ or β' indicates a product channel). In the same equation E is the total energy while Q_λ and q_λ are the atom to the diatom center of mass and the diatom internuclear distances. Q_λ and q_λ distances are related to the physical R_λ and r_λ distances by the following mass scaling relationships:

$$Q_\lambda=\left(\frac{\mu_\lambda}{\mu}\right)^{\frac{1}{2}}R_\lambda;\ q_\lambda=\left(\frac{m_\lambda}{\mu}\right)^{\frac{1}{2}}r_\lambda \quad (2)$$

being μ an arbitrary mass and μ_λ and m_λ the reduced mass of the triatom and diatom in the λ arrangement, respectively.

Equations (1) have the same structure as those for collinear atom-diatom scattering and therefore can be solved by adapting well established computational procedures.[12] Therefore, for each value of γ_λ the integration of equations (1) is carried out starting from the central line separating reactant and product channels. To this purpose the reactant α and the product (β or β') channels are divided into two regions. The first region begins at the central line which connects the zero of the Q_λ, q_λ axes to a point (P^*, the turning center) located on the ridge sufficiently far from the coordinate origin to ensure that, even at the largest energy considered for the calculation, it is crossed by a negligible interchannel flux. The point P^* is also taken as the center of the circular coordinates. As will be discussed later, these coordinates are used for formulating the scattering equations in the inner region. The upper limit of this region is given by the perpendicular from P^* to the Q_λ axis. The second region goes from the end of the inner region to a Q_λ value for which the potential cut is almost coincident with the asymptotic potential.

For each arrangement channel the actual numerical integration of equations (1) is performed by partitioning all regions of the potential energy into several small sectors. Inside each sector i, $\Phi(Q_\lambda, q_\lambda; \gamma_\lambda)$ is expressed in terms of products of the translational $\chi_{\lambda v}(\gamma_\lambda; Q_\lambda)$ and vibrational $\phi_{\lambda v}(q_\lambda; Q_\lambda^i)$ functions. The vibrational functions $\phi_{\lambda v}(q_\lambda; Q_\lambda^i)$ are calculated at the sector midpoint value (Q_λ^i) of the propagation coordinate Q_λ. To this purpose in the original version of the program the one dimensional bound state problem

$$\left[-\frac{\hbar^2}{2\mu}\frac{\partial^2}{\partial q_\lambda^2} + V(q_\lambda; \gamma_\lambda^c, Q_\lambda^i) - \varepsilon_{\lambda v}^i\right] \phi_{\lambda v}^i(q_\lambda; Q_\lambda^i) = 0 \tag{3}$$

is solved using a numerical technique[13] for a chosen reference value γ_λ^c of the angle of approach γ_λ.

At each value of γ_λ equations (1) are solved by premultiplying both their lhs and rhs terms by $\phi_{\lambda v'}^{*i}(q_\lambda; Q_\lambda^i)$ and averaging over q_λ. In this way one obtains a set of coupled differential equations of the type

$$\left[\frac{d^2}{dQ_\lambda^2} - \underline{D}_\lambda^i\right] \underline{\chi}_\lambda(\gamma_\lambda; Q_\lambda) = 0 \tag{4}$$

where $\underline{D}_\lambda^i$ is the coupling matrix whose elements read as

$$\begin{aligned} D_{\lambda v v'}^i &= \frac{2\mu}{\hbar^2} < \phi_{\lambda v}^i \left| V(q_\lambda; \gamma_\lambda, Q_\lambda^i) - V(q_\lambda; \gamma_\lambda^c, Q_\lambda^i) \right| \phi_{\lambda v'}^i > \\ &+ < \phi_{\lambda v}^i \left| \frac{l_\lambda(l_\lambda+1)}{Q_\lambda^2} + \frac{j_\lambda(j_\lambda+1)}{q_\lambda^2} \right| \phi_{\lambda v'}^i > + \frac{2\mu}{\hbar^2}(\varepsilon_{\lambda v}^i - E)\delta_{vv'} \end{aligned} \tag{5}$$

As already mentioned, inside the internal region of each channel the circular coordinates u_λ and w_λ are preferred to the Jacobi coordinates for ensuring a smooth connections between different arrangement channels. In fact, the bound circular coordinates of the two channels (w_α and w_β) coincide at the intermediate line making the switch from the reactant to the product arrangement continuous. The only difference between w_α and w_β on the separating line is a shift of their origin (the distance of the origin from the turning center P^* is σ_λ). The introduction of the circular coordinates has little effect on the complexity of the calculation. Its impact on the computational procedure is mainly relevant to the construction of those parts of the $\underline{D}$ matrix elements which are calculated, as will be discussed in more detail in the next section, once for ever in the first section of the computer code. In fact, when using circular coordinates the $\underline{D}$ matrix elements read as

$$\begin{aligned} D_{\lambda v v'}^i &= \frac{2\mu}{\hbar^2} < \phi_{\lambda v}^i \left| \left[V(w_\lambda; \gamma_\lambda, u_\lambda^i) - V(w_\lambda; \gamma_\lambda^c, u_\lambda^i)\right] \eta_\lambda^2 \right| \phi_{\lambda v'}^i > \\ &+ < \phi_{\lambda v}^i \left| \frac{l_\lambda(l_\lambda+1)}{Q_\lambda^2} + \frac{j_\lambda(j_\lambda+1)}{q_\lambda^2} \right| \phi_{\lambda v'}^i > \\ &+ \frac{2\mu}{\hbar^2}\left(\frac{\varepsilon_{\lambda v} + \varepsilon_{\lambda v'}}{2} - E\right) < \phi_{\lambda v} | \eta_\lambda^2 | \phi_{\lambda v'} > + \frac{3}{4\sigma_\lambda}\delta_{vv'} \end{aligned} \tag{6}$$

where η_λ is a function of w_λ and σ_λ and where u_λ^i is the value of the propagation coordinate at the sector midpoint.

For non-collinear calculations (as is the case of the IOS approach) the matching between different arrangement channels is to a large extent arbitrary. A recipe has been given for making unique the mapping of the entrance onto the exit channel at the dividing line.[14] The IOS approach we have followed is designed to deal with systems having a central atom much heavier than the external ones. This ensures that reactants and products coordinates are exchangeable ($Q_\alpha = q_\beta; q_\alpha = Q_\beta; \gamma_\alpha = \gamma_\beta = \gamma$) and the matching between the reactant and product channels is uniquely defined.[8] Continuity of the solution propagated from the central line outwards to the asymptotes through the different sectors, regions and arrangement channels is ensured by mapping the solution at the border between adjacent sectors via overlap matrices of the related vibrational eigenfunctions.[12] By imposing the appropriate boundary conditions[8] at both reactant and product asymptotes the detailed IOS fixed γ S matrix elements ($S_{jj'vv'}(\gamma, E)$) for the reactive process can be derived. From these values the fixed γ reactive cross section from a given initial vib-rotational state ($S_R^{vj}(\gamma, E)$) of interest for our applications can be calculated using the relationship

$$S_R^{vj}(\gamma, E) = \sum_{v'j'} \frac{\pi}{k_{vj}^2}(2j'+1)\left|S_{jj'vv'}(\gamma, E)\right|^2 \tag{7}$$

where as usual k_{vj}^2 is $2\mu/(\hbar(E - E_{vj}))$ and E_{vj} is the energy associate with the considered reactant vibrational state. Then the corresponding global IOS 3D reactive cross section for a given total energy can be obtained by integrating over $\cos(\gamma)$

$$S_R^{vj}(E) = \frac{1}{2}\int_{-1}^{1} S_R^{vj}(\gamma, E)\, d\cos\gamma \tag{8}$$

In the BCRLM approach the equations to be integrated are

$$\left[-\frac{\hbar^2}{2\mu}\left(\frac{1}{\eta^2}\frac{\partial^2}{\partial u^2} + \frac{1}{\eta}\frac{\partial}{\partial \nu}\eta\frac{\partial}{\partial \nu} - \frac{l_\lambda(l_\lambda+1)+1}{R^2}\right) + V(u, \nu; \gamma_\alpha = 180^o) - E^{BC}\right]$$
$$*\Phi(u, \nu) = 0 \tag{9}$$

where u and ν are natural coordinates, $R = (Q_\alpha^2 + q_\alpha^2)^{1/2}$, $\eta = 1 + \sigma(u)\nu$ (σ is the inverse radius of the local curvature and E^{BC} is the total energy subtracted of the amount lodged in the (adiabatically evolving) bending motion.[5] Once performed in the usual way a partial wave expansion, the resulting coupled differential equations have the same form of equation (4) provided that the local E^{BC} is estimated at each step of the integration. After performing the propagation of the solution from the central line to large Q_λ values, using the same computational machinery of the IOS program, detailed BCRLM S matrix elements ($S_{vv'}^l(E)$) of the reactive process can be derived by applying the appropriate asymptotic conditions.[5] The detailed reactive probability $P_{vv'}$ can then be estimated using the relationship

$$P_{vv'}(E) = \sum(2l+1)P_{vv'}^l(E) \tag{10}$$

where $P_{vv'}^l$ is the square modulus of the $S_{vv'}^l$ (E) matrix element. Finally the detailed reactive rate constants $k_{vv'}$ (T) of interest for our applications can be calculated using the relationship

$$k_{vv'}(T) = \left[\frac{2\pi\hbar^4 N^2}{k_B^3 T^3 \mu^3}\right]^{\frac{1}{2}} \int_0^\infty P_{vv'}(E) e^{-E/k_B T} dE \tag{11}$$

3. The programs' structure

The general structure of both the BCRLM and the IOS (at a given value of γ) programs is

Section I

```
LOOP on sectors
    Calculation of sector vibrational eigenvalues and eigenfunctions
    Evaluation of the l, j and energy independent components of the D matrix elements
END of the sector-loop
```

Section II

```
LOOP on energies
    Fixed energy initializations
    LOOP on l values
        LOOP on arrangement channels
            LOOP on propagation sectors
                Map of the border values of previous sector quantities
                Construction of the D matrix
                Within sector diabatic propagation
            END of the sector-loop
        END of the channel-loop
        Asymptotic analysis and S matrix elements evaluation
    END of the l values-loop
    Evaluation of the fixed angle reactive cross section
    Storage of the quantities needed for off line integration
END of the energy-loop
```

The goal of these programs is the calculation of the fixed angle quantum cross section. Such a value is the final result for the BCRLM treatment while for the IOS program several fixed γ results are at first calculated and then stored on disk for an off-line integration over the angle of approach. As apparent from the scheme, the programs are divided into two separate sections.

In the first section, the potential energy channel connecting reactant and product arrangements is cut into several sectors. On the one-dimensional cuts of the potential energy channels taken at a fixed value (the sector midpoint) of the propagation coordinate vibrational eigenfunctions and eigenvalues as well as other quantities independent from both the quantum number and energy (such as overlap integrals and part of the coupling matrix) are calculated for use in the second section. In the original version of the program this is done only at γ_c (which is also the first value of γ_λ considered during the calculation). Calculated coupling matrix elements are, in this case, read-in from the disk for subsequent runs.

The second section of the program is devoted to the propagation of the solution and to the calculation of the final quantities for an array of energies. This section contains four nested DO loops. The outer DO loop runs over the specified energy values. The next internal DO loop runs over the quantum number l. The innermost DO loop runs over the various sectors to perform the propagation. Inside this DO loop several calls to the routines performing the different matrix manipulations (add, multiply, transpose, invert) are made.

Parts of the computer code eligible for vectorizing are the routines called inside the DO loop of the first section evaluating the potential energy as well as those called for manipulating large matrices. As a consequence, it is important that the potential energy routine is designed to vectorize. This is not always the case. As will be discussed in the next section, the restructure of the potential energy functional representation can be considered as an example of the possibility of obtaining a speed enhancement when a detailed revisit of the theoretical approach and a careful redesign of the computer code is performed. For both the first and the second section further time saving can be obtained by making use of library routines explicitly designed for manipulating matrices in vector mode.

For parallelizing the BCRLM and IOS programs it can be taken advantage of two facts. One is that in the first section of the code the calculation of the vibrational eigenfunctions is an independent computational task for each sector of the potential energy channel. Therefore, it can be carried out in parallel for different sectors. The other advantage is that in the second section of the program several DO loops are concerned with independent computational tasks and are therefore eligible for parallel execution.

4. Software restructuring

As already mentioned the most radical restructure was concerned with the redesign of the potential energy routine. In fact, the high quality of the ab initio potential energy values[15] and the fine detail of the experimental data[16] obtainable nowadays set so severe constraints on the characteristics of the potential energy surface that its analytical formulation has to be very flexible. For the systems we have investigated, in fact, the reactivity is highly sensitive not only the height and location of the surface barrier but also to several other features. This is the case, for example, of the well associate to the formation of an intermediate complex either inside the entrance or the exit channel.[17] For this reason, the functional form adopted for interpolating the ab initio points has to be highly flexible. Flexibility is usually achieved at the expenses of simplicity and, therefore, of the possibility of vectorizing. This is the case of functional forms assembled by patching local interpolators or of global interpolators requiring different local corrections.

An approach which couples flexibility to a fairly simple analytical formulation is that suggested by Murrell and coworkers[18] decomposing the potential energy in two and three body terms. These terms are then represented using polynomial expansions in physical coordinates[19] damped by exponential or hyperbolic tangent factors. We have suggested elsewhere expansions in terms of bond order (BO)[20] or mixed physical and bond order[21] variables. In these cases there is no need for superimposing damping factors because a map of the potential energy into the BO space ensures a smooth evolution of the polynomial expansion to the correct asymptotic limit. For simplicity, our study has been confined to the BO PES whose generalized functional representation is

$$V(r_1,r_2,r_3)=\sum_j^J\sum_k^K\sum_l^L a_{jkl}n_1^j n_2^k n_3^l \tag{12}$$

where $n_i = e^{-b_i(r_i-r_{e_i})}$ is the BO variable of the ith diatom, and r_i and r_{e_i} are its actual and equilibrium distance and b_i a parameter. The simplicity of the BO potential makes it an ideal case for investigating to what extent the design of a computer code is important for determining the speed-up obtainable on a supercomputer. In order to investigate in depth such a problem, we have isolated the potential energy calculation from the program context and designed several algorithms for its evaluation.[22]

For a scalar execution the polynomial of equation (1) can be evaluated using three nested DO loops. In this case (subroutine POTJKL) the time consumption is 0.80 ms for the IBM 3090 and 0.35 ms for the CRAY XMP-12. A restructuring of this routine for vectorizing can be performed in several different ways[22]. The simplest choice is to determine n_1, n_2 and n_3 and run a loop over the terms obtained by rising the n_is to the appropriate powers. A more elaborate algorithm takes advantage of the properties of products of exponentials and run a DO loop over single exponential terms whose argument is the sum of those of the three BO variables. Another algorithm makes use of indirect addressing for multiplying array elements containing powers of n_i once that these have been determined by running a separate DO loop over the corresponding values of the exponent. Because of the unavailability of a hardware indirect addressing feature on the CRAY XMP-12, calls to GATHER and SCATTER firmware utilities have been invoqued in this case.

For each of the three proposed methods we wrote a separate routine (POTPOW, POTEX, POTCUR, respectively). To take advantage of the vector facility, the coefficient and exponent matrices were transformed into vectors of appropriate dimensions. As a results, for our BO potentials the sum of the polynomial terms can be performed by running a simple vectorizable DO loop. For the potential energy surfaces used in our calculations the length of these vectors is 77. The average cpu time per call of these routines are in scalar mode 0.47, 0.37 and 0.13 ms, for the IBM and 0.21, 0.21 and 0.06 ms, for the CRAY. These values cannot be considered for a direct comparison of the performance of the two machines because the used time analysis procedures were not standard. Therefore, strictly speaking, the comparison can be performed only between scalar and vector performances of the same machine. For these routines speed-up factors (SUR), defined as the ratio of the vector over the scalar velocity, are, respectively, 1.3, 3.7 and 3.3 on the IBM and 3.5, 9.2 and 1.3 on the CRAY.

As shown from numbers reported above, all routines restructured for vectorizing (independently upon whether they are run in scalar or in vector mode) are more efficient than POTJKL. This means that a DO loop reduction and, in general, a reorganization of the code for vector execution on a supercomputer frequently pays off also in terms of scalar speed-up. Moreover, times measured for POTPOW on the IBM 3090 show that this may be true even

Table I: Detailed run times (in ms) for the IOS program.

Routine	Scalar NV=15	Vector NV=15	Scalar NV=30	Vector NV=30
COOLEY	38.9	31.7	25.2	11.3
MAIN	72.0	49.1	390.7	228.8
MATINV	21.4	15.6	12.6	6.8
MMULO	408.9	57.2	3163.4	333.0
MMULI	202.5	28.6	1575.4	163.1
MMUL2	200.5	29.2	1588.0	167.6
OVLP	210.3	150.5	113.4	13.6
POTBO	7.8	1.5	12.6	2.3
RMATRX	29.2	11.2	100.8	27.2
SDMATQ	183.0	47.9	1424.1	373.1
TQL2	366.0	113.8	2709.6	543.6
TRED2	157.7	66.6	1323.3	364.7

when overheads for the actual vectorizing make vector routines more complex than scalar ones. The POTPOW routine performs much better on the CRAY XMP-12. In fact, even in scalar mode, time spent per call by POTPOW is about half that of the original routine. On top of that a quite large SUR value (3.5) is achieved when running in vector mode. The POTEX routine is more efficient than POTPOW. On the IBM 3090 the routine POTEX is a factor of two faster than POTJKL. In addition, when running in vector mode the SUR value is 3.7 leading to an overall speed gain larger than seven. Efficiency of the CRAY is not that great when the routine is run in scalar mode (time gain with comparison to the original routine is lower than a factor of two as in the POTPOW case). However, the strength of the restructured code can be fully appreciated when it runs in vector mode. In fact the SUR factor is 9.2 making the POTEX routine one order of magnitude faster than POTJKL. In the POTCUR case the IBM 3090 reaches its top performance. In fact, although the value of the SUR factor is similar to that obtained for POTEX (3.3 against 3.7 of POTEX) the overall performance is higher because the scalar speed is already quite large. A similar situation has been reproduced on the CRAY machine by using the GATHER and SCATTER subroutines. Regardless of the additional DO loops and routine calls indirect addressing shows to be a good way of organizing the calculation of a BO potential on the CRAY. Additional test runs have been carried out on more recent models of these machines. They indicate that POTEX and in general routines using indirect addressing have more comparable performances on both machines.

Vector restructuring for the second section of the program was not as deep as for the first section. In fact, the main task of this section is the appropriate manipulation of matrices for carrying out the propagation of the solution. As a matter of facts, all algorithms were left substantially unaltered and vectorizable library routines were used. Most of the restructuring effort was, therefore, dedicated to optimizing some DO loop indexes in a suitable way for vectorizing.

To evaluate the overall speed-up obtained as a result of the program restructuring we have performed test runs of the IOS program on the IBM 3090. Calculations have been carried out for the $Li + HF$ reaction at one value of γ and using two different values of the partial wave set dimension (NV). Detailed run times obtained for two values of the collision energy are reported in Table I.

Cpu times requested for scalar runs are respectively 1947s and 12603s for NV=15 and NV=30. For vector runs the need for cpu time reduces to 622s and 2265s respectively leading to SUR values of 3.1 and 5.6. These speed-up factors, although smaller than those obtained for test runs on the potential energy routine alone, still lead to a significant saving of computer time. The time saving for vector runs is more effective when dimensions are large. The detailed results reported in the Table confirm also that the crucial parts of the program are those dedicated to matrix manipulations. The routine mentioned in the Table perform the following tasks:

COOLEY	calculates the sector vibrational eigenfunctions;
MATINV	inverts complex matrices;
MMUL0, MMUL1 and MMUL2	perform matrix times matrix, transposed matrix times matrix and matrix times transposed matrix multiplications respectively;
OVLP	evaluates the overlap matrix between different sets of vibrational eigenfunctions;
POTBO	evaluates the BO potential energy value using the indirect addressing algorithm;
RMATRIX	propagates the R matrix along the potential energy channel;
SDMATQ	solves a system of real linear algebraic equations;
TQL2 and TRED2	find eigenvalues by tridiagonalizing the interaction matrix;

The table evidences the non-negligible demand of cpu time associate to the potential energy routine (when using the unrestructured POTJKL routine the time consumption needed for the evaluation of the potential is more than six times larger than for POTBO) and the large quantity of time consumed by routines performing matrix manipulations when the program runs in scalar mode. Time consumption for vector runs of the whole programs confirm the large SUR factor estimated for benchmark runs of the potential energy routine and single out the importance of vectorizing matrix multiplications. Quite large speed-up factors are also obtained for the other routines devoted to matrix manipulations. The only apparent exception is the OVLP routine that saves only a 25% of its time share when running in vector mode. This is due to the fact that the compiler vectorizes automatically the outer DO loop running over NV rather than the innermost DO loop running over the grid points of the vibrational eigenfunctions. Obviously, when NV is small most of the speed gain is counterbalanced by the vectorizing overheads. However, for larger NV values the time gain becomes appreciable. This is the case of NV=30 for which the Table gives a SUR value larger than 8.

Absolute cpu values reported in the Table indicate that the use of this program is quite convenient. A first important point to be emphasized is that (as will be discussed in greater detail below) the program has been modified to take advantage of the memory space available on the machine to avoid I/O operations to/from disk to save them. An estimate of these times is 50s and 170s saved in scalar runs for NV=15 and 30 respectively.

We turned the above mentioned time benefits into an improvement of the computational procedure. In the improved procedure[9] the $\underline{D}$ matrix elements are recalculated at each γ value by using the proper vibrational basis set. This has been easily implemented in the program by executing the first section of the program also for $\gamma \neq \gamma_c$.

As far as restructuring for parallel execution was concerned, a comment is in order before discussing it in detail. Our discussion on the efficiency of a parallel restructuring is based on the use of the Macro Tasking Facility (MTF). The computer used for our calculations, in fact, was not equipped with a facility for fine grain parallelism. Therefore, the parallel reorganization of the program was focussed on the large grain parallelism. This is ideally suited for those sections of the program which are sufficiently large to render negligible the overhead associate to a parallel restructuring.

In the first section of the program the obvious choice for a MTF approach is to execute in parallel the DO loop running over different sectors being the integration of equation (3) a independent task for each of them. A more difficult problem is the reorganization of the second section program which is responsible for the bulk of the cpu time consumed by the program. When the restructure is carried out using the MTF, several different approaches can be followed. Our choice has been to parallelize the l DO loop. The main reason for this choice is related to the fact that the number of cycles of the outermost DO loop (running over the energy values) might be too small for taking full advantage of a multiprocessor machine. The case of single energy runs, in fact, is not infrequent. However, even the most usual case of a few energy run does not allow a fair share of the tasks when several processors are available. At the same time, the use of the parallelism at an innermost level makes the demand of memory associate to every dispatched process lighter. In our case, the most internal DO loop suitable for parallelizing is that on the reaction channel index which for these methods never exceeds 2. Unfortunately, this is the worst situation for a parallel organization of a computer code. In fact, in this case there is no advantage for using computer architectures based on more than two processors. On top of that, even when the machine has only two processors this DO loop cannot fully profit of the parallelism because of the synchronization delays for tasks not having the same length. For the two channels, in fact, the number of sectors in which they are divided may be largely different when the reactive system is asymmetric. The most internal DO loop runs over the different sectors of the potential energy channels accumulating the propagated R matrix. Because of this, it has a strictly sequential organization and cannot be considered for a parallel execution.

The above discussion confirms that the DO loop best suited for parallelizing is the intermediate one running over the l number. Each l cycle, in fact, is fully independent from others. On top of that, this DO loop is ideally located in an intermediate position of the computer code so that the amount of memory involved is sufficiently small. Such a characteristic is vital for a parallel organization of the program. When using MTF, in fact, each independent task has to be confined into a separate routine connecting (without communicating to the external environment) two sets of input and output variables. The MTF utility dispatches a copy of the routine to the first available processor jointly with the reference of the input and output sets of variables (for machines having more than two processors other copies of the routine are dispatched according to the user requests and to the installation management parameters). When a dispatched process comes to an end the related processor is disengaged and results returned to the appropriate memory location. It is, therefore, of great advantage if the dispatched process avoids as much as possible paging faults and I/O operations. By parallelizing the *l* DO loop the program occupies 14 megabytes of the memory space during execution. In turn, as already mentioned, a significant speed-up was obtained. Furthermore, being in general the maximum value of *l* to be considered for getting a reasonable convergency of the calculated reactive probabilities at least of a few tens, the number of parallelized processes is sufficiently large. Time consumptions obtained for test runs indicate that in our case the parallelism is quite effective and that the relate time gain is a factor very close to the number of processors used.

5. Two prototype atom-diatom calculations

In recent years we have become interested in the study of some prototype reactive processes. Our goal is a comparison of the dynamical behaviour of reactions governed by a bent transition state with that of reactions occurring through collinear intermediates. The investigation has been carried out using the IOS and BCRLM programs illustrated in the previous sections. As already discussed, the reduction of the cpu time demand obtained because of the program restructure has made possible extended calculations of the reactive properties of several realistic atom-diatom prototype reactions. We shall discuss in the remainder of this paper the two theoretically most investigated collinearly and non collinearly dominated reactions.

The most celebrated prototype for collinearly dominated reactions is $H + H_2$. Several theoretical studies of this reaction have already been reported in the literature.[24] For this system, most of the theoretical work has been focussed on the calculation of reactive properties at low reactant vibrational states. Only a limited number of theoretical investigations have been concerned with the calculation of reactivity of the vibrationally excited H_2. However, after the advent of supercomputers it has been possible to undertake a systematic study of the effect of increasing the initial internal energy of the reactant molecule. A practical motivation for our calculations has been supplied by plasma chemistry studies. In particular, the modeling of magnetic multicusp H^- ion sources[25] needs among other detailed information the vibrational excitation and deexcitation rate constants of the H_2 molecules in collisions with hydrogen atoms. From the theoretical side this kind of studies is motivated by the need for understanding the efficiency of the reactant vibrational energy in promoting a redistribution of the energetic content of a chemical system among the various vibrational states when a collision occurs.[26] The first part of the investigation has been concentrated on calculating state to state rate constant for the temperatures 300, 500, 1000, and 4000 K using quasiclassical trajectories.[27] The advantage of using quasiclassical trajectories for high energy calculations is well known. In our case the calculation was made faster by the use of a program restructured for running on a supercomputer.[22,28] The potential energy of ref. 29 was used for the calculation. The state to state rate constants were calculated for the set of reactant vibrational levels ranging from 1 to 9. Results obtained from these calculations indicate that the vibrational deactivation is always much more efficient than excitation even at the largest temperature of the investigated interval. They also evidence that while at low temperature the most efficient reactive process is that conserving the initial vibrational number at higher temperatures the most efficient process is deexcitation. An interesting finding of these calculations is also the fact that the detailed rate constants of all the processes involved in a $H + H_2$ collisions can be parametrized using simple functionals.

The large saving in computer time obtained for the BCRLM restructured computer program has made competitive such a quantum treatment with comparison to classical techniques. Previous calculations based on the BCRLM method were performed by R.B. Walker for total energies lower than 30 kcal/mol.[5] In our case we have extended the energy interval to about 100 kcal/mol. In this way it has been possible to determine the reactive probability for the range of the collision energy important for evaluating detailed rate constants up to 4000 K and $v < 8$. For convergency reason l values up to 30 have been considered. Typical evolutions of the fixed l state to state probabilities $P^{l}_{vv'}(E)$ are reported in figure 1 as a function of the total energy.

The figure shows that the main effect of increasing l is a shift of the probability curves to a higher total energy. At the same time as shown by the figure, the structure of the curves becomes smoother and the total amplitude decreases. When considering more averaged quantities such as the $P_{vv'}$ reactive probability or the $k_{vv'}$ rate constant the structures are expected to compensate each other and results to be parametrizable in terms of simple functionals as quasiclassical results. From our results we find that this is indeed the case of the $H + H_2$ reaction. In fact, they indicate that the average amplitude of the reactive probabilities decrease with the quantum vibrational jump.

To give a more quantitative description of this property we have reported in figure 2 the dependence from the initial vibrational state of the value of the integral over energy of some state to state reactive probabilities

$$I(v,\delta) = \int_0^{\infty} P_{vv'} dE \tag{13}$$

In particular the curves plotted in the figure refer to deexcitation processes at different values of the vibrational state jump δ. Four different δ values are considered in the figure. From the figure it can be clearly seen that the value of $I(v,\delta)$ smoothly decreases when moving from low to high v values. A similar decrease occurs also when considering increasingly higher vibrational jumps. The plotted curves can clearly be interpolated using simple functionals indicating that simple rules can model the evolution of the state to state reactive probability $P_{vv'}(E)$. The possibility of performing extended calculations on a supercomputer in a reasonable time allows such a large collection of computed data that the calculated information can be mapped into appropriate empirical rules covering homogeneous region of the reaction phenomenology. This type of modeling, based on the formulation of system specific rules, is of great importance for approaches to the treatment of chemical systems based on artificial intelligence techniques. As an example, expert sys-

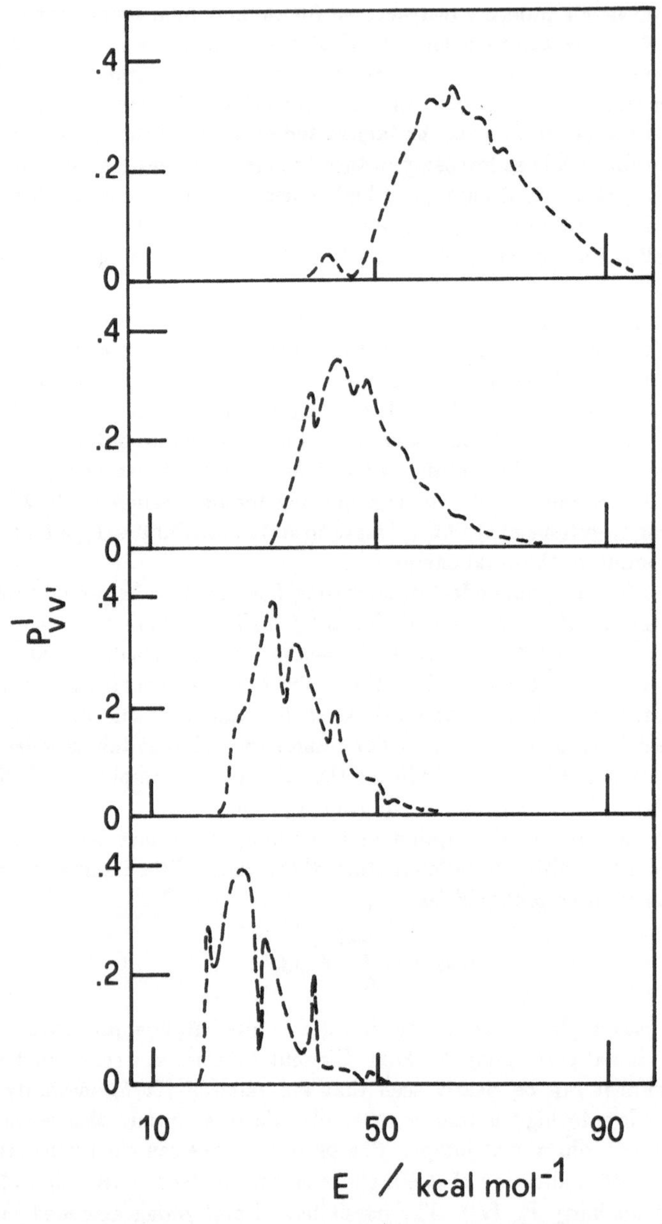

Figure 1: Fixed l state to state reactive probabilities for the $H + H_2$ reaction plotted as a function of the total energy. Reported values have been calculated at l=1, 10, 20, and 30 (from the lower to the higher panel) and v=0. Because of the fine grid used values are reported as a single line.

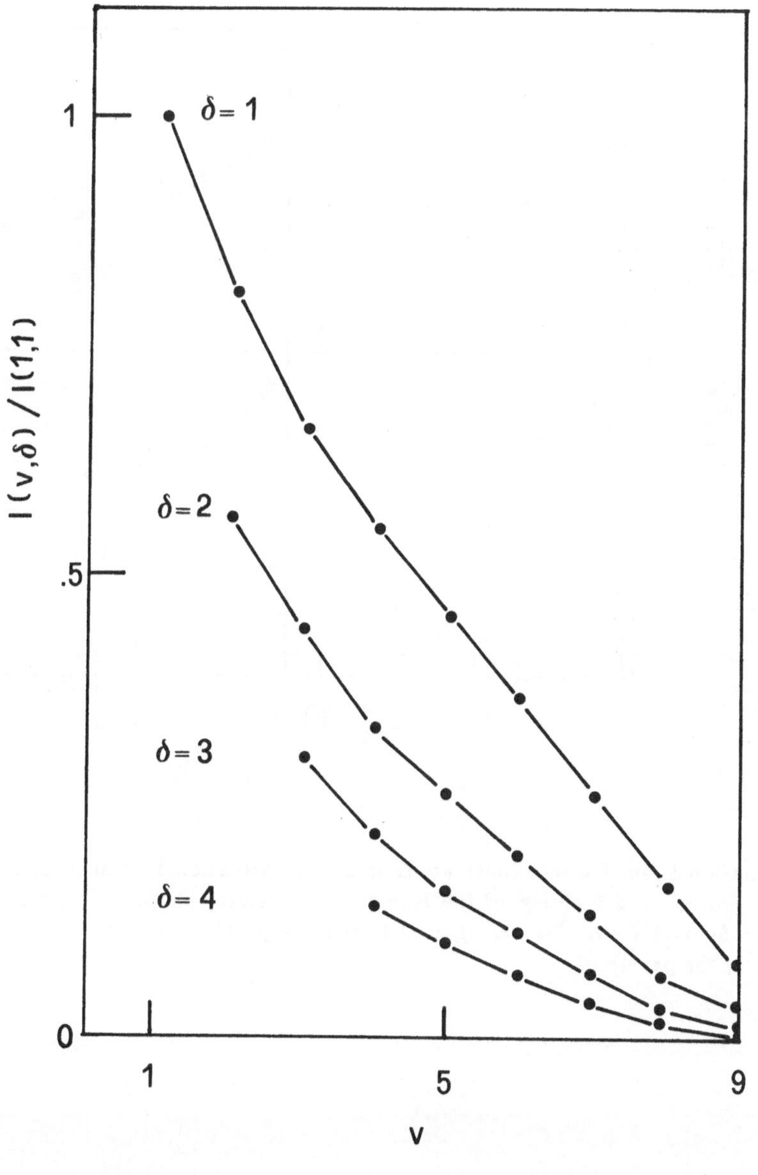

Figure 2: Integral of the state to state reactive probabilities over total energy $I(v,\delta)$ normalized to I(1,1) for the $H + H_2$ reaction plotted as a function of the initial vibrational number v at different values of δ.

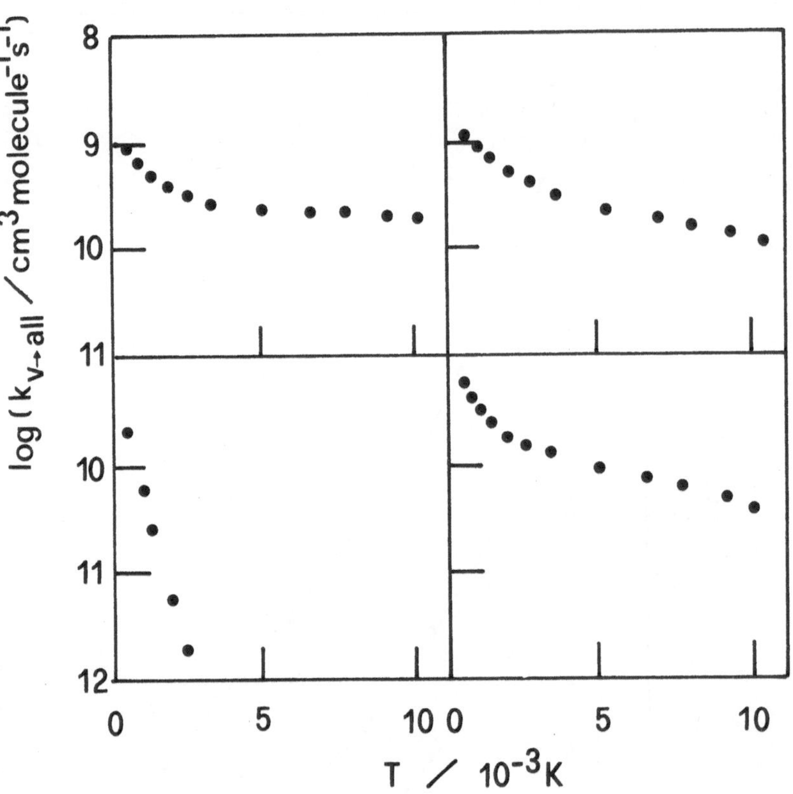

Figure 3: Global reactive rate constants from a given vibrational number for the reaction $H + H_2$ plotted as a function of the inverse temperature. Reported values have been calculated for v=1 (lower lhs panel), v=3 (lower rhs panel), v=5 (upper lhs panel), and v=9 (upper rhs panel).

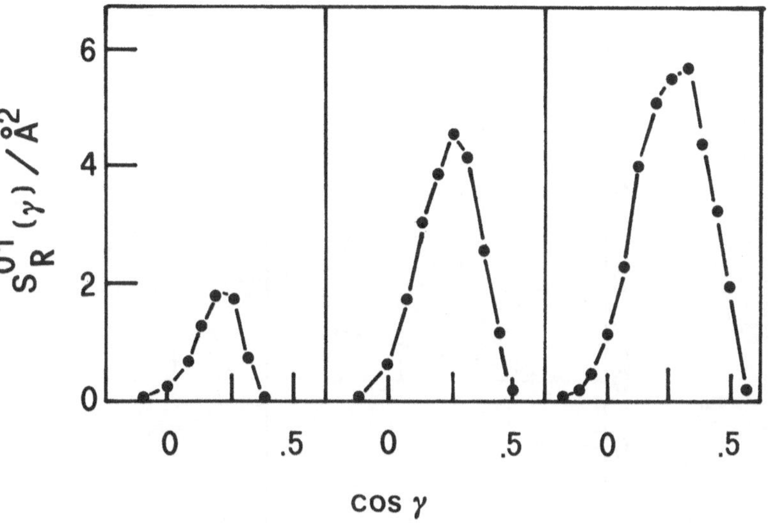

Figure 4: Fixed angle reactive cross sections for the $Li + HF$ reaction at v=0 and j=1 plotted as a function of the angle of approach γ.

tems devoted to the individuation of elementary processes occurring in a complex chemical system base their search strategy on a set of easy to handle rules which select the type of steps more likely to occur.[30] Examples of expert systems aimed at defining the possible reactive evolution of a complex chemical system are given elsewhere in this book.[31]

From the reactive probabilities we have calculated the global rate constants for a given initial vibrational number (by summing over all accessible final vibrational states) for a large variety of initial vibrational states and an extended interval of temperatures. Some of these results are reported in figure 3 where the temperature evolution for rate constants calculated at v= 1, 3, 5 and 7 is reported. The figure shows that at T=4000 K and large v these rate constants are one order of magnitude larger than those at low v. The difference gets much larger at lower temperature because the decrease with the inverse of the temperature of these rate constants is dramatic at low v values while it is smoother at large v values.

As a prototype for reactions occurring through a bent transition state we have considered the $Li+HF$ reaction. A great deal of investigation has been focussed on this system. Measurements of the reactive cross section and of the detailed properties of the products have

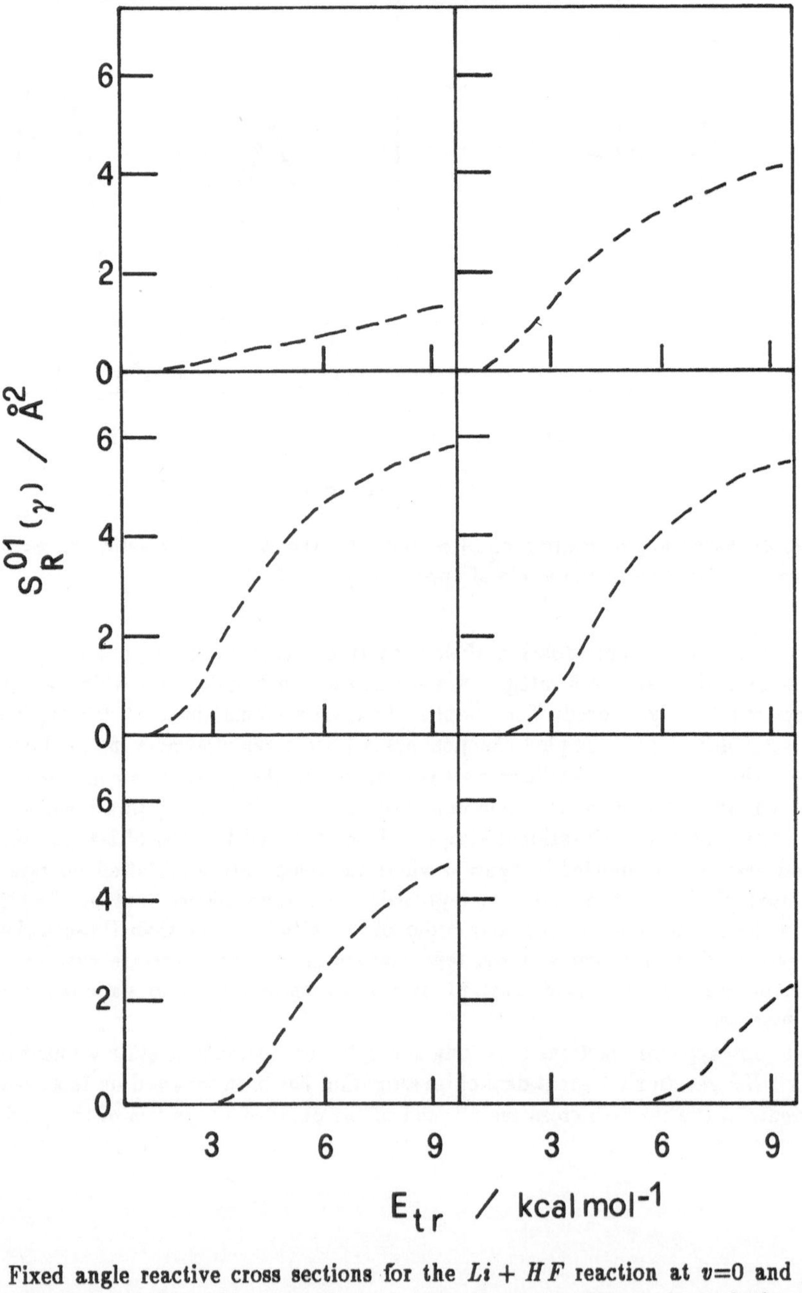

Figure 5: Fixed angle reactive cross sections for the $Li + HF$ reaction at v=0 and j=1 plotted as a function of the collision energy for γ= 68, 76, and 90 degrees from the lower to the upper lhs panel and γ= 60, 72, and 83 degrees from the lower to the upper rhs panel.

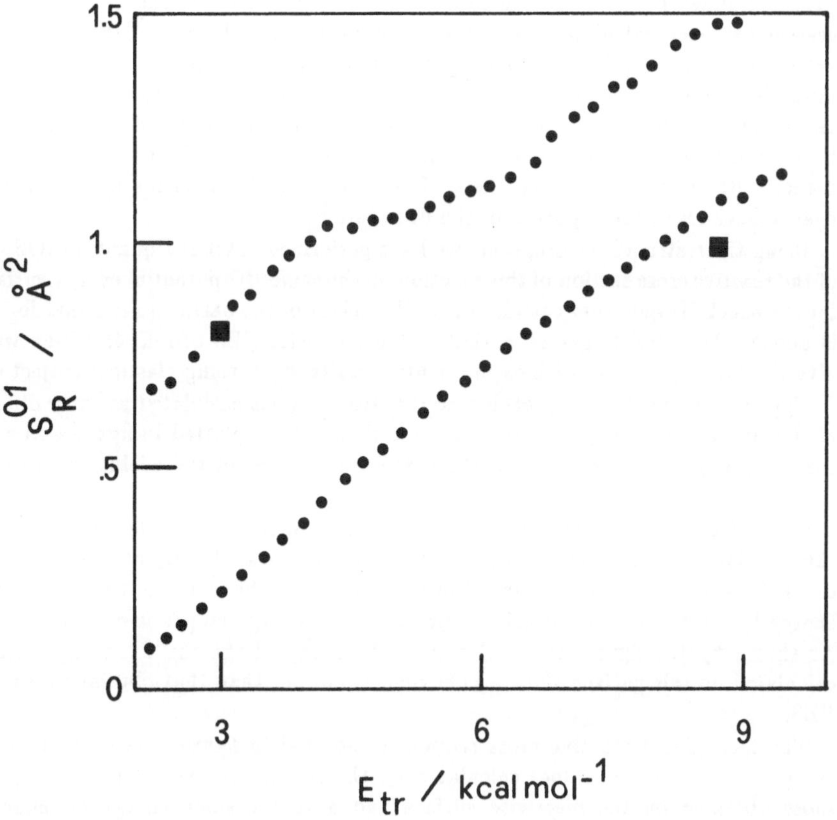

Figure 6: Integral reactive cross sections for the $Li+HF$ reaction at v=0 and j=1 plotted as a function of the collision energy. Reported values have been calculated on the BO potential energy surface of ref. 36 (solid circles, lower curve). For comparison values calculated on a piecewise potential energy surface[38] (solid circles, upper curve) as well as experimental values[32] are also shown.

been performed by Y.T. Lee and collaborators.[32] The potential energy surface of the *LiFH* system has been calculated using high level ab initio techniques.[33] Fits to the calculated values have been performed using different functional forms.[34–36] Empirically corrected potential energy surfaces have been used for trajectory calculations of the reactive cross section starting from initial conditions mimicking the experimental situation. Results have been compared with experimental values to understand the effectiveness of the corrections made.[37] Recently, preliminary exact 3D quantum calculations for a restricted interval of energy have also been reported in the literature.[2]

Using the restructured program. we have performed extended quantum IOS calculations of the reactive cross section of this reaction on the same BO potential energy surface adopted for the exact 3D quantum calculations. The height of the barrier to reaction for this surface is about 2 kcal/mol larger than that of the piecewise PES of ref. 35 which was found to give the best agreement with experimental results when using classical trajectories.[37]

Typical fixed angle of approach reactive cross sections calculated at three different values of the collision energy (3.0, 6.0, and 8.8 kcal/mole) are plotted in figure 4 as a function of $\cos\gamma$. As expected, the shape of the curve mimics that of the evolution of the barrier to reaction.

A few fixed angle of approach cross sections are reported in figure 5 as a function of the collision energy. The investigated energy interval was slightly more extended than that covered by the experiment. In addition, thanks to the reduced time consumption, the energy interval was scanned using a fine grid (the energy step was of about 0.2 kcal/mol). As shown by the figure, the evolution with energy of the the fixed angle cross sections calculated on this surface show a behaviour smoother than that obtained on the piecewise PES.

The final global reactive cross section is reported in figure 6 as a function of collision energy. As expected, values calculated on the BO surface are systematically lower than those obtained on the piecewise surface and have the same energy dependence. With comparison to the experiment, however, the BO results show also a good agreement. They fall, in fact, almost entirely within the error bars suggested in ref. 32. The main difference between the two sets of calculated cross sections consists in the fact BO results better agree with the experiment at higher collisional energy while the others better reproduce the low energy point.

6. Conclusions

When restructuring two different reduced dimensionality programs to run on supercomputers we have operated at various levels. At the first level we have reformulated part of the theoretical approach to better suit vector and parallel architectures. At the second level we have invoqued the automatic vectorizing and made use of the library routines optimized to run in vector mode. At the third level we have made use of the Macro Tasking Facility for running in parallel those sections of the program that could take effective advantage of the availability of more than two processors.

At the first level, the potential energy functional has been redesigned to obtain the maximum speed-up while maintaining a very high flexibility in representing ab initio potential energy values. At the second level, automatic vectorizing facilities have been invoqued to make vectorizable DO loops run in vector mode. To improve performances at this level some DO loops have been reorganized and vector routines linked. At the third level, parallelism for the l DO loop has been introduced obtaining the advantage of optimizing the balance between the size of the memory involved and the extent of concurrence in dispatched processes.

The large quantity of saved time has been partly reinvested in the restructured version of the program for considering explicitly the angular dependence of the partial wave expansion. After all these changes, the time consumption for a single energy run is still low enough to allow a systematic theoretical quantum investigation of the energy dependence of the cross section of several elementary chemical reactions. In this paper, we have reported the results of a detailed investigation of two prototype systems ideally suited for the two different theoretical approaches we have considered. The wealth of results produced by these calculations could be used for simple modeling of chemical reactivity to incorporate into rule-based deciding procedures typical of some expert systems.

7. Acknowledgments

Calculations have been performed at IBM ECSEC (Roma, Italy), CNUCE (Pisa, Italy), CNUSC (Montpellier, France) and CINECA (Casalecchio di Reno, Italy). Partial financial support from CICYT (Spain, grant no. PB85-0316) and CNR (Italy) is acknowledged. We also thank the Ministry of Education of Spain and Italy for financial support within a binational collaborative agreement.

References

1. M. Mishra, J. Linderberg and Y. Ohrn, Chem. Phys. Letters **111**,439(1984); K. Haug, D.W. Schwenke, Y. Shima, D.G. Truhlar, J.Z.H. Zhang, D.J. Kouri, J. Phys. Chem. **90**,6757(1986); A. Kuppermann, P.G. Hypes, J. Chem. Phys. **84**,5962(1986); J. Linderberg, Int. J. Quantum Chemistry, **S19**,467(1986); F. Webster, and J.C. Light, J. Chem. Phys. **85**,4744(1986); K. Haug, D.W. Schwenke, D.G. Truhlar, Y. Zhang, J.Z.H. Zhang, D.J. Kouri, J. Chem. Phys. **87**,1892(1987); P.G. Hypes, A. Kuppermann, Chem. Phys. Letters **133**,1(1987); J. Linderberg and B. Vessal, Int. J. Quantum Chemistry, **31**,65(1987); R.T Pack, G.A. Parker, J. Chem. Phys. **87**,3888(1987); G.A. Parker, R.T Pack, B.J. Archer, R.B. Walker, Chem. Phys. Letters **137**,564(1987); M. Baer, J. Phys. Chem. **91**,5846(1987); D.W. Schwenke, K. Haug, D.G. Truhlar, Y. Sun, J.Z.H. Zhang, D.J. Kouri, J. Phys. Chem. **91**,6080(1987); J.Z.H. Zhang, W.H. Miller, Chem. Phys. Letters **140**,329(1987); J.Z.H. Zhang, W.H. Miller, J. Chem. Phys. **88**,4549(1988); J.M. Launay, B. Lepetit, Chem. Phys. Letters **144**,346(1988); M. Mladenovic, M. Zhao, D.G. Truhlar, D.W. Schwenke, Y. Sun, D.J. Kouri, Chem. Phys. Letters **146**,358(1988); Y.C. Zhang, J.Z.H. Zhang, D.J. Kouri, K. Haug, D.W. Schwenke, D.G. Truhlar, Phys. Rev. Letters **60**,2367(1988); J.Z.H. Zhang, D.J. Kouri, K. Haug, D.W. Schwenke, Y. Shima, D.G. Truhlar, J. Chem. Phys. **88**,2492(1988); M. Baer, J. Chem. Phys. **91**,5846(1988).

2. A. Laganà, R.T Pack and G.A. Parker, Faraday Discuss. Chem. Soc. **84**(1987); G.A. Parker, R.T Pack, A. Laganà, B.J. Archer, J.D. Kress and Zlatko Bačic "*Exact Quantum Results for Reactive Scattering Using Hyperspherical (APH) Coordinates*" in this book.

3. J.M. Bowman and A.F. Wagner, in "*The theory of Chemical Reaction Dynamics*" D.C. Clary Ed.; Reidel, Dordrecht, 1986, p.47.

4. R.B. Walker and E.F. Hayes, J. Phys. Chem., **87**,1255(1983); T.P Tsien and R.T Pack, Phys. Letters **6**,54,400(1970);

5. R.B. Walker and E.F. Hayes, J. Phys. Chem. **88**,1194(1984); E.F. Hayes and R.B. Walker, J. Phys. Chem. **88**,3318(1984).

6. D.L. Miller, and R.E. Wyatt, J. Chem. Phys. **86**,5557(1987).

7. T.P Tsien and R.T Pack, Chem. Phys. Letters **6** 54, 400(1970); *ibid.* **8**, 579 (1971); *ibid.* **14**, 393 (1972).

8. D.C. Clary and G. Drolshagen, J. Chem. Phys. **76**,5027(1982); D.C. Clary, Chem. Phys. **81**,379(1983)

9. A. Laganà and E. Garcia, Chem. Phys. Letters **139**,140(1987).

10. S.I. Drodzov, Zh. Exp. Theor. Fiz. **28**,2734(1955) (English translation Sov. Phys. JETP 1,591(1955); D. Chase, Phys. Rev. **104**,838(1956).

11. R.T. Pack, J. Chem. Phys. **60**,653(1974); P.M. McGuire and D.J. Kouri, J. Chem. Phys. **60**,2488(1974); D.J. Kouri in "Atom Molecule Collision Theory: A Guide for the Experimentalist" R.B. Bernstein Ed. (Plenum, New York,1979); D.J. Kouri and D.E. Fitz, J. Phys. Chem. **86**,2224(1982).

12. J.C. Light and R.B. Walker, J. Chem. Phys. **65**,1598(1976); ibid. **65**,4272(1976).

13. J.W. Cooley, Math. Comput. **15**,363(1961).

14. G.Grossi, J. Chem. Phys. **81**,3355(1984);

15. see for example: N.C. Handy, "*Modern electronic structure calculations"; T.H. Dunning, Jr, Calculation and characterization of potential energy surfaces for chemical reactions* " in this book; C.W. Bauschlicher, Jr, S.R. langhoff and P.R. Taylor, "*The calculation of accurate potential energy surfaces* " in this book; K. Morokuma, "*Potential energy surfaces of several elementary chemical reactions* " in this book.

16. R.B. Bernstein, "*Chemical Dynamics via molecular beams and laser techniques* " (Oxford Univ. Press, New York, 1982).

17. R.L. Jaffe, M.D. Pattengill, F.G. Mascarello, and R.N. Zare, J. Chem. Phys., **86**, 6150 (1987).

18. S. Carter, I.M. Mills, J.N. Murrell, and A.J.C. Varandas, Mol. Phys. **45**, 1053 (1982).

19. J.N. Murrell, S. Carter, S.C. Farantos, P. Huxley, and A.J.C. Varandas "*Molecular Potential Energy Functions* " (Wiley, New York, 1984).

20. E. Garcia and A. Laganà, Mol. Phys. **56**, 621 (1985); *ibid.*, **56**, 629 (1985).

21. M. Dini, A. Laganà, and M. Paniagua, to be published.

22. E. Garcia, L. Ciccarelli, and A. Laganà, Theor. Chim. Acta, **72**,253(1987).

23. L. Ciccarelli and A. Laganà, J. Phys. Chem. **92**,932(1988).

24. See for example G.C. Schatz in "*The theory of chemical reaction dynamics*, D.C. Clary, ed. (Reidel, Dordrecht, 1986). Pag. 1.

25. C. Gorse, M. Capitelli, M. Bacal, J. Bretagne, and A. Laganà Chem. Phys. **117**,177(1987)

26. Topics in Current Physics, "*Nonequilibrium Vibrational Kinetics* M. Capitelli ed., Springer, Berlin, 1986.

27. A. Laganà Int. J. Chemical Kinetics **18**,1009(1986)

28. J.M. Alvariño, E. Garcia and A. Lagana, "*Quasiclassical Calculations for alkali/alkaline earth + hydrogen halide chemical reactions using supercomputers*" in this book.

29. D.G. Truhlar and C.J. Horowitz, J. Chem. Phys. **68**, 2466 (1978); ibid. **71**, 1514 (E) (1979).

30. Nikolaos Kaniadakis, Tesi di Laurea "*Modellizzazione di processi chimici reattivi mediante tecniche di intelligenza artificiale*", Perugia, 1988.

31. G.M. Come and G. Scacchi, "*The modeling of complex gas phase reactions: from expert systems to supercomputers*" in this book.

32. C.H. Becker, P. Casavecchia, P.W. Tiedemann, J.J. Valentini and Y.T. Lee, J. Chem. Phys., **73**, 2833 (1980).

33. M.M.L. Chen and H.F. Schaefer, III, J. Chem. Phys., **72**,4376(1980).

34. S. Carter and J.N. Murrell, Mol. Phys. **41**,567(1980).

35. E. Garcia and A. Laganà, Mol. Phys. **52**,1115(1984).

36. A. Laganà, O. Gervasi and E. Garcia, Chem. Phys. Letters **143**,174(1988).

37. J.M. Alvariño, P. Casavecchia, O. Gervasi, and A. Laganà, J. Chem. Phys. **77**, 6341 (1982); A. Laganà, M.L. Hernandez, and J.M. Alvariño, Chem. Phys. Lett., **106**,41(1984); E. Garcia, A. Laganà, J.M. Alvariño, and M.L. Hernandez, J. Chem. Phys. **84**, 3059 (1986).

38. A. Laganà, E.Garcia, and O. Gervasi, J. Chem. Phys. (in press).

APPROXIMATE QUANTUM MECHANICAL CALCULATIONS ON MOLECULAR ENERGY TRANSFER AND PREDISSOCIATION

D. C. Clary
University Chemical Laboratory
Lensfield Rd
Cambridge CB2 1EW
UK

ABSTRACT. Various quantum mechanical scattering methods for performing calculations on the vibrational and rotational energy transfer in three dimensional collisions involving polyatomic molecules are described. Furthermore, calculations of the lifetimes for the vibrational and rotational predissociation in van der Waals molecules are also discussed. Examples of computations include rotational excitation in Ne+HF and He+CH_3CN, vibrational excitation in collisions of rare gases with N_2, CO_2, $C_2O_2H_2$ and para-difluorobenzene(pDFB), rotational predissociation and bound states in Ne-HF and Ne-HCl and vibrational predissociation in rare gas-C_2H_4, $[C_2H_4]_2$ and Ar-pDFB complexes.

I. INTRODUCTION

There has been a resurgence of interest in the collision-induced vibrational and rotational excitation of gas-phase molecules in recent years[1]. This is partly due to the fact that new experimental methods such as pulsed molecular beams, supersonic jets, dispersed laser-fluorescence and high-resolution infrared spectroscopy are providing exciting new results which are probing the mode selective aspects of these collisional processes in more detail than has been possible previously[2]. It is now proving possible for the first time to measure cross sections as a function of collisional energy for the excitation of state-selected vibrational and rotational states in polyatomic molecules at low collision energies[3]. Furthermore, the attention of several spectroscopists has turned, very recently, to the reverse process of vibrational energy transfer, namely the infrared photodissociation of van der Waals molecules[4]. Here, linewidths for vibrational and rotational predissociation are measured and are dependent on the same potential energy surface and, essentially, the same dynamics as that which controls vibrational and rotational energy transfer.

The advances in experimental techniques have been matched by theoretical and

A. Laganà (ed.), Supercomputer Algorithms for Reactivity, Dynamics and Kinetics of Small Molecules, 295–325.

computational developments, with the availability of supercomputers and mini-supercomputers playing a significant role(for a recent review with many references see Ref.5). Furthermore, the significant efforts from *ab initio* quantum chemists in calculating potential energy surfaces of good quality has done much to stimulate scattering theoreticians to develop methods, algorithms and computer programs that are capable of yielding results for comparison with these new experiments. The quantum mechanical scattering calculations often involve the numerical solution of large numbers of coupled equations and, since many matrix manipulations are involved, computers with vector processing capabilities are becoming essential when realistic calculations are made. However, the number of vibrational-rotational states in a polyatomic molecule is often too large for highly accurate quantum scattering calculations and, therefore, the use of approximate scattering methods remains a very significant area from the point of view of comparing with the experimental results. However, the advent of supercomputers has enabled more accurate quantum mechanical computations to be performed and these benchmark results have enabled us to assess with more certainty the accuracy of the more approximate theories.

The aim of this article is to give the reader a feel for the developments in this field through a description of some of the practical quantum mechanical theories for inelastic scattering. Furthermore, applications of these methods are illustrated through the presentation of several calculations performed recently on vibrational and rotational energy transfer and predissociation in our laboratory. Examples have been chosen to illustrate the various types of calculations that can be performed, ranging from atom-diatom systems to polyatomic dimers and atom-aromatic molecule complexes. A particular emphasis is placed on comparing approximate calculations with more accurate ones and on comparing calculated results with experiments. Through these comparisons we hope to illustrate the power of the three-dimensional quantum scattering computations in providing results that not only can compare well with experiment, but can make some useful predictions also. It is our opinion that many energy transfer process in polyatomic molecules involve perhaps only two or three active degrees of freedom(certain vibrations and rotations) and by treating these motions accurately, and averaging over all other motions, very realistic models can be obtained even for quite complicated polyatomic systems.

In Section II we outline the coupled-channel approach to quantum scattering, which forms the basis for almost all the calculations we will be describing. We discuss the numerical methods for solving the coupled channel equations. Furthermore, we describe the various sudden approximations to the exact theory that have enabled coupled-channel computations to be performed on vibrational and rotational relaxation in polyatomic molecules. We also describe how the coupled channel method can be used to calculate the spectra for vibrational and rotational predissociation in van der Waals molecules. In Section III we describe results for rotational and vibrational energy transfer. Systems discussed include rotational excitation in Ne+HF and He+CH_3CN and vibrational excitation in collisions of rare gases with N_2, CO_2, glyoxal and para-difluorobenzene(pDFB). Rotational predissociation and bound states in Ne-HF and Ne-HCl and vibrational predissociation in rare gas-C_2H_4, [C_2H_4]$_2$ and Ar-pDFB complexes are discussed in Section IV. Conclusions are in Section V in which the future prospects for computations in this field are outlined.

II. QUANTUM THEORY OF VIBRATIONAL-ROTATIONAL EXCITATION

IIa. The Coupled-channel method

We here outline the coupled-channel approach to inelastic scattering for atom-molecule or molecule-molecule collisions[6]. It is necessary to compute a wavefunction Ψ^J with fixed total angular momentum J and collision energy E. An appropriate expansion is

$$\Psi^J = \sum_i f_i^J(R)\, g_i \tag{1}$$

where $f_i(R)$ is a translational wavefunction that depends on the distance R between the centres of mass of the colliding particles and $\{g_i\}$ is a suitably chosen set of internal basis functions which are often constructed from the vibrational-rotational states of the molecule or molecules involved and are combined in such a way that they are eigenfunctions of the total angular momentum of the total system. For molecule-molecule interactions, the basis functions would normally involve products of the eigenstates of each individual molecule. The calculations are usually performed using either a set of space-fixed coordinates[6] or with a body-fixed set of coordinates[7] in which the z axis is placed along the vector **R** between the centres of mass of the colliding particles. The hamiltonian for the problem can be written as

$$H = -\hbar^2/(2\mu R)\partial^2/\partial R^2 R \; + H_{mol} \; + \ell^2/(2\mu R^2) \; + V \tag{2}$$

where H_{mol} represents the hamiltonian or hamiltonians of the isolated molecule or molecules, ℓ^2 is the orbital angular momentum operator associated with the rotation of **R** and V is the intermolecular potential that vanishes as R gets large and becomes repulsive as R gets very small. Substitution of equation (1) into the Schrödinger equation , multiplication by one of the basis functions and integration over all the internal coordinates produces the coupled-channel equations for the $f_i(R)$

$$d^2 f_i(R) / dR^2 = \sum_{i'} C_{ii'}(R)\, f_{i'}(R) \, , \tag{3}$$

where the $C_{ii'}(R)$ are matrix elements which contain the vital potential coupling terms

$$< g_i \mid V \mid g_{i'} >, \qquad (4).$$

The coupled-channel equations (3) can be solved numerically using procedures discussed below by starting the integration in the classically forbidden region at small R and integrating out until the effect of the intermolecular potential can be neglected. Application of boundary conditions

$$f_{ii'}(R \rightarrow 0) \rightarrow 0, \tag{5}$$

$$f_{ii'}(R \rightarrow \infty) \rightarrow (k_i)^{-1/2} \{\delta_{ii'} \exp[-i(k_i R - J\pi/2)] - S^J_{ii'} \exp[\, i(k_{i'} R - J\pi/2)]\} \tag{6}$$

produces the S matrix element $S^J_{ii'}$, the absolute square of which gives the probability for the transition between the quantum states i and i'. In equation (6), k_i is the wavenumber for the initial state i. The computations are repeated for many values of J, and this enables the state-selected integral cross sections

$$\sigma(i \rightarrow i') = \pi/(k^2_i)\ \Sigma_J\, (2J+1)\ |\ S^J_{ii'}|^2 \tag{7}$$

to be obtained. Maxwell-Boltzmann averaging of these cross sections gives the rate coefficients $k(i \rightarrow i', T)$ which are the temperature-dependent quantities produced in certain types of bulb experiments.

IIb. Numerical solution of the coupled-channel equations

Much attention has been diverted to the numerical solution of the coupled-channel equations (3). Two of the most widely used and stable methods are the log-derivative approach introduced originally by Johnson[8] and the R-matrix propagator method developed by Light and Walker[9]. The log-derivative approach propagates the ratio between the derivative of the wavefunction (with respect to R) and the wavefunction, while the R matrix propagates the inverse of this quantity. Thus the log derivative matrix is defined by

$$\mathbf{Y}(R) = \mathbf{F}'(R)\, \mathbf{F}^{-1}(R) \tag{8}$$

while the R matrix is

$$\mathbf{R}(R) = \mathbf{F}(R)[\mathbf{F}'(R)]^{-1} \tag{9}$$

In the log-derivative method, the scattering coordinate R is divided into sectors and the propagator matrices $\{\mathcal{Y}_i(a,b)\}$ for the interval R=a to R=b are defined by

$$\begin{bmatrix} \mathbf{F}'(a) \\ \mathbf{F}'(b) \end{bmatrix} = \begin{bmatrix} \mathcal{Y}_1(a,b) & \mathcal{Y}_2(a,b) \\ \mathcal{Y}_3(a,b) & \mathcal{Y}_4(a,b) \end{bmatrix} \begin{bmatrix} -\mathbf{F}(a) \\ \mathbf{F}(b) \end{bmatrix} \tag{10}$$

The recurson relationship for the log derivative matrix is then

$$Y(b) = \mathcal{y}_4(a,b) - \mathcal{y}_3(a,b)\,[\,Y(a) + \mathcal{y}_1(a,b)]^{-1}\,\mathcal{y}_2(a,b) \qquad (11).$$

Formulae for the propagator matrices $\{\mathcal{y}_i(a,b)\}$ can be derived by starting from the solution to the simple homogeneous problem on the interval [a,b]

$$\Phi''(R) = C_{ref}(R)\,\Phi(R) \qquad (12)$$

where the reference potential can be freely chosen. Johnson set $C_{ref}(R)$ to zero, while, in a significant modification, Manolopoulos[10] used a constant diagonal reference potential which can be obtained, for example, from the diagonal elements of the close-coupling matrix C(R). The algorithm then involves treating equation (3) as an integral equation which can be discretised with a Simpson's rule quadrature. This requires evaluating the residual coupling matrix

$$U(R)= C(R)-C_{ref}(R) \qquad (13)$$

at the points a,b and the midpoint of the sector at R= c.

In the simplest form of the R-matrix propagator method[9], the scattering coordinate is again split up into sectors and the close-coupling matrix at the midpoint of each sector i is diagonalised ro give an eigenvector matrix $\mathbf{T}_i$. In sector i with width h_i, the local scattering wavefunction is then approximated as

$$G_i(R) = g_{i1}(R) + g_{i2}(R)\,[\mathbf{W}_i]^T\mathbf{R}_{i-1}\mathbf{W}_i \qquad (14)$$

where the diagonal matrices $g_{i1}(R)$ and $g_{i2}(R)$ are constructed from sine and cosine functions, $\mathbf{W}_i$ is the sector overlap matrix

$$\mathbf{W}_i = [\mathbf{T}_{i-1}]^T\mathbf{T}_i \qquad (15)$$

and $\mathbf{R}_{i-1}(R)$ is the R matrix for sector i which is obtained from

$$\mathbf{R}_{i-1} = G_{i-1}(R_{i-1} + h_{i-1}/2)\,[dG_{i-1}(R_{i-1}+h_{i-1}/2)/dR]^{-1} \qquad (16).$$

There are several other techniques for solving the coupled-equations, some of which involve modifications and hybrids of the above approaches. For example, Alexander and Manolopoulos[11] combine the modified log-derivative method for the shorter-range region with a linear reference potential algorithm for the long-range region, where it is often appropriate to take sectors with large widths. A general program incorporating this technique can be obtained from Alexander[12]. The program that is most widely used is the MOLSCAT code originally developed by Green and more recently modified by Hutson to incorporate the Manolopoulos algorithm and several other features including the detection and search of scattering resonances, the calculation of bound states using close-coupling methods and surface scattering using a basis set of diffraction states[13]. A thorough examination of the relevant merits of the R matrix propagator and modified log-derivative approaches has been made by Manolopoulos in his Ph.D. thesis[14]. The number of N^3 operations required for the R-matrix, log derivative and modified log-derivative algorithms have the ratio $4^2/3$, $1^1/2$ and $1^1/2$ respectively at the first energy, and 2, $1^1/2$ and 1 at subsequent energies. These ratios are reflected directly in the computer time needed for the three approaches with the same number of sectors. Furthermore, in the limit of a very small sector width h, the numerical error in the calculated S matrix elements normally depends on h^2 and h^4 for the R-matrix and log-derivative methods respectively. Thus the log-derivative method gives a superior convergence with respect to the number of sectors. This is shown in Figure 1 were the calculated root mean square error in the S matrix elements for the 9-channel Lester Bernstein rotational excitation close-coupling problem are plotted for the R-matrix, log-derivative and modified log-derivative methods. This will often make the modified log-derivative method the most efficient approach. It should be emphasised that most of the numerical work in these propagator methods goes into the inversions and multiplications of matrices. These operations are, of course, ideal for supercomputers or minisupercomputers with vector processing capabilities.

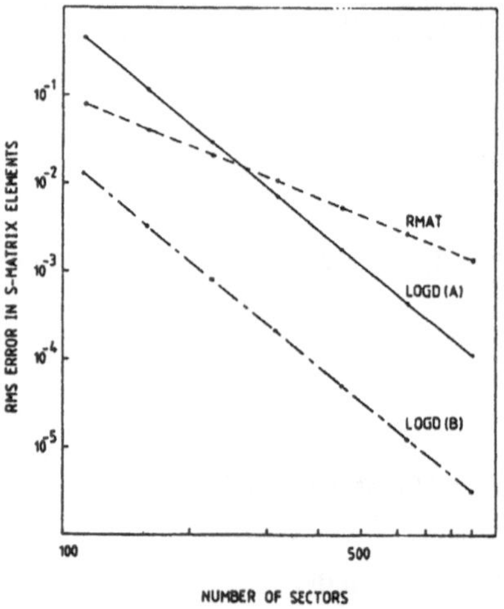

Figure 1. Convergence of S matrix elements for 9-channel Lester-Bernstein rotational excitation problem[10]. LOGD(A) refers to method of Johnson[8], LOGD(B) to method of Manolopoulos[10].

Furthermore, the modified log-derivative method has also been shown to be a very useful method for calculating the bound vibrational-rotational states of van der Waals molecules[14]. Here, the log-derivative matrix is integrated outward from the classically forbidden region and inward from the asymptotic region, where exponentially decaying boundary conditions are applied. The energy can be varied until the determinants of the two log-derivative matrices are equal at a suitably chosen intermediate point and an eigenvalue is thus obtained as the wavefunction and its derivative will then be continuous. A detailed account of this procedure, including the complicated aspects of multichannel node counting, is given in the thesis of Manolopoulos[14]. As is discussed below, this method is also very useful for determining the positions of scattering resonances relevant to vibrational and rotational predissociation. The R-matrix method has also been adapted for coupled-channel bound state computations by Danby[15], but , for the reasons discussed above, the modified log-derivative method would normally be expected to be more efficient.

Another important application of the coupled-channel approach is in the calculation of integrals that arise in molecular photodissociation. In the case of photodissociation, it is necessary to calculate integrals of the form[16]

$$|\langle \Psi_I^{AB} | \mu | \Psi_i^{[A+B]} \rangle|^2 \tag{17}$$

where Ψ_I^{AB} is the wavefunction for the original molecule in state I, μ is the transition dipole operator and $\Psi_i^{[A+B]}$ is the scattering wavefunction for the dissociation into A+B in quantum state i. Because of the simple expansion of the local translation wavefunction in each sector in terms of sine and cosine functions as shown in equation (14), these integrals are straightforward to evaluate in the R-matrix propagator method[17,18]. Formulae for their evaluation within the framework of the log-derivative method have also been derived by Mrugala[19] and, in a more compact fashion, by Manolopoulos[14]. A numerical comparison of the R-matrix propagator and log-derivative approaches has been carried out by the author and Manolopoulos for the dissociation of CF_3I into CF_3+I, with CF_3 in different vibrational states[20]. A three-dimensional combined vibrational close-coupling, rotational infinite order sudden method was used[18].

Table 1. Photodissociation probabilities[20] for $CF_3I \rightarrow CF_3$+I with laser frequency of ω=40323cm^{-1}. For more details see Refs 18 and 14. RM=R matrix, LD=Log-derivative methods.

	Ground state		C-I Stretch excited		C-I Bend excited	
Sectors	RM	LD	RM	LD	RM	LD
100	0.5455(-2)	0.3129(-3)	0.7346(-2)	0.1825(-2)	0.4261(-2)	0.3767(-3)
200	0.5304(-2)	0.5319(-2)	0.7520(-2)	0.7685(-2)	0.4123(-2)	0.4134(-2)
300	0.5304(-2)	0.5303(-2)	0.7549(-2)	0.7548(-2)	0.4123(-2)	0.4122(-2)

Table 1 compares the convergence of the photodissociation integrals of equation (17) for CF_3I calculated for the two methods with respect to the number of equally spaced sectors. Here it can be seen that, unlike the case of scattering S matrix elements, a superior convergence is obtained for the R-matrix propagator method. This is because this approach performs a local expansion in terms of oscillatory sine and cosine functions across a sector, the integrals involving which can be evaluated analytically. In contrast, the log-derivative approach involves evaluating the integrals on a quadrature grid, and the oscillatory nature of the scattering wavefunction might then not be approximated so accurately at the higher energies often considered in photodissociation calculations. However, at very low collision energies, the log-derivative approach will be more competitive when the scattering wavefunction is not so highly oscillatory.

Yet another useful application of the coupled-channel wavefunctions is in the evaluation of integrals arising in applications of perturbation theory (eg. coupled-channel distorted wave Born approximation calculations(DWA)). For example, we have recently been working on an approach for calculating vibrational-rotational cross sections that involves performing separate rotational coupled-channel calculations for each vibrational state of a molecule involved in a collision and calculating matrix elements over the vibrational coupling matrix elements with these scattering wavefunctions. To take an atom-diatom system for example, coupled-channel calculations with a spherical harmonic rotational basis set $\{Y^{j\Omega}\}$ can be used, where Ω represents the projection of the rotational and total angular momentum along the scattering coordinate R. The vibrational state v is held fixed for each rotational coupled-channel calculation and the coupled-channel expansion is of the form

$$\Psi^{Jv} = \sum_{j\Omega} f_{Jj\Omega v}(R)\, Y^{j\Omega} \quad (18).$$

The S matrix for the transition $(J, v, j, \Omega \rightarrow v', j', \Omega'\}$ is then calculated from

$$S_{Jvv'} = [i/2][2\mu/\hbar^2] \int_0^\infty f_{Jv}(R)\, V_{vv'\Omega\Omega'}(R)\, f_{Jv'}(R)\, dR \quad (19)$$

where $V_{vv'\Omega\Omega'}(R)$ is the close-coupling matrix with elements labelled by $\{v,j,\Omega\}$. It is straightforward to calculate these integrals using both the R-matrix propagator[21,22] and log-derivative techniques[19]. For example, it can be seen from equation (14) that the local expansion of the wavefunction in terms of sine and cosine functions in the R-matrix approach enables the contribution from a particular sector to the integral of equation (19) to be evaluated analytically and the integrals can then be accumulated in a propagator fashion. Only minor modifications of existing close-coupling codes are needed to do this and the extra numerical work largely requires just matrix multiplies, which are ideally suited to vector computers. Surprisingly, although this method has been used for molecule-surface scattering[21] and reactive scattering calculations[23], it has not been applied before to vibrational-rotational energy transfer computations. Since cross sections for pure rotational transitions are normally large and those involving vibrational transitions are small, this would seem to be a natural method for vibrational-rotational energy transfer. This will be particularly true for polyatomic molecules, where there are many different vibrational states. By performing rotational coupled-channel calculations for each of the individual vibrational states {v}, and storing the necessary numerical information, it is then possible to evaluate the vibrational-rotational cross sections for all the transitions between the states (v,j→v',j'). In Section III, we describe an unpublished application of this rotational coupled-channel, distorted-wave approach to the He+N_2(v,j→v',j') problem.

IIc. Close-coupling calculations

The term "close-coupling"(CC) is often used to describe coupled-channel calculations that apply no approximations to the kinetic energy part of the hamiltonian (the potential energy surface will always be an approximation). Even with the best algorithms and the most modern supercomputer or minisupercomputers that are available, it is not comfortable to do computations with more than about 200 basis functions and this makes it very difficult to obtain converged close-coupling calculations of cross sections for most polyatomic molecule problems. Truhlar and coworkers[24] have made very extensive close-coupling calculations on the HF+HF vibrational-rotational relaxation problem and, through an intensive use of CRAY computers, and a very thorough vector-optimisation of their R-matrix propagator computer code, they have managed to obtain converged transition probabilities for J=0 with more than a thousand basis functions. However, to obtain converged cross sections for this system, in which the calculations have to be repeated for several larger J values, seems out of the reach of even the most powerful computational facilities. Because of the matrix-based nature of the numerical algorithms, the computer time depends roughly on N^3, where N is the number of basis functions. Furthermore, many more basis functions are needed for larger values of J than for J=0, with the number needed for atom-diatom sytems scaling roughly linearly with J. Thus the main use of "exact" close-coupling calculations in the near future will be to provide benchmark results which can test the accuracy of approximate methods which can more readily be applied to more complicated problems.

II d. The Centrifugal Sudden Approximation

A variety of approximations have been developed that approximate or neglect certain angular momentum terms in the kinetic energy part of the hamiltonian for the scattering problems. These are often termed sudden approximations(for a list of references see Ref.5). Their derivation from the exact close-coupling theory is best understood by making used of the Body-Fixed(BF) systems of coordinates in which the z axis is placed along the vector **R** joining the atom to the centre of mass of the molecule, and both the total and rotational angular momenta **J** and **j** have the projection Ω along this axis. For an atom-diatom system, the angular momentum operators in the exact hamiltonian are

$$H_{ang} = j^2/(2\mu R^2) + | \mathbf{J} - \mathbf{j} |^2/(2\mu R^2) \tag{20}$$

In the close-coupling expansion, basis functions with different Ω states are coupled together and this can lead to an enormous basis set being required. Expansion of the $|\mathbf{J}\text{-}\mathbf{j}|^2$ operator gives

$$|\mathbf{J}\text{-}\mathbf{j}|^2 = J^2 + j^2 - 2j_z^2 - [J^- j^+ + J^+ j^-] \tag{21}$$

and it is the $[J^- j^+ + J^+ j^-]$ operators that couple basis functions together that differ by one in Ω. Neglect of these particular operators removes the coupling in basis functions with different Ω quantum numbers and drastically reduces the size of the basis set[7,25]. This is the centrifugal sudden approximation (CSA) and is often known as the coupled-states approximation, although that is a term we prefer not to use as it is easily confused with "close-coupling". The basis set expansion for a CSA calculation of vibrational-rotational excitation cross sections would thus be

$$\Psi^{J\Omega} = \Sigma_{jv} \; f_{jv}^{J\Omega}(R) \, Y^{j\Omega} \, h_{vj} \tag{22}$$

where h_{vj} is an appropriate vibrational wavefunction for the isolated diatomic moelcule for a particular j value. The S matrix elements calculated for fixed Ω are thus $S^{J\Omega}{}_{vjv'j'}$ and the CSA integral cross secions are given by

$$\sigma(vj \rightarrow v'j') = \pi/([2j+1]k^2{}_{vj}) \; \Sigma_J \Sigma_\Omega (2J+1) \; | \; S^{J\Omega}{}_{vjv'j'} |^2 \tag{23}$$

Most CSA calculations of integral rotational or vibrational-rotational cross sections have given very good agreement with close-coupling results when these have been available[26]. However, for subtle magnetic properties or calculations involving high J or j, the CSA must be used with care. This is illustrated in Section III where unpublished calculations we have performed of cross CSA cross sections for the relaxation of HF(j=13) by Ne atoms are shown to underestimate the close-coupling results.

CSA calculations can be performed of rotational excitation cross sections for atom-asymmetric top collisions quite easily at low collision energies. To date, the only application of the CSA to an atom-polyatomic molecule vibrational-rotational problem has been to $He+CO_2(01^10,j \rightarrow 00^00,j')$[27] and some of these calculations are described in Section III. Unfortunately, most polyatomic molecules have rotor constants that are too small to enable converged CSA cross sections to be obtained for vibrational-rotational excitation as the required rotational basis sets will be large since so many rotational channels will be energetically open at typical collision energies. Thus further approximations are required.

II e. The Azimuthal approximation

A symmetric top molecule has the molecular rotational hamiltonian

$$H_{rot} = bj^2 + (a-b)j_z^2 \tag{24}$$

where a and b are rotor constants and j_z is the operator associated with the projection of the molecular angular momentum along the symmetry axis z of the symmetric top. If the molecule is allowed to vibrate, a set of molecular-fixed axes can be defined by the Eckart conditions and a scattering atom will then have spherical polar angles of orientation (θ, ϕ) with respect to these axes. It so happens that a large number of polyatomic molecules are near prolate symmetric tops, and those containing hydrogen atoms often have a b rotor constant much smaller than the a rotor constant. Thus the rotational levels described by the j rotor constant are often quite closely spaced while those associated with the k rotor constant can have quite wide spacings. This suggests[28] the sudden approximation of setting the small b rotor constant to zero, while retaining the $(a-b)j_z^2$ term in H_{rot}. Thus the coupled-channel calculations can be performed with a basis set that depends on the azimuthal angle ϕ, coupled with a suitable vibrational basis for the molecule. These calculations are done for fixed values of the polar angle θ. The appropriate coupled-channel expansion is[28]

$$\Psi^J = \Sigma_{vk}\, f_{vk}^J(R;\theta)\, h_v(q) \exp(ik\,\phi) \tag{25}$$

where $h_v(q)$ is a vibrational wavefunction that depends on the normal coordinates q of the isolated molecule. The S matrix elements associated with this basis are $S^J{}_{vkv'k'}(\theta)$. By repeating the calculations for a range of θ values, the vibrational cross sections

$$\sigma(v,k \rightarrow v') = \pi/(k^2{}_v)\ \Sigma_J\ \Sigma_{k'}\,(2J+1)\ \int^{\pi}_0 \sin\theta\ d\theta\ |S^J{}_{vkv'k'}(\theta)|^2 \tag{26}$$,

summed over all product (j',k') states and averaged over all initial j states, are obtained. It is also possible to calculate S matrix elements for the transition (v,j,k,$\rightarrow$v',j', k') with fixed Ω by taking the matrix elements[29]

$$S^J{}_{vjkv'j'k'} = \langle d^{j\Omega}{}_k(\theta)\ |\ S^J{}_{vkv'k'}(\theta)\ |\ d^{j'\Omega}{}_{k'}(\theta)\rangle \tag{27}$$

where $d^{j\Omega}{}_k(\theta)$ is the Wigner d function suitably normalised. We have given this method the rather lengthy title[28], azimuthal and vibrational close-coupling, infinite order sudden method(AVCC-IOS), although for this particular review we will call it the Azimuthal method. First application of the method was to He+C_2H_4 vibrational-rotational relaxation[28] and the method has also been applied to rotational excitation in He+CH_3CN[29] (see section III). New unpublished results[30] for the collisional relaxation in He and Ar collisions with glyoxal($C_2H_2O_2$) are also presented in section III. Furthermore, the azimuthal method can be adapted to account for strong coriolis terms in the molecular hamiltonian that can mix together v and k states in the isolated molecule and calculations have been performed with this procedure on He+D_2CO[31], which is a system of considerable experimental interest[32]. A very useful application of the Azimuthal method has been in the vibrational predissociation of atom-ethylene van der Waals molecules[33] and the ethylene dimer[34]. This is described in more detail in Section IV. The power of the azimuthal method is that it allows important rotational effects to be treated explicitly in scattering calculations on polyatomic molecules, which could not otherwise be tackled using fully quantum-mechanical techniques.

IIf. The infinite-order sudden approximation

In the infinite-order sudden approximation(IOSA)[35-40], all the rotational operators in the molecular hamiltonian are set to an average value (quite commonly zero). This is known as the energy sudden approximation(ESA). Furthermore the CSA is applied. The approximation thus assumes that the rotational energy levels of the molecule are degenerate during the collision and is clearly appropriate when the rotational constants of the molecules are small, which is the case with many polyatomic molecules. Indeed, it would seem that the larger the molecule is, the more appropriate is the IOSA. The vibrtaional coupling can still be treated with no approximations with this theory and we then have the vibrational coupled-channel, IOSA(VCC-IOSA)[41,42]. Here the wavefunction expansion is

$$\Psi^J = \sum_v \; f_v^J(R; \theta, \phi)\, h_v(q) \qquad (28)$$

and the calculations are performed for fixed angles of the spherical polar angles (θ, ϕ). Solution of the coupled equations yields the S matrix elements $S^J_{vv'}(\theta,\phi)$ from which the vibrational cross sections[42]

$$\sigma(v \rightarrow v') = 1/(4\,k^2_v)\; \Sigma_J\,(2J+1) \int^{\pi}_0 \sin\theta\, d\theta \int^{2\pi}_0 d\phi \;|\; S^J_{vv'}(\theta,\phi)\,|^2 \qquad (29)$$

are obtained. For many energy transfer problems involving polyatomics it is only necessary to couple together two or three different vibrational normal modes at a time and this makes it straightforward and automatic to perform VCC-IOSA calculations on polyatomic molecules as large as aromatics providing a suitable potential energy surface is available[43,44].We have developed a general computer code[43] which automatically yields vibrational relaxation cross sections and rate coefficients from a supplied potential energy surface for atom-polyatomic systems. Furthermore, this code has now been arranged so that it can directly take harmonic force-fields calculated using the ab initio CADPAC program of Amos[45] which uses the method of SCF-analytical gradients to obtain the molecular geometry and harmonic force-field. This force-field is needed as it is necessary to calculate the VCC-IOSA matrix elements[42]

$$\langle h_v(q) \;|\; V(R,q,\theta,\phi) \;|\; h_{v'}(q)\rangle \qquad (30),$$

where we note that $h_v(q)$ is the full vibrational wavefunction for the polyatomic molecule and q represents all the normal modes in the molecule. However, for those modes that are not being considered explicitly, the normal coordinates can be set to zero. It is straightforward to include anharmonicities in the molecular potential when calculating the $\{h_v(q)\}$. However, for many low-energy problems, it is just excitation or relaxation involving v=1 levels that are relevant and here harmonic oscillators with the correct experimental frequencies will usually suffice. The integrals of equation (30) are evaluated using a Gauss-Hermite quadrature and, therefore, it is necessary to transform the normal coordinates to the distances and angles in which the intermolecular potential is usually expressed. The required matrix that transforms the normal modes to cartesian coordinates is obtained from the *ab initio* gradient code[45]. Table 2 summarises the various sudden acronyms.

Table 2. Summary of acronyms

Acronym	Meaning	Basis Set*
CC	Close Coupling	v x j x k x Ω
CSA	Centrifugal Sudden Approximation (Coupled States Approximation)	v x j x k; Fixed Ω
IOSA	Infinite Order Sudden Approximation	Fixed θ and ϕ
VCC-IOSA	Vibrational Coupled Channel IOSA	v; Fixed θ and ϕ
AVCC-IOSA	Azimuthal and VCC-IOSA	v x k ; Fixed θ

* Basis set refers to the collision of an atom with a symmetric top molecule having the vibrational quantum number v, rotational quantum numbers j and k, and rotational projection quantum number Ω along the body-fixed z axis.

Our VCC-IOSA computer code has been applied to a variety of problems including the interactions of rare gases with CO_2, SO_2, O_3, SF_6, C_2H_4, C_3H_6 and $C_6H_4F_2$ (see Ref.5 for a review of these computations). It has also been used in calculations on the vibrational excitation of CO_2 and H_2O by O atoms[46-48]. A careful examination of the accuracy of the approach has been carried out by Banks in his Ph.D. thesis[49] by comparing with CSA cross sections for atom+CO_2[27] and He+N_2[50] vibrational-rotational excitation. Generally speaking, the VCC-IOSA cross sections are accurate providing the rotor constants and vibrational frequencies are not too large.

IIg. Vibrational and rotational predissociation of van der Waals complexes

The theories described above can all be adapted to perform calculations of linewidths for the rotational and vibrational predissociation of van der Waals molecules. These quantities can now be measured in high resolution infrared and far-infrared spectroscopy experiments[4]. The weakly-bound complex absorbs a photon with an energy such that the complex can dissociate. However, this process takes a certain time and, through the Heisenberg principle, the lifetime is inversely related to the linewidth. A complete theory would normally require evaluation of a photodissociation integral such as (17). However, the dominant contribution to the photodissociation integral comes from resonances in the scattering wavefunction for the dissociation process and , at energies close to a scattering resonance, the phase of the S matrix element sharply increases by 2π. For a multichannel problem, the S matrix is diagonalised in the form[51]

$$S(E) = \mathbf{B}(E)\, \Lambda^2(E)\, \mathbf{B}^T(E) \tag{31}$$

and the diagonal matrix Λ has the elements

$$\Lambda_i(E) = \exp[\, i\lambda_i(E)\,] \tag{32}$$

The eigenphase sum

$$e(E) = \Sigma_i\, \lambda_i(E) \tag{33}$$

is then constructed and, at energies close to a resonance, can be fitted to a Breit-Wigner function

$$e(E) = e_b(E) + \tan^{-1}\{\, \Gamma_r/[2(E_r-E)]\, \} \tag{34}$$

where $e_b(E)$ is a slowly varying background term, Γ_r is the resonance width and E_r is the resonance position. A problem with resonances associated with vibrational predissociation is that the linewidths are often very narrow($< 10^{-4}cm^{-1}$) and the resonances are very hard to detect numerically. However, a good guess at the resonance position can be acheived by performing a bound state calculation in which all channels that are open asymptotically are removed from the basis set expansion. This bound state calculation can either be done by using the bound-state coupled-channel method as described above, or by using a basis set expansion in the R coordinate together with a suitable basis for the internal coordinates[52]. In calculations of the rotational predissociation on molecules such as Ne-HF, we have found the latter basis set approach particularly simple to apply[53]. Furthermore, this procedure also enables the intensity of a transition to be easily calculated and this quantity, together with the position and width of the resonance enables a Lorenztian or Voigt profile lineshape to be fitted to the resonance. By repeating this procedure for all available resonances, the complete absorption spectrum can be predicted. An example of this for Ne-HF is shown in Section IV.

For such a rigid rotor atom-diatom system, we have the hamiltonian

$$H = -\hbar^2/(2\mu R)\partial^2/\partial R^2 R \quad +B\,\mathbf{j}^2 \quad +\boldsymbol{\ell}^2/(2\mu R^2) \quad +V(R,\theta) \tag{35}$$

in space-fixed coordinates. The intermolecular potential is expanded as a Legendre series on a chosen grid in R

$$V(R,\theta) = \Sigma_{n=0}\, c_n(R)\, P_n(\cos\theta) \tag{36}$$

The coefficients { $c_0(R)$} are then taken as an effective stretching potential and a basis set of n_t distributed Gaussian functions[54] equally spaced on the grid $\{R_i\}$ are used to obtain the eigenfunctions

$$\Psi^k{}_{str}(R) = \Sigma^{n_t}{}_{i=1}\; d_{ki}\,(2\alpha/\pi)^{1/4}\exp[-\alpha(R-R_i)^2] \tag{37}$$

of the "stretching" hamiltonian

$$H = -\hbar^2/(2\mu R)\partial^2/\partial R^2 R + c_0(R) \tag{38}.$$

With the basis set for the atom-diatom "stretching" motion obtained in the above way, the next step is to find an optimum angular basis set for each value of J. This is done with respect to the stretching function $\Psi^{k=1}{}_{str}(R)$ with lowest energy. The hamiltonian

$$H = B\,\mathbf{j}^2 + < \Psi^1{}_{str}(R) \;|\boldsymbol{\ell}^2/(2\mu R^2) + \Sigma_{n=1}\, c_n(R)\, P_n(\cos\theta)\,|\, \Psi^1{}_{str}(R) > \tag{39}$$

is diagonalised with the angular basis

$$y^{JM}_{\ell j}\ (\theta',\phi',\theta'',\phi'') = \Sigma_{mj}\Sigma_{m\ell}\ C(j,\ell,J,m_j,m_\ell,M)\ Y^{m_j}{}_j(\theta',\phi')\ Y^{m_\ell}{}_\ell(\theta'',\phi'') \qquad (40)$$

where $C(j,\ell,J,m_j,m_\ell,M)$ is a Clebsch-Gordan coefficient. Here, M is the projection of J along the space-fixed z axis and is set to zero in the energy level calculations. The orientation angles of the diatomic vector **r** and scattering coordinate **R** with respect to the space-fixed set of axes are (θ',ϕ') and (θ'',ϕ'') respectively. All asymptotic rotational states with energies less than the energy of interest are not included in the basis set expansion of equation (40) (ie. all open channels to which the molecule can dissociate are neglected). The angular eigenfunctions are then

$$\psi^{kJM}{}_{ang}\ (\theta',\phi',\theta'',\phi'') = \Sigma_j\ \Sigma_\ell\ d^k{}_{\ell j}\, y^{JM}{}_{\ell j}\ (\theta',\phi',\theta'',\phi'') \qquad (41).$$

The final wavefunction is expanded in the "configuration-interaction" form

$$\Psi_f^{nJM}\ (R,\theta',\phi',\theta'',\phi'') = \Sigma_k\ \Sigma_{k'}\ d^{nJM}{}_{kk'}\Psi^{k'JM}{}_{ang}(\theta',\phi',\theta'',\phi'')\Psi^k{}_{str}(R) \qquad (42)$$

and the full hamiltonian of equation (35) is diagonalised. This is also a very efficient procedure for obtaining accurate bound-state eigenvalues. If the dipole moment for the complex is known, it is then very straightforward to calculate the intensity for the transition. For a complex such as Ne-HF, it is a very good approximation[55] to take the dipole moment of the Ne-HF complex along the HF molecule and the transition probability

$$|\langle\Psi_f^{nJM}(R,\theta',\phi',\theta'',\phi'')|\ Y^0{}_1(\theta',\phi')\ |\Psi_f^{n'J'M}(R,\theta',\phi',\theta'',\phi'')\rangle|^2$$

is then trivial to evaluate.

Many calculations of linewidths for the vibrational and rotational predissociation have been performed by various researchers in the last few years and both highly accurate and approximate methods have been used. In Section IV, we describe a complete calculation we have carried out of the spectrum for rotational predissociation in Ne-HF[53], and also show results of AVCC-IOS calculations on the vibrational predissociation in rare gas-C_2H_4 complexes[33] and in $[C_2H_4]_2$[34] together with VCC-IOSA calculations of vibrational predissociation in the Ar-paradifluorobenzene complex[56].

IIh. Potential energy surfaces

Other papers in this book will deal in more detail with the problems associated with the calculation of potential energy surfaces. Here we briefly refer to some of the difficulties associated particularly with polyatomic scattering. First of all, it is necessary to obtain a suitable intramolecular force-field for the individual molecules involved in the collision. As was discussed in Section IIf, these can now conveniently be calculated using *ab initio* computer programs that exploit gradient techniques.

A much harder problem is to obtain a suitable potential energy surface for the intermolecular interaction. For atom-diatom and some simple diatom-diatom systems it is now possible to use high-quality *ab initio* techniques, such as the correlated electron pair approximation(CEPA)[57] or Many Body Perturbation Theory(MBPT)[58] with large basis sets to obtain quite accurate potential energy surfaces. There have been several such calculations and, to give two examples connected with our own work,the CEPA method has been used to obtain a potential, including angular and vibrational motion appropriate for the He+N_2 vibrational relaxation[50] and has also been applied to obtain a potential suitable for calculating the high-resolution spectrum for the rotational predissociation of the Ne-HF van der Waals molecule[59]. Calculations with both these potentials are reported in Section III and IV and, providing the dynamics computations are of good quality, good agreement with experimental data can be obtained. For calculations on the intermolecular potential for larger polyatomic systems, it is very difficult to obtain potentials of a good accuracy and it is normally necessary to use SCF potentials for the short-range interaction with C_6 coefficients for pair interactions for the long-range interaction. This approach has been used, for example, in our calculations on the Ar,He+glyoxal interactions described in Section III.

Even when points on a potential energy surface have been computed using *ab initio* techniques it is still a major problem to obtain a suitable functional fit to those points. For atom-diatom systems, flexible functions such as that due to Maitland and Smith[60,61] can be used to fit points with a minimal number of parameters but, even for diatom-diatom systems, the functional fits can be very unwieldy. If vibrational dependences are required it is normally necessary to resort to using sums of suitably parameterised atom-atom pair potentials[43]. This has the advantage that the pair functions can be easily transferred between similar systems. However, these types of functions have come in for some criticism as being too simplistic and more research on practical functional forms is required[62]. Nevertheless, despite all of these deficiencies, the many exciting experimental studies on polyatomic molecule scattering and predissocation problems do demand that calculations should be attempted using the best available potentials.

III. CALCULATIONS ON ROTATIONAL AND VIBRATIONAL ENERGY TRANSFER

IIIa. Rotational energy transfer in Ne+HF

A very large number of calculations have been performed on rotational energy transfer in the past and this review will concentrate on vibrational energy transfer. However, it is instructive to discuss two recent calculations we have performed on rotational energy transfer to illustrate some of the points discussed above.

Rotational relaxation of HF by collision with rare gas atoms is of particular interest as a recent laser double-resonance measurement of rate coefficients for the relaxation of HF(j=13) has been made by Taatjes and Leone at room temperature[63]. Since a potential surface suitable for calculations on van der Waals dynamics in Ne-HF has recently become available[59], it is of interest to calculate the rotational relaxation cross sections and rates for Ne+HF(j=13→j'). We have done this using both the CC and CSA methods. The calculations were performed on the Convex C1 computer and it was necessary to include basis functions with j having a maximum value of 15 and minimum value of 9 in the CC computations (amounting to a 84 basis function calculation).

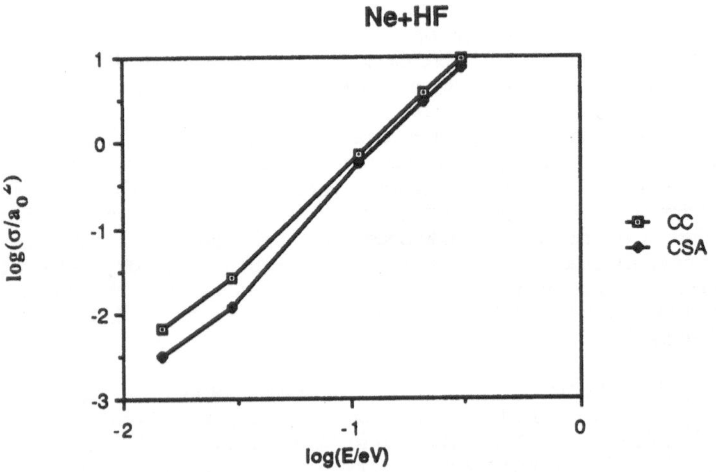

Figure 2. Comparison of relaxation cross sections for Ne+HF(j=13)

The j=13 CC and CSA relaxation cross sections summed over all j' states apart from j=13 are shown in Figure 2. It is interesting to note that even the CSA underestimates the integral rotational cross sections by as much as a factor of two at lower energies. The most likely transition is j=13→12 which has the relatively large energy difference of 0.064eV and, consequently, the cross sections are relatively small at low collision energies (less than $10^{-2}a_0^2$) and are sensitive to the coriolis terms left out of the hamiltonian in the CSA approximation. These coriolis terms depend on $[j(j+1)]^{1/2}$ and become more significant for large j values. The calculated room temperature rate coefficients in units of $cm^3s^{-1}molec^{-1}$ are $3.9x10^{-13}$(CSA) and $5.4x10^{-13}$(CC) which are to be compared with the value of $(2.2\pm0.4)x10^{-12}$ obtained in the experiment[63]. Thus the CC rate coefficient calculated with the *ab initio* surface is a factor of three lower than experiment.

This disagreement is a little surprising considering the quality of the calculation.

Furthermore, the anisotropy in the potential, which largely controls the degree of rotational excitation, is expected to be accurate as the frequency of the van der Waals bending mode of the Ne-HF complex is obtained very accurately[53](see Section IV). However, it should be remembered that the Ne-HF potential was specially set up for the van der Waals spectrum of the complex, which is involved with a somewhat longer-range part of the potential than that likely to be most important for Ne+HF(j=13) relaxation. It is normally considered that the CSA is a highly accurate method for obtaining intregral cross sections but these results show that the method must be used with great care when large j values are considered and the cross sections are small.

IIIb. Rotational relaxation in He+CH_3CN

The rotational relaxation problem He+CH_3CN(j=0,k=0→j',k'=3) presents an excellent case for testing the accuracy of the IOSA and of the Azimuthal method for a symmetric top, CH_3CN, which has one rotor constant(a=5.25cm^{-1}) much larger than the other (b=0.307cm^{-1}). Figure 3 shows such a comparison[29] against results obtained using the CSA. It can be seen that the Azimuthal method(ACC-IOS) does give a much better agreement with the CSA than the IOSA, thus suggesting that it will be a useful approach for studying k changing transitions in symmetric top or near symmetric top molecules.

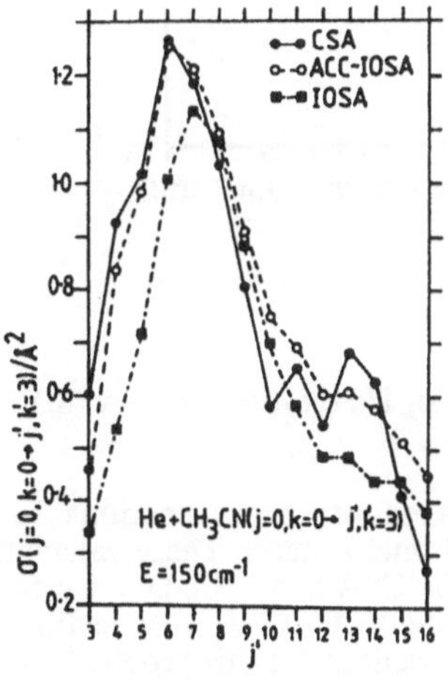

Figure 3. Comparison of cross sections for He+CH_3CN(j=0,k=0→j',k'=3)

IIIc. Vibrational relaxation in He+N_2

The vibrational-rotational relaxation process He+N_2(v=1,j→v=0,j') provides a good benchmark system for testing various methods for calculating cross sections because a good potential energy surface can be used[50] (see section IIh) and good quality vibrational relaxation rate coefficients have been measured down to 100K[64]. We have shown before[50] that the CSA gives excellent agreement with this experimental data. We have recently implemented a CSA version of the coupled-channel distorted-wave Born Approximation to calculate cross sections for this problem(called CSA-DWA). Thus, separate rotational coupled-channel calculations are performed for the v=0 and v=1 states and scattering matrix elements with these wavefunctions over the coupling term that depends on the vibrational coordinate are then computed to give the vibrational-rotational relaxation cross sections.

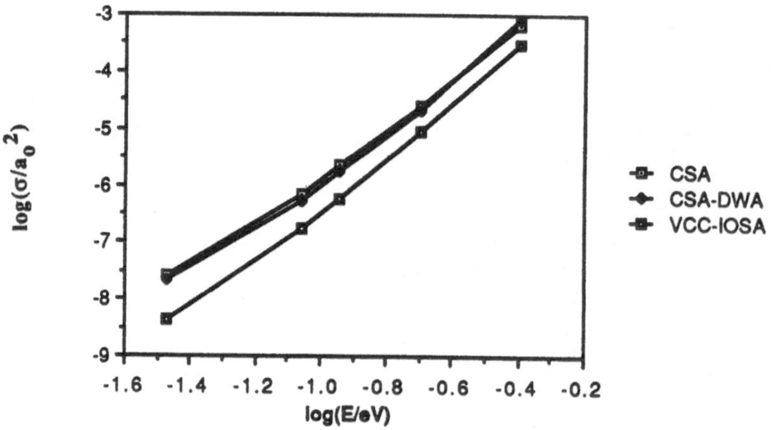

Figure 4. Comparison of cross sections for He+N_2(v=1,j=0→v'=0).

Figure 4 gives a comparison of cross sections for the relaxation process He+N_2(v=1,j=0→v'=0) summed over all final j' states. These calculations were performed on a CRAY-1 computer. It can be seen that the CSA and CSA-DWA cross sections are in excellent agreement for all collision energies while the VCC-IOSA cross sections are too low by as much as a factor of 5 at lower energies. The cross sections for this problem are extremely small (ranging from 10^{-8} to 10^{-3} a_0^2) and are very sensitive to the effect of rotational energies narrowing the effective energy gap between the vibrational levels. Since the VCC-IOSA explicitly neglects rotational energies it does not give nearly such accurate results for this system. The CSA-DWA does accurately account for the rotational energies and gives much more accurate results. Furthermore, as the cross sections are so small, it is accurate to treat the vibrational coupling using perturbation theory as is done in the DWA.

To our knowledge, this represents the first application of the CSA-DWA to gas-phase non-reactive vibrational-rotational relaxation and the results suggest that the method will be a very accurate one for treating vibrational relaxation in molecular collisions. The technique should be particularly useful for calculations on polyatomic molecule systems with large rotor constants and many vibrational levels open. It has the advantage that only one rotational coupled-channel calculation need be done for each vibrational state to get the whole manifold of vibrational-rotational cross sections involving all vibrational states.

IIId. Vibrational relaxation in He+CO_2

The VCC-IOSA is a method generally applicable to vibrational relaxation in polyatomic molecule scattering, but the disappointing perfomance of the method for He+N_2 vibrational relaxation puts a question-mark against its utility. To examine this in more detail, a series of CSA and VCC-IOSA computations have been carried out on the vibrational relaxation of the bending mode in He+$CO_2(01^10,j \rightarrow 00^00,j')$[27,49].

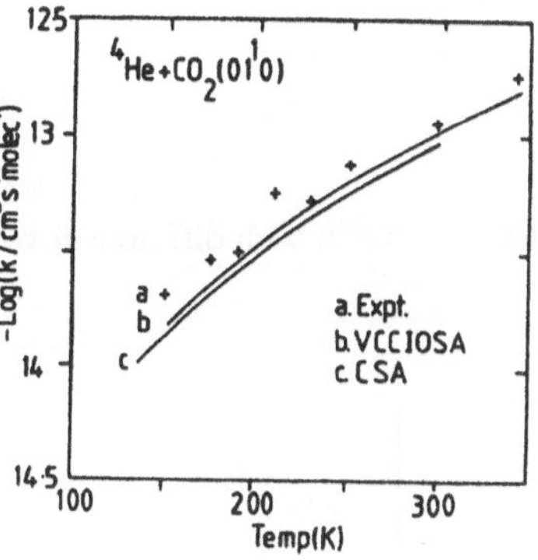

Figure 5. Rate coefficients for He+$CO_2(01^10 \rightarrow 00^00)$.

Figure 5 gives a comparison of the these calculated vibrational relaxation rates, summed over all j' states, with those obtained in a photoacoustic experiment. It can be seen that the CSA and VCC-IOSA rate coefficients are in good agreement with each other and with experiment. The reasons why the VCC-IOSA works so well for He+$CO_2(01^10 \rightarrow 00^00)$ and so poorly for He+$N_2(v=1,j=0 \rightarrow v'=0)$ is that the vibrational frequencies and rotor constants in the two systems are very different. In N_2 these constants are ω=2359cm^{-1} and B=2.00cm^{-1} respectively while the values are ω=667cm^{-1} and B=0.389cm^{-1} for CO_2. A small vibrational frequency gives larger vibrational relaxation rate coefficients which will be less sensitive to effects of rotational energies. Furthermore, the smaller the rotor constant, the less important the rotational energy effects will be.

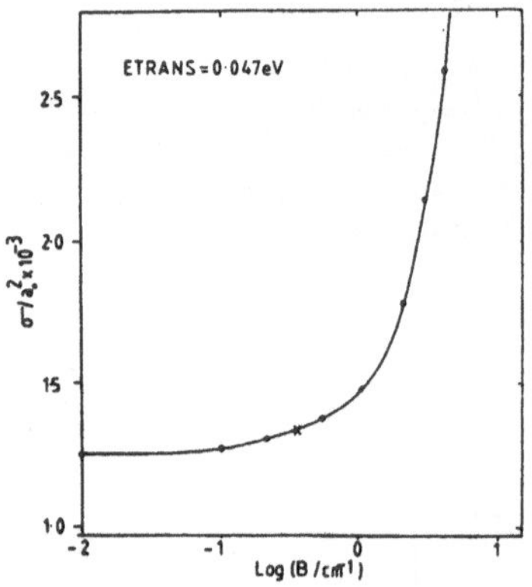

Figure 6. CSA cross sections for He+CO_2(01^10,j=1→00^00) with different B values.

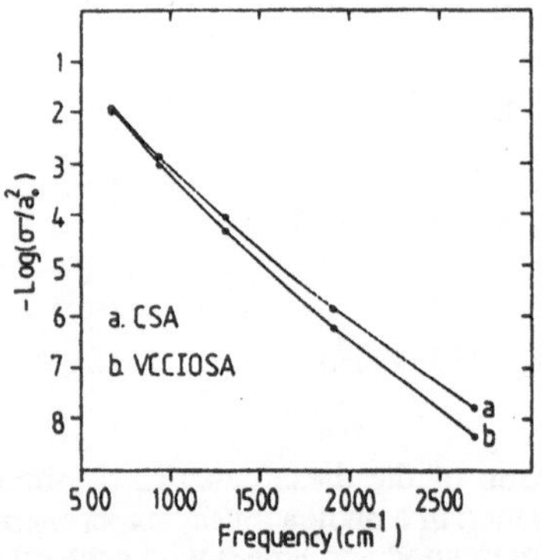

Figure 7. Cross sections for He+CO_2(01^10→00^00) with different vibrational frequencies.

In Figure 6 the rotor constant dependence of the cross sections is examined in more detail by showing CSA cross sections for He+$CO_2(01^10,j=1\rightarrow00^00)$, summed over all j' states, with the rotor constant artificially varied. The CSA and VCC-IOSA results are formally identical for B=0. It can be seen that providing B is less than 1cm^{-1} there is excellent agreement between the VCC-IOSA and CSA results. In Figure 7 a similar comparison is performed for variation of the vibrational frequency. Here it is seen that if the vibrational frequency is less than 1000cm^{-1} the VCC-IOSA and CSA cross sections are in excellent agreement. Thus, providing the rotor constants and vibrational frequencies are in these limits, the VCC-IOSA is expected to be a reliable technique. Fortunately, this is the case for most polyatomic molecules and, indeed, the larger the polyatomic, the more likely it is to be true! When one of the rotor constants is large while the others are small it might be more appropriate to use the Azimuthal technique. This is the case for glyoxal and new results for relaxation of this molecule are presented in the next section.

IIIe. Vibrational and rotational relaxation in glyoxal

Calculations on the vibrational and rotational relaxation in glyoxal (trans $C_2O_2H_2$) are of interest because Parmenter and co-workers[65] have very recently performed rotationally state-selected crossed molecular beam experiments to study the competition between rotational and vibrational excitation in glyoxal(S_1) in collisions with H_2. They find that the vibrational-rotational cross sections for excitation of the lowest frequency mode (the v_7 torsion with ω=233cm^{-1}) are of the same order of magnitude as those for pure rotational excitation(v_7=0) of levels with similar energies to the v_7=1 energy. Glyoxal is a near prolate symmetric top with an a rotor constant of 1.8cm^{-1} and a b rotor constant of 0.15cm^{-1}. Therefore, this molecule is ideal for using the Azimuthal method combined with vibrational close-coupling for the v_7=0 and v_7=1 levels. Calculations have been performed on both He and Ar scattering off glyoxal(S_1).

The intramolecular force-field for the S_1 state was determined by using the SCF gradient method. The intermolecular potential was taken as a sum of atom-atom pair functions with parameters obtained as a fit to *ab initio* data for the rare gas-formaldehyde system[66]. A total of 18 and 11 k basis functions were needed in the v_7=0 and v_7=1 states respectively to obtain cross sections converged to within 10% or better.

Calculations[30] of the vibrational-rotational cross sections for the process He and Ar+$C_2H_2O_2(v_7=0;k=0\rightarrow v_{7'}=0,1;k')$ are presented in figures 8a and 8b. In the case of He scattering it is seen that the cross sections for vibrational excitation of the v_7=1 mode are quite close in magnitude to those for excitation of the (v_7=0,k') levels with similar energies to those for the v_7=1 level. Substitution of the mass of H_2 for He , with the He+glyoxal potential, gives the cross sections into $v_{7'}$=1 very similar to the rotational excitation cross sections , in good agreement with experiment. However, for the Ar collisions, it is found that the rotational excitation cross sections are all much larger than the vibrational relaxation cross sections, which is a prediction that should be verifable in the experiments. This difference occurs because rotational excitation is much more efficiently produced by atoms with heavier masses. Calculations have also recently been carried out on collisions at extremely low collision energies(~1K) and preliminary results[67] suggest that orbiting resonances can enhance the cross sections for the process He+$C_2H_2O_2(v_7=1,k=0\rightarrow v_{7'}=0,k')$.

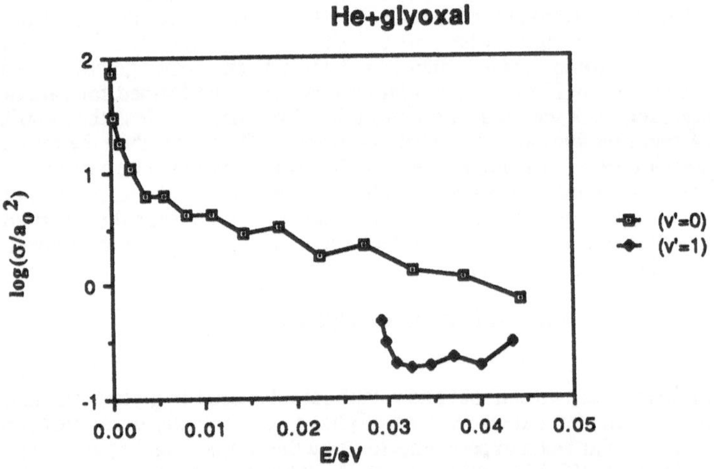

Figure 8a. Cross sections for He+glyoxal(v_7=0,k=0→v_7',k') against final state energy.

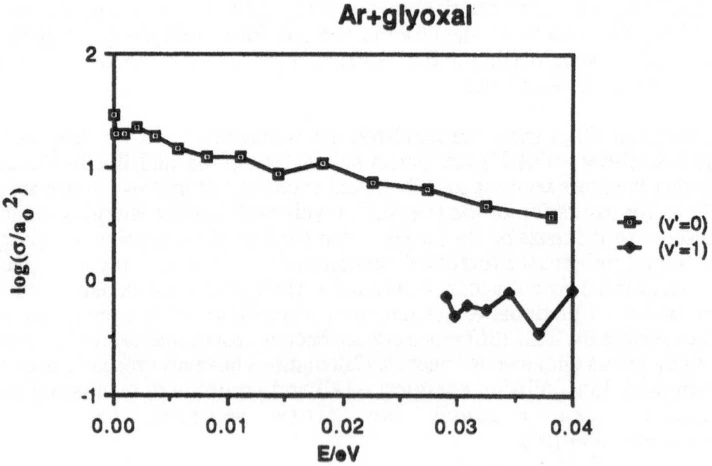

Figure 8b. Cross sections for Ar+glyoxal(v_7=0,k=0→v_7',k') against final state energy.

IIIf. Vibrational excitation in difluorobenzene

It is possible to apply the VCC-IOSA to molecules as large as para-difluorobenzene. As this system has very small rotor constants, this should be an accurate method for this molecule. An extensive series of VCC-IOSA computations have been reported elsewhere for the He+pDFB vibrational relaxation and excitation process involving several different modes[44]. This problem is of particular interest because molecular beam[3] and bulb experiments[68] have all shown a pronounced propensity for excitation of the mode of lowest frequency, v_{30}, which is an out-of-plane bending mode. Our calculations confirm these experimental findings and also show that the v_{30} mode is preferentially excited over V-V energy transfer even when other vibrational channels lie very close in energy to the particular vibrational mode already excited. The calculations also show that ,when modes have similar frequencies, those involving motion out of the molecular plane are preferentially excited over in-plane modes.

IV ROTATIONAL AND VIBRATIONAL PREDISSOCIATION

IVa Rotational predissociation in Ne-HF

The methods described in Section IIg have been applied to calculate the spectrum[53] for excitation of the low-lying vibrational-rotational states of the Ne-HF van der Waals complex using a CEPA *ab initio* potential energy surface[59]. This potential gives a dissociation energy D_0 of only 22cm^{-1} and, before the spectrum was predicted, the complex was unobserved. Since the coupling is so weak in the Ne-HF interaction, the energy levels for the hindered rotor states in Ne-HF(v=0) will be very similar to those for Ne-HF(v=1) and this approximation enables the infrared spectrum to be predicted, even though the *ab initio* points were calculated with the HF bond distance held fixed.

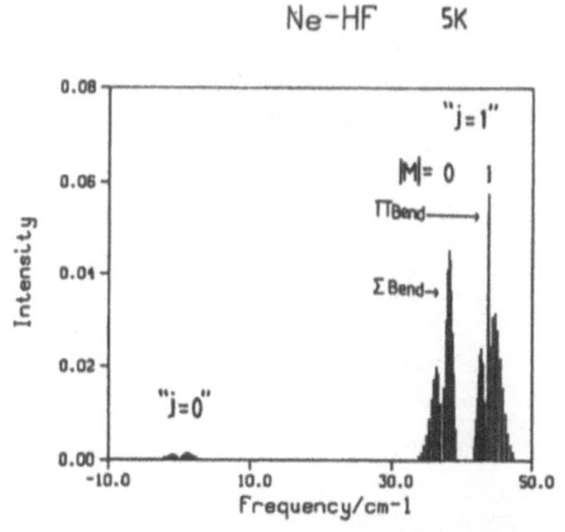

Figure 9. Spectrum for NeHF with no linewidths given to lines.

Figure 9 shows the predicted spectrum for a temperature of 5K which is typically accessed in supersonic jet experiments with the original Ne-HF molecule in its ground state. In this figure, the calculated lines have not been given linewidths. To obtain the infrared spectrum it is necessary to add the fundamental frequency for NeHF($v=0\rightarrow1$) to the reported energies. The lines close to $0cm^{-1}$ correspond to transitions into the fundamental energy level which correlates with HF($j=0$). However, the anisotropy of the potential mixes in a little $j=1$ character with $j=0$ and this gives the fundamental band just enough intensity to be observed. Indeed, soon after this prediction was made it was verified to be true in experiments[53]. The lines around 40 cm^{-1} correspond to excitation of levels correlating with HF($j=1$) which is an energy above the dissociation limit in Ne-HF. The so called Σ bend states have a projection quantum number of zero along the intermolecular axis. These states are dissociative and close-coupling calculations of resonances give linewidths $>0.3cm^{-1}$ for these states ,which makes them too broad to be observed. However, the Π bend state also correlates with $j=1$ and has angular momentum projection quantum numbers of +1 or -1 along the intermolecular axis, appropriate linear combinations of which give Π^+ and Π^- states. The Π^+ state is linked by relatively weak coriolis coupling to the Σ states and has a linewidth for $J=1$ of $0.003cm^{-1}$ which increases as $J(J+1)$ for higher J states. The Π^- wavefunction is made up of basis functions with $j\geq1$ and thus this state cannot dissociate to HF($j=0$), its linewidth being determined purely through the experimental conditions. Thus the P and R branches of the spectrum have linewidths that increase dramatically with J(corresponding to excitation into the Π^+ state) while the Q branch is very narrow and corresponds to excitation into the Π^- state.

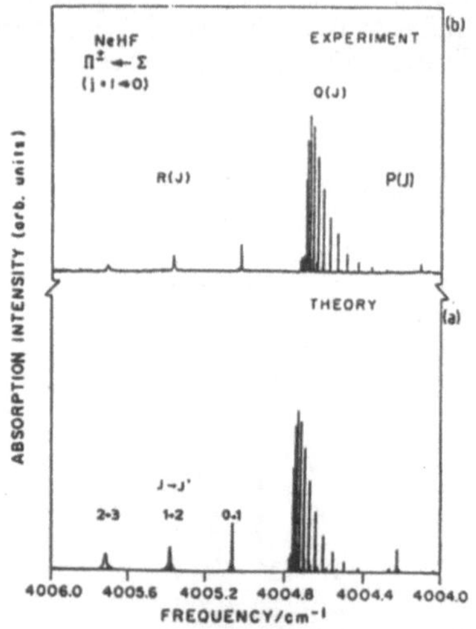

Figure 10. Experimental and theoretical spectrum for portion of the Π bending mode excitation

in NeHF, including linewidths in calculation.

A comparison of a portion of the predicted spectrum[53], including calculated linewidths, for the Π bend of Ne-HF is shown in Figure 10. Also shown is the experimental spectrum that was measured after the first predictions of the spectrum had been made[53]. It can be seen that the agreement for the frequency of the Π bend, the intensities and widths of the lines is excellent. This illustrates well the predictive power that can now be achieved from performing accurate bound state and dynamics calculations on a good *ab initio* surface for small systems.

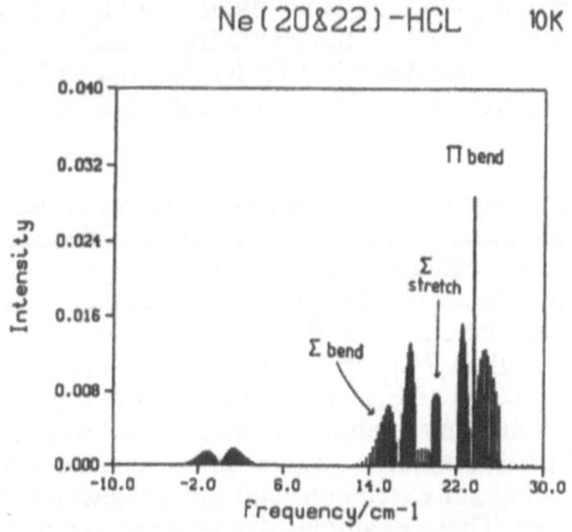

Figure 11. Predicted spectrum for Ne-HCl

Figure 11 shows the equivalent spectrum for the Ne-HCl van der Waals complex[52]. Here the well depth in the potential is deep enough for both the Σ bend and the Π bend levels to be bound. The potential energy surface for this system was obtained by Hutson and Howard through a multiproperty fit. The lines differ from the measured infrared spectrum by less than $1.3cm^{-1}$ and the intensities of all the bands compare well also with experiment[69]. Predictions of the spectra for Ar,Kr and Xe complexes with HCl have also been made in a similar way[52].

IVb Vibrational predissociation in ethylene complexes

Given that the quantum mechanical methods can be used very effectively to predict the spectrum for rotational predissociation for a syatem such as Ne-HF it is of interest to see if the methods can be extended to more complicated polyatomic systems. Here the vibrational frequencies are normally much lower than those for HF which makes the study of vibrational predissociation relevant and interesting. In recent years, we have been making an intensive study of the vibrational predissocation associated with the excitation of the out-of-plane bending mode ν_7 (ω=$946cm^{-1}$) of ethylene complexed with rare gases[33] or another ethylene molecule[34]. A detailed description of most of these calculations can be found in the Ph.D. thesis of Peet[70]. There is a lower lying vibrational level at ω=826cm^{-1}, the in-plane bending v_{10} level, and very efficient predissociation can occur into this level provided the D_0 of the complex is smaller than 120cm^{-1}. This is likely to be the case for the Ne-C_2H_4 complex , but the well depths will probably be too big for the Ar and Kr complexes[33], and will certainly be too big for the ethylene dimer[34].

The combination of the Azimuthal method of Section IIe with close-coupling expansions in the v_7=0, v_7=1 and v_{10}=1 vibrations enables vibrational predissociation resonances to be predicted. For the reasons given above these predicted resonances are quite broad for the Ne complex(~ 10^{-2}cm^{-1}) but are much narrower for Ar(~10^{-8}cm^{-1}). This prediction remains to be verified in experiments. Furthermore, methods similar to those described for Ne-HF, but with the inclusion of the sudden approximation, have been used very recently to predict the full infrared spectrum for Ne-C_2H_4[71].

The vibrational predissociation in the ethylene dimer has been the subject of many experiments, the first of which suggested[72] that this system should have a linewidth of the order of 10cm^{-1}. Our first Azimuthal-vibrational close-coupling calculations[73], which required basis sets expansions of the order 200 and extensive utilisation of the CRAY-1 computer, gave linewidths of the order of 10^{-6}cm^{-1}, in strong disagreement with the existing experiments. However, new experiments at high resolution[74,75] did then give narrower linewidths(~10^{-3}cm^{-1}). The most recent calculations[70,76] take account of the fact that the v_7 vibration in the ethylene dimer will be strongly mixed with many rotationally excited v_{10} levels and thus it is more appropriate to calculate vibrational predissociation resonances associated with rotationally excited v_{10} rather than v_7 with very little rotational excitation. This gives linewidths of the order 10^{-3}cm^{-1} in good agreement with experiment. These examples demonstrate again the important role that calculations can play in this developing field.

IVc Vibrational predissociation in Ar-pDFB

Vibrational predissociation lifetimes have been measured by Parmenter and coworkers for the system Ar-pDFB(S_1)[77]. They have observed linewidths for several vibrational states including the in-molecular-plane v_6 and v_5 modes. Very recently, we have been doing VCC-IOSA calculations of the linewidths for these vibrational states[56]. In our calculations, the linewidths $\Gamma(\theta,\phi)$ and line positions $E_r(\theta,\phi)$ are calculated and the $E_r(\theta,\phi)$ are used as an effective potential for the hamiltonian describing the atom-molecule bending motion to produce a bending wavefunction which can then be used to average over $\Gamma(\theta,\phi)$. One surprising result obtained in the experiment is that the v_5 state has a large linewidth (~10^{-3}cm^{-1}) for dissociation into the ground vibrational state, even though the v_5 =1 frequency is 818cm^{-1} which gives a very large translational energy release implying very inefficient predissociation. Our calculations so far have failed to give large linewidths for this particular predissociation and more theoretical work is required.

V CONCLUSIONS

In this overview of recent calculations that we have performed on polyatomic rotational-vibrational energy transfer and predissociation we have attempted to illustrate the many types of calculations and systems that can now be studied using quite rigorous quantum mechanical techniques. We have also tried to give the reader a feeling for the difficulties involved in performing the calculations and have assessed the accuracy of the many approximations that can be used. Furthermore, we have chosen particular examples that illustrate the close interaction between theory and experiment in this field that has lead to the confirmation of theoretical predictions in experiments and a deeper understanding of experimental results and the mechanisms of molecular interactions through theory. The main feature of our approach is that most problems of experimental interest in this field will require the detailed quantum-mechanical treatment of only a small number of degrees of freedom in a polyatomic system, while it is possible to average over all other degrees of freedom.

This is an area that has benefited enormously from the use of supercomputers and mini-supercomputers. The theories described in this review are almost all based on a coupled-channel expansion of the wavefunction and the numerical work involved in the calculation is largely concerned with matrix manipulations and the computations of multidimensional integrals. These are all operations that are made efficient through the use of computers with vector processing capabilities. Furthermore, the power of these computers has enabled more accurate calculations to be performed on more complicated systems which have provided benchmark results that have allowed the accuracy of the many approximate methods to be properly calibrated. Also, the new computers have now made it possible to perform approximate computations on quite large molecules of real chemical interest such as aromatic molecules. The impact of these types of calculations on experimental chemistry is already being realised and will become even more significant in the years to come.

ACKNOWLEDGMENTS

Some of the calculations reported here were initiated when the author was a Visiting Fellow at the Joint Institute for Laboratory Astrophysics, National Bureau of Standards and University of Colorado, Boulder, Colorado, USA.

REFERENCES

1. D.J.Krajnovich,C.S Parmenter and D.L.Catlett, Chem.Rev.,1987,**87**,237.
2. R.B.Bernstein and A.H.Zewail, J.Phys.Chem., 1986 , **90**, 3467 .
3. G.Hall, C.F.Giese and W.R.Gentry, J.Chem.Phys., 1985, **83**, 5343 .
4. R.E.Miller, J.Phys.Chem., 1986, **90**, 3301 .
5. D.C.Clary, J.Phys.Chem.,1987,**91**,1718.
6. A.M.Arthurs and A.Dalgarno, Proc.Roy.Soc.London Ser.A,1960,**256**,540.
7. R.T Pack, J.Chem.Phys., 1974 ,**60**, 633 .
8. B.R.Johnson, J.Comput.Phys.,1973, **13**, 445.
9. J.C.Light and R.B.Walker, J.Chem.Phys.,1976, **65**, 4272 .
10. D.E.Manolopoulos, J.Chem.Phys., 1986,**85**,6425.
11. M.H.Alexander and D.E.Manolopoulos, J.Chem.Phys., 1987,**86**,2044
12. M.H.Alexander, Department of Chemistry, University of Maryland.
13. J.M.Hutson, MOLSCAT computer code version 9(1986), distributed by Collaborative Computational Project No.6 of the Science and Engineering Research Council,UK.
14. D.E.Manolopoulos, 1988,Ph.D. Thesis, Department of Theoretical Chemistry, University of Cambridge.
15. G.Danby, J.Phys.B.,1983,**16**,3393.
16. G.G.Balint-Kurti and M.Shapiro, Chem.Phys., 1981, **61**,137.
17. K.C.Kulander and J.C.Light, J.Chem.Phys., 1980, **73**, 4337 .
18. D.C.Clary, J.Chem.Phys., 1986, **84**, 4288 .
19. F.Mrugala, J.Comp.Phys., 1985, **58**,113.
20. D.C.Clary and D.E.Manolopoulos, unpublished calculations 1987, see also Ref.14.
21. K.B.Whaley and J.C.Light, J.Chem.Phys.,1984,**87**,3334.
22. D.C.Clary, J.Chem.Phys.,1985,**83**,4470.
23. G.C.Schatz, L.M.Hubbard, P.S.Dardi and W.H.Miller, J.Chem.Phys., 1984, **81**, 231 .
24. D.W.Schwenke and D.G.Truhlar, Theor. Chim. Acta, 1986, **69**, 175 .
25. P.McGuire and D.J.Kouri, J.Chem.Phys., 1974, **60**, 2488 .
26. D.J.Kouri and D.E.Fitz, J.Phys.Chem., 1982, **86**, 2224 .
27. A.J.Banks and D.C.Clary, J.Chem.Phys., 1987,**86**,802.
28. D.C.Clary, J.Chem.Phys., 1984, **81**, 4466 .
29. D.C.Clary and S.Green, Chem.Phys., 1987,**112**,15.
30. D.C.Clary and C.E. Dateo, to be published.
31. A.C. Peet and D.C.Clary, Molec.Phys., 1986,**59**,529.
32. C.P.Bewick, A.B.Duval and B.J.Orr, J.Chem.Phys., 1985, **82**, 3470.
33. A.C.Peet, D.C.Clary and J.M.Hutson, J.Chem.Soc.,Faraday Trans. 2,1987,**83**,1719.
34. A.C.Peet, D.C.Clary and J.M.Hutson, Faraday Discuss.Chem.Soc.,1986,**82**,327.
35. R.T Pack, Chem.Phys.Lett., 1972 ,**14**, 393 .
36. M.A.Brandt and D.G.Truhlar, Chem.Phys.Lett., 1973 , **23**, 48 .
37. D.Secrest, J.Chem.Phys., 1975 , **62**, 710 .
38. L.W.Hunter, J.Chem.Phys., 1975, **62**, 2855 .
39. S.Green, J.Chem.Phys., 1976, **64**, 3463 .
40. J.M.Bowman and S.Leasure, J.Chem.Phys.,1977, **66**, 288 .
41. D.C.Clary, J.Chem.Phys., 1981, **75**, 209 .
42. D.C.Clary, J.Chem.Phys., 1981, **75**,2899 .
43. D.C.Clary, J.Am.Chem.Soc., 1984 ,**106**,970 .

. D.C.Clary, J.Chem.Phys., 1987,**86**,813.
. R.D.Amos, Department of Theoretical Chemistry, University of Cambridge, private communication.
. N.M.Harvey, Chem.Phys.Lett., 1982, **88**, 553 .
. B.C.Garrett, Chem.Phys., 1984, **87**, 63 .
. B.R.Johnson, J.Chem.Phys., 1986, **84**, 176 .
. A.J.Banks, 1986, Ph.D. Thesis, Department of Theoretical Chemistry, University of Cambridge.
50. A.J.Banks, D.C.Clary and H.J.Werner, J.Chem.Phys., 1986, **84**, 3788 .
51. M.S.Child, Molecular Collision Theory, (Academic, London, 1974); C.J.Ashton, M.S.Child and J.M.Hutson, J.Chem.Phys., 1983, **78**, 4025 .
52. D.C.Clary and D.J.Nesbitt, to be published.
53. D.C.Clary, C.M.Lovejoy, S.V.Oneil and D.J.Nesbitt, Phys.Rev.Lett., 1988, in press.
54. I.P.Hamilton and J.C.Light, J.Chem.Phys., 1986,**84**,306.
55. P.W.Fowler and A.D.Buckingham, Molec.Phys., 1983,**50**,1349.
56. A.R.Tiller, A.C.Peet and D.C.Clary, Chem.Phys.,1988,in press.
57. W.Meyer, J.Chem.Phys., 1973,**58**,1017.
58. N.C.Handy, P.J.Knowles and K.Somasundram, Theor. Chim. Acta.,1985,**68**,87.
59. S.V.ONeil, D.J.Nesbitt, H.-J.Werner, P.Rosmus and D.C.Clary, to be published.
60. G.C.Maitland and E.B.Smith, Chem.Phys.Lett.,1973,**22**,443.
61. J.M.Hutson and B.J.Howard, Molec.Phys.,1982,**45**,769.
62. A.J.Stone and S.L.Price, J.Phys.Chem., 1988,**92**,3325.
63. C.A.Taatjes and S.R.Leone, J.Chem.Phys., 1988,in press.
64. M.M.Maricq, E.A.Gregory, C.T.Wickham-Jones, D.J.Cartwright and C.J.S.M.Simpson, Chem.Phys., 1983, **75**, 347.
65. K.W.Butz, H. Du, D.J.Krajnovich and C.S.Parmenter, J.Chem.Phys., 1987,**87**,3699.
66. S.Brode, private communication.
67. G.Kroes, C.E.Dateo and D.C.Clary, to be published.
68. D.L.Catlett,Jr., K.W.Holtzclaw, D.Krajnovich, D.B.Moss, C.S.Parmenter, W.D.Lawrance and A.E.W.Knight, J.Phys.Chem., 1985, **89**, 1577 .
69. C.M.Lovejoy and D.J.Nesbitt, Chem.Phys.Lett.,1988,**147**,490.
70. A.C.Peet, 1988,Ph.D. Thesis, Department of Theoretical Chemistry, University of Cambridge.
71. A.Tiller, 1988,unpublished work.
72. K.C.Janda, Adv.Chem.Phys., 1985, **60**, 201 .
73. A.C.Peet, D.C.Clary and J.M.Hutson, Chem.Phys.Lett.,1986, **125**,477 .
74. M.Snels, R.Fantoni, M.Zen, S.Stolte and J.Reuss, Chem.Phys.Lett., 1986, **124**, 1 .
75. K.G.H.Baldwin and R.O.Watts, Chem.Phys.Lett., 1986, **129**, 237 .
76. A.C.Peet, D.C.Clary and J.M.Hutson, Faraday Discuss. Chem.Soc.,1986,**82**, 373.
77. K.W.Butz, D.L.Catlett, G.E.Ewing,D.Krajnovich and C.S.Parmenter, J.Phys.Chem.,1986, **90**,3533.

Temperature-Dependent Rate Constants For Ion-Dipole Reactions: $C^+(^2P) + HCl(X\,^1\Sigma^+)$

Christopher E. Dateo and David C. Clary
University Chemical Laboratory,
Lensfield Road, Cambridge CB2 1EW
United Kingdom

ABSTRACT. Calculation of the rate constant at several temperatures for the reaction $C^+(^2P)$ + $HCl(X\,^1\Sigma^+)$ are presented. A quantum mechanical dynamical treatment of ion-dipole reactions which combines a rotationally adiabatic capture and centrifugal sudden approximation is used to obtain rotational state-selective cross sections and rate constants. *Ab initio* SCF (TZ2P) methods are employed to obtain the long- and short-range electronic potential energy surfaces. This study indicates the necessity to incorporate the multi-surface nature of open-shell systems. The spin-orbit interactions are treated within a semiquantitative model. Results fare better than previous calculations which used only classical electrostatic forces, and are in good agreement with CRESU and SIFT measurements at 27, 68, and 300°K .

1. Introduction

The field of gas-phase ion chemistry has undergone dramatic growth during the past decade and has received increasing attention from both experimentalists and theoreticians. The rapid development in this area is primarily due to advances in experimental methods employed in the preparation and detection of highly reactive species. Many exothermic ion-molecule reactions proceed very fast, typically having little or no barrier to reaction. Hence, nearly every collision results in reaction. For these cases, it is often accurate to assume that the reaction is dominated by the long-range intermolecular forces. A variety of approximate theories have been developed based on a classical capture approximation.[1] This approximation assumes that if there is enough translational energy along the reaction coordinate to surmount any existing centrifugal barrier then "capture", i.e., reaction will occur. For systems in which the interaction potential can be described as that between a charge (the ion) and the induced-dipole of the molecule, the capture approximation yields the well-known Langevin formula predicting temperature independent rate constants. A recent experiment[2] involving reaction of atomic cations with nonpolar neutral molecules confirmed this prediction to temperatures as low as 8°K . For systems where the interaction potential is highly anisotropic the application of capture theories becomes somewhat more complicated. For example, if the molecule has a significant permanent dipole moment, its rotation will be hindered by the presence of the ion. The anisotropy of the interaction potential couples together different rotational states of the target molecule and a rigorous treatment of the problem would require a multichannel scattering theory.

A large number of approximate theories have been proposed for ion-dipole reactions. Some of these include the average dipole orientation (ADO) approximation[3,4] and its extension to include conservation of angular momentum (the AADO method[5]), various transition-state theories involving variational[6] and statistical[7] modifications, the semiclassical perturbed rotational state (PRS) approximation,[8,9] classical trajectory studies,[6,10] the adiabatic invariance method,[11] and the statistical adiabatic channel model[12,13] (SACM).

A. Laganà (ed.), Supercomputer Algorithms for Reactivity, Dynamics and Kinetics of Small Molecules, 327–338.

All of these approaches employ the classical capture approximation and use a charge-dipole interaction potential of the form

$$V(R,\theta) = -\frac{\alpha q^2}{2R^4} - \frac{q\mu_D \cos\theta}{R^2}. \tag{1}$$

Here **R** is the vector between the centers of mass of the ion, with charge q, and the molecule, with isotropic polarizability α and dipole moment μ_D, and θ is the angle that **R** makes with the direction of the dipole. If θ can be assumed constant (the Locked Dipole approximation[1]), then simple capture theory results in a rate constant expression which increases with decreasing temperature according to $T^{-1/2}$. This phenomenon of a negative temperature dependence for rate constants of ion-dipole reactions has been observed experimentally.[14,15]

In the present study, we employ the ACCSA method[16,17] which involves a combined rotationally adiabatic capture and centrifugal sudden approximation. The method enables the computation of cross-sections and rate constants which are state-selective in the initial rotational states of the reactant molecule. The ACCSA approach has been successful for ion-dipole reactions which involve proton transfer. Predicted rate constants[15,17] for the reactions $H_3^+ + HCN \rightarrow H_2CN^+ + H_2$ and $HCO^+ + HCN \rightarrow H_2CN^+ + CO$ are in excellent agreement over the temperature range at which experimental data are available. On the other hand, for reactions in which heavy ion or atom exchange occurs, ACCSA theory appears to work well for some systems and not for others. Good agreement with experiment exists for reactions of C^+ or N^+ with H_2O, but poorer results are obtained for the respective reactions with NH_3.[18] At present, therefore, it remains unclear whether theory is capable of making reliable predictions. However, more experimental data will be necessary to thoroughly test the available theories.

A main goal of the present work is to investigate those reactions for which the ACCSA and other methods do not fare as well in the hope of developing a modified theory which will be able to give reliable temerature dependent rate constants for a wider base of systems. A possible source of breakdown could lie in the assumption that only long-range forces are important. The ACCSA theory is not limited to interaction potentials of the form given in Eq.(1), but is easily extended to incorporate more complicated potentials. We report here a study of the $C^+(^2P) + HCl(X\,^1\Sigma^+) \rightarrow CCl^+(X\,^1\Sigma^+) + H(^2S)$ reaction, exothermic by 0.91eV.[19] Previous ACCSA studies with the charge-dipole interaction potential of Eq.(1) did not agree well with experimental measurements. *Ab initio* SCF methods are employed to obtain a more accurate description of the intermolecular forces for both short- and long-range regions. A brief outline of the ACCSA theory is given section 2, followed by description of the calculations and results in section 3, including a discussion on the effect of spin-orbit coupling interactions. A final summary is given in section 4.

2. Theoretical Approach

In this section we very briefly review the basis of the ACCSA method. We concentrate here on the treatment of a bimolecular reaction between an atomic ion and a diatom with a significant permanent dipole moment. However, the general theory has been described in detail elsewhere.[16,17] The theory is applicable to systems in which the long-range intermolecular forces are dominant, such as strongly exothermic reactions which have no barrier in the potential energy surface. Under these conditions it is reasonable to neglect the possibility of back reflected flux from any collision complex, and only the entrance channel need be considered in the calculation. We start from the partial-wave hamiltonian for the entrance channel of an atom-diatom reaction, written in body-fixed (BF) coordinates[20] as,

$$H = -\frac{1}{2\mu R}\frac{\partial^2}{\partial R^2}R + B\mathbf{j}^2 + \frac{|\mathbf{J}-\mathbf{j}|^2}{2\mu R^2} + V(R,\theta). \qquad (2)$$

Here μ is the reduced mass of the collisional system, B is the rotational constant of the diatom (rigid rotor approximation assumed), **j** and **J** are the operators for rotational angular momentum of the diatom and total angular momentum, respectively, and V is the interaction potential. The BF z-axis is chosen to lie along the ion-diatom center of mass vector **R**, such that Ω is the projection of both **J** and **j** along the BF z-axis.

Employing the centrifugal sudden approximation (CSA[20,21]), we replace $|\mathbf{J}-\mathbf{j}|^2$ by the diagonal value $[J(J+1)+\mathbf{j}^2-2\Omega^2]$, which combined with the potential depending only on R and θ, eliminates the coupling between states of different Ω. The CSA approximation should be valid for ion-molecule systems having a large collisional reduced mass where the neglected off-diagonal coupling terms should be small. For a fixed value of R, the hamiltonian of Eq.(2) is then diagonalized using a basis set of spherical harmonics to yield a set of rotationally adiabatic potential energy curves $V^J_{j,\Omega}(R)$. We assume that the ion-dipole reaction occurs on a single adiabatic potential energy curve $V^J_{j,\Omega}(R)$ which correlates to the initial (J,j,Ω) asymptotic rotational state of the reactants. Thus any non-reactive rotational energy transfer is neglected. Furthermore, as a large number of J values will generally contribute to the reaction cross section, tunnelling for high J values will not make a significant contribution. The reaction probability is given a value of unity provided the energy is above the centrifugal barrier.

The assumption that reactions dominated by long-range intermolecular forces can be treated on a single rotationally adiabatic potential energy surface has been tested in previous calculations on the $O(^3P)+OH$ (quadrupole-dipole) and $He^+ + HCl$ (ion-dipole) reactions. For these systems,[14,22] the ACCSA method gives excellent agreement with cross sections and rate constants obtained using a 'detailed quantum transition-state theory' (DQTST),[23] which involves solving the coupled equations in the entrance channel and therefore does take into account inelastic rotational energy transfer between different rotational states.

3. Calculations and Results

3.1. THE INTERACTION POTENTIAL

In this study we have investigated the ion-dipole reaction $C^+(^2P) + HCl(X\,^1\Sigma^+) \rightarrow H(^2S) + CCl^+(X\,^1\Sigma^+)$ and the dependence of the calculated rate constants on the interaction potential, $V(R,\theta)$. We compare results for two forms of the interaction potential, which we label as $V_2(R,\theta)$ and $V_{SCF}(R,\theta)$. Both of these can be conveniently expressed in terms of a Legendre polynomial expansion. $V_2(R,\theta)$ is similar to Eq.(1) but includes the contributions from the permanent quadrupole moment, Θ, and static anisotropic polarizability, α_2, of the HCl molecule,

$$V_2(R,\theta) = -\frac{q^2\alpha}{R^4}P_0(\cos\theta) - \frac{q\mu_D}{R^2}P_1(\cos\theta) + \left[\frac{q^2\Theta}{R^3} - \frac{q^2\alpha_2}{2R^4}\right]P_2(\cos\theta). \qquad (3)$$

The final interaction potential, $V_{SCF}(R,\theta)$, is obtained from the fitting of *ab initio* SCF calculated energy points in the long and short-range regions of the potential energy surface.

Previous studies which employed the ion-dipole potential show that the capture approximation overestimates the experimentally measured rate constants for this system.[14] One possibility for this discrepancy is that the long-range intermolecular forces are less attractive than modelled by $V_2(R,\theta)$. The approach from the attractive side of the dipole, $\theta = 0$, is energetically favored by the ion-dipole term. The additional terms included in $V_2(R,\theta)$ will add a repulsive contribution for collinear approach. These terms are not expected to have a significant effect at long-range as they are proportional to R^{-3} and R^{-4} versus R^{-2} for the ion-dipole term, but become more important in the intermediate and short-range regions of the potential.

The asymptotic expansion $V_2(R,\theta)$ is only a valid description for the long-range intermolecular interaction. The definition of long-range will vary with different systems and will be dependent on the dominant R^{-n} term in the expansion. If the reaction can be described by a single determinant everywhere on the surface, it is expected that *ab initio* SCF calculations with a good basis set should give a qualitative and semiquantitative accurate description of the true potential surface for all regions.

3.2. SCF CALCULATIONS

Ab initio SCF calculations have been carried out over an extended range of intermolecular separations R, and orientations θ, to obtain a potential surface that describes the interaction between the $C^+(^2P)$ atom with $HCl(X\ ^1\Sigma^+)$ treated as a rigid rotor. The triple-zeta plus double polarization (TZ2P) quality contracted gaussian basis set employed consists of (11s7p2d/7s5p2d) for chlorine[24,25] with α_d's = 0.75 and 0.25, (5s2p/3s2p) for hydrogen[24,26] with α_p's = 1.5 and 0.5, and (10s6p2d/5s4p2d) for carbon[24,26] with α_d's = 1.2 and 0.4. As indicated in Table I, this basis set yields reasonable values of the molecular properties of $HCl(X\ ^1\Sigma^+)$.[27] The dipole and quadrupole moments are each within 10% of the experimental values, and hence the SCF surface should give an accurate description of the long-range intermolecular interaction.

Table I. Molecular Properties of $HCl(X\ ^1\Sigma^+)$

Property	SCF†	Expt.‡
r_e (Å)	1.2702	1.2746
B (cm^{-1})	10.66	10.59
μ_D (D)	1.225	1.109
Θ (10^{-26} esu)	3.67	3.80
α ($Å^3$)	1.88	2.60
α_2 ($Å^3$)	0.380	0.209

†Present study, see text.
‡See reference 27.

The interaction potential $V_2(R,\theta)$ assumes the C^+ atom to be a point charge, neglecting any anisotropy in the electronic charge distribution. In reality, the carbon cation ground state is an open-shell ^2P electronic state which asymptotically is triply degenerate, corresponding to the unpaired electron occupying one of the carbon $2p_x$, $2p_y$, or $2p_z$ atomic orbitals. In the presence of the HCl molecule, the degeneracy is lost and the potential surface splits into two $^2A'$ and a $^2A''$ surfaces. For comparison with experimental measurements, the final total cross sections and rate constants must be averaged over all three surfaces.

Comparison of $V_2(R,\theta)$ with the *ab initio* SCF potentials V^i_{SCF} gives an indication to what extent the asymptotic expansion of Eq.(3) is accurate. The agreement between $V_2(R,\theta)$ and all three SCF surfaces is within 10% up until an interfragment separation of 15 bohr, at which point the asymptotic expansion breaks down and $V_2(R,\theta)$ diverges significantly from the others. All three *ab initio* SCF potentials are similar, in agreement to within 5% up until about $R = 10$ bohr, where $2-{}^2A'$ diverges dramatically from the other two and becomes less attractive. The $1-{}^2A'$ and $1-{}^2A''$ surfaces separate as well, but qualitatively behave in a similar fashion. By $R = 3.5$ bohr, the $2-{}^2A'$ surface becomes strongly repulsive at all orientations, unlike the two lower surfaces which remain strongly attractive at near perpendicular geometries. The total cross sections and rate constants must be averaged over these three surfaces and their behavior at short range could possibly alter the results obtained assuming only the long-range forces are important.

The SCF potential energy surfaces, calculated on a grid between $R = 75$ and 3.5 bohr and $\theta = 0$ to 180° (in increments of 45°), were least-squares fit to a Legendre polynomial expansion for each fixed value of R. As R becomes smaller, the increasing anisotropy in the SCF interaction potentials is reflected in the increasing size of the coefficents for the higher terms. At $R = 5.0$ and 3.5 bohr, the potentials are strongly anisotropic and additional energy points were necessary to obtain a more reasonable fit (13 angles at 15° increments).

The SCF method completely neglects dispersion contributions to the interaction. However, long-range interfragment correlations should be negligible, and the main corrections will come from intrafragment correlations. Thus, we expect the general shape of the long-range SCF interaction to be maintained in more sophisticated correlated wavefunction methods, though perhaps shifted somewhat in energy. As the SCF molecular parameters dominant in the asymptotic expansion are within 10% of the experimental values, we expect the SCF potentials we have calculated in the present work to be accurate for the long-range region. For the short-range regions of the potential, correlation corrections will be much more important. However, as discussed above, the qualitative and semiquantitative accuracy of these SCF potentials should be adequate for these ion-dipole systems within the ACCSA theory.

For low energy collisions, the only energetically allowed product channel is ground state $CCl^+(X\,{}^1\Sigma^+) + H({}^2S)$ which is exothermic by 0.91 eV. The $1-{}^2A''$ and $2-{}^2A'$ surfaces correlate to a higher $CCl^+({}^2\Pi)$ surface, which is 5.3 eV above. Asymptotically all three surfaces are equally probable but it appears only one results in products, hence the total reaction cross sections should be one-third those obtained for the lowest $1-{}^2A'$ surface. However, both the $1-{}^2A'$ and $1-{}^2A''$ are strongly attractive and show bonding character along the entrance channel. Indeed, along a collinear approach, they are degenerate and comprise the two components of the ${}^2\Pi$ ground state surface. If there exists a mechanism, such as strong nonadiabatic couplings, which allows fast and efficient transitions between these surfaces, then the $1-{}^2A''$ surface could also lead to products. Then, the total reaction cross section would approach the limit of one-third the sum of the cross sections for the two surfaces.

3.3. CROSS SECTIONS

A basis set of spherical harmonics including a maximum of 20 for j and Ω gives excellent convergence for $V^J_{j,\Omega}(R)$ for the range of interfragment distances, R, between 75 and 3.5 bohr. Figure 1 shows a plot of the (J=0, j=0, Ω=0) rotationally adiabatic potential curves for each of the surfaces being considered. The four curves resulting from the SCF potentials and $V_2(R,\theta)$, essentially lie on top of each other within the scale of the diagram until $R < 10$ bohr. At this point the asymptotic expansion breaks down and chemical bonding forces within the $1-{}^2A'$ and $1-{}^2A''$ surfaces cause the corresponding attractive rotationally adiabatic potential curves to drop faster than those of the $V_2(R,\theta)$ surface, whereas the repulsive nature of the $2-{}^2A'$ surface yields a rotationally adiabatic curve which turns over. This secondary repulsive barrier exists for every (j,Ω) curve belonging to the $2-{}^2A'$ surface. Thus, in the adiabatic capture approximation there will be no contribution to the total cross section from this surface.

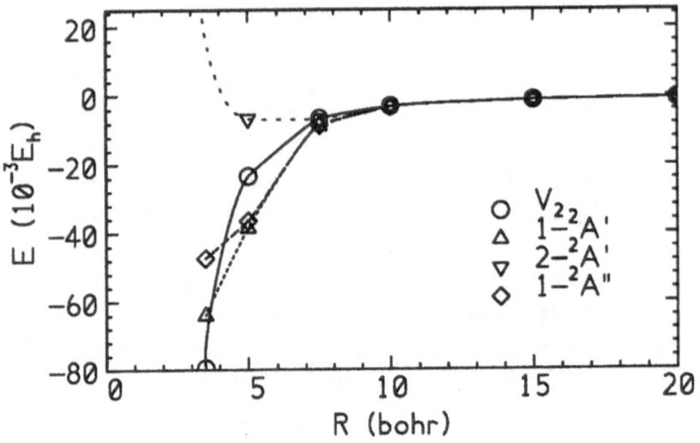

Figure 1. Rotationally Adiabatic Potential Curves, $V^J_{j,\Omega}(R)$, for J=0, j=0, Ω=0

Figure 2 plots the rotationally adiabatic potential curves for several (J=0, j, Ω) states obtained from the SCF $1-{}^2A'$ surface. Similar behavior is observed for the other reactive surfaces as well. Asymptotically each $V^J_{j,\Omega}(R)$ curve will reach an energy $E_j = Bj(j+1)$. (Note that for calculations on all surfaces the experimental value for the rotational constant B was employed.) For a given value of Ω, the $V^J_{j,\Omega}(R)$ curves are well separated and the interfragment separation corresponding to the maximum of the rotational barrier, R^*, decreases for increasing j. For a given value of j, the $V^J_{j,\Omega}(R)$ curves are also well separated near the rotational barrier, but asymptotically converge to the same energy, E_j. Increasing Ω has an opposite effect, tending to reduce the height of the barrier and to increase R^*. However, it is possible that $V^J_{j,\Omega}(R)$ curves with both j and Ω differing can cross. The ACCSA approximation, which assumes the reaction occurs on an individual $V^J_{j,\Omega}(R)$ curve, will be valid for systems in which the CSA approximation neglecting the coupling between states of different Ω applies. As discussed previously, for systems of large collisional reduced mass as considered here, the neglected coupling terms should be small.

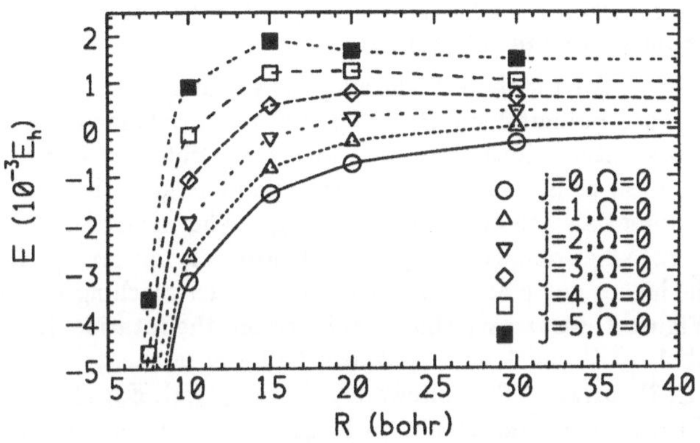

Figure 2. $V_{SCF}^{1-{}^2A'}$ Rotationally Adiabatic Potential Curves, $V_{j,\Omega}^{J}(R)$, for J=0

Table II. $C^+ + HCl$ ACCSA Cross Sections†

j	E_c	V_2	$V_{SCF}^{1-{}^2A'}$	$V_{SCF}^{1-{}^2A''}$
	1.0(-4)	9762	10152	10350
	1.0(-3)	1288	1266	1310
0	5.0(-3)	370	442	442
	1.0(-2)	236	328	324
	2.0(-2)	166	238	251
	1.0(-4)	6590	6784	6984
	1.0(-3)	1216	1237	1273
1	5.0(-3)	356	440	438
	1.0(-2)	230	321	318
	2.0(-2)	163	234	248
	1.0(-4)	4061	4079	4061
	1.0(-3)	1120	1140	1123
2	5.0(-3)	339	430	417
	1.0(-2)	223	316	309
	2.0(-2)	158	230	245

†Collisional energies given in hartrees, Cross sections given in bohr^2. Numbers in parentheses refer to powers of 10.

To determine the state-selected cross sections for a given collisional energy E_c, the maximum value of the total angular momentum $J_{max}(j,\Omega,E_c)$, for which $V_{j,\Omega}^{J}(R)$ for all R is less than $E = E_c + E_j$ must be found. An analytic form for $V_{j,\Omega}^{J}(R)$ is obtained from fitting $V_{j,\Omega}^{J}(R)$ with $J = 0$ to an exponential-spline function over the grid of R values with the addition of the centrifugal term $BJ(J+1)/(2\mu R^2)$. Standard algorithms for maximization of functions can be used to find $J_{max}(j,\Omega,E_c)$ and R^* for each (j,Ω) state and E_c. The centrifugal term is repulsive for all R and is proportional to R^{-2}, hence, for increasing J, the barrier height increases and R^* decreases.

Table II gives results for the calculated cross sections for several (j, Ω) states at different energies for each potential surface. The SCF potentials for the $1-{}^2A'$ and $1-{}^2A''$ surfaces yield cross sections which agree well with those obtained using the asymptotic expansion $V_2(R,\theta)$ for low collisional energies. At higher energies, the cross sections obtained for the full SCF potentials are considerably larger. The explanation is found in the R^* values which indicate what range of the potential surface is important. Our calculations indicate that for the higher energies E_c, R^* becomes smaller reaching values between 7.5 and 5.0 bohr. From Figure 1, we can see that in this region, the rotationally adiabatic potential curves of the SCF $1-{}^2A'$ and $1-{}^2A''$ surfaces lie lower in energy than the corresponding curve of the $V_2(R,\theta)$ surface. This allows higher $J_{max}(j,\Omega,E_c)$ effective potential curves satisfying $V^J_{j,\Omega}(R) < E_c + E_j$ and results in larger cross sections for the SCF potentials at these energies.

3.4 RATE CONSTANTS

The state-selected rate constants $k_j(T)$ are obtained from Maxwell-Boltzmann averaging over the calculated cross sections, which can then be Boltzmann averaged over j yielding the temperature dependent total rate constant, $k(T)$. The results and trends obtained for the calculated state-selected and total rate constants are both qualitatively and quantitatively similar for each reactive surface. Figure 3 plots the calculated total rate constants obtained over a temperature range between 2 and 400°K. The strong negative temperature dependence is observed for each surface, also in qualitative agreement with experiment.

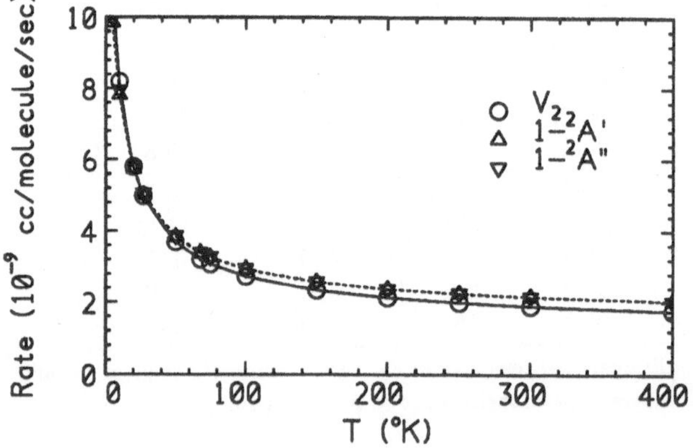

Figure 3. ACCSA Rate Constants, $k(T)$

Another important feature which arises from the theory is the sensitivity of the rotationally state-selected rate constants $k_j(T)$ to the initial rotational state j. Figure 4 illustrates this point for the calculated $k_j(T)$ of the SCF $1-{}^2A''$ surface for $j = 0$ to 3. It is seen that $k_j(T)$ decreases strongly as j is increased. This effect is due to the fact that initially at long-range the energetically favored approach of the ion is collinear with the direction of the dipole. As the molecule rotates, the ion feels all orientations of the potential. Therefore, the ion spends less time along the minimum energy path and the reaction probablility is reduced. For j $\geq$ 3, the temperature dependence of $k_j(T)$ has essentially vanished. This is due to the anisotropy of the potential being averaged out by the rotation. The total rate constant $k(T)$ is the Boltzmann average of $k_j(T)$ over j. At low temperatures, only small values of j will contribute and there is a strong temperature dependence. As T increases, higher j rotational states will eventually dominate and the temperature dependence becomes weaker approaching the Langevin limit.

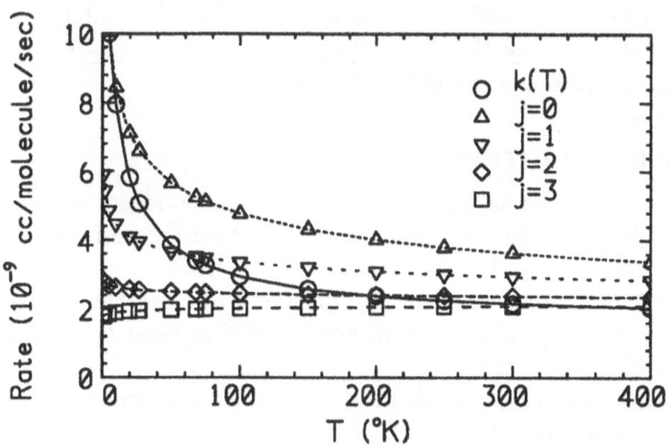

Figure 4. $V_{\mathrm{SCF}}^{1-{}^2A''}$ ACCSA State-Selected Rate Constants, $k_j(T)$

Comparison between the $k(T)$ obtained from the $V_2(R,\theta)$ and the full SCF potential surfaces, shows good agreement at low temperatures, but larger rate constants are obtained for the $1-{}^2A'$ and $1-{}^2A''$ SCF surfaces at higher temperatures. This is consistent with the arguments given previously to explain the behavior of the calculated cross sections. At the higher temperatures, larger collisional energies will contribute and the shorter range region of the potential is more important. The attractive bonding nature of the short range interaction for the $1-{}^2A'$ and $1-{}^2A''$ surfaces tends to yield larger rate constants for higher temperatures.

From Figure 3 and Table III, we can see that the $V_2(R,\theta)$, $1-{}^2A'$ and $1-{}^2A''$ surfaces give similar results for the total reaction rate constants, overestimating the experimental values. The multi-surface character of open-shell systems has often been overlooked in previous studies which employ the asymptotic expansion to model the long-range interactions. For the present system, only the lowest $1-{}^2A'$ surface correlates to the energetically accessible products, although the probability of starting on each surface is the same. Hence, for a proper comparison, the cross sections for the $1-{}^2A'$ surface should be divided by three, resulting in roughly one-third the reaction rate constants reported in Table III and an underestimation of the experimental values. Allowing for an efficient transition mechanism for the reaction to occur on the $1-{}^2A''$ surface, the averaging procedure would entail one-third the sum of the cross sections for each surface. These averaged rate constants are given in Table III under the column labeled AVE. The rate constants are in good agreement with experiment being within the error bars at low T and just slightly larger at 300°K. However, as discussed above, the higher temperature results are more sensitive to the short-range potential. It is likely that inclusion of electronic correlation via more sophisticated *ab initio* methods would be needed to attain greater accuracy in this region.

Table III. $C^+ + HCl$ ACCSA Rate Constants†

T (°K)	V_2^{SCF}	$V_{SCF}^{1-{}^2A'}$	$V_{SCF}^{1-{}^2A''}$	AVE	SO	Expt.‡
27	4.98	5.00	5.06	3.35	4.84	3.8±1.1
68	3.21	3.38	3.40	2.26	2.80	1.9±0.6
300	1.89	2.15	2.15	1.43	1.51	1.0±0.2

†Rate constants are given in units 10^{-9} cc/molecule/sec.

‡See reference 14.

3.5 SPIN-ORBIT INTERACTIONS

In the preceeding analysis, the electronic potential surfaces were obtained by solving the nonrelativistic Schrödinger equation within the Born-Oppenheimer approximation. In reality, the spin-orbit interactions within the $C^+({}^2P)$ atom causes a splitting of 63.42 cm^{-1} between the ${}^2P_{1/2}$ doubly degenerate ground state and the ${}^2P_{3/2}$ excited quartet.[28] We adopt a simple model employed by Handy *et al.*[29] in their treatment of the spin-orbit interaction in the $F + H_2$ system.

We can incorporate the spin-orbit interaction via the angular momentum Hamiltonian for the collinear system, given by

$$H = \Sigma + (\Pi - \Sigma)\mathbf{L_z}^2 + \frac{2}{3}\Delta(\mathbf{L}\cdot\mathbf{S}). \tag{4}$$

Here Σ and Π represent the three SCF potentials determined as before without regard to spin-orbit coupling. The components of the Π electronic ground state correspond to the $1-{}^2A'$ and $1-{}^2A''$ surfaces and Σ to the $2-{}^2A'$ in a collinear arrangement. The observed ${}^2P_{1/2}$ - ${}^2P_{3/2}$ C^+ splitting is given by Δ. The model assumes that the spin-orbit coupling is the same as that for the free C^+ atom, although it varies with the nuclear geometry as the SCF electronic wavefunctions will have varying mixtures of the C^+ atomic orbitals. As we are only considering the entrance channel, which is predominantly reactant-like and in which a collinear approach is energetically favored, this simple model should give a semiquantitative estimate of the importance of spin-orbit coupling.

The electronic potential surfaces which include the spin-orbit coupling are obtained from diagonalization of the Hamiltonian (4) within a six-dimensional spin-angular momentum basis $|L, M_L, S, M_S>$, with $L = 1$ and $S = 1/2$, and $M = M_L + M_S$ is a good quantum number. The eigenvalues obtained are given by

$$\begin{aligned} V^{(3/2)} &= \Pi + \frac{\Delta}{3}\,; \quad |M| = \frac{3}{2} \\ V_{\pm}^{(1/2)} &= \frac{1}{2}(\Sigma + \Pi) - \frac{\Delta}{6} \pm \frac{1}{2}\left[(\Sigma - \Pi)^2 + \frac{2\Delta}{3}(\Sigma - \Pi) + \Delta^2\right]^{1/2}\,; \quad |M| = \frac{1}{2}. \end{aligned} \tag{5}$$

Of the new potential surfaces, $V_{-}^{(1/2)}$ asymptotically correlates to the ${}^2P_{1/2}$ doubly degenerate ground state surfaces and both $V^{(3/2)}$ and $V_{+}^{(1/2)}$ go to the four-fold degenerate ${}^2P_{3/2}$ surfaces which lie 63.42 cm^{-1} higher in energy. The relative shapes of the new surfaces are found to be very similar to the original nonrelativistic SCF surfaces. Indeed, employing the ACCSA theory, essentially identical cross sections are obtained for the $V_-^{(1/2)}$ and $V^{(3/2)}$ surfaces as were found for the $1-{}^2A'$ and $1-{}^2A''$ surfaces. Similar to the $2-{}^2A'$ surface, the $V_+^{(1/2)}$ surfaces become strongly repulsive and hence nonreactive in the adiabatic capture approximation. The magnitude of Δ is sufficiently large to assume that coupling between the excited quartet surfaces and the doublet ground state surfaces is not important.

Assuming a Boltzmann probability of occupying each level, the average rate constants are obtained from the following relation,

$$\bar{k} = \frac{2k(V_-^{(1/2)}) + 2\exp\left(-\frac{\Delta}{KT}\right) k(V^{(3/2)})}{2 + 4\exp\left(-\frac{\Delta}{KT}\right)}. \qquad (6)$$

In the limit of very small T, the final rate constant will be dominated by reaction occuring on the $V_-^{(1/2)}$ surfaces. At higher T, however, there will be a more significant contribution from the $V^{(3/2)}$ surfaces, as for example, at room temperature 42% of the C^+ population will exist in the ${}^2P_{3/2}$ state. The rate constants obtained from these spin-orbit (SO) surfaces are also given in Table III. The calculated rate constants still slightly overestimate experiment, but are much closer than those obtained from employing the asymptotic expansion which ignores the multi-surface and spin-orbit interaction aspects. Further improvement could possibly be attained from a more rigorous treatment of the spin-orbit interaction in three dimensions.

4. Summary

The ACCSA theory is applied to the $HCl(X\ {}^1\Sigma^+) + C^+({}^2P) \rightarrow CCl^+(X\ {}^1\Sigma^+) + H({}^2S)$ reaction to detemine rotational state-selected cross sections and reaction rate constants. *Ab initio* SCF calculations were performed to obtain a more accurate potential energy surface. For this open-shell system, three surfaces, $1-{}^2A'$, $1-{}^2A''$, and $2-{}^2A'$, must be considered. Both the $1-{}^2A'$ and $1-{}^2A''$ surfaces are strongly attractive, especially for short $C-Cl$ distances at perpendicular geometries. The $2-{}^2A'$ surface is repulsive at short-range and hence does not contribute to the reaction. Comparisons of the results with those obtained using an asymptotic multipole expansion for the long-range interaction indicate good agreement for low-energy cross sections and low-temperature rate constants. For the higher energy and temperature results, the short-range region of the potential becomes more important and the more attractive SCF potentials yield higher cross sections and rate constants. Spin-orbit coupling is also shown to have an important effect on the final rate constants. Good agreement is obtained with CRESU experiments at 27, and 68 °K, and SIFT measurements at 300°K.

The ACCSA theory has been shown in the past to be reliable for predicting low temperature rate constants for exothermic proton-transfer reactions. A main goal of this study was to investigate the nature of other types of reaction for which the theory was not so reliable. From this study on the $HCl + C^+$ system, it is clear that much more careful consideration must be taken when dealing with open-shell systems. The multi-surface nature of the problem is not adequately described by an asymptotic multipole expansion. However, reasonable results within the ACCSA framework have been obtained by including more detailed information of the pertinent surfaces. Further investigation of other systems appears necessary before generalizations can be made concerning the reliability of the ACCSA for open-shell systems.

Acknowledgements.

This research is supported by the Science and Engineering Research Council. The *ab initio* calculations were performed using the CADPAC program developed by Dr. R.D. Amos and Dr. J.E. Rice at Cambridge University.

References

1. See for example, T. Su and M.T. Bowers, in *Gas Phase Ion Chemistry*, edited by M.T. Bowers (Academic Press, New York, 1979) Vol. 1, p.84.
2. B.R. Rowe, J.B. Marquette, G. Dupeyrat, and E.E. Ferguson, Chem. Phys. Lett. **113**, 403 (1985).
3. T. Su and M.T. Bowers, J. Chem. Phys. **58**, 3027 (1973).
4. T. Su and M.T. Bowers, Int. J. Mass. Spectro. Ion Phys. **17**, 211 (1975).
5. T. Su, E.C.F. Su, and M.T. Bowers, J. Chem. Phys. **69**, 2243 (1978).
6. W.J. Chesnavich, T. Su, and M.T. Bowers, J. Chem. Phys. **72**, 2641 (1980).
7. R.A. Barker and D.P. Ridge, J. Chem. Phys. **64**, 4411 (1976).
8. K. Sakimoto and K. Takayanagi, J. Phys. Soc. Jpn. **48**, 2076 (1980).
9. K. Sakimoto, Chem. Phys. Lett. **116**, 86 (1985).
10. T.Su and W.J. Chesnavich, J. Chem. Phys. **76**, 5183 (1982).
11. D.R. Bates, Proc. R. Soc. London, Ser. A **384**, 289 (1982).
12. M.Quack and J. Troe, Ber. Bunsenges. Physik. Chem. **78**, 240 (1974).
13. J. Troe, Chem. Phys. Lett. **122**, 425 (1985).
14. C. Rebrion, J.B. Marquette, B.R. Rowe and D.C. Clary, Chem. Phys. Lett. **143**, 130 (1988).
15. D.C. Clary, D. Smith, and N.G. Adams, Chem. Phys. Lett. **119**, 320 (1985).
16. D.C. Clary, Mol. Phys. **53**, 3 (1984).
17. D.C. Clary, Mol. Phys. **54**, 605 (1985).
18. D.C. Clary, J. Chem. Soc., Faraday Trans. 2 **83**, 139 (1987).
19. Determined from ionization potentials and thermochemical data, in *CRC Handbook of Chemistry and Physics, 68th Edition*, edited by R.C. West (CRC Press, Boca Raton, Florida, 1987).
20. R.T. Pack, J. Chem. Phys. **60**, 633 (1974).
21. P. McGuire and D.J. Kouri, J. Chem. Phys. **60**, 2408 (1974).
22. D.C. Clary and J.P. Henshaw, Faraday Disc. Chem. Soc. **84**, 333 (1987).
23. J.C. Light and A. Altenberger-Siczek, Chem. Phys. Lett. **30**, 195 (1975).
24. S. Huzinaga, J. Chem. Phys. **42**, 1293 (1965).
25. T.H. Dunning and P.J. Hay, in *Modern Theoretical Chemistry,* edited by H.F. Schaefer (Plenum, New York 1977) Vol. 3, pp. 1-27.
26. T.H. Dunning, J. Chem. Phys. **53**, 2823 (1970).
27. C.G. Gray and K.E. Gubbins, *The Theory of Molecular Fluids,* (Clarendon Press, Oxford 1984) Vol. 1, p. 579.
28. C.E. Moore, Natl. Bur. Std. Ref. Data Ser. **34** (1970).
29. N.C. Handy, T.J.Lee, and W.H. Miller, Chem. Phys. Lett. **125**, 12 (1986).

CLASSICAL PATH APPROACH TO INELASTIC AND REACTIVE SCATTERING

Gert Due Billing
Chemistry Laboratory III
H.C. Ørsted Institute
University of Copenhagen
DK-2100 Copenhagen Ø
Denmark

ABSTRACT. Exact quantum mechanical calculations on energy transfer processes are possible or numerical feasible only for a few mainly hydrogen containing systems. However, often the semiclassical (classical path) method will be sufficiently accurate to allow a quantitative determination of cross sections and rate constants for inelastic as well as reactive processes. A brief review of the classical path method is given and several numerical aspects are discussed.

1. INTRODUCTION

In the classical path approach a certain number of degrees of freedom are treated classically whereas the remaining ones are quantized. In order to decide upon how to divide the phase space in a classical and a quantum part one utilizes that the various degrees of freedom often are well separated on the energy scale, i.e. have different de Broglie wave lengths. As a rule of thumb a given degree of freedom may be treated classically if the available energy for that particular degree of freedom is of the order 10 to 20 times the quantum mechanical level spacing or more. However, we know for instance from classical S-matrix theory that the classical treatment is only adequate if a certain amount of averaging is introduced, i.e. the classical degrees of freedom should somehow be averaged out before comparison with either experiment or less approximate methods is carried out. Thus we expect for instance that a classical mechanical treatment of the rotational motion of a diatomic molecule is better for the rotational averaged quantitites as e.g. rate constants or cross sections for vibrational transitions than the state to state vib/rot cross sections since the latter can only be obtained by using the socalled bin or boxing methods in classical phase space. Likewise if the rotation and vibration are quantized

A. Laganà (ed.), Supercomputer Algorithms for Reactivity, Dynamics and Kinetics of Small Molecules, 339–356.

whereas translation and rotational projection states are treated classically it is the m-state averaged cross sections which are expected to be the most accurate. Thus the classical path method implicitly assumes an average over part of phase-space namely that part which is treated classically - although more detailed information is in principle available from the trajectories.

Once it has been decided how to divide the phase space it is possible from the time dependent Schrödinger equation to derive a set of equations for the time evolution of the amplitude of the quantum mechanical part of the system. These equations may quite generally be written as:

$$i \hbar \dot{\underline{\underline{A}}}(t,t_o) = \underline{\underline{H}}\, \underline{\underline{A}}(t,t_o) \tag{1}$$

The squared amplitudes are either quantum state probabilities or they may represent quantum operators if the hamiltonian is expanded in quantum operators rather than quantum states. Aside from a set of equations of the type (1) we have some classical equations of motion for the classical part of the system. In order to couple the classical and quantum subsystems one introduces an average hamiltonian, i.e. an average over the total hamiltonian of the system such that the quantum variables only appear with some average or expectation value in the hamiltonian governing the classical motion. Thus we also obtain a set of equations of motion of the type:

$$\dot{Q} = \frac{\partial H_{av}}{\partial P_i} \tag{2a}$$

$$\dot{P}_i = - \frac{\partial H_{av}}{\partial Q_i} \tag{2b}$$

Since H_{av} depends upon the solution to eq. (1) and vice versa the equations (1,2) are solved simultanelusly. The equations (2) may be solved by standard Runge-Kutta or Predictor Corrector methods and efficient algorithms for the solution of eqs. (1) have been developed using the matrix diagonalization technique.

Improvements in the classical path method may be introduced by using a Gaussian wave packet description for some of the degrees of freedom which are treated classically. This "quantum trajectory" approach has recently been used to establish the validity of the velocity symmetrization usually introduced in the classical path method. Other ways of improving the classical path method uses multitrajectory approaches and eventually include interference effects. However, for a number of processes the "primitive" classical path method is sufficiently accurate.

2. THE CLASSICAL PATH METHOD

In the classical path formulation of collision problems one divides the phase space in two sets: a classical {x} and a quantum {y}. The dynamical motion in the classical phase space is as mentioned in the introduction governed by an effective or average potential. However, it has been known for many years that it is necessary also to introduce a "symmetrization procedure" of the S-matrix elements or transition probabilities obtained by solving the equations (1, 2). This symmetrization procedure which has been motivated by results obtained using first order perturbation technique (for further details see ref. [1]) has recently been rationalized by the quantum trajectory derivation by Muckerman et al. [2].

The results of ref. [2] will be briefly reviewed here. Introducing for the system under consideration the trial wave function of the type:

$$\psi(x,y,t) = \Phi(y,t)\ F(x,t) \tag{3}$$

i.e. separability in the two sets of variables is assumed. More generally one could introduce an expansion in a number of product functions:

$$\psi(x,y,t) = \sum_{\alpha} \Phi_{\alpha}(y,t)\ F_{\alpha}(x,t) \tag{4}$$

But a particular simple approach obtains if the system may be described by a single trial function (3). Inserting (3) in the time dependent Schrödinger equation

$$i\hbar \frac{\partial\psi}{\partial t} = \left[H_o(y) - \frac{\hbar^2}{2\mu}\frac{\partial^2}{\partial x^2} + V(x,y)\right]\psi \tag{5}$$

yields:

$$i\hbar(\langle\Phi|\frac{\partial\Phi}{\partial t}\rangle F + \frac{\partial F}{\partial t}) = \langle\Phi|H_o+V|\Phi\rangle F - \frac{\hbar^2}{2\mu}\frac{\partial^2 F}{\partial x^2} \tag{6}$$

where we have chosen to consider a system with 'internal' coordinate y, external coordinate x and where μ is a reduced mass, V(x,y) the interaction between the external and internal system and $H_0(y)$ the hamiltonian for the isolated 'y-system'. The brackets denote integration over y-space and we have used that $\langle\Phi|\Phi\rangle = 1$. If we now introduce a simple Gaussian wave packet for the 'external' degree of freedom:

$$F(x,t) = \exp\{ \frac{i}{\hbar}(\alpha(t)\ (x-\bar{x}(t))^2 + p(t)\ (x-\bar{x}(t))\ \gamma(t))\} \tag{7}$$

i.e. a form suggested by Heller [3], we obtain by expanding the interaction potential V(x,y) through second order power of $(x-\bar{x}(t))^k$ (k=0,1,2) the following set of equations:

$$\dot{\alpha}(t) = -\frac{2}{\mu}\,\alpha(t)^2 - \frac{1}{2}\,\langle\Phi|V''|\Phi\rangle\ , \tag{8a}$$

$$\dot{\bar{x}}(t) = \frac{1}{\mu}\,p(t)\ , \tag{8b}$$

$$\dot{p}(t) = -\,\langle\Phi|V'|\Phi\rangle\ , \tag{8c}$$

$$\dot{\gamma}(t) = i\hbar\,\frac{\alpha(t)}{\mu} + \frac{p(t)^2}{2\mu} \tag{8d}$$

Introducing also the eigenfunctions to $H_0(y)$:

$$H_0\ \phi_n = E_n\ \phi_n \tag{9}$$

and

$$\Phi(y,t) = \sum_n a_n(t)\ \phi_n(y)\ \exp(-\frac{i}{\hbar}\,E_n t) \tag{10}$$

we furthermore obtain from eq. (6) a set of coupled equations for the amplitudes $a_n(t)$:

$$i\hbar\ \dot{a}_m = \sum_n V_{mn}(\bar{x}(t))\ a_n\ \exp(i\omega_{mn}t) \tag{11}$$

where $\omega_{mn} = \hbar^{-1}(E_m - E_n)$ and $V_{mn} = \langle\phi_m|V|\phi_n\rangle$. We notice that the eqs. (8b, 8c) and (10) may be solved independently of (8a, 8d) and that eqs. (8b,c) are the usual classical trajectory equations governed by the Ehrenfest potential $\langle\Phi|V|\Phi\rangle$. Eqs. (8a,d) determine the width parameter α and the phase of the wavepacket F(x,t). We notice that the eqs. (8b,c) and (10) are specific examples of the general ones (1,2) mentioned in the introduction. The parameters of the eqs. (8-10) are the initial momentum $p(-\infty)$ and width $\alpha(-\infty)$. However, the width α_o at the turning point of the external motion is more convenient to work with as a parameter. In ref. [2] it was demonstrated that:

a) the transition probabilities $P_{IF} = |a_F(\infty)|^2$, where $a_n(-\infty) = \delta_{nI}$, were independent of α_o for a given value of $p(-\infty) = p^*(-\infty)$.
b) Good agreement with quantum results is obtained with

   $p(-\infty) = \frac{1}{2}(p_I+p_F)$ and $\alpha_o = \alpha_o^{min}$ where α_o^{min} defines the

   width parameter which minimizes the error associated with the Gaussian wave packet approach. Since the Gaussian wave packet is exact for harmonic potentials the error term involves the cubic derivative of the interaction potential.
c) At low energies $p^*(-\infty) = 1/2(p_I+p_F)$, where the momenta p_n are defined by

$$\frac{p_n^2}{2\mu} = E - E_n \qquad (12)$$

where E is the total 'quantum' energy.

Thus the quantum trajectory method essentially derives the symmetrized Ehrenfest approach suggested many years ago [4] and it introduces a correction factor [2] which, however, turns out to modify only slightly the results obtained with the 'primitive' classical path method. More important the derivation shows that the classical path method can be expected to be accurate when the separability assumption (3) is fulfilled. This is the case in the 'weak coupling' limit which typically extends to energies up to 10 to 20 times the energy level spacing. However, for higher energies we may as mentioned in the introduction introduce a classical mechanical description also of the y-system. Thus the classical path and the classical mechanical description will for many systems be complementary, i.e. when the system enters the strong coupling region the x and y degrees of freedom should be treated evenhandedly and classically. However, there may be situations where we are in a strong coupling region and where classical mechanics is not valid. Such situations occur e.g. in the tunnelling region of reactive scattering, but as we shall see later, we may deal with this situation by changing to coordinates which do not discriminate between the reaction channels in the way the cartesian coordinate system does. Thus before the classical path method is applied, the problem should be examined in terms of strong and weak coupling regimes, and this analyses eventually require a change in coordinates before the separability assumption (3) can be introduced. If this is still not sufficient, one may have to use the more general expression (4) eventually just with a few terms [5]. However, for a large class of problems it will be possible to use the simple classical path scheme, and we have in table 1 shown how it is possible to reduce the dimensionality of the $\underline{\underline{H}}$ matrix. But often the dimension of $\underline{\underline{H}}$ is large and the integration of eqs. (1) will then be the timedetermining part of the classical path method. We shall below give a numerically convenient method for solving the eqs. (1).

3. NUMERICAL METHODS

As mentioned in the introduction the classical equations of motion (2) may be solved using standard Runge-Kutta (RK) or Predictor-Corrector (PC) methods. If the rhs of eq. (2) is cumbersome to calculate, it is advantageous to use the PC method since it calculates the derivative only twice in each integration step. Furthermore it is desirable with a large step length which then requires a high order PC-method. Thus for exploratory calculations it can be recommended to use a variable order/variable step size method suggested by Krogh [14]. If the problem under consideration happens to be such that the method rarely changes the order/stepsize this may then be utilized by switching to a PC or RK method which does not have this facility. However, since the quantum part of the problem is the time consuming part it is necessary to use a steplength in the trajectory part which takes full advantage of the matrix diagonalization technique given below. If the eq. (2a) yields the result

$$\dot{Q}_i = P_i \qquad (14)$$

one might consider combining eqs. (2b) and (14) to the algorithm often used in molecular dynamics calculations:

$$Q_i(t+h) = 2\,Q_i(t) - Q_i(t-h) + C_i(t)\,h^2 + O(h^3) \qquad (15)$$

where $C_i(t) = -\frac{\partial}{\partial Q_i} H_{av}$.

However, we can hardly recommend to use this method in scattering calculations since there is nothing in eq. (15) which prevents propagation beyond the true turning point, and since the probability for inelastic or reactive events are extremely sensitive to the propagation being correct in this region, the algorithm (15) may lead to large errors if it is not accompanied by an accurate turning point predictor which predicts the correct step length h to be used in (15) in order to hit the correct turning point. As mentioned the propagation of the classical trajectories is usually not the time consuming part of the classical path method. But in cases where the integration has to be carried out over a large asymptotic region of the intermolecular interaction it is advantageous to switch to a classical mechanical perturbation theory [15] at a given separation R_o and then propagate in variables which changes more slowly in time than do the cartesian coordinates and momenta. Such variables are the action-angle variables which at least for diatomic molecules are well defined.

Also the time dependent matrix equation (1) may be integrated using the standard methods mentioned above. Here, however, the size of the matrix is usually so large that it is advantageous to introduce an alternative technique. Furthermore, the following method which is based upon a matrix diagonalization and multiplication technique can utilize the vector facilities offered by modern machines. The eqs. (1) where $\underline{\underline{H}}$ is a hermitian matrix should be rewritten as

$$i\hbar \dot{\underline{\underline{B}}} = (\underline{\underline{W}} + \underline{\underline{E}})\, \underline{\underline{B}} \tag{16}$$

where $\underline{\underline{W}}$ is a real symmetric and $\underline{\underline{E}}$ a diagonal matrix and where we have introduced $\underline{\underline{B}} = \exp(-\frac{i}{\hbar}\underline{\underline{E}}t)\underline{\underline{A}}$. Following ref. [16] we expand the matrix $\underline{\underline{U}} = \underline{\underline{W}} + \underline{\underline{E}}$ in the time interval $[t_n;\ t_{n+1}]$:

$$\underline{\underline{U}}(t) = \underline{\underline{U}}(T_n) + \left.\frac{dU}{dt}\right|_{T_n} \tau + \frac{1}{2} \left.\frac{d^2U}{dt^2}\right|_{T_n} \tau^2 \tag{17}$$

where $T_n = \frac{1}{2}(t_n + t_{n+1})$ and $\tau = t - T_N$.

We now introduce the transformation matrix $\underline{\underline{A}}$ such that

$$\underline{\underline{A}}\ \underline{\underline{U}}(T_n)\ \underline{\underline{A}}^T = \underline{\underline{D}} \text{ (diagonal matrix)} \tag{18}$$

and define

$$\underline{\underline{b}} = \underline{\underline{A}}\ \underline{\underline{B}} \tag{19}$$

Inserting (17) in (16) and using eqs. (18,19) we obtain the

following differential equation for $\underline{\underline{b}}$:

$$i\hbar\, \dot{\underline{\underline{b}}} = (\underline{\underline{D}} + \underline{\underline{F}}\tau + \underline{\underline{G}}\tau^2)\underline{\underline{b}} \tag{20}$$

where $\underline{\underline{F}} = \underline{\underline{A}}\, \dot{\underline{\underline{U}}}(T_n)\, \underline{\underline{A}}^T$ and $\underline{\underline{G}} = \frac{1}{2}\, \underline{\underline{A}}\, \ddot{\underline{\underline{U}}}(T_n)\, \underline{\underline{A}}^T$.

We now define a matrix $\underline{\underline{C}}$ as the solution to the differential equation

$$i\hbar\, \dot{\underline{\underline{C}}} = [\underline{\underline{D}} + \underline{\underline{F}}_D\, \tau + \underline{\underline{G}}_D\, \tau^2,\, \underline{\underline{C}}] + (\underline{\underline{F}}_o\, \tau + \underline{\underline{G}}_o\, \tau^2)\, \underline{\underline{C}} \tag{21}$$

where $\underline{\underline{F}}_D$ and $\underline{\underline{G}}_D$ denote the diagonal of $\underline{\underline{F}}$ and $\underline{\underline{G}}$ respectively, [,] a commutator and where

$$\underline{\underline{F}}_o = \underline{\underline{F}} - \underline{\underline{F}}_D, \tag{22}$$

and

$$\underline{\underline{G}}_o = \underline{\underline{G}} - \underline{\underline{G}}_D \tag{23}$$

are the off diagonal part of $\underline{\underline{F}}$ and $\underline{\underline{G}}$. We furthermore define the matrix b_o by:

$$i\hbar\, \dot{\underline{\underline{b}}}_o = (\underline{\underline{D}} + \underline{\underline{F}}_D\, \tau + \underline{\underline{G}}_D\, \tau^2)\, \underline{\underline{b}}_o \tag{24}$$

The solution to eq. (20) can now be shown to be:

$$\underline{\underline{b}} = \underline{\underline{C}}\, \underline{\underline{b}}_o \tag{25}$$

Since eq. (24) only contains a diagonal 'coupling' it is easily solved, i.e.

$$\underline{\underline{b}}_o(t) = \underline{\underline{H}}(t)\, \underline{\underline{b}}_o(t_n) \tag{26}$$

where

$$[\underline{\underline{H}}(t)]_{lk} = \exp\{-\frac{i}{\hbar}\, (D_{kk}(\tau + 1/2\, \Delta t_n) + \frac{1}{2}\, F_{kk}(\tau^2 - \frac{1}{4}\, (\Delta t_n)^2)$$

$$+ \frac{1}{3}\, G_{kk}(\tau^3 + \frac{1}{8}\, (\Delta t_n)^3))\}\, \delta_{lk} \tag{27}$$

and $\Delta t_n = t_{n+1} - t_n$. The boundary condition for the matrix $\underline{\underline{C}}$ may be chosen such that

$$\underline{\underline{C}}(T_n) = \underline{\underline{I}}(\text{unit matrix}) \tag{28}$$

Since $\underline{\underline{C}}$ is presumably slowly varying in the interval $[t_n, t_{n+1}]$ we can obtain a good estimate of it by substituting $\underline{\underline{C}} = \underline{\underline{I}}$ on the r.h.s. of (21) we then obtain:

$$\underline{\underline{C}} = \underline{\underline{I}} - \frac{i}{\hbar} \{\frac{1}{2} \underline{\underline{F}}_o \tau^2 + \frac{1}{3} \underline{\underline{G}}_o \tau^3\} \quad (29)$$

From eqs. (19), (25), (26) and (29) we obtain the final propagator suggested in ref. [16] by Billing and Baer:

$$\underline{\underline{B}}(t_{n+1}) = \underline{\underline{A}}^T \underline{\underline{C}}(t_{n+1}) \underline{\underline{H}}(t_{n+1}) \underline{\underline{C}}^{-1}(t_n) \underline{\underline{A}} \underline{\underline{B}}(t_n) \quad (30)$$

where

$$\underline{\underline{C}}(t_{n+1}) = \underline{\underline{I}} - \frac{i}{\hbar}\{\frac{1}{8} \underline{\underline{F}}_o(\Delta t_n)^2 + \frac{1}{24} \underline{\underline{G}}_o (\Delta t_n)^3\} , \quad (31)$$

$$[\underline{\underline{H}}(t_{n+1})]_{lk} = \exp\left[- \frac{i}{\hbar} (D_{kk} \Delta t_n + \frac{1}{12} G_{kk} (\Delta t_n)^3)\right] \quad (32)$$

and consistent with the approximation leading to (29) we obtain:

$$\underline{\underline{C}}^{-1}(t_n) \simeq \underline{\underline{C}}^*(t_n) = \underline{\underline{I}} + \frac{i}{\hbar} \{\frac{1}{8} \underline{\underline{F}}_o(\Delta t_n)^2 - \frac{1}{24} \underline{\underline{G}}_o(\Delta t_n)^3\} \quad (33)$$

The propagation of the solution from t_n to t_{n+1} involves one diagonalization and ten matrix multiplications if the complex matrices are expressed in terms of a real and an imaginary part. The propagator method is capable of using step lengths which are 10-20 times those of conventional PC methods, and since it is a matrix method it utilizes the advantages offered by vector computers. The speed may be increased further by introducing a 'dynamical' basis set (see below) or for large matrices the RRGM-method [17] instead of the diagonalization technique. The possibility of combining the RRGM and the above propagator method is presently being investigated at our laboratory.

4. DYNAMICAL BASIS SET

In the classical path method one usually propagates a number of vectors rather than a full matrix. The reason being that the effective potential which couples the quantum and classical subsystems depends upon the initial state or even upon the specific transition under consideration. However, in an approximation one may use a trajectory which is governed only by the elastic part of the interaction potential. Here the propagator becomes independent of the initial quantum state and it is convenient to propagate a full matrix. One should keep in mind, however, that the convergence properties are not independent of the initial state, and a basis set which is large enough for a specific initial state I may not be sufficient for another state. Thus the number of converged transition probabilities is always a subset of those obtained from the B-matrix which means that in any case it could be advantageous to tie the size of the B-matrix to the initial state. If this is the case, one may then as mentioned introduce a dynamical basis. Since the initial state (I) through the matrix elements of $\underline{\underline{W}}$ (eq. (16)) is to first order coupled only to a relatively small number of states one may reduce the dimension of $\underline{\underline{W}}$ (and $\underline{\underline{B}}$) in the beginning of the integration. As the propagation is continued, more states should be added according to the way the probability distribution upon the quantum levels changes, i.e. if the probability for being in a given state n exceeds a certain value, the basis set should be enlarged with those states which for first order are connected to it. This method has recently been used [18] to reduce the dimensions of the rotational-vibrational basis for the HF-HF system. The method may be used without any problems for the 'inward' integration, but for the 'outward' integration one must decide upon the probabilities one wants to determine accurately before truncating the basis set.

5. REACTIVE SCATTERING.

Before introducing the classical path approximation in reactive systems it is necessary to switch to a coordinate system which does not discriminate between the reaction channels. This condition is fulfilled in the so-called hyperspherical coordinates in which the atom-atom distances for the three-body system are expressed in terms of the hyperradius ρ and the two hyperangles θ, ϕ by [19]:

$$R_1 = \frac{\rho}{\sqrt{2}} d_3[1 + \sin\theta \cos(\phi+\varepsilon_3)]^{1/2}, \tag{34}$$

$$R_2 = \frac{\rho}{\sqrt{2}} d_1 [1 + \sin\theta \cos\phi]^{1/2}, \tag{35}$$

and

$$R_3 = \frac{\rho}{\sqrt{2}} d_2[1 + \sin\theta \cos(\phi-\varepsilon_2)]^{1/2} \tag{36}$$

where $d_1^2 = m_1(m_2+m_3)/\mu M$, $d_2^2 = m_2(m_1+m_3)/\mu M$,

$d_3^2 = m_3(m_1+m_2)/\mu M$, $M = m_1+m_2+m_3$, $\mu = (m_1m_2m_3/M)^{1/2}$

and

$$\epsilon_2 = 2\ \mathrm{arctg}(m_3/\mu) \tag{37}$$

$$\epsilon_3 = 2\ \mathrm{arctg}(m_2/\mu) \tag{38}$$

The coordinates ρ, θ, ϕ determine the shape and size of the molecular plane. The rotation of the molecular plane in space is specified by the Euler angles α, β and γ. In the CP-approximation suggested by Billing and Muckerman [11] the Euler angles and the motion along the hyper radius ρ are treated classically, i.e. only the θ,ϕ space is quantized. The reason for this 'choice' is that it may be shown that the asymptotic eigenstates $\psi_n(\theta,\phi;\rho)$ i.e. the diatomic fragment wavefunctions for $\rho\to\infty$ are then labelled by the vibrational/rotational quantum numbers v, j, but not by the projection quantum numbers m_j. Thus the suggested CP-method reduces the dimensionality of the quantum problem to a two dimensional one and decouples the m_j-states without making a decoupling approximation. In reality the m_j-states are treated classically! The mixed quantum-classical hamiltonian obtained in this manner is [11]:

$$\hat{H}_m = \frac{1}{2\mu} P_\rho^2 + \hat{H}_o + \frac{P_\gamma(P_\gamma - 4\cos\theta\ \hat{P}_\phi)}{2\mu\ \rho^2\ \sin^2\theta} + \frac{J^2 - P_\gamma^2}{\mu\rho^2\cos^2\theta}(1 + \sin\theta\ \cos^2\gamma) \tag{39}$$

where

$$\hat{H}_o = -\frac{1}{2\mu}\frac{4}{\rho^2}\hbar^2\left[\frac{\partial^2}{\partial\theta^2} + \frac{1}{\sin^2\theta}\frac{\partial^2}{\partial\phi^2}\right] + \Delta V(\rho,\theta) + V(\rho,\theta,\phi) \tag{40}$$

The total angular momentum is $P_\alpha = J$ and the angle β is defined by $P_\gamma = J\cos\beta$ [20]. ΔV is an additional potential arising from the coordinate transformation and is given by [13]:

$$\Delta V(\rho,\theta) = -\frac{\hbar^2}{2\mu\rho^2}\left[\frac{1}{4} + \frac{4}{\sin^2 2\theta}\right] \tag{41}$$

The adiabatic eigenstates are now defined by

$$\hat{H}_o(\theta,\phi;\rho)\ \psi_n(\theta,\phi;\rho) = \varepsilon_n(\rho)\ \psi_n(\theta,\phi;\rho) \tag{42}$$

and the total wave function $\psi(\theta,\phi,t;J)$ is expanded in these functions, i.e.

$$\psi(\theta,\phi,t;J) = \sum_n a_n(t;J)\ \exp\left[-\frac{i}{\hbar}\int_{-\infty}^{t} dt'\varepsilon_n(\rho(t'))\right]\psi_n(\theta,\phi,\rho) \tag{43}$$

From the timedependent Schrödinger equation we obtain a set of coupled equations for the amplitudes

$$\dot{a}_m(t;J) = -\sum_n a_n(t;J)\ \exp(i\ \Delta_{mn}(t))\ \left[\dot{\rho}\langle\psi_m|\frac{\partial}{\partial\rho}|\psi_n\rangle + \right.$$

$$\left. + \frac{i}{\hbar}\langle\psi_m|\hat{H}_1^{\ J}|\psi_n\rangle\right] \tag{44}$$

where

$$\Delta_{mn}(t) = \frac{1}{\hbar}\int_{-\infty}^{t} dt'[\varepsilon_m(\rho(t')) - \varepsilon_n(\rho(t'))] \tag{45}$$

and

$$\hat{H}^J = \frac{P_\gamma\ [P_\gamma - 4\ \cos\theta\hat{P}_\phi]}{2\mu\rho^2\ \sin^2\theta} + \frac{J^2-P_\gamma^2}{\mu\rho^2\cos^2\theta}\ (1+\sin\theta\ \cos^2\gamma) \tag{46}$$

The classical equations of motion obtain by applying Hamilton's principle upon the effective Hamiltonian:

$$\hat{H}_{SC}(\rho,P_\rho,\gamma,P_\gamma,t;J) = \frac{1}{2\mu}\ P_\rho^2 +$$

$$+ \sum_n |a_n(t;J)|^2\ \varepsilon_n(\rho) + \langle\psi|\ \hat{H}_1(\rho,\gamma,P_\gamma,t;J)|\psi\rangle \tag{47}$$

Thus we get:

$$\dot{\gamma} = \frac{\partial H_{SC}}{\partial P_\gamma} \; ; \; \dot{P}_\gamma = - \frac{\partial H_{SC}}{\partial \gamma} \tag{48}$$

and

$$\dot{\rho} = \frac{\partial H_{SC}}{\partial P_\rho} \; ; \; \dot{P}_\rho = - \frac{\partial H_{SC}}{\partial \rho} - \frac{\mu}{P_\rho} \frac{\partial H_{SC}}{\partial t} \tag{49}$$

The eqs. (48,49) assure conservation of the average energy, i.e. $\dot{H}_{SC} = 0$. The above equations have recently been programmed for the $A + B_2$ system and used to calculate cross sections and rates for the reaction

$$D + H_2(vj) \rightarrow DH(v'j') + H \tag{50}$$

The numerical work associated with the CP-method for reactive collisions consists of:

a) Solution of eq. (42) at a number of ρ-distances $\rho\epsilon[\rho_{min};\rho_{max}]$ where ρ_{min} is determined by $\epsilon_n(\rho_{min}) > E_{max}-\epsilon_n(\rho_{max})$ where E_{max} is the maximum value of the total energy to be considered. The value of ρ_{max} should be so large that the coupling for reactive channels vanishes. For $\rho \rightarrow \infty$ the physically interesting (θ,ϕ) region shrinks to zero, one therefore switches to an alternative method at $\rho = \rho_{max}$ and propagates from ρ_{max} to infinity in a single step (see refs. [11,20]).

b) Calculation of the coupling matrix elements $\langle\psi_m|\frac{\partial}{\partial\rho}|\psi_n\rangle$ and the four Coriolis like coupling elements: $\langle\psi_m|\hat{O}_i|\psi_n\rangle$ (i=1,2,3,4)

   where

$$\hat{O}_1 = \cos^{-2}\theta \tag{51a}$$

$$\hat{O}_2 = - \frac{2\cos\theta\,\hat{P}_\phi}{\sin^2\theta} \tag{51b}$$

$$\hat{O}_3 = \sin\theta\,\cos^{-2}\theta \tag{51c}$$

   and

$$\hat{O}_4 = \sin^{-2}\theta \tag{51d}$$

The matrix elements are splinefitted in the range $[\rho_{min};\rho_{max}]$ and used when integrating the coupled equations (44, 48, 49).

c) The amplitudes for inelastic and reactive events are obtained by integrating eqs. (44, 48, 49) with the initial conditions:

$$a_n(-\infty;J) = \delta_{nI} \tag{52a}$$

$$\rho = \rho_{max} \tag{52b}$$

$$\gamma \in [0;2\pi] \tag{52c}$$

$$P_\rho = -(2\mu(E_{kin} - V_{eff}(\rho_{max})))^{1/2} \tag{52d}$$

$$P_\gamma = J\cos\beta \tag{52e}$$

where $\cos\beta = \pm\left[1 - \frac{j^2}{J^2}\right]^{1/2}$.

Finally the cross sections obtain as:

$$\sigma_{In}(E) = \frac{\pi}{k_I^2}\sum_J (2J+1)\langle|a_n(\infty;J)|^2\rangle \tag{53}$$

where the brackets indicate an average over the random initial angles in (52), i.e. γ and $\Delta = \cos^{-1}(m_j/j)$.

For obtaining state to state cross sections for the $D+H_2$ system including 50 (vj) states in the expansion (43), we have estimated that the CP method is about a factor 5-10 times more time consuming than a purely classical mechanical trajectory method but at least an order of magnitude faster than a full quantum mechanical treatment of the problem. Furthermore the fact that the quantum states are only labelled by the vibrational and rotational quantum numbers makes it possible to use the approach for heavier A + BC systems. The classical treatment of the ρ-motion may underestimate the tunneling in ρ and thereby the reaction probabilities at low energies. In this energy region it may therefore be necessary to introduce quantum corrections in the ρ-coordinate. This can be done by using a wave packet or a WKB approach instead of a purely classical, mechanical treatment of this degree of freedom. As mentioned previously the separability assumption inherent in the CP-method is expected to break down in the strong coupling limit. At present we do not know precisely where this limit is for reactive systems, but we expect that a complete classical mechanical description of the system becomes adequate in this energy regime. Our results on the $D+H_2$ system show that the classical mechanical cross sections are too small by a factor of 2-3 at E_{kin} ~ 0.25 eV but only by 30% at 0.45 eV. But before complementarity between the classical path and the classical mechanical results can be established, the CP-calculations have to be extended to energies E_{kin} ~ 1 eV or more. Such calculations are presently being carried out on the Siegbahn-Liu potential surface [22].

TABLE I

An overview of the classical path approach to energy transfer and reactive processes.

System	Quantum	Classical	Dimension	Phenomena
A+BC	n	Translation	1D	Inelastic [4]
$He+NH_3$	jm	Translation	3D	Inelastic [6]
A+BC	nj	Translation,m	3D	Inelastic [7]
AB+CD	$n_1j_1n_2j_2$	Translation,m_1m_2	3D	Inelastic [8]
A+BC	n	Translation + Rotation	3D	Inelastic [9]
AB+CD	n_1n_2	Translation + Rotation	3D	Inelastic [10]
A+BC	nj	Eulerangles + ρ-motion	3D	Reactive [11]
A+surface	$\{n_i\}$	Translation of A	3D	Phonon excitation [12]
A+surface	$\{n_i\},k_xk_y$	z-motion of A	3D	+diffraction [13]

n vibrational quantum number
j rotational quantum number
m rotational projection
$k_{x,y}$ diffraction channels.

REFERENCES

[1] Billing, G.D. 1984, Comp.Phys.Rep. 1, pp. 237.

[2] Muckerman, J.T., Kanfer, S., Gilbert, R.D. and Billing, G.D. 1988, J.Chem.Phys. (in press).

[3] Heller, E.J. 1975, J.Chem.Phys. 62, pp. 1544.

[4] Billing, G.D. 1975, Chem.Phys.Lett. 30, pp. 391; 1976, J.Chem.Phys. 64, pp. 908; Muckerman, J.T., Rusinek, I., Roberts, R.F. and Alexander, M., 1976, J.Chem.Phys. 65, pp. 2416.

[5] Makri, N. and Miller, W.H. 1987, J.Chem.Phys. 87, pp. 5781.

[6] Billing, G.D. and Poulsen, L.L. 1984, J.Chem.Phys. 81, pp. 3866.

[7] Billing, G.D. 1976, J.Chem.Phys. 65, pp. 1.

[8] Billing, G.D. 1986, J.Chem.Phys. 84, pp. 2593.

[9] Billing, G.D. 1975, Chem.Phys. 9, pp. 359.

[10] Poulsen, L.L., Billing, G.D. and Steinfeld, J.I. 1978, J.Chem.Phys. 68, pp. 5121.

[11] Muckerman, J.T., Gilbert, R.D. and Billing, G.D. 1988, J.Chem.Phys., 88, pp. 4779; Billing, G.D. and Muckerman, J.T. 1988, J.Chem.Phys. (to be published).

[12] Billing, G.D. 1982, Chem.Phys. 70, pp. 223.

[13] Billing, G.D. 1988, Surf.Sci. (in press).

[14] Krogh, F.T. 1973, SIAM J. Numer.Anal. 10, pp. 949.

[15] Billing, G.D. and Poulsen, L.L. 1979, Chem.Phys.Lett. 66, pp. 177.

[16] Billing, G.D. and Baer, M. 1977, Chem.Phys.Lett. 48, pp. 342.

[17] Haydock, R. 1980, Solid State Physics 35, pp. 215; Nauts, A. and Wyatt, R.E. 1983, Phys.Rev.Lett. 51, pp. 228; Moiseyev, N., Friesner, R. and Wyatt, R.E. 1986, J.Chem.Phys. 85, pp. 331.

[18] Billing, G.D. 1987, Chem.Phys. 112, pp. 95.

[19] Johnson, B.R. 1980, J.Chem.Phys. 73, pp. 5051.

[20] Johnson, B.R. 1983, J.Chem.Phys. 79, pp. 1906.

[21] Gilbert, R.D. and Muckerman, J.T. 1985, J.Chem.Phys. 82, pp. 3132.

[22] Liu, B. 1973, J.Chem.Phys. 58, pp. 1925; Siegbahn, P. and Liu, B. 1978, J.Chem.Phys. 68, pp. 2457.

INTRAMOLECULAR ENERGY TRANSFER IN HC AND HO OVERTONE EXCITED MOLECULES

J. Santamaría, A. García Ayllón, C. Getino and P.A. Enríquez
Departamento de Química Física. Facultad de Químicas
Universidad Complutense, 28040 Madrid
Spain

ABSTRACT. The mechanisms for energy flow from overtone excited HC and HO local modes have been elucidated in two mode model Hamiltonians of benzene and trihalomethanes and in a six mode model of HOOH molecule. Intramolecular vibrational relaxation (IVR) from the excited 2:1 Fermi resonance is shown to be very sensitive to the stretch-bend potential energy coupling in connection with the stability of the HC stretch periodic orbit. The overtone induced dissociation of HOOH, which is a slow process in comparison with the initial HO overtone relaxation, is explained in terms of the details of the potential energy surface.

I. INTRODUCTION

Detailed understanding of intramolecular vibrational relaxation (IVR) is of crucial importance in interpreting the behavior of highly vibrationally excited states of molecules, for example the overtone spectra of HC and HO stretching modes(1,2,3). State to state measurements of chemical processes are now almost routine(4). As experimental probes bring out increasing details in chemical reactions(5)and spectroscopic transitions(6) it becomes necessary to understand the underlying dynamics on a finer scale. It is widely accepted today, first, that the energy redistribution in a molecule occurs from an excited mode to the small set of strongly coupled modes and not simultaneously to most, if not all, vibrations; and second, that the specific pathways for energy flow are a consequence of low order frequency commensurabilities between zeroth-order modes together with the presence of appropiate resonant kinetic and potential energy coupling terms(7).

The methods of Nonlinear Dynamics(8) can be applied to gain new theoretical insight into the underlying dynamics in terms of the molecular phase space structure. In particular, the existence of low order Fermi resonances between vibrationally anharmonic local (or normal) modes or between bending vibrations and rotations, can cause dramatic changes in phase space structure, manifest in the breakdown

A. Laganà (ed.), Supercomputer Algorithms for Reactivity, Dynamics and Kinetics of Small Molecules, 357–366.

of traditional normal mode/rigid rotor model and appearance of large resonance zones indicating extensive energy exchange between zero-order modes. Moreover, recent research has shown that both the barriers in phase space, which prevent the classical flow from exploring energetically allowed regions in near integrable systems, and the partially penetrable barriers as the perturbation increases (cantori or broken tori) are significant for the quantized systems in the sense that they determine the accesible regions of quantum phase space (Wigner representation)(9).

The advent of new experimental techniques such as overtone spectroscopy has provided information on polyatomic molecules which challenge theoretical methods to directly deal with large amplitude vibrational motions. Recently there has been a large amount of experimental research on the overtone spectroscopy of HC(1,2) and HO(3) local modes. Assuming that the observed linewidths are due to homogeneous broadening the time scale of the initial vibrational relaxation can be inferred; in both cases this relaxation is fast, taking place within a fraction of one picosecond. Subsequent relaxation, in the case of HO overtone of HOOH, leading to the rupture of O-O bond (overtone induced dissociation) takes place on a time scale one to two orders of magnitude more slowly. Also the measured product energy distributions show non thermal rotational state distributions for HO. Much recent theoretical work(10,11,12)) has been focused on the extraction of information on especific pathways that govern the intramolecular vibrational relaxation from overtone spectroscopy. Of particular importance is the coupling between HC (HO) stretching vibrations and adjacent bending modes; the initial energy transfer from the excited stretch to the adjacent modes has been shown to proceed via a strong 2:1 Fermi resonance for the HC stretch in benzene and weak resonance in the case of HOOH for the HO.

Quantum and corresponding quasiclassical calculations of Sibert et al.(10) for HC overtone relaxation in benzene yielded subpicosecond lifetimes consistent with the linewidths observed in the room temperature experiments of Reddy et al.(1). The model Hamiltonian used by these authors incorporates only kinetic coupling between curvilinear HC stretch and HCC bend modes. It is clear, however, that the HCC bend force constant must be attenuated by stretching of the HC bond, and Lu et al.(10) have shown that the nature and rate of short time IVR are sensitive to the magnitude of this attenuation. In this work we summarize a recent classical trajectory study of a two mode model of stretch-bend interaction in a HC_3 fragment of benzene(13). Our aim is to understand the sensitivity of short time IVR to potential coupling in terms of the classical phase space structure of the stretch-bend Hamiltonian.

In trihalomethanes several resolved absortion bands appear in each overtone region (2). These bands result from a very strong coupling between the HC stretch and the bend and are characteristic of Fermi resonances. The measured splittings allow the bend-stretch anharmonic couplings to be determined quantitatively, and demonstrate that HC stretching energy flows first to the bending modes and subsequently on a longer time scale to the heavy atom vibrations. It has been shown that these Fermi resonances in CHF_3, CHD_3 and $CH(CF_3)_3$ can be analysed via effective tridiagonal Hamiltonian matrices. The off diagonal Fermi resonance coupling matrix elements depend on an effective coupling constant(11).

We have also studied the overtone induced dissociation of HOOH by classical trajectory calculations using different Hamiltonians (12). Our results illustrate the importance of bending modes and provide evidence for incomplete relaxation of energy. The experimental data suggest a strong coupling of the torsional vibration to the HO stretch mode due to the dependence of the torsional barrier heights on the level of HO excitation, but the trajectory studies are not yet conclusive. The importance of rotation has been stressed by Sumpter and Thompson(12).

In our study we investigate the sensitivity of the unimolecular decay dynamics of HOOH to potential energy surface features, initial condition selection and overall rotation.

In section II, we describe the results of a classical trajectory calculation on two mode model for benzene relaxation in conection with the sensitivity of IVR to potential coupling in terms of the classical phase space structure. The same kind of approach is used in section III for a Hamiltonian model of fluoroform. Preliminary results of classical trajectory calculations of overtone induced dissociation in HOOH using two types of potential surface are presented in section IV. We end with a brief discussion of some possible extensions of our work and outstanding problems.

II. SENSITIVITY OF IVR TO POTENTIAL ENERGY COUPLING IN HC_3 FRAGMENT OF BENZENE

To establish a connection between the stretch-bend potential coupling and the decay rate constant from overtone excited HC bonds in benzene, we have studied a two-mode model system describing a HC_3 fragment of benzene. The Hamiltonian, which is expressed in internal curvilinear coordinates, includes both kinetic and potential stretch-bend coupling(13):

$$H = 1/2G_{rr}p_r^2 + 1/2G_{\theta\theta}p_\theta^2 + D[1-\exp(-\beta r)]^2 + 1/2S(r)f_{\theta\theta}(\Delta\theta)^2 \qquad (2.1)$$

Here, the HC bond is taken to be a Morse oscillator (r is the stretch coordinate) while the HCC bending potential (θ is the HCC wag coordinate) is a harmonic oscillator with r-dependent force constant. The CC bond stretch coordinate R and the CCC bond angle are frozen at their equilibrium values. The stretch-bend G-matrix elements are:

$$G_{rr} = 1/m_H + 1/m_C \qquad (2.2a)$$

$$G_{\theta\theta} = 1/r^2m_H + (1/r^2 + 1/rR + 3/4R^2)/m_C \qquad (2.2b)$$

Stretch-bend kinetic coupling occurs through the r-dependence of $G_{\theta\theta}$, while the potential coupling is determined by the r-dependence of the switching function S(r) in the form(10)

$$S(r) = 1 \qquad r<r_o \qquad (2.3a)$$
$$S(r) = \exp[-a(\Delta r)^2] \qquad r>r_o \qquad (2.3b)$$

The strength of the potential coupling is determined by the magnitude of the parameter a. We have studied the effect of three values of this attenuation parameter, a=0.0 $Å^{-2}$(no coupling), a= 0.825 $Å^{-2}$(moderate coupling) and a= 2.0 $Å^{-2}$(strong coupling).

The phase space structure of this model has been studied at different energies ($J_\theta = n_\theta + 1/2, n_\theta = 0$; $J_r = n_r + 1/2$, n_r=1-12) for the three cases of potential coupling via the surfaces of section (sos). Each sos has been obtained by plotting (θ, p_θ) every time r=0,p_r>0 for several trajectories at a particular constant energy. Representative sos are shown in reference 13. The sos can be catalogued according to the stability of the central fixed point of the return map, which is by symmetry, the periodic orbit in which all available energy resides in the HC stretch mode. At low excitation energy the central fixed point is stable (elliptic) and the two islands (top left and bottom right) of the sos are a consequence of the particular cut used to define the sos and are not associated with a stretch-bend resonance. Consequently we have a single family of regular non resonant trajectories and the interchange of energy is small. When one increases the excitation energy, the central fixed point becomes unstable (hyperbolic with reflection) and, as a result of the bifurcation of the HC stretch periodic orbit, we have two distintic families of trajectories divided by a separatrix: resonant (with large interchange of energy between stretch and bend) and non-resonant. Finally when we turn on the stretch-bend potential energy coupling for the same excitation level, the central point becomes stabilized, i.e., the HC stretch periodic orbit no longer lies inside the 2:1 resonance zone. There are now three families of trajectories, two of which correspond to the previous case. The third family is associated with stable quasiperiodic motion of predominantly HC stretch character and no overtone relaxation occurs for those trajectories.

The previous analysis is confirmed when one measures the quasiclassical probability Pn(t) for remaining in the initial state (n_r,0) for an ensemble of two hundred trajectories defined with initial conditions: $J_r = n_r + 1/2$, ϕ_r = random; $J_\theta = 1/2$, ϕ_θ = random. The decay is close to exponential when the central fixed point is unstable and on the contrary it is distinctively non exponential with prominent oscillations ("beats") when the fixed point is stable. Thus, the sensitivity of short time relaxation to potential energy coupling derives from the drastic effect of a change from instability to stability on the stretch-bend energy flow in quasiperiodic trajectories. In this way, we have shown that the short time overtone decay dynamics of the two mode model exhibit the same sensitivity to potential energy coupling as does the trajectory calculations for the full planar benzene Hamiltonian of Hase and coworkers(10).

Moreover, short time decay rate constants can be estimated very accurately by linearization of the dynamics(8) around the HC periodic orbit(14). The eigenvalues of the Jacobian matrix propagator of the return map are $\lambda_1 = \exp(i\sigma)$, $\lambda_2 = \exp(-i\sigma)$ for elliptic (stable) fixed point, and $\lambda_1 = \exp(\sigma)$, $\lambda_2 = \exp(-\sigma)$ for hyperbolic (unstable) fixed point (σ real), where the exponents are the (real or complex) phases acquired by the eigenvectors after τ_s (the period of the orbit). In the unstable case it is possible to estimate the decay constant k for the stretch-bend relaxation since in this case $\sigma = k\,\tau_s$. The method requires the numerical calculation of the Jacobian matrix by determining the action of the return map on initial conditions displaced by small amounts (0.001) along the θ and p_θ directions

TABLE I. RATE CONSTANTS (PS^{-1}) FOR SHORT TIME OVERTONE DECAY IN MODEL HAMILTONIAN OF EQ. (2.1) FOR BENZENE.

	Type 1		Type 2		Type 3	
n_r	k^a	k^b	k^a	k^b	k^a	k^b
1	N[c]	N	N	N	N	N
2	9	18	N	4	N	N
3	29	28	23	19	N	N
4	34	33	35	25	N	N
5	31	36	38	27	N	N
6	39	37	35	28	N	2
7	37	36	33	28	N	N
8	28	34	28	26	N	N
9	25	30	21	24	N	N
10	28	22	18	21	N	N
11	31	5	15	17	N	N
12	N	N	13	11	N	N

[a]Decay rate constant calculated by least squares fit of ln $P_n(t)$.

[b]Decay rate constant calculated by linear stability analysis of CH periodic orbit.

[c]N Denotes nonexponential decay.

respectively. The relaxation rate k is estimated by diagonalizing the resulting Jacobian matrix. The values of k obtained by linear stability analysis are given in Table I and are in good accord with the quasiclassical trajectory results.

These results, together with the dynamical behavior described above, show that there is a cancellation of the kinetic and potential stretch-bend coupling terms in the Hamiltonian.

III. IVR VS. KINETIC AND POTENTIAL COUPLINGS IN FLUOROFORM.

The magnitude and relative importance of kinetic and potential couplings are better well known in trihalomethanes, where the structure of the overtone spectra is well resolved.

Our model Hamiltonian is a reduced two mode model in terms of symmetrized curvilinear internal coordinates and their conjugate momenta. The two modes are the HC stretch and a doubly degenerate bend(15). The Hamiltonian includes both kinetic and potential stretch-bend coupling and is formally identical to eqn (2.1). Nevertheless the θ angle is a linear combination (doubly degenerate) of the three HCF wag coordinates(ϕ angles). Consequently the G matrix elements $G_{\theta\theta}$ is defined as

$$G_{\theta\theta} = G_{\phi\phi} - G_{\phi\phi'} \qquad (3.1)$$

where

$$G_{\phi\phi} = 1/r^2 m_H + 1/R^2 m_F + (1/r^2 + 1/R^2 - 2\cos\phi/rR).1/m_C \qquad (3.2a)$$

$$G_{\phi\phi'} = \cos\psi/r^2 m_H + [\cos\psi/r^2 - 2\cos\phi\cos\psi/r.R + (\sin^2\phi.\sin^2\psi + \cos\alpha\cos\psi)/R^2].1/m_C \qquad (3.2b)$$

The angle α is the equilibrium FCF angle and the ψ angle is the dihedral angle defined by α and ϕ (16). The stretch-bend kinetic coupling occurs, as usual, through the r dependence of $G_{\theta\theta}$. The potential coupling appears in the r-dependence of the switching function S(r). Its expression given by the expansion of f_θ in terms of $y = 1-\exp(-\beta r)$ is

$$S(r) = 1 + \frac{1}{\beta}\frac{Fr\theta\theta}{f_{\theta\theta}}\, y + \frac{1}{2}\left(\frac{1}{\beta^2}\frac{Frr\theta\theta}{f_{\theta\theta}} + \frac{1}{\beta}\frac{Fr\theta\theta}{f_{\theta\theta}}\right)y^2 \qquad (3.3)$$

where $Fr\theta\theta$ and $Frr\theta\theta$, defined in ref. 15, are obtained from the fit to the experimental spectrum. We have found that the expression (3.3) can be very well approximated with the formula

$$S(r) = 1-\tanh(\xi\beta r) \qquad (3.4)$$

which reproduces the experimental energy levels of the polyads up to $N= V_S+1/2V_b = 6$ (V_S and V_b are the quantum numbers of the stretch and bend modes) for $\xi = 0.316136$ and at the same time gives the correct assimptotic behaviour. The parameters for the Hamiltonian are given in Tables I and III of ref. 15.

Following the method described for benzene, we are currently investigating the effect of stretch-bend couplings for this hamiltonian in terms of phase space structure and stability vs. instability of HC periodic orbits. Representative surfaces of section are shown in Fig. 1 for the Hamiltonian previously described including both kinetic and potential couplings. The next step is to expand the $G_{\theta\theta}$ matrix element in terms of y, and observe the effect of the approximate Hamiltonian on the stability of the HC periodic orbits in each polyad. At the same time one can study the effect of the inclussion and suppression of the potential coupling. In this respect it is interesting to note that Green et al.(17) have studied

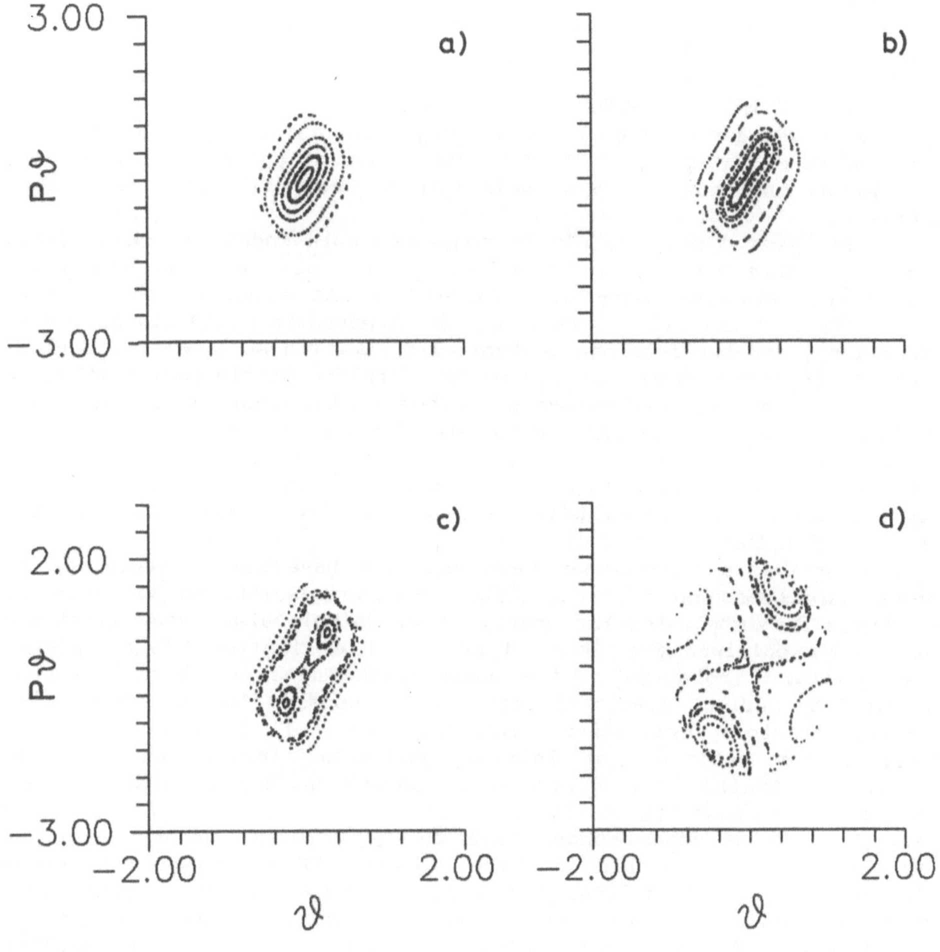

Figure 1.
Surfaces of section (θ, p_θ) for the exact $G_{\theta\theta}$ matrix element and potential coupling switching function proposed by us, at energies corresponding to states:
$(n_r=2,\ n_\theta=0)$ for case a),
$(n_r=3,\ n_\theta=0)$ for case b),
$(n_r=4,\ n_\theta=0)$ for case c),
$(n_r=8,\ n_\theta=0)$ for case d).

stretch-bend couplings in trihalomethanes, and have shown that the kinetic coupling in their model Hamiltonian is larger than the coupling experimentally observed, suggesting that potential and kinetic coupling terms contribute to the Hamiltonian with opposite sign.

IV. OVERTONE INDUCED DISSOCIATION OF HOOH

We have carried out classical trajectory calculations of overtone induced dissociation in HOOH using two types of potential surface(18) to explore the sensitivity of unimolecular decay lifetimes to surface details.

The first type is a simple near-separable model, in which local bond stretches are described by Morse potentials, bending potentials are taken to be harmonic with force constant attenuated in several ways (BEBO fashion(19), gaussian or hyperbolic tangent switching function), and the torsional potential is written as a Fourier series. The second type of potential is an empirical/ab initio surface(20), in which local bond stretches are described by switched Morse functions, bends are taken as harmonic attenuated by a gaussian function of the HO stretch, and a switching function of the central O-O bond length; the torsion is a truncated Fourier series with coefficients attenuated by the OO stretch, and suitable scaled ab initio results are used for the off-diagonal F-matrix elements.

Approximate curvilinear mode energies have been calculated for several picoseconds following initial HO bond excitation to determine pathways for intramolecular energy flow. Unimolecular decay lifetimes have been obtained from trajectory ensemble lifetime distributions. The effect of rotation on the unimolecular decay dynamics has been studied by adding rotational energy corresponding to T= 300 K after inital overtone excitation. The influence of different sampling schemes for selection of initial conditions (quasiclassical mode sampling with and without zero point energy vs. Wigner distribution) has also been investigated(18).

In all the cases, short time energy transfer occurs from the initially excited HO bond into the adjacent HOO bend, and the second HO bond remains vibrationally unexcited throughout the dissociation. We have found dissociation lifetimes ranging from 1.2 ps to approximately 50 ps for v=6, depending on the potential and sampling method. For example, lifetimes for the first type of surface are very sensitive to the precise way in which the HOO bending force constant is attenuated with OO bond distance, increasing from ca. 1 ps to 10 ps on going from a BEBO function $\exp(-\alpha\Delta r_{OO})$ to a hyperbolic tangent switching function. Rotation has very little effect on the decay lifetimes for the first type of surface. Lifetimes are considerably longer for the empirical/ab initio surface, and here, rotation has more effect, uniformly tending to increase decay rates. Wigner sampling yields slighly longer lifetimes than quasiclassical initial conditions with zero point energy.

Our current best estimate for the unimolecular decay lifetime of a hydrogen peroxide molecule excited with v = 6 quanta in the HO bond is 18.9 ps.

V. DISCUSSION AND CONCLUSIONS

In our study of HC overtone decay in benzene, we have found that the short time probability decay of the two mode stretch-bend Hamiltonian of a HC_3 benzene fragment shows the same trends with changes in the potential coupling parameters as the full benzene Hamiltonian: This result is consistent with the local nature of the stretch-bend interactions responsible for short time overtone decay in benzene. We have identified a mechanism whereby changes in a coupling parameter can have significant effect on the dynamics by inducing only local changes in the phase space structure, i.e. a change from instability to stability of the HC periodic orbit. Our results are, on the whole, in cualitative accord with the room temperature experiment of Reddy et al(1) and with the quantum calculation of Sibert et al. Nevertheless the most recent experiment by Lee and coworkers(21) suggests that the true homogeneus linewidths may, in fact, be considerable narrower than those obtained via quasiclassical simulations (ca. 3 cm^{-1} for $n_T = 3$), indicating a much smaller decay rate. It is not clear at present how much of the gross discrepancy between quasiclassical theory and experiment is due to inadequacy of the model potential surface, and how much to a fundamental failure of classical mechanics to yield a correct degree of adiabaticity in the dynamics.

With the method outlined for fluoroform, we can make useful predictions of the order of magnitude of the anharmonic kinetic couplings and compare them with the potential one, in order to evaluate the rates of energy transfer between modes(17).

In the case of HOOH, recent experimental results(22) indicate 6 ν_{HO} dissociation lifetimes longer than previously reported(3) and of the order of the values that we have found with the empirical/ab initio surface. The sensivity of unimolecular decay lifetimes to details of the potential surface, in particular stretch-bend coupling, is clearly demonstrated. Further work on effects of rotations and interaction with torsional modes is in progress.

Acknowledgements

This work was supported in part by the CICYT project nºPB86-0540. We gratefully acknowledge helpful comments by Prof. G.S. Ezra.

REFERENCES

1. K.V. Reddy, D.F. Heller and M.J. Berry, J. Chem. Phys., 76, 2814 (1982).
2. H.R. Dübal and M. Quack, J. Chem. Phys., 81, 3779 (1984). J.S. Wong, W.H. Green, C. Cheng and C.B. Moore, J. Chem. Phys., 86, 5994(1987).
3. F. Crim, Ann. Rev. Phys. Chem., 35, 657 (1984). T.R. Rizzo, C.C. Hayden and F.F. Crim, Farad. Discuss. Chem. Soc., 75, 223 (1983); J. Chem. Phys., 81, 4501 (1984). L.J. Butler, T.M. Ticich, M.D. Likar and F.F. Crim, J. Chem. Phys., 85, 2331 (1986). M.D. Likar, J.E. Baggot, T.M. Ticich, R.L. Vander Wal and F.F. Crim, J. Chem. Soc. Farad. Trans., 2, (1988), (in press).
4. P.J. Vaccaro, J.L. Kinsey, R.W. Field and H.L. Dai, J. Chem.

Phys., 77, 573 (1982). K.K. Lehman, G.J. Scherer and W. Klemperer, J. Chem. Phys., 77, 2853 (1982). E.E. Marinero, C.T. Retlner and R.N. Zare, J. Chem. Phys., 80, 4142 (1984).

5. R.D. Levine and R.B. Bernstein, "Molecular Reaction Dynamics and Chemical Reactivity", (Oxford University Press, 1987). S. Leone, Ann. Rev. Phys. Chem., 35, 109 (1984).
6. R. Lefebvre and S. Mukamel (Editors), "Stochasticity and Intramolecular Redistribution of Energy", (Kluwer Academic Publishers, 1987).
7. E.L. Sibert, J.T. Hynes and W.P. Reinhardt, J. Phys. Chem., 87, 2032 (1983).
8. A.J. Lichtenberg and M.A. Lieberman, "Regular and Stochastic Motion" (Springer-Verlag, New York, 1983).
9. G. Radons, T. Geisel and J. Rubner, Adv. Chem. Phys., (in press).
10. E.L. Sibert, W.P. Reinhardt and J.T. Hynes, J. Chem. Phys., 81, 1115 (1984), ibid., 81, 1135 (1984). R.J. Wolf, D.S. Bathla and W.L. Hase, Chem. Phys. Lett., 132, 493 (1986). D-H. Lu, W.L. Hase and R.J. Wolf, J. Chem. Phys., 85, 4422 (1986).
11. H.R. Dübal and M. Quack, Mol. Phys., 53, 257 (1984).
    A. Amrein, H.R. Dübal and M. Quack, Mol. Phys., 56, 727 (1985).
    M. Lewerenz and M. Quack, Chem. Phys. Lett., 123, 197 (1986).
    S. Peyerimhoff, M. Lewerenz and M. Quack, Chem. Phys., Lett., 109, 563 (1984).
    J.E. Baggot, M.C. Chuang, R.N. Zare, H.R. Dübal and M. Quack, J. Chem. Phys., 82, 1186 (1985).
    G.A. Voth, R.A. Marcus, A.H. Zewail, J. Chem. Phys., 81, 5494 (1984).
12. T. Uzer, J.T. Hynes and W.P. Reinhardt, J. Chem. Phys., 85, 5791 (1986). B.G. Sumpter, D.L. Thompson, J. Chem. Phys., 82, 4557 (1985), ibid, 86, 2805 (1987)
13. A. García-Ayllón, J. Santamaría and G.S. Ezra, J. Chem. Phys., (in press).
14. E.J. Heller, E.B. Stechel and M.J. Davis, J. Chem. Phys., 73, 4720 (1980).
15. L. Halonen, T. Carrington and M. Quack, J. Chem. Soc. Faraday Trans 2, 84, 000 (1988).
16. E.B. Wilson, J.C. Decius and P.C. Cross, "Molecular Vibrations" (Dover, New York, 1980).
17. W.H. Green, W.D. Lawrance and C.B. Moore, J. Chem. Phys., 86, 6000 (1987).
18. C. Getino, B.G. Sumpter, J. Santamaría and G. Ezra, 1988, (unpublished work).
19. H.S. Johnston, "Gas Phase Reaction Rate Theory", (Ronald Press, New York, 1966).
20. B.G. Sumpter, 1986, (unpublished work).
21. R.H. Page, Y.R. Shen and Y.T. Lee, Phys. Rev. Lett., 59, 1293 (1987), ibid., J. Chem. Phys., (in press).
22. T. Rizzo, (private communication). A. Zewail, (private communication).

CLASSICAL TRAJECTORY STUDIES OF GAS PHASE REACTION DYNAMICS AND KINETICS USING *AB INITIO* POTENTIAL ENERGY SURFACES

Richard L. Jaffe
NASA Ames Research Center
Moffett Field, CA 94035

Merle D. Pattengill
Department of Chemistry
University of Kentucky
Lexington, KY 40506
and
David W. Schwenke
Eloret Institute
Sunnyvale, CA 94087

ABSTRACT. Strategies for constructing global potential energy surfaces from a limited number of accurate *ab initio* electronic energy calculations are discussed. Generally these data are concentrated in small regions of configuration space (e. g., in the vicinity of saddle points and energy minima) and difficulties arise in generating a potential function that is globally well-behaved. We also describe efficient computer codes for carrying out classical trajectory calculations on vector and parallel processors. Illustrations are given from recent work on the following chemical systems: $Ca + HF \rightarrow CaF + H$, $H + H + H_2 \rightarrow H_2 + H_2$, $N + O_2 \rightarrow NO + O$ and $O + N_2 \rightarrow NO + N$. The dynamics and kinetics of metathesis, dissociation, recombination, energy transfer and complex formation processes will be discussed.

1. INTRODUCTION

One of the tasks undertaken by the Computational Chemistry Branch at NASA Ames Research Center is to provide critically needed chemical and physical data for NASA hypersonics projects such as the aeroassisted orbital transfer vehicle[1] and the scramjet propulsion system in the National Aerospace Plane[2,3]. In order to meet this goal, we have embarked on theoretical studies of the reaction kinetics and dynamics of high temperature air ($T \approx 5000-50000$ K) and moderately high temperature hydrogen-air mixtures ($T < 3000$ K).

In order to reliably predict the rates for these processes, it is necessary to perform accurate dynamical calculations using accurate potential energy surfaces (PES). However, one is faced with difficulties realizing these ideals. Quantum mechanical scattering methods are not yet advanced to the point where 3-dimensional calculations can be performed on three and four atom systems at elevated temperatures and accurate PES information is rarely available. Fortunately it is often possible to ameliorate these difficulties and in this paper we discuss recent advances in this area.

A. Laganà (ed.), Supercomputer Algorithms for Reactivity, Dynamics and Kinetics of Small Molecules, 367–382.

The reaction dynamics will be treated using classical mechanics. This should not be a cause of significant error, especially for heavy particle collisions well above threshold. Furthermore, in spite of the fact that systems containing hydrogen atoms may have important quantum effects, such as tunneling[4], we expect that the calculated rate constants should still be accurate at high temperatures. These calculations are implemented on a vector pipeline supercomputer, and strategies for doing this in an efficient manner will be discussed.

Accurate PES for scattering can come from two sources, experiment and *ab initio* theoretical calculations. Unfortunately, experimentally derived PES are restricted to relatively simple systems and then usually only reflect small regions of the PES. Thus we usually rely on theoretical calculations. The use of an *ab initio* PES requires first the calculation of the electronic energies for a set of atomic geometries followed by the generation of a suitable interpolating function. This second step is one of the more difficult aspects of the process of predicting rate constants and is required because of the expense of calculating accurate PES data. Usually *ab initio* quantum chemical calculations of "chemical accuracy" (i.e., relative energies to within 1 kcal/mol) are so computationally intensive that the necessary computer resources are not available to generate energies for the hundreds or thousands of geometries needed to map out the global PES required for the dynamics calculations[5,6]. Thus one is usually supplied only with calculations of the energy and its second derivatives at stationary points on the PES (e.g. saddle points, reaction intermediates) or limited grids of energies clustered around the reaction coordinate.

The choice of the interpolating function is not an easy task and depends, to a certain extent, on the amount of *ab initio* data available as well as the system studied. For reactive systems, several methods have been used to represent the global PES, including: the valence-bond derived LEPS[7] and diatomics-in-molecules (DIM)[8] methods, the rotated Morse oscillator-spline (ROMS) expression[9] and the Sorbie-Murrell (SM)[10] approach. These all rely on diatomic fragment potentials for ground and excited electronic states which are used to describe the correct asymptotic limits (separated reactants and products). The diatomic potentials can be obtained from spectroscopic data or quantum chemical calculations. In the LEPS and DIM methods, the excited state potential energy curves are treated as adjustable fitting functions to control the topography of the PES. The parameters are further adjusted to reproduce the saddle point geometry and height, if it is known, and/or some experimental measure of the reaction dynamics such as a product scattering angle or product internal state distribution. All of these methods are generally satisfactory for simple atom-diatom exchange reactions occurring on purely repulsive potential energy surfaces, especially when the reaction coordinate favors collinear geometries, as is the case in most hydrogen-halogen reactions. However, if a collinear approach between reactants is not favored, or if intermediate complex formation is possible, the PES cannot usually be represented by the LEPS and ROMS methods. In principle, the DIM method should give satisfactory results for these complicated chemical systems, however, a large number of diatomic fragment electronic states may be required to serve as basis functions in order to build the proper features into the PES.

Another difficulty in using these methods arises from the use of arbitrary grids of *ab initio* points to determine the PES parameters. This is because the adjustable parameters in these formulations generally do not provide enough flexibility to allow complete adjustment of the PES to build in the precise locations and shapes of features determined by quantum chemical calculations. One solution of this problem has been the introduction of geometry dependent Sato parameters[11] to extend the flexibility of the LEPS formulation. However, these parameters have an indirect effect on the PES topography and it is difficult to design a Sato function to fit specific features. However, we have used a different modification of the LEPS formulation to construct PES for $N + O_2$ and $O + N_2$.

Of the methods discussed, perhaps the most promising for general systems is the SM method, which is based on a sum of a series of terms describing all possible 2-body to n-body interactions. The increased flexibility of treating each term separately is a powerful feature of the method. We have used variations of this method to construct PES for $Ca + HF$ [12] and $N + O_2$ [13].

For systems which are primarily nonreactive, such as $H_2 + H_2$, a wider variety of approaches are available. However, because of the high temperatures considered and because at least one of the molecules must dissociate properly, aspects similar to those present in reactive systems must be addressed. This has led to the development of hybrid methods[14].

The organization of the rest of the paper is as follows: in the next section we discuss the implementation of the dynamics calculations, and in Sections 3-5 we discuss the specific reaction systems $Ca + HF$, $H_2 + H + H$, $N + O_2$ and $O + N_2$. Finally in Section 6 we give our conclusions.

2. ALGORITHMS FOR CLASSICAL TRAJECTORY CALCULATIONS

The classical trajectory method[15] has been a standard approach for the calculation of the dynamics and kinetics of gas phase collisions of small molecules for more than 30 years. The calculations consist of integrating the classical equations of motion, a set of $6n - 6$ coupled first order differential equations for an n-atom system after elimination of the center of mass:

$$dQ_i/dt = \partial\mathcal{H}/\partial P_i, \tag{1}$$

$$dP_i/dt = -\partial\mathcal{H}/\partial Q_i, i = 1, ..., 3n - 3, \tag{2}$$

where Q_i and P_i are the generalized coordinates and their conjugate momenta, respectively, and $\mathcal{H}$ is the classical Hamiltonian (sum of kinetic and potential energies relative to the center of mass of the system). For conservative systems, the potential energy, $\mathcal{V}$, depends only on the generalized coordinates, Q_i, and the kinetic energy only depends on the conjugatemomenta. Thus the right hand side of Eq. (2) is $-\partial\mathcal{V}/\partial Q_i$. Our calculations use, as the Q_i, linear combinations of the atomic cartesian coordinates, thus the evaluation of $\partial\mathcal{H}/\partial P_i$ only requires multiplication of the P_i by mass factors. The solution of these equations for a set of initial coordinates and momenta is called the calculation of a single trajectory and corresponds to the motion of the colliding species over a time period of ≈ 1 ps. Through the application of appropriate random sampling techniques, the behavior of an ensemble of molecules with a collision frequency of $10^{20} s^{-1}$ can be simulated by on the order of 10^4 trajectories.

A variety of numerical integrators have been used for the calculation of trajectories, and both fixed and variable step size algorithms are popular. The optimal integrator depends both on the components of the colliding system as well as the conditions under which the collisions take place. An additional consideration is that not all algorithms are equally suited for implementation on specialized computers. Thus it may be more economical to use an algorithm which requires more operations than competing methods if it completes these operations more efficiently. In the applications considered here, we perform the calculations on the NASA Ames Research Center Cray XMP/48, which is a 4 processor vector pipeline supercomputer. However, we use only one of the processors for our calculations. Our general strategy is to take advantage of the fact that each trajectory in an ensemble is independent of all others so that many trajectories can be integrated simultaneously. This technique can provide significant speedups compared to integrating the trajectories sequentially provided that enough of the operations required are identical for each trajectory. This can be a difficulty for variable step size algorithms, thus for our calculations on $N + O_2$ and $Ca + HF$ we use the fixed step size sixth order Gear integrator[16]. In contrast, for our calculations on $H_2 + H + H$, we use a variable step size algorithm which is a modification[14] of the Bulirsch-Stoer method[17]. The motivation for using a variable step size algorithm in this case is that the velocities of the recombining atoms can vary over a very wide range and the savings obtained by using a variable step size more then compensates for the inefficiencies incurred adjusting the step size. Despite these differences, our implementation of the two different integrators is similar.

The basic flow of our calculations is as follows. The calculation is initiated by specifying the initial conditions of N_V trajectories. Then these trajectories are simultaneously integrated forward in time. Periodically the trajectories are checked for completion, and as trajectories finish, new initial conditions are generated so that there remain N_V trajectories to be simultaneously integrated. Each integration step requires the evaluation of the gradient of the potential, $\partial\mathcal{V}/\partial Q_i$. This process continues until at total of N_{Total} trajectories are completed. The code using the Gear integrator uses N_V fixed at 64 for all time so that some incomplete trajectories are discarded after N_{Total} trajectories are completed[18]. In contrast, the modified Bulirsch-Stoer integrator code does not discard any incomplete trajectories but rather when necessary decreases N_V from its maximum value of 500. Once all trajectories are completed, a final state analysis is carried out.

The steps which are not vectorizable in this process are the specification of the initial conditions, the end checks, the final state analysis, and for the modified Bulirsch-Stoer integrator, the step size adjustment. All other steps vectorize, with vector length N_V, and since most of the time is spent on the vectorizable steps, a very high rate of execution is obtained. For typical runs using the code with the Gear integrator, about 5% of the central processing unit (CPU) time is spent on the numerical integration routines and about 90% of the CPU time is spent evaluating the gradient of the potential. For $N + O_2$ (using a modified LEPS potential energy surface), this means an overall performance of 65 million floating point operations per second (MFLOPS) as measured by the hardware performance monitor on the Ames Cray XMP/48. For the code using the modified Bulirsch-Stoer integrator, typical runs for $H_2 + H_2$ spend about 95% of the CPU time evaluating $\partial\mathcal{V}/\partial Q_i$, and 3% of the CPU time in the integration routine and have an overall performance of about 96 MFLOPS as measured by the hardware performance monitor on the Ames Cray XMP/48.

3. $Ca + HF \rightarrow CaF + H$

The reactions of alkaline earth atoms (M) with hydrogen halides (HX) have been widely studied in molecular beam experiments owing in part to the ease of detection (by laser induced fluorescence) of the MX products. One interesting example of this class of reaction is the formation of CaF by collision of Ca with HF[19,20]. The reaction is endoergic by 14.1 kcal/mol and has only been observed for HF in excited vibrational levels. Experiments have been carried out in which HF is prepared in a specific ro-vibrational level ($v = 1, J = 0 - 7$) and the product CaF vibrational distribution is measured[19]. In addition, comparison has been made between the product distribution for $Ca + DF(v = 2, J = 1)$ and $Ca + HF(v = 1, J = 7)$ which are nearly isoenergetic[20]. These studies indicate that the energy disposal for CaF formation is statistical and consistent with the presence of a long-lived intermediate complex.

The *ab initio* potential energy surface computed for the $Ca - F - H$ system[12] is dominated by a 39.0 kcal/mol deep well corresponding to a linear but floppy $H - Ca - F$ molecule. The favored angle of approach of the Ca atom to the HF molecule is 75°. The energy barrier for the abstraction pathway is 16.1 kcal/mol (relative to separated reactants) and is located at $R_{CaF} = 2.12Å$ and $R_{HF} = 1.39Å$. This barrier is also the saddle point for the insertion process forming the triatomic complex. At this geometry, the $Ca - F$ and $H - F$ bonds are 0.15Å and 0.47Å longer than the diatomic r_e values, respectively. For other angles of approach, the abstraction barrier height increases markedly and has a value of 30.0 kcal/mol for the collinear reaction. These PES calculations were carried out for 175 $Ca-F-H$ geometries with the MOLECULE-SWEDEN system of computer codes[21] and utilized the complete active space SCF algorithm (CASSCF) followed by a multireference configuration interaction procedure (MRCI). Fairly large contracted gaussian basis sets were used[22] and the calculation was designed to recover all important valence correlation effects. While it is known that higher level treatments are needed to accurately reproduce the $Ca - F$ and $H - F$ bond energies, the PES calculations are balanced in that the computed endoergicity (14.1 kcal/mol) agrees quite well with the experimental value. A schematic diagram of the potential energy profile is shown in Figure 1.

The fit of the *ab initio* PES data was achieved by partitioning the problem into sections. Simple 2x2 valence bond treatments were used to describe the ionic-covalent interactions for the diatomic fragments $Ca - F$ and $Ca - H$[23] and a Morse potential was used for HF. The next phase of the fitting procedure involved determination of separate polynomial representations (in R_{CaF}, R_{HF} and R_{CaH}) to the collinear and 75° data. These terms were multiplied by appropriate attenuation factors and hyperbolic tangent switching functions to limit their contributions to the desired geometries. Finally, additional terms were added to describe the $H - Ca - F$ well. The resulting potential energy function[12] has 41 adjustable parameters and reproduces the *ab initio* data points to a mean square deviation of 1.2 kcal/mol.

From the initial trajectory runs, it was discovered that the outcome of some individual trajectories was sensitive to the magnitude of the integration time step. Even though the trajectories were stable to back integration and conserved energy and angular momentum, the final energy partitioning and scattering angle would change if a different integration time step were used. In certain cases these *unstable* trajectories would be reactive for one step size and non-reactive for another! Closer

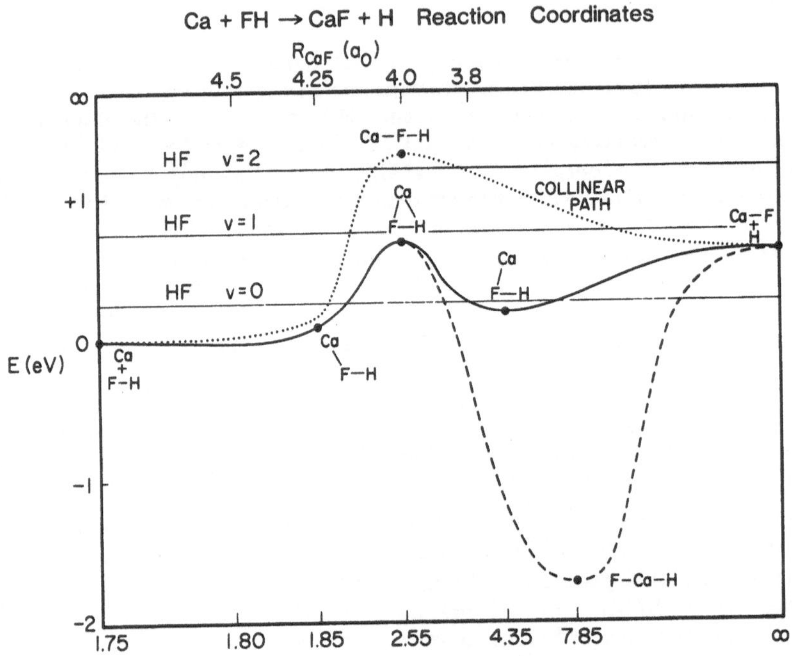

Figure 1. Schematic representation of the potential energy profiles (in eV) along the $Ca + HF \rightarrow CaF + H$ reaction coordinates.

analysis indicated that the changes occurred while the system was in the $H-Ca-F$ potential well and seemed to result from an interplay between round off errors due to the finite numerical precision available and errors in the numerical integration algorithm. Further reductions in the time step were not feasible as the likelihood of round off errors was greater when the number of integration steps was increased. The strategy we chose to employ involved running the identical set of trajectories at two or three different time steps in the range between 1.25 and 5.0×10^{-17}s and comparing the outcomes. If the trajectory endpoints agreed for two different time steps, that trajectory was classified as *converged*. If the endpoints did not agree for two of three different time steps, the trajectory was considered *nonconverged*. In general, the lifetimes of nonconverged trajectories were in excess of 1 to 2 ps, however the trajectory lifetime itself did not constitute a reliable criterion for identifying nonconverged behavior.

Trajectories were computed mainly for HF in v=1 with J=0,1,..10 and relative translational energies E_{rel} between 5 and 40 kcal/mol. For the lower collision energies, the number of nonconverged trajectories was comparable to the number of converged reactive trajectories, while at higher collision energies the ratio of nonconverged to converged reactive trajectories was small ($\approx 0.1 - 0.2$). Further tests revealed that nearly all reactive and nonconverged trajectories sampled the

$H-Ca-F$ well for at least several $Ca-F$ vibrational periods and thereby resulted in complex formation. For $E_{rel} = 5$ kcal/mol, nearly all intermediate complexes ultimately went on to $CaF+H$ products. However, at $E_{rel} = 10$ kcal/mol, 25-50% of the complexes dissociated to form reactants. Thus the fate of the $H-Ca-F$ complexes seems to depend on the reactant energy partitioning! On the other hand, the energy partioning in the products is quite insensitive to the reactant rotational and translational energy. Approximately 50% of the total energy ended up in translation, 25-30% ended up in vibration and 20-25% ended up in rotation. This energy partitioning agrees well with the recent experiments of Zhang *et al.*[20] for $E_{rel} = 2.6$ kcal/mol who found the following energy partitioning: translation 48-51%, vibration 20-26% and rotation 23-32%. They also found virtually no difference in the CaF vibrational distribution for $Ca + HF(v = 1, J = 7)$ and $Ca + DF(v = 2, J = 1)$ which have almost identical vibration-rotation energies. These results are completely consistent with those from the trajectory calculations. Unfortunately, it was not possible to reproduce the experimental collision energy in the trajectory calculation. Both the small cross sections and the large numbers of nonconverged trajectories at the lower collision energies made such calculations impractical.

In general, there is complete agreement between the trajectory calculations and state-to-state molecular beam experiments for the $Ca+HF$ reaction. This indicates that the *ab initio* PES is accurate and that the reaction does indeed proceed via an insertion pathway forming a transient $H-Ca-F$ complex. It also provides justification for our approach of computing *ab initio* PES and using classical trajectories to study the dynamics of chemically reacting systems. Interestingly, the related reactions $Sr + HF$ and $Ba + HF$ exhibit different dynamics. The former appears to undergo both direct abstraction and insertion while the latter undergoes mainly abstraction even though $H-M-F$ complexes are expected to exist in both cases. These differences cannot be explained in terms of the mass differences alone and must be due to differences in the respective PES.

4. $H_2 + H + H \rightarrow H_2 + H_2$

At NASA, the interest in this system is in determining the rate of three-body recombination of hydrogen atoms under fuel rich conditions in the hydrogen burning scramjet. This reaction is thought to be important in determining the amount of thrust generated by the partially burnt fuel as it exits the nozzle of the engine.

An analytic representation of the PES for this system should be able to accurately reproduce a wide range of features present in the accurate PES. The approach we have taken to reach this goal was to partition configuration space into the geometries which reflect these different features. We consider the configurations $H_2...H_2$ (non-interacting hydrogen molecules), H_2-H_2 (strongly interacting molecules with bond lengths near their equilibrium values) and H_2-H-H (one molecule distorted far from equilibrium). Potential functions which describe these various regions are then combined with switching functions to produce an overall description of the system. An additional question for this system is the problem of assigning the bonds. That is, if we label the four classically distinguishable hydrogen atoms as A through D, then there are the three possibilities $AB+CD$, $AC+BD$ and $AD+BC$. We have defined a procedure for switching between these choices. Thus our PES is unaltered by the interchange of any atoms. Full details are given in Ref. (14).

Some novel aspects of our PES concern the various configurations. For the asymptotic system $H_2...H_2$, we reproduce the accurate H_2 potential curve including corrections for relativistic and nuclear motion effects. This is done using an analytic function which is the sum of a damped dispersion term and a short range contribution. The short range contribution is written in the form of a Morse function but with a more complicated exponent which depends nonlinearly on bond length. Because the long range part of the potential is treated separately, the exponent is approximately a linear function of the bond length, thus it is fairly easy to represent the exponent in a manner in which all spurious behavior at large and small bond length is avoided.

The PES in the region of $H_2 - H_2$ is also broken up into a long range contribution and a short range one. The long range part includes damped dispersion and quadrupole-quadrupole interactions which accurately reproduce extensive calculations of the van der Waals minimum. The short range part is represented as a product of the sum of nonbonding pairwise interactions and a correction factor which is expanded in terms of Jacobi coordinates. By this means only a few orientations of the molecules are required to parameterize the full anisotropy of the system. The parameters in this part of the analytic representation are determined by fitting *ab initio* points which include the displacement of one or both molecules from their equilibrium bond length. Thus we accurately reproduce the force along the molecular bonds, which is expected to be important for controlling the amount of vibrational energy transfer[24].

The final region consists of configurations of the type $H_2 - H - H$. Here the potential is built from the the accurate H_3 potential[25]. The H_3 potential is the sum of a three-body LEPS function and a correction term. This is extended to $H_2 - H - H$ by using a four-body LEPS function and a scaled sum of the correction terms. In the LEPS function, the H_2 potential discussed above is used instead of the original spline fit, and the triplet curve is modified so that it never crosses this new H_2 potential. This does not significantly alter the H_3 potential. The sum of the correction terms consists of the four ways H_4 can be broken down to $H_3 + H$. By scaling this sum, it is possible to smoothly match this region to the $H_2 - H_2$ region. Thus this PES is consistent with all known properties of H_4 and should be useful for simulations of a wide variety of properties.

Perhaps the most undesirable aspect of this potential is the presence of the many switching functions which can cause problems if one is not careful. The rapid switching between functions of differing size can give rise to spurious humps and valleys. Although it is difficult to make sure that no such features exist on a six dimensional function such as this one, we have endeavored to do so and have not detected any undesirable properties of this analytic representation.

The quasiclassical trajectory method was used to study this system, and the variable step size modified Bulirsch-Stoer algorithm was specially developed for recombination problems such as this one. Comparisons were made with the fourth order Adams-Bashforth-Moulton predictor-corrector algorithm[14], and the modified Bulirsch-Stoer method was always more efficient, with the relative efficiency of the Bulirsch-Stoer method increasing as the desired accuracy increased. We measure the accuracy by computing the rms relative difference between the initial coordinates and momenta and their back-integrated values. For example, for a rms relative difference of 0.01, the ratio of the CPU times for the two methods was 1.6, for a rms relative difference of 0.001 it was 2.0, and for a rms relative difference of 10^{-5} it was 3.3. Another advantage of the variable step size method is that the errors in individual trajectories are more similar, e.g. a test run of ten trajectories yielded rms errors which differed by a factor of 53 when using the modified Bulirsch-Stoer method and a factor of 3200 when using the Adams-Bashforth-Moulton method. In these two runs, both methods used parameters which gave about the same overall rms error. In typical runs, the modified Bulirsch-Stoer algorithm varies the step size over about a factor of ten.

We have used this PES along with a quasiclassical trajectory implementation of the energy transfer mechanism of the resonance complex theory to predict the three body recombination rate over the temperature range 100 – 5000 K. The energy transfer mechanism,

$$H + H \leftrightarrow H_2^*$$

followed by

$$H_2 + H_2^* \rightarrow H_2 + H_2,$$

H_2^* metastable hydrogen, is expected to be the dominant pathway in the temperature regime under study. The results are shown in Figure (2)

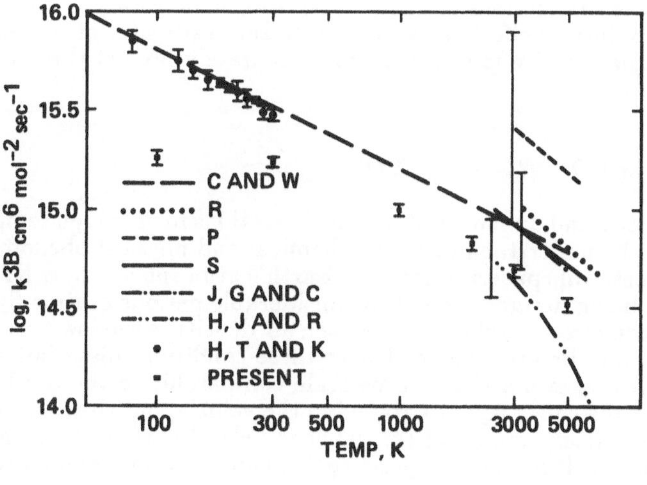

Figure 2. A comparison of the theoretical and experimental recombination rate constants for $H_2 + H + H \rightarrow H_2 + H_2$. C and W is from Ref. (32), R from Ref. (26), P from Ref. (27), S from Ref. (28), J, G and C from Ref. (29), H, J and R from Ref. (30), and H, T and K from Ref. (31).

along with selected experimental results[26−31] and a recommendation based on the critical evaluation of the available experimental data[32]. Over the temperature range of 300 – 5000 K, the theoretical results closely parallel the experimental rate constants, but are about a factor of two smaller than the recommended value in Ref. (32). At lower temperatures, below 300 K, the theoretical results are even smaller than experiment. The increasing underestimate of the calculations at the lowest temperature is probably due to the neglect of the chaperon mechanism,

$$H + H_2 \leftrightarrow H_3^*$$

followed by

$$H + H_3^* \rightarrow H_2 + H_2,$$

which is expected to be important only at low temperature. At the highest temperatures, the underestimate of the present results should be considered suggestive rather than definite, for the experiments show very large error bars in the 3000–5000 K temperature range. More important is the underestimate at 300 K, where the experiments are more reliable. Since the theoretical and recommended values are parallel, this underestimate at 300 K probably indicates that the underestimate at high temperature is real.

At this time the primary cause of this underestimate is not known. However, it is more likely due to the dynamics and/or kinetics calculations rather than the uncertainties in the PES. We conclude this because although it is possible to produce an analytic representation of the H_4 PES which does reproduce the magnitude of the experimental rates when using the resonance complex mechanism, this PES is inconsistent with *ab initio* calculations[14]. It probably will be necessary to improve the treatment of nonequilibrium effects as well to introduce estimates of tunneling into the resonance complex theory in order to more accurately predict three body recombination rates.

5. $N + O_2 \rightarrow O + NO$ and $O + N_2 \rightarrow N + NO$

Studies of the kinetics and dynamics of chemical reactions in very high temperature air are critical to the understanding of the chemical and physical phenomena occuring during the re-entry of spacecraft into the Earth's atmosphere. New NASA hypersonic vehicles are being designed to spend considerable periods of time flying through the upper atmosphere (75-125 km) where the air density is too low for complete thermal equilibrium to be maintained. Under these conditions dissociation of the N_2 and O_2 molecules takes place on a time scale comparable to the residence time of the air species in the shock layer surrounding the vehicle, but the excitation of the vibrational and electronic modes of the gas lags behind that of the translational and rotational modes. It is common practice to define separate temperatures for translation, vibration, rotation and electronic degrees of freedom[1,33]. Under flight conditions envisioned for the AOTV[1], the translational temperature, T_{trans}, might reach values as high as 50000 K while the vibrational temperature, T_{vib}, will be under 15000 K. It is generally assumed that $T_{rot} \approx T_{trans}$ and $T_{elec} \approx T_{vib}$ for the rotation and electronic temperatures, respectively. The rate constants for air exchange reactions

$$N + O_2 \rightarrow NO + O$$

and

$$O + N_2 \rightarrow NO + N$$

have been measured over a wide temperature range[34,35], but their dependence on differing amounts of translational, vibrational and rotational energy is not known.

Walch and Jaffe[36] have completed calculations of the *ab initio* PES for these systems. In both, cases there are two low-lying surfaces connecting the ground electronic states of reactants and products. For the $N + O_2$ reaction, which is exoergic by 33.2 kcal/mol, these are the $^2A'$ and $^4A'$ surfaces with saddle points of 10.2 and 18.0 kcal/mol, respectively, above the minimum in the $N + O_2$ and $O + N_2$ asymptotes. It was estimated that the calculated $^2A'$ saddle point energy is too high by ≈ 2-3 kcal/mol. The reaction paths for both surfaces favored a 110° approach of the reactants. An additional complicating factor for the $^2A'$ PES is the presence of a deep potential well corresponding to the ground electronic state of the NO_2 molecule.

The $O + N_2$ reaction is endothermic by 75 kcal/mol and has no additional energy barrier on the lowest PES ($^3A''$). The second surface ($^3A'$) has a barrier that is 14.5 kcal/mol higher than the product $NO + N$ asymptote. Reactions occurring on both of these surfaces favor $\angle O - N - N = 110°$. No potential energy wells corresponding to stable intermediates were located on these surfaces.

Fits to both $N - O - O$ surfaces and the ground state $O - N - N$ surface were obtained using a modification[37] of the LEPS approach. An additional angle dependent term ($\mathcal{V}_\theta$) was added to the standard LEPS function[7] ($\mathcal{V}_{LEPS}$) to shift the favored angle of approach from collinear to 110°. We first adjusted the Sato parameters such that the energy barrier for the LEPS function matched the location and energy of the saddle point as determined by Walch and Jaffe[36] when $\angle N-O-O$ or $\angle O - N - N$ was constrained to be 110° . We next determined the additive correction to $\mathcal{V}_{LEPS}$ required to match the calculated variation of barrier height with angle and fit this correction factor to a polynomial in $\cos^2\theta$ where θ is $\angle O - N - O$ or $\angle N - O - N$ for the $N - O - O$ and $O - N - N$ systems, respectively. This angle was chosen to ensure that the resulting potential energy function would be invariant to interchange of the two equivalent atoms. Finally, a hyperbolic tangent switching function was used to attenuate $\mathcal{V}_\theta$ as the system approached the atom-diatom or atom-atom-atom asymptotic limits. The attenuation was accomplished using the repulsive part of the fragment diatomic Morse potential ($V_{rep}(R_{AB}) = D_e^{AB} exp\{-2\beta_{AB}(R_{AB} - r_e^{AB})\}$). The full definition of $\mathcal{V}_\theta$ for the general case of $A + BC \to AB + C$ is as follows:

$$\mathcal{V}_\theta = (A_0 + A_2\cos^2\theta + A_4\cos^4\theta + \ldots) \times \frac{1}{2}\left(\tanh\{0.4(\sqrt{W} - 7.0) + 1\}\right) \qquad (3)$$

where

$$W(R_{AB}, R_{AC}, R_{BC}) = V_{rep}(R_{AB})V_{rep}(R_{AC}) + V_{rep}(R_{AB})V_{rep}(R_{BC}) + V_{rep}(R_{AC})V_{rep}(R_{BC}). \qquad (4)$$

The actual parameters employed will be given elsewhere[13]. The resulting potential energy functions $\mathcal{V} = \mathcal{V}_{LEPS} + \mathcal{V}_\theta$ provided a satisfactory representation of the *ab initio* PES for the $^4A'$ surface of $N - O - O$ and the $^3A''$ surface of $O - N - N$ where only a small amount of data was available. It was not, however, satisfactory for the $^2A'$ PES of $N-O-O$. For that case, a more elaborate formulation based on the SM method has been developed[13]. As of this writing, the improved $N - O - O$ doublet potential energy function has not yet been used for trajectory calculations.

Before presenting the results of trajectory calculations using the above PES for $O + N_2$ and $N + O_2$ it is necessary to discuss the problem of selecting initial conditions for diatomic molecules in very highly excited ro-vibrational levels. Porter, Raff and Miller[38] have determined a quasiclassical proceedure for a rotating Morse oscillator. However, their work is based on a power series expansion and is not suitable for describing the coordinates of diatomic molecules in energy levels near or above the rotationless dissociation limit. An alternative classical formulation given by Clarke and Burns[39] can be applied to any bound or quasibound energy level for an arbitrary diatomic potential. In their scheme, the internal energy E_{int} is provided and randomly partitioned into rotational (angular) and vibrational (radial) components. First the range of the square of the rotational angular momentum (q_{min}, q_{max}) is found such that for q_{min} the value of the effective diatomic potential[40] at the centrifugal barrier is greater than or equal to E_{int} and for q_{max} the energy minimum on the effective potential is equal to E_{int}. For each trajectory, q_i is then randomly selected from a uniform distribution between q_{min} and q_{max}. The choice of q_i also serves to specify a unique effective diatomic potential curve and vibrational energy which is fixed by the original value of E_{int}. Each trajectory in an ensemble so chosen has a weight of $\Delta q_i \tau_i / \sum_i \Delta q_i \tau_i$, where Δq_i is $(q_{max} - q_{min})$ for the ith trajectory and τ_i is the vibrational period for the particular values of q_i and E_{int}. If the initial energies E_{int} are sampled from a Boltzmann distribution at high temperatures, the weighted sample closely approximates a thermal distribution of $E(v, J)$ based on a Dunham expansion.

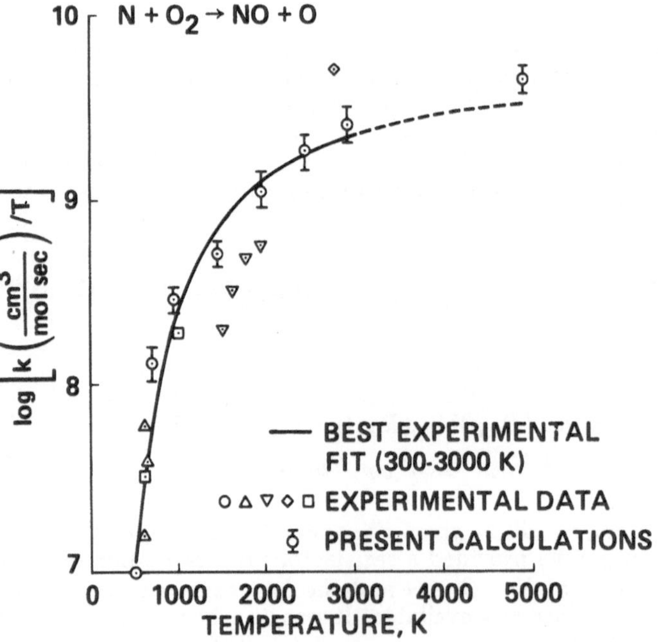

Figure 3. A comparison of the theoretical and experimental rate constants for $N + O_2 \rightarrow NO + O$. Experimental data are from Ref. (34).

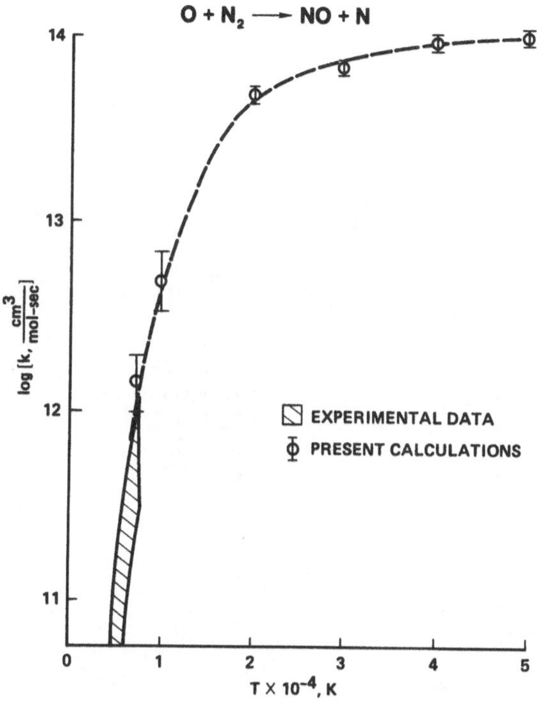

Figure 4. A comparison of the theoretical and experimental rate constants for $O + N_2 \rightarrow NO + N$. Experimental data are from Ref. (34).

Using the Clarke and Burns sampling scheme[39] for a Boltzmann distribution of E_{int}, thermal rate constants were computed for both the $N + O_2$ and $O + N_2$ reactions. For these calculations E_{rel} was also chosen from a thermal distribution at the same temperature. The results shown in Figures 3 and 4 are the combined rate constants for exchange and dissociation.

For $N + O_2$ (figure 3), rate constants for reaction occurring on the $^2A'$ and $^4A'$ surfaces have been combined with the proper degeneracy factors of 1/6 and 1/3, respectively. Below 2000 K reaction on the quartet surface makes little contribution to the overall rate constant because of the higher energy barrier. However, the larger degeneracy factor causes this process to become dominant at temperatures greater than 3000 K. Dissociation does not contribute significantly at temperatures below 5000 K. As can be seen from the figure, the agreement between the calculated and measured rate constants is quite good.

For $O + N_2$ (figure 4), only reaction occurring on the lowest PES was considered. The large endoergicity made calculations of the thermal rate constants difficult at the lower temperatures. Even though the calculations used stratified sampling[15] and large ensembles of trajectories (N_{Total}=5000 for most runs), the rate constants for temperatures less than 2000 K have large statistical errors. Nevertheless, the agreement between the experimentally determined and computed rate constants is good in the narrow range of temperatures where they overlap. As was found for the $N + O_2$ reaction, the dissociation channel does not contribute significantly for temperatures below 5000 K. Perhaps a better approach would have been to study the reverse reaction or to use a phase space sampling scheme and start the trajectories in the interaction region[41]. However, in such studies it would be much more difficult to extract rate constants for nonequilibrium conditions.

Table I. Effect of Reactant Energy Partitioning on Total Reaction Cross Sections ($\AA^2$) for $N + O_2 \rightarrow NO + O$.

f^a_{trans}	$\sigma^b_{E_{int}}$	$\sigma^c_{E_{vib}}$
$E_{total} = 32.4$ kcal/mol		
.92	1.14±.08	1.17±.07
.77	1.10±.08	0.75±.06
.64	0.80±.06	0.35±.04
.51	0.48±.05	0.05±.02
.38	0.08±.02	0.00±.02
$E_{total} = 58.0$ kcal/mol		
.95	3.51±.13	3.83±.13
.66	3.54±.13	3.52±.13
.39	3.19±.12	2.41±.11
.14	0.45±.05	0.07±.02

[a] Fraction of total energy of reactants in translation. Energy is referenced to the minimum in the reactants asymptote.
[b] Using the sampling scheme of Ref (39). E_{int} is randomly distributed between vibration and rotation.
[c] Using the sampling scheme of Ref (38). E_{rot} is selected from a 300 K thermal distribution and E_{vib} is chosen such that $E_{vib} + E_{trans} = E_{total}$. The values of E_{vib} correspond to integral vibrational quantum numbers v.

We have begun to study the role different reactant energy modes play in promoting the exchange and dissociation reaction. Using the Clarke and Burns[39] sampling method we cannot vary the degree of rotational and vibrational excitation. However, we can compute reaction cross sections or rate constants for ensembles of trajectories with fixed E_{int} and E_{rel}. In addition, we have used the rotating Morse oscillator sampling scheme[38] for the same values of E_{int} to delineate rotational and vibrational effects. In these calculations, the diatomic rotational quantum numbers were chosen from a 300 K thermal distribution with the balance of the internal energy being in vibration. Typical results for $N+O_2$ at a fixed total energy are given in Table I. It can be seen that placing a larger fraction of the energy in translation enhances the reaction. It also appears that rotational energy is more effective than vibrational, because the cross sections for the Clarke and Burns sampling method are slightly larger than those for the rotating Morse oscillator sampling method. The former distributes the internal energy between rotation and vibration while the later was used to place it predominately in vibration. For $O + N_2$ the opposite trend is found. Cross sections are much larger when most of the available energy is present as E_{int}. Larger cross sections were also obtained when most of the internal energy was present as vibration.

Finally, we have also studied the effect of varying E_{int}. Batches of trajectories were run using the Clarke and Burns[39] sampling scheme for fixed values of E_{int} and with E_{rel} selected from a thermal distribution at a fixed value of T_{trans}. As expected, the rate constants for exchange (k_{exch}) and dissociation (k_{dissoc}) increased with increasing internal energy. However, at energies very near D_e, the rotationless dissociation energy, k_{dissoc} increased markedly and k_{exch} decreased. At higher energies (i. e., for quasibound diatomic energy levels), the rate constants returned to their lower E_{int} behavior. This resulted in a resonance-like spike in k_{dissoc}. Closer examination of these results revealed that the increase in k_{dissoc} is due to the enhanced dissociation cross sections for diatomic molecules with very high v and very low J. This effect does not carry over to quasibound molecules where rotation quantum numbers are higher. It is suggested that the nature of the outer turning point is quite different for these cases and that for the low J and high v combination, molecules spend more time near that outer turning point than do molecules with other (v, J) combinations. It is not certain that this effect will be observable. However, it can play an important role in enhancing the rate constants for dissociation. Further studies of the dynamics of these dissociation and exchange reactions are in progress.

6. CONCLUSIONS

These studies demonstrate the utility of classical trajectory calculations in probing the kinetics and dynamics of elementary gas phase reactions. When accurate potential energy surfaces based on *ab initio* quantum chemistry calculations are used, quantitative agreement with experimental results can be obtained. However, these studies required lengthy trajectory calculations because of the complicated nature of the potential energy functions used.

References:

1. D. M. Cooper, R. L. Jaffe and J. O. Arnold, J. Spacecraft and Rockets **22**, 60 (1985).
2. G. Y. Anderson, "An Outlook on Hypersonic Flight", AIAA Paper 87-2074, AIAA/SAE/ASME/ASEE 23rd Joint Propulsion Conference, San Diego, CA, June 1987.
3. "Pioneering the Space Frontier", Report of the National Commission on Space, (Bantam Books, New York, 1986).
4. B. C. Garrett and D. G. Truhlar, J. Phys. Chem. **83**, 2921 (1979).
5. See, for example, C. W. Bauschlicher, S. R. Langhoff and P. R. Taylor, paper presented at this workshop.
6. C. W. Bauschlicher, S. P. Walch, S. R. Langhoff, P. R. Taylor, and R. L. Jaffe, J. Chem. Phys. **88**, 1743 (1988).
7. J. C. Polanyi and J. L. Schreiber, The dynamics of Bimolecular Reactions, in "Physical Chemistry - An Advanced Treatise", Vol. VI, Kinetics of Gas Reactions, H. Eyring, W. Jost and D. Henderson, ed. (Academic Press, New York, 1974), p. 383.
8. J. C. Tully, Adv. Chem. Phys. **42**, 63 (1980).

9. See, for example: J. M. Bowman, A. F. Wagner, S. P. Walch, and T. H. Dunning, J. Chem. Phys. **81**, 1739 (1984)
10. K. S. Sorbie and J. N. Murrell, Mol. Phys. **29**, 1387 (1975).
11. R. Steckler, D. G. Truhlar and B. C. Garrett, J. Chem. Phys. **83**, 2870 (1985).
12. R. L. Jaffe, M. D. Pattengill, F. G. Mascarello, and R. N. Zare, J. Chem. Phys. **86**, 6150 (1987).
13. R. L. Jaffe, M. D. Pattengill, T. Halicioglu, and S. P. Walch, manuscript in preparation.
14. D. W. Schwenke, J. Chem. Phys., in press
15. D. G. Truhlar and J. T. Muckerman, in "Atom-Molecule Collision Theory", ed. by R. B. Bernstein (Plenum Press, New York, 1979), chap. 16.
16. C. W. Gear, J. SIAM Numer. Anal. Ser. B **2**, 69 (1964).
17. R. Bulirsch and J. Stoer, Numer. Math. **8**, 1 (1966).
18. Care must be taken to ensure that the set of discarded trajectories is not statistically different from the set of completed trajectories. For example, it might contain mainly trajectories that formed long-lived complexes. One way to check this is to compare batches of trajectories with different N_V or N_{Total}.
19. R. I. Altkorn, F. E. Bartoszek, J. DeHaven, G. Hancock, D. S. Perry and R. N. Zare, Chem. Phys. Lett. **98**, 212 (1983).
20. R. Zhang, D. J. Rakestraw, K. G. McKendrick and R. N. Zare, submitted to J. Chem. Phys.
21. P. E. M. Siegbahn, C. W. Bauschlicher, Jr., B. Roos, A. Heiberg, P. R. Taylor and J. Almlöf, SWEDEN, a vectorized SCF-MCSCF-direct CI program.
22. The calculations used a $(12s, 8p, 5d/10s, 7p, 1d/5s, 1p)$ primitive basis for (Ca/F/H) which was contracted to $(9s, 6p, 3d/5s, 4p, 1d/3s, 1p)$.
23. R. Grice and D. R. Herschbach, Mol. Phys. **27**, 159 (1974).
24. D. G. Truhlar, F. B. Brown, D. W. Schwenke, R. Steckler, and B. C. Garrett in "Comparison of Ab Initio Quantum Chemistry with Experiment for Small Molecules", ed. by R. J. Bartlett (Reidel, 1985), p. 95.
25. B. Liu, J. Chem. Phys. **58**, 1924 (1973); P. Siegbahn and B. Liu, J. Chem. Phys. **68**, 2457 (1978); D. G. Truhlar and C. J. Horowitz, J. Chem. Phys. **68**, 2466 (1978); **71**, 1514 (1979).
26. J. P. Rink, J. Chem. Phys. **36**, 262 (1962).
27. R. W. Patch, J. Chem. Phys. **36**, 1919 (1962).
28. E. A. Sutton, J. Chem. Phys. **36**, 2923 (1962).
29. T. A. Jacobs, R. R. Giedt, and N. Cohen, J. Chem. Phys. **47**, 54 (1967).
30. T. A. Hurle, A. Jones, J. L. J. Rosenfeld, Proc. Roy. Soc. **A 310** , 253 (1969).
31. D. O. Ham, D. W. Trainor, and F. Kaufman, J. Chem. Phys. **53**, 4395 (1970).
32. N. Cohen and K. R. Westberg, J. Phys. Chem. Ref. Data **12**, 531 (1983).
33. R. L. Jaffe, "Rate Constants for Chemical Reactions in High-Temperature Nonequilibrium Air", AIAA Progress in Astronautics and Aeronautics: Thermophysical Aspects of Re-entry Flows, Vol. 103, edited by J. N. Moss and C. D. Scott, New York, 1986, p. 123; R. L. Jaffe, "The Calculation of High-Temperature Equilibrium and Nonequilibrium Specific Heat Data for N_2, O_2 and NO", AIAA Paper 87-1633, AIAA 22nd Thermophysics Conference, Honolulu, HI, June 1987.
34. D. L. Baulch, D. D. Drysdale, and D. G. Horne, "Evaluated Kinetic Data for High Temperature Reactions, Vol. II Homogeneous Gas Phase Reactions of the $H_2 - N_2 - O_2$ System" (Butterworth, London, 1973).
35. R. K. Hanson and S. Salimian, "Survey of Rate Constants in the $N/H/O$ System", in Combustion Chemistry, edited by W. C. Gardiner, Jr., (Springer-Verlag, New York, 1984), p. 361.

QUASICLASSICAL CALCULATIONS FOR ALKALI AND ALKALINE EARTH + HYDROGEN HALIDE CHEMICAL REACTIONS USING SUPERCOMPUTERS

J.M. Alvariño
Departamento de Quimica Fisica
Universidad de Salamanca
Salamanca
Spain

E. Garcia
Departamento de Quimica Fisica
Universidad del Pais Vasco
Bilbao
Spain

A. Laganà
Dipartimento di Chimica
Università di Perugia
Perugia
Italy

ABSTRACT. In quasiclassical trajectory programs for calculating atom-diatom reactivity, routines dedicated to the evaluation of the potential energy and its derivatives are in general a computational bottleneck. A significant speed-up of these calculations has been obtained by running trajectory programs in vector mode on supercomputers after expressing the potential energy as a polynomial in the Bond-Order variables. Other computational advantages have been obtained by allowing trajectories to run in parallel. Applications to the systematic study of detailed properties of the reactions of the alkali/alkaline earth + hydrogen halide family have been made.

1. Introduction

Quasiclassical trajectory (QCT) programs are nowadays routinely used for 3D studies of the atom-diatom reactions. In this way, it has been possible not only to compare theoretical results with experimental findings, but also to investigate in detail some properties of the atom-diatom reactive collisions. However, the amount of computer time needed to carry out these quasiclassical investigations even for small chemical systems is so large, when performed on usual scalar computers, to make a systematic study difficult. This is especially true when detailed (state to state or vector-to-vector) rather than global (integral) quantities need to be calculated.

A. Laganà (ed.), Supercomputer Algorithms for Reactivity, Dynamics and Kinetics of Small Molecules, 383–393.

In the last few years we have afforded the problem of performing a systematic study of the reactive properties of alkali/alkaline earth + hydrogen halide reactions[1-4]. Our aim has been the investigation of the reactivity of these systems by paying a particular effort for establishing its relationships with the shape of the potential energy surface (PES), the amount of energy supplied as reactants' rotation, the reorientation of the collision partners during the collision and the effect of competing reaction paths on detailed reaction attributes (such as vector alignment).

In all these cases, to have a good statistics (i.e. to have a significant sample of reactive events) one has to run a large number of trajectories. Therefore to obtain results in a reasonable time the cpu consumption per single trajectory has to be very short. This is possible only when using supercomputers provided that the programs have been restructured to take advantage of the vector and parallel features of these machines.

2. Vector performances of the BO potential

The schematic structure of a classical trajectory program is as follows:

```
LOOP on trajectories
    Random numbers generation for setting initial conditions
    LOOP on time integration steps
        LOOP on differential equations
            Calculation of potential energy values and derivatives
        END of differential equations loop
        Check of the constants of motion
    END of time integration loop
    Evaluation of the final quantities of the given trajectory
    Statistical analysis update
END of the trajectory loop
```

The program consists of an initial part in which the initiator of the random sequence as well as the physical constants of the problem are read in. The outermost DO loop running over the trajectory index is then opened. This determines the set of initial values to be selected randomly from a uniform distribution. Nested inside the main loop are respectively the DO loop over time integration and the DO loop over the 12 projections of the conjugated quantities on the Space Fixed Cartesian frame. Within the innermost DO loop the routines POTBO and POTDER are called to evaluate respectively the potential energy and its derivatives with respect to the internuclear distances. In the same DO loop the integration (INTEGR), chain rule (CHAINR), and coordinate transformation (COORDT) routines are also called.

It turns out that most of the overall cpu time needed for integrating a trajectory is taken by the innermost DO loop, i.e. the loop in which the potential energy and its derivatives are calculated. This fact singles out the importance of designing an efficient way of calculating both quantities, i.e. of finding an adequate functional representation of the ab initio potential energy data.

Simple formulations of the PES, such as the LEPS or the three term DIM[5] functionals are suited for fast trajectory integrations. However, these analytical representations of the PES have been shown[6,7] to be unappropriate fot the reactions of our interest. In fact, the wealth of published theoretical and experimental information concerning the electronic structure of these systems can be fitted only by a complex functional form.

Among the various analytical representations of a PES, those advantageous for use on supercomputers are polynomial expansions. For this reason, following Murrell et al.[8] it is useful to divide the potential energy of a triatomic ABC system into its multi-body components

$$V(r_1,r_2,r_3) = V2_1(r_1) + V2_2(r_2) + V2_3(r_3) + V3_{123}(r_1,r_2,r_3) \tag{1}$$

For simplicity, the AB, BC and AC pairs in eq. (1) have been numbered as 1, 2, and 3, respectively. The $V2_i$ terms can be represented as polynomials in the related bond order (BO) variable n_i

$$V2_i(r_i) = \sum_k^K a_{ik} n_i^k \tag{2}$$

In eq.(2) $n_i = exp(-b_i(r_i - r_{e_i}))$, b_i is a parameter, r_{e_i} is the equilibrium distance of the diatom and K (the degree of the polynomial) is usually 4. In this way, b_i and a_{ik} parameters can be obtained by solving a system of algebraic equations generated by the requirement of reproducing the spectroscopic force constants of the diatoms. This functional form is highly flexible and, at the same time, easy to compute. On top of that it naturally converges to the correct asymptotic limits[9a] A polynomial expansion in BO variables has been successfully adopted also for the three body term[9b] although in this case it does not converge to the right limit at short distances.

To have a more compact design of the routine, the BO potential can be reformulated in the more general way

$$V(r_1,r_2,r_3) = \sum_j^J \sum_k^K \sum_l^L A_{jkl} n_1^j n_2^k n_3^l \tag{3}$$

where diatomic terms have been incorporated into the three-body expansion by setting equal to zero the power of the BO variables other than that relative to the considered diatom. Accordingly, A_{j00}, A_{0k0} and A_{00l} are set equal to the corresponding diatomic coefficients. Usually we choose $J \leq 5$, $K \leq 5$, $L \leq 5$, $0 < j + k + l \leq 6$.

It can be easily seen by inspection, that for a BO potential the routines calculating partial derivatives with respect to the internuclear distances have the same structure as the routine calculating the potential energy value. For this reason, we shall refer to ref. 10 for a discussion of the different ways of coding a BO potential on a supercomputer. We merely recall here that a global speed-up factor ranging from 15 to 18 can be obtained when

Table I: Absolute and relative cpu times in a test QCT program.

Routine	Scalar time	Scalar time%	Vector time	Vector time%
POTDER	100.0	72.9	24.7	57.1
INTEGR	22.6	16.5	5.8	13.4
CHAINR	9.6	7.0	9.3	21.6
COORDT	3.0	2.2	2.2	5.0

restructuring the original potential energy subroutine (POTJKL) designed for running on a scalar computer.

A more complete test of the efficiency of the vector restructuring of this kind of computer codes has been carried out on the IBM 3090/200 VF by running sample quasiclassical calculations using a typical classical trajectory program[11]. For these test runs, input parameters typical of realistic reactive systems having the potential energy surface expressed as a BO functional have been used. The BO potential and potential derivatives routines adopted on the program were the ones best performing on the IBM machine. In them the BO potential was coded as a single DO loop using single exponential terms vectorized by making use of indirect addressing (POTCUR).

The first batch of calculations were performed to assess to what extent the speed-up obtained for test calculations is transferred to runs of the whole trajectory program. For these runs input parameters were selected as those of the

$$Li + HCl(v,j) \rightarrow LiCl(v',j') + H \qquad \text{(R1)}$$

reaction for which a BO PES is available[9c].

As expected, these runs confirm that the evaluation of the potential energy derivatives is the critical step of a QCT calculation. As shown by Table I, such a step takes about 73% of the global cpu time when the QCT program is run in scalar mode after substituting a POTCUR- to a POTJKL-type routine for calculating the potential energy and its derivatives.

It is worth mentioning here that in the original program containing POTJKL the fraction of computer time spent for calculating potential energy derivatives was much larger. The speed-up for a vector execution of the POTDER routine in its restructured form is an additional factor of 4. The importance of vectorizing POTDER is confirmed by the small speed-up factors obtained for the remaining routines. On a per cent basis this means a worsening of the performance of the smaller routines such as INTEGR and COORDT.

As apparent from the sketch of the program structure given in the previous section, paralellism is quite easily implemented in a trajectory program. In fact, apart from a few preliminary and final assignment instructions, the whole body of the program can be executed in parallel. As a consequence, when the number of trajectories to be computed is large the execution of non-parallel sections as well as the load of synchronization delays negligibly contribute to the overall time consumption. Accordingly, in our calculations we found a speed-up factor very close to the number of available processors.

3. Trajectory calculations of some atom diatom reactions

Restructured programs have been used for a systematic investigation of the reactive properties of some atom-diatom systems for which accurate ab initio PESs are available. Some illustrative results obtained for the alkali/alkaline earth atoms + hydrogen halide family of reactions are reported in this paper. In addition to reaction R1 the following systems are considered

$$Li + HF(v,j) \rightarrow LiF(v',j') + H \tag{R2}$$

$$Be + HF(v,j) \rightarrow BeF(v',j') + H \tag{R3}$$

$$Mg + HF(v,j) \rightarrow MgF(v',j') + H \tag{R4}$$

have been investigated.

For reaction R1 the goal of our investigation was to reproduce experimental cross sections measured on a crossed molecular beam apparatus[12]. Unfortunately, the original ab initio potential energy values[2] could not be used directly for this purpose. They give, in fact, a barrier to reaction too high for allowing reactivity at the lowest experimental collision energy. For this reason, they were modified by successive lowering of the transition state height while the overall shape of the PES was left unaltered. Two different BO surfaces (BO2 and BO3) were fitted to these modified ab initio values. The transition states of the BO2 and BO3 surfaces are respectively 3.6 and 2.0 kcal/mol higher than the entrance channel asymptote (the ab initio value is 11.4 kcal/mol). Cross sections calculated starting from initial values of the reactant parameters simulating the experimental situation are reported in figure 1. For both surfaces the calculated quasiclassical cross sections are much smaller than the corresponding experimental values. A better comparison with experimental results is found for BO3. In this case, in fact, not only the absolute value of the cross section is better reproduced but also its energy dependence is more accurately mimicked.[2]

Extended quasiclassical calculations have also been carried out for reaction R2[1,7,13]. Results of these calculations have been used for estimating empirical corrections to the ab initio PES[14] and for understanding the importance of quantum effects for determining reactivity. The detail of the calculated results has allowed the study of the dependence of chemical reactivity from the atom-diatom angle of attack. An example of such a dependence

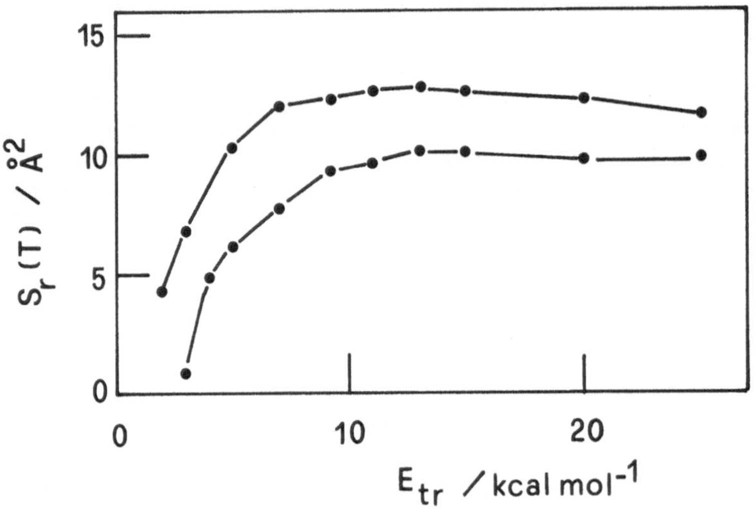

Figure 1: Reactive cross sections plotted as a function of the translational energy for reaction R1 at v=0 and a rotational temperature of 60 K. Reported values have been calculated on the BO2 surface (lower curve) and on the BO3 surface (upper curve).

for v=j=0 is given in Figure 2 where the reactive cross section is reported as a function of the angle of attack for all four isotopic variants of reaction R2.

For $X = ^1H$ we had previously found[15] an "anomalous" high reactivity for attacks occurring from the light particle end of the target molecule. On the contrary, attacks from the F-end of ^1HF (F is the final product partner of Li) were essentially unreactive for a cone of approach of 30 degrees around collinearity. Such an "anomaly" was rationalized in ref. 1 in terms of reorientation of the target molecule during the collision when the F partner is light. This hypothesis is supported by the variation of the angular dependence of the cross section along the isotopic series as well as by a detailed graphical analysis of the individual trajectories. The reorientation was found to be favoured by the anisotropy of the PES and by the angular momentum transfer typical of reactions of a heavy atom with a target molecule having a small moment of inertia[1].

For reaction (R3) one of the goals of our calculations was the investigation of the effect of the reactants' rotational energy on the efficiency of the reactive process. Calculations were carried out on a BO potential energy surface[9b] at several initial vibrational and translational energy values of the reactants. We found that the reactive cross section increases monotonously with the HF rotational excitation. Results of this type are shown in figure

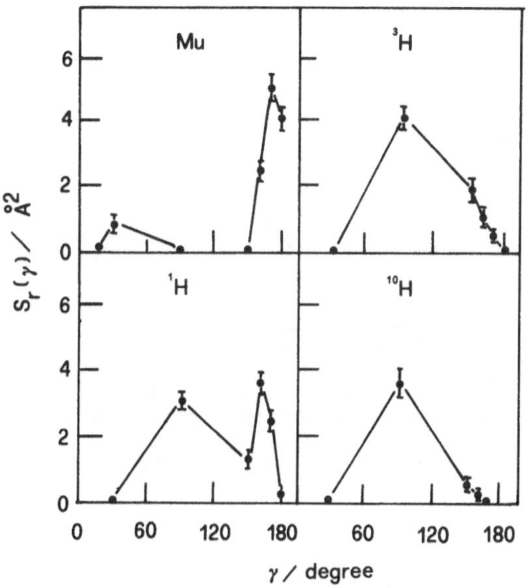

Figure 2: Reactive cross sections plotted as a function of the angle of approach for all the isotopic variants of reaction R2 at v=j=0 and 15 kcal/mol of collision energy.

3 where the reactive cross section calculated for HF in its ground vibrational state and a translational energy (T) of 50 kcal/mol is reported as a function of j.

The fact that the rotational energy promotes the reactivity of this system more efficiently on the BO PES than on the other PES[16] has been interpreted as a result of the greater anisotropy of the BO surface. The detailed study of the characteristics of the reactive trajectories of this system has also pointed out the importance of events taking a complex (C) path to products. These C trajectories form in the exit channel an extremely short-lived FBeH linear complex. The breaking-up of this complex leads very often to the BeF product.

A more systematic investigation of the importance of the microscopic branching between reactive events following a straight path (S) to products after overtaking the saddle and those going through the formation of a linear intermediate complex has been carried out for reaction R4. In this case too an extended interval of reactant collision energies (from threshold up to 40 kcal/mol) as well as of vibrational (v=0-5) and rotational (j=0-20) quantum states of HF has been investigated. Calculations were performed on the BO PES[17] fitted to the ab initio values of ref. 18. Examples of the results obtained for this reaction are given in figures 4 and 5 where the reactive integral cross section ($S_r(T)$) calculated, respectively, at (v=5, j=0) and at (v=5, j=4) is plotted as a function of the collision energy.

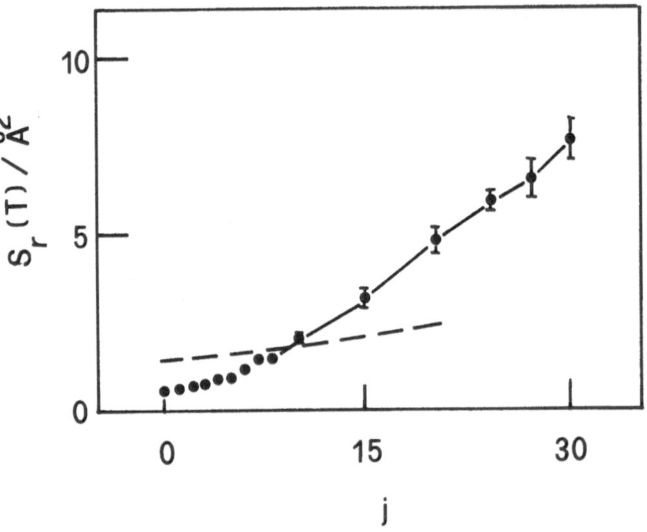

Figure 3: Reactive cross sections plotted as a function of j for reaction R3 at v=0 and a collision energy of 50 kcal/mol. Also shown (dashed line) are the results obtained on a different PES[16] under the same initial conditions.

To emphasize the different roles played by C and S events C and S contributions to the total (C+S) cross section are separately plotted in both figures.

The $S_r(T)$ curves for complex trajectories exhibit a slowly increasing trend from threshold at both j=0 and j=4. Such a behaviour qualitatively agrees those of Be+HF[4] and Ca+HF[19].On the other hand, the corresponding $S_r(T)$ curves for S trajectories have a clearly different trend. In fact, while both cases show a post-threshold sudden rise (more evident for j=0) as is typical of late barrier reactions when the reactant vibrational energy is larger than the barrier height, their behaviour is very different at slightly larger energies. As an example, for j=0 $S_r(T)$ exhibits a maximum at about T=4 kcal/mol while for j=4 it increases monotonously.

4. Conclusions

A careful restructuring of a quasiclassical trajectory code and the adoption of a polynomial representation of the potential energy has made it possible to take profit from both vector and parallel features of the IBM 3090/200 VF supercomputer to perform extended calculations of the reactivity of some alkali/alkaline earth + hydrogen halide systems. Thanks to the reduction of the needed cpu time it has thus been possible to examine in detail the dynamics of these reactions. In particular, it has been possible to improve ab initio potential

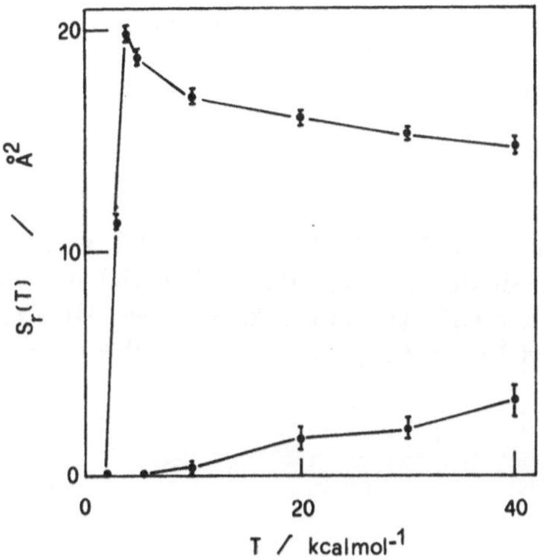

Figure 4: C (filled circles, lower curve) and S (open circles, upper curve) contributions to the reactive cross section plotted as a function of the collision energy for reaction R4 at v=5, j=0. C contributions have been multiplied by a factor 10.

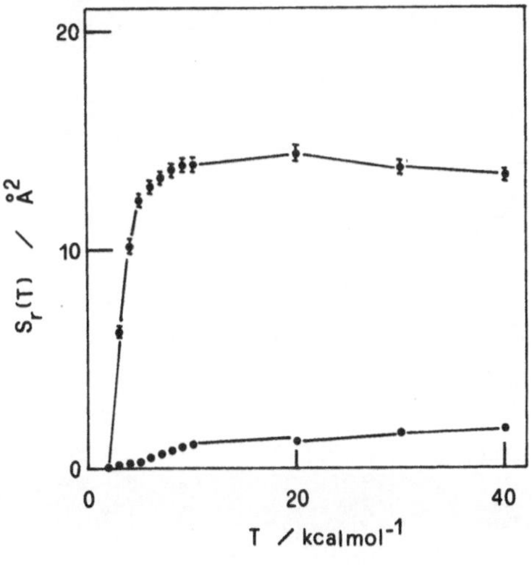

Figure 5: As figure 4 for v=5, j=4. The C curve has not been scaled.

energy values by a trial and error comparison of calculated cross sections and experimental data. Other detailed investigations carried out included the influence of initial rotational energy on the efficiency of the reactive process, the reorientation of the reactive system during the collision and the competition between direct and complex mechanisms.

5. Acknowledgments

Calculations have been performed at the IBM ECSEC (Roma, Italy), at CNUCE (Pisa, Italy), and at CINECA (Casalecchio di Reno, Italy). Partial financial support from CICYT (Spain, grant no. PB85-0316) and CNR (Italy) is acknowledged. We also thank the Ministries of Education of Spain and Italy for support within a collaborative agreement scheme.

References

1. J.M. Alvariño, F.J. Basterrechea, and A. Laganà, Mol. Phys., 59,559(1986).
2. P. Palmieri, E. Garcia, and A. Laganà, J. Chem. Phys., 88,181(1988).
3. M. Paniagua, J.C. Sanz, J.M. Alvariño, and A. Laganà, Chem. Phys. Lett., 126, 330 (1986).
4. J.M. Alvariño and A. Laganà, Chem. Phys. Letters,144, 558 (1988).
5. Y. Zeiri and M. Shapiro, Chem. Phys., 31, 217 (1978); M. Shapiro and Y. Zeiri, J. Chem. Phys., 70, 5264 (1979).
6. I. Noorbatcha and N. Sathyamurthy, Chem. Phys., 77, 67 (1983).
7. E. Garcia, A. Laganà, J.M. Alvariño, and M.L. Hernandez, J. Chem. Phys., 84, 3059 (1986).
8. J.N. Murrell, S. Carter, S.C. Farantos, P. Huxley, and A.J.C. Varandas in *Molecular Potential Energy Functions* (Wiley, New York, 1984).
9. a) E. Garcia and A. Laganà, Mol. Phys., 56, 621 (1985); b) ibid., 56, 629 (1985); c) E. Garcia, A. Laganà, P. Palmieri, and J.M. Alvariño, in preparation.
10. E. Garcia, L. Ciccarelli, and A. Laganà, Theor. Chim. Acta, 72, 253 (1987); E. Garcia, O. Gervasi, and A. Laganà, in "*Approximate Quantum Techniques for Atom Diatom Reactions*" in *Supercomputer Algorithms for Reactivity, Dynamics and Kinetics of Small Molecules*, A. Laganà, Ed. (Kluwer, Dordrecht, 1989).
11. A program derived from QCPE no. 273.
12. C.H. Becker, P. Casavecchia, P.W. Tiedemann, J.J. Valentini and Y.T. Lee, J. Chem. Phys., 73, 2833 (1980).
13. J.M. Alvariño, P. Casavecchia, O. Gervasi, and A. Laganà, J. Chem. Phys., 77, 6341 (1982); A. Laganà, M.L. Hernandez, and J.M. Alvariño, Chem. Phys. Lett., 106, 41 (1984).
14. M.M.L. Chen and H.F. Schaefer, III, J. Chem. Phys., 72, 4376 (1980).
15. J.M. Alvariño, F.J. Basterrechea, M.L. Hernandez, and A. Laganà, J. Molec. Struct. (Theochem) , 120, 187 (1985).
16. S. Chapman, J.Chem. Phys., 81, 262 (1984).
17. M. Dini, A. Lagana, and M. Paniagua, to be published.
18. M. Paniagua, M. Garcia de la Vega, J.R. Alvarez Collado, J.M. Alvariño and A. Laganà, Chem. Phys., 101, 55 (1986).
19. R.L. Jaffe, M.D. Pattengill, F.G. Mascarello, and R.N. Zare, J. Chem. Phys., 86, 6150 (1987).

DYNAMICS OF THE LIGHT ATOM TRANSFER REACTION: Cl+HCl → ClH+Cl

J.N.L. CONNOR
Department of Chemistry
University of Manchester
Manchester M13 9PL
England

W. JAKUBETZ
Institut für Theoretische Chemie und Strahlenchemie
Universität Wien
A-1090 Wien
Austria

ABSTRACT. Recent research on the Cl+HCl → ClH+Cl reaction is reviewed, covering the period 1983 to mid-1988. Topics discussed include: accurate collinear quantum reaction probabilities, cross sections calculated by the centrifugal-sudden distorted-wave method and results from three dimensional (3D) quasiclassical trajectory (QCT) computations. In addition, we report and discuss new 3D QCT calculations at total energies of E_{total} = 0.50, 0.70 and 1.183 eV for the extended London-Eyring-Polanyi-Sato potential energy surface No. 3 of Persky and Kornweitz.

1. Introduction

The dynamics of reactions in which a light atom (**L**) is exchanged between two heavy ones (**H′** and **H**), *i.e.*

H′+LH → H′L+H,

are of particular theoretical and experimental interest at the present time [1-3]. In this Chapter, we shall be concerned with the dynamics of the isoergic reaction

Cl+HCl(v,j) → ClH(v′,j′)+Cl,

where v is a vibrational quantum number and j is a rotational quantum number. This reaction has been the focus of a considerable amount of research since 1983 [4-32].

In Section 2, we first review some of the theoretical work on the Cl+HCl reaction. In particular, we will discuss accurate quantum collinear (1D) reaction probabilities, cross sections calculated by the centrifugal-sudden distorted-wave (CSDW) method, and results obtained from three dimensional (3D) quasiclassical trajectory (QCT) computations. Two potential energy surfaces (PES) have been used in these

A. Laganà (ed.), Supercomputer Algorithms for Reactivity, Dynamics and Kinetics of Small Molecules, 395–411.

calculations: an extended London–Eyring-Polanyi-Sato (LEPS) surface [4], which will be denoted BCMR, and a scaled and fitted *ab initio* one (denoted sf-POLCI) [5,28].

In Section 3, we present some new QCT results for a second LEPS surface, namely the PES No. 3 of Persky and Kornweitz (denoted PK3) [17]. The 3D dynamical properties of BCMR and PK3 are quite different, although for collinear geometries the two surfaces are very similar, with both of them having a symmetric barrier of height 0.37 eV [17]. The QCT results for PK3 in Section 3 complement CSDW calculations that are currently being performed for this surface [22]. We will also compare with our earlier QCT results for the BCMR surface [21].

2. Recent Research on the Cl+HCl Reaction

2.1. COLLINEAR QUANTUM CALCULATIONS

We begin our brief review of recent research on the Cl+HCl reaction with Ref. [4], which reported the first accurate 1D quantum calculation of reaction probabilities for Cl+XCl→ClX+Cl (X=Mu,H,D), using the BCMR surface. Reference [4] also contains many citations to pre-1983 research on the Cl+HCl reaction.

Plots of the 1D reaction probabilities versus total energy, E_{total}, showed the following striking properties [4]:

(a) Diagonal vibrational transitions, $v \to v'=v$, dominate over off-diagonal, $v \to v' \neq v$, transitions.
(b) A spectrum of resonances is present in the reaction probability curves.
(c) The $v \to v$ reaction probabilities oscillate in a sinusoidal manner, which is particularly pronounced for the $v=0 \to v'=0$ transition.

These properties can be understood by making an adiabatic separation of variables in mass-weighted Delves' polar coordinates [4]. It is then found that the radial polar coordinate corresponds physically to the slow relative motion of the Cl atoms, whilst the angular polar coordinate represents the vibrational motion of the light H atom [4].

The quantum reaction probabilities can also be used to compute accurate 1D thermal rate coefficients. If a transmission coefficient κ (at a temperature T) is defined as the ratio of the accurate quantum rate coefficient divided by the rate coefficient calculated from variational transition state theory (VTST), with the reaction path motion treated classically, then it is found for Cl+HCl that $\kappa(200\ K) = 93$ and $\kappa(300\ K) = 16$, *i.e.* tunnelling is very important for the 1D reaction [6]. These large transmission coefficients can be modelled by VTST if tunnelling corrections are incorporated that allow for the large curvature of the reaction path [5-7,9].

An interesting question to ask is the following: To what extent do the 1D properties (a) - (c) survive in 3D ? Preliminary experimental evidence for resonances [*i.e.* property (b)] in ClHCl and ClDCl has recently been obtained by Metz *et al.* [27]. They probed the transition state of the unstable ClHCl molecule by photodetachment of the corresponding stable anion $ClHCl^-$, an approach suggested in Ref. [33]. The positions of the peaks in the photoelectron spectra are in qualitative accord with the 1D quantum calculations [4].

Further theoretical support for the existence of resonances in 3D ClHCl has been obtained recently by Schatz [30]. He calculated 3D reaction probabilities for total angular momentum quantum number, J=0, for Cl+HCl using a new coupled-channel hyperspherical (CCH) quantum method [29]. Both the BCMR and PK3 surfaces were used, and narrow resonance-like structures were found in the energy dependence of the J=0 reaction probabilities.

The possible occurence of properties (a) and (c) in 3D will be discussed in the next subsection.

2.2. THREE DIMENSIONAL CALCULATIONS

The Cl+HCl reaction is the first **H+LH** system for which accurate 3D quantum cross sections have been calculated (the BCMR surface was used) [14,16,21]. The quantum calculations were done by the CSDW technique [34], which treats the reaction as a perturbation on the inelastic scattering in the entrance and exit channels. For collinearly dominant reactions with high barriers, this perturbative limit is expected to be valid when the total energy, E_{total}, is close to, or below, the QCT threshold for reaction. For a review of DW theories of chemical reactions, see Ref. [35]

A more precise estimate for the reliability (to within *ca.* 30%) of the CSDW method is for the reaction probability summed over final states at each total angular momentum, J, to be less than about 0.1 [34]. For Cl+HCl, this corresponds to E_{total} < 0.45 eV [14,16,21]. This estimate has also been verified for J=0 by Schatz [29], whose CCH cumulative reaction probability agreed with CSDW at low energies, but there was disagreement above 0.45 eV.

The CSDW degeneracy-averaged integral state-to-state cross sections at five energies in the range, 0.30 eV $\leq E_{total} \leq$ 0.50 eV, exhibited the following properties [14,16,21]:

(a) Reactant rotational energy very effectively promotes the reaction. For example, at E_{total} = 0.40 eV, the cross sections, $\sigma_{vj \to v'j'}(E_{total})$, increase by nearly three orders of magnitude for each j′ as j changes from j=0, v=0 to j =9, v =0 (*n.b.*, v=v′=0 for $E_{total} \leq$ 0.50 eV).

(b) The product ClH(v′=0,j′) molecules are highly rotationally excited, with 40-50% of the available energy, E_{avail}, going into rotation. Note that E_{avail} equals E_{total} minus the zero point energy of ClH(v′=0, j′=0), which is 0.183 eV.

(c) The partitioning of rotational energy in the products is only weakly dependent on initial j, in contrast to simple kinematic propensity rules which predict that j→j′=j transitions should dominate for **HLH** reactions.

(d) The surprisial function is nonlinear and is neither functionally simpler than $\sigma_{0j \to 0j'}$ nor independent of j.

The integral CSDW cross sections in Ref. [14] were calculated from reaction probabilities computed at every fifth partial wave, up to the maximum needed for convergence (typically J_{max} = 100). Linear interpolation was then used for the remaining reaction probabilities at J values that had not been computed.

A more extensive calculation at E_{total} = 0.40 eV, in which all partial waves were calculated in the range J = 0(1)70, has been reported in Ref. [16]. This calculation allowed accurate differential cross sections to be computed, which are very sensitive to the phases of the S matrix elements, in contrast to the integral cross sections, which depend only on the modulus of the S matrix. All the differential cross sections were found to be backward peaked [16], which is typical of a rebound mechanism.

The accurate differential cross sections at E_{total} = 0.40 eV have also been used to test a simple semiclassical optical model for reactive angular distributions, that was introduced by Herschbach [36,37]. This model treats the Cl+HCl reaction as an elastic hard sphere collision between the two Cl atoms, with an impact parameter distribution that is determined by the reaction probability (the model therefore avoids the calculation of S matrix phases). The optical model differential cross sections were found to agree well with those computed by the CSDW method [16] (see also Section 3 for further discussion).

An important question suggested by the 1D quantum calculations is whether the differential cross sections at fixed scattering angle θ, oscillate as a function of energy [38]. This question was also investigated in Ref. [16] for the BCMR surface using the semiclassical optical model together with information on the energy dependence of the J=0 reaction probability. For the energy range, 0.30 eV $<$ E_{total} $<$ 0.45 eV, the differential cross sections were found to be backward peaked and to increase monotonically with energy at fixed θ. This does not rule out the possibility of oscillations at higher energies - indeed oscillatory behaviour has been seen for the PK3 surface in QCT calculations [17,31] (see also Refs. [10-12,15] for additional results on this topic).

The CSDW cross sections calculated in Refs. [14,16,21] for the BCMR surface are nearly exact at low energies. They can therefore be employed to assess the accuracy of more approximate theories, in particular the widely used QCT method. In both 1D and 3D, it was found that the QCT thresholds are much higher in energy than the corresponding 1D and 3D quantum thresholds [4,21]. This suggests that tunnelling through the 0.37 eV potential barrier is an important process in the Cl+HCl reaction.

In order to explore the dynamics of the Cl+HCl reaction in an energy regime where the CSDW method breaks down, QCT cross sections have been computed in Ref. [21] at E_{total} = 0.50 , 0.60, 0.70 and 1.183 eV. In all, 441330 trajectories were calculated, of which 8447 reacted. At all four energies it was found that reactant rotational energy is very effective in promoting the reaction and that the products are formed in highly excited rotational states [21]. This behaviour is similar to that discussed above for the CSDW cross sections at lower values of E_{total}.

The properties of individual trajectories during the course of the reaction (including vector correlations) have also been studied [21], in order to gain insight into the dynamics of the reaction. For rotationally excited reactants, trajectories undergoing a "figure of eight" motion were found to form an important part of the reaction mechanism at E_{total}=0.70 and 1.183 eV. The term "figure of eight" motion means one in which the H atom in the incident rotating HCl molecule passes quickly between the approaching Cl′...Cl atoms, with the product Cl′H molecule spinning off in the opposite sense to that of the initial HCl molecule. In addition, the configuration of the three atoms Cl′HCl at the crossover distance is approximately collinear.

The CSDW and QCT calculations discussed so far have all used the BCMR surface. However Garrett *et al.* [5] have reported *ab initio* POLCI calculations in order to produce a surface which is expected to be more realistic. This surface differs from the LEPS/BCMR one in that the reaction path is noncollinear, being bent by an angle of 18.6°.

CSDW calculations have also been performed for a scaled and fitted adaptation of this POLCI surface [28]. The resulting PES will be denoted sf-POLCI. This is the first time an accurate 3D quantum scattering calculation has been carried out for any reaction involving a noncollinear reaction path. The quantities computed by the

Table 1. QCT integral cross sections and energy partitioning results for the PK3 surface at E_{total} = 0.50 eV, which corresponds to E_{avail} = 0.32 eV. The numbers in parentheses are powers of 10 by which the whole entry must be multiplied, *e.g.* 0.61±0.07(-1) ≡ 0.061±0.007.

v	j	$E^{v,j}_{trans}$/eV	$\sigma_{vj}/Å^2$	$\langle j' \rangle_{vj}$	$\langle E_j' \rangle_{vj}$/eV	$\langle f_j' \rangle_{vj}$	$\langle v' \rangle_{vj}$
0	0	0.317	0.46±0.02	1.4±0.1	0.7±0.1(-2)	0.22±0.02(-1)	-0.04±0.01
0	1	0.315	0.35±0.02	2.1±0.1	0.12±0.01(-1)	0.38±0.04(-1)	-0.05±0.01
0	2	0.309	0.19±0.02	2.2±0.2	0.13±0.02(-1)	0.41±0.05(-1)	-0.05±0.01
0	3	0.30	0.61±0.07(-1)	2.7±0.2	0.18±0.03(-1)	0.56±0.08(-1)	-0.04±0.01
0	4	0.29	0.10±0.02(-1)	1.7±0.2	0.8±0.2(-2)	0.25±0.05(-1)	0.05±0.02
0	5	0.28	0.43±0.1(-2)	1.6±0.2	0.7±0.1(-2)	0.21±0.04(-1)	0.14±0.02
0	6	0.26	0.4±0.2(-3)	1.6±0.6	0.6±0.3(-2)	0.20±0.10(-1)	0.29±0.06
0	7	0.24	<0.6(-5)	a	a	a	a

[a] No reactive trajectories were obtained for this case.

Table 2. QCT integral cross sections and energy partitioning results for the PK3 surface at E_{total} = 0.70 eV (E_{avail} = 0.52 eV).

v	j	$E^{v,j}_{trans}$/eV	σ_{vj}/Å^2	$\langle j' \rangle_{vj}$	$\langle E_j' \rangle_{vj}$/eV	$\langle f_j' \rangle_{vj}$	$\langle v' \rangle_{vj}$
0	0	0.517	0.88±0.06	3.4±0.2	0.29±0.03(-1)	0.57±0.06(-1)	-0.04±0.01
0	1	0.515	0.85±0.06	4.1±0.2	0.37±0.03(-1)	0.71±0.06(-1)	-0.07±0.01
0	2	0.509	0.73±0.06	4.1±0.3	0.38±0.04(-1)	0.74±0.08(-1)	-0.09±0.02
0	3	0.50	0.71±0.05	4.6±0.2	0.46±0.04(-1)	0.88±0.07(-1)	-0.09±0.01
0	4	0.49	0.44±0.03	5.1±0.2	0.51±0.04(-1)	0.99±0.07(-1)	-0.07±0.01
0	5	0.48	0.38±0.03	6.0±0.3	0.70±0.05(-1)	0.14±0,01	-0.06±0.02
0	6	0.46	0.33±0.03	6.5±0.3	0.76±0.05(-1)	0.15±0.01	-0.02±0.02
0	7	0.44	0.30±0.03	6.5±0.3	0.75±0.05(-1)	0.15±0.01	0.01±0.01
0	8	0.42	0.31±0.03	7.3±0.3	0.94±0.06(-1)	0.18±0.01	0.03±0.02
0	9	0.40	0.30±0.03	8.7±0.3	0.122±0.005	0.235±0.010	-0.01±0.02
0	10	0.38	0.23±0.02	8.9±0.3	0.126±0.006	0.243±0.012	0.04±0.02
0	11	0.35	0.27±0.02	9.3±0.3	0.139±0.006	0.269±0.012	0.07±0.02
0	12	0.32	0.16±0.02	10.3±0.3	0.159±0.007	0.308±0.014	0.09±0.02
0	13	0.28	0.11±0.01	9.7±0.3	0.150±0.007	0.290±0.015	0.20±0.02
0	14	0.25	0.29±0.03(-1)	8.0±0.5	0.111±0.010	0.215±0.020	0.43±0.03
0	15	0.21	<0.3(-4)	a	a	a	a

a No reactive trajectories were obtained for this case.

CSDW technique included [28]: Partial wave reaction probabilities, integral cross sections and product rotational distributions. In addition differential cross sections were obtained from the reaction probabilities by applying Herschbach's optical model.

The sf-POLCI results are generally similar to the LEPS/BCMR results, although there are some important differences. In particular the integral cross sections in the threshold region were found to increase more slowly with energy for the sf-POLCI surface. Both the sf-POLCI and LEPS cross sections exhibit high product rotational excitation with the sf-POLCI products more excited than those from the LEPS surface [28].

Accurate 3D thermal rate coefficients have also been computed for both surfaces. The activation energy is smaller for the LEPS surface, even though it has a higher barrier than the sf-POLCI surface. The reactant rotational state that contributes most to the thermal rate coefficient in the temperature range 300-400 K is j=11 for sf-POLCI and j=10 for LEPS, in contrast to the thermally most probable state which is j=3 [28]. Despite these (small) differences, the CSDW rate coefficients for both surfaces agree with measured values within the experimental uncertainties.

The CSDW calculations described above require substantial computer resources. A description of the algorithms and structure of the five programs that comprise the CSDW suite can be found in Ref. [16], which also reports the effect of basis set variation on CPU times for CDC 7600, CDC Cyber 176, CDC Cyber 205, Cray X-MP and Cray-2 computers.

No experimental measurements of product state distributions or the effect of reagent rotational excitation have so far been reported for Cl+HCl. However, measurements have been made for the related H′+LH reactions, $O(^3P)$+HCl → OH+Cl [39] and $O(^3P)$+HBr → OH+Br [40,41]. These experiments found effects similar to properties (a) - (d) discussed above for Cl+HCl.

3. QCT Calculations for the PK3 Potential Energy Surface

In this Section, we report the results of QCT computations for the PK3 surface. These calculations have been undertaken in order to (a) explore the dynamics of Cl+HCl on PK3 in an energy regime where CSDW is no longer valid, and (b) to compare with our earlier QCT results on the BCMR surface [21].

The Monte Carlo QCT calculations were carried out in the same way as has been described previously [21]. Results have been obtained for three values of the total energy, namely, E_{total} = 0.50, 0.70 and 1.183 eV. The QCT cross sections σ_{0j} at these energies are reported in Tables 1-3. Also given are the average product rotational and vibrational quantum numbers, $\langle j'\rangle_{0j}$ and $\langle v'\rangle_{0j}$ respectively, the average product rotational energy $\langle E_j'\rangle_{0j}$, and the average fraction of available energy in the products $\langle f_j'\rangle_{0j}$. These quantities are defined in the same way as in our earlier calculations [21]. Note that $\langle j'\rangle_{0j}$, $\langle E_j'\rangle_{0j}$ and $\langle v'\rangle_{0j}$ are computed using continuous product vibrational and rotational energies, *i.e.* no "half-integer box quantization" is involved. The QCT errors in the Tables (and Figures) correspond to ±one standard deviation. Altogether, 236845 trajectories were run, of which 8739 reacted.

Figures 1-3 show plots of σ_{0j} versus j at E_{total} = 0.50, 0.70, 1.183 eV respectively for the PK3 and BCMR surfaces. The behaviour for BCMR is similar at all three energies: σ_{0j} increases rapidly with j reaching a maximum at relatively large values of j; then a steep decline sets in as the reaction threshold is approached.

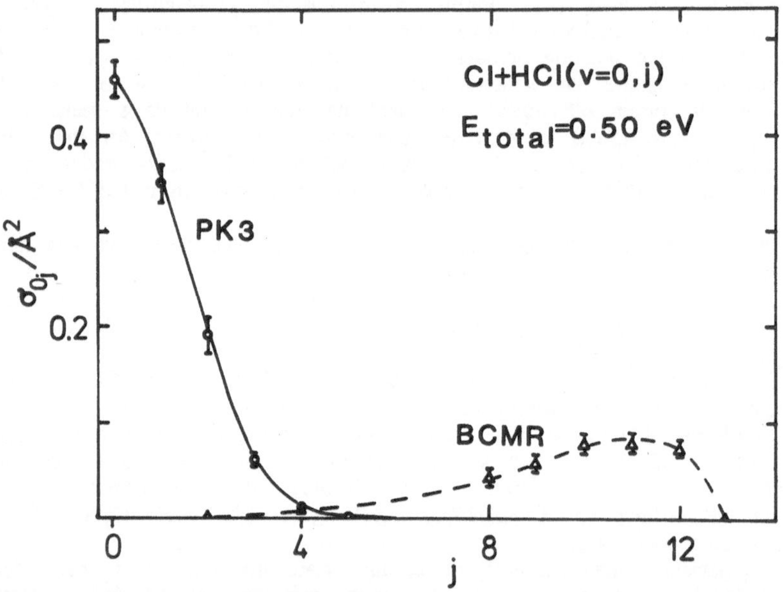

Figure 1. QCT integral cross sections σ_{0j} versus j at E_{total} = 0.50 eV. Circles (0): PK3. Triangles (Δ): BCMR. Vertical lines denote ±one standard deviation in the QCT data (when these exceed the size of 0 or Δ). Full (PK3) and dashed (BCMR) lines are visual fits to the QCT data.

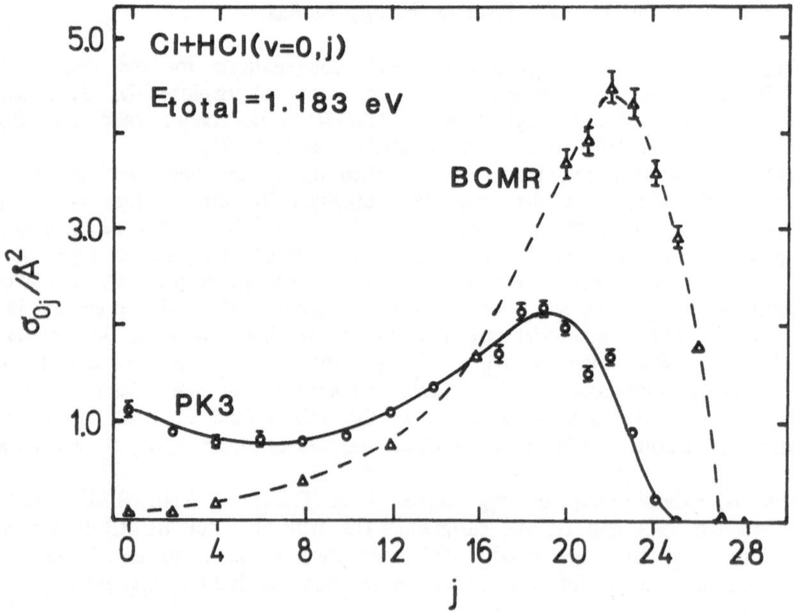

Figure 2. Same as Fig. 1, except for E_{total} = 0.70 eV.

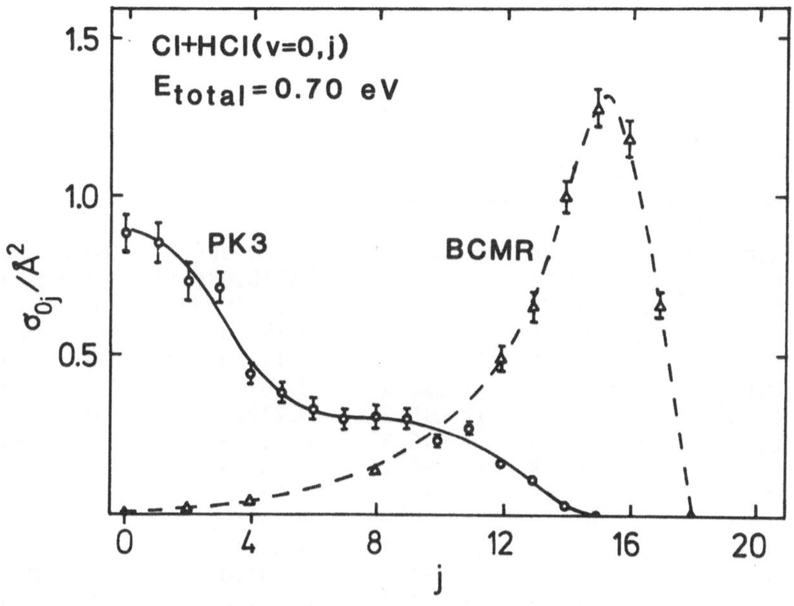

Figure 3. Same as Fig. 1, except for E_{total} = 1.183 eV.

The behaviour for PK3 is quite different. At E_{total} = 0.50 eV (Fig. 1), the largest cross section occurs for j=0, which is followed by a rapid decline as j increases. A similar trend is evident at E_{total} = 0.70 eV (Fig. 2), with the new feature that a shoulder has appeared for j=7-9. At E_{total} = 1.183 eV (Fig. 3), the shoulder has become much more pronounced and the distribution is bimodal.

Figure 4 shows a plot of the average product rotational quantum number, $\langle j' \rangle_{0j}$, versus j at the same three energies for BCMR and PK3. On the BCMR surface, the products are always highly rotationally excited and $\langle j' \rangle_{0j}$ does not vary rapidly except at high j. On the other hand for PK3, the products are always colder than for BCMR, except at j=21,22 for E_{total} = 1.183 eV, where $\langle j' \rangle_{0j}$ is comparable for both surfaces.

Although vibrationally adiabaticity is not enforced in the QCT calculations for PK3, Tables 1-3 (for $\langle v' \rangle_{0j}$) show that it is approximately obeyed, except at high values of j. In general, simple kinematic propensity rules for **HLH** reactions are obeyed better on PK3 than on BCMR.

The differences described above for the dynamical behaviour of Cl+HCl on BCMR and PK3 are dramatic. To understand these differences better, we have also examined individual trajectories at different times as the reaction proceeds on PK3 (results for BCMR are in Ref. [21]). In addition, Persky and Kornweitz [17] have also used QCT to study the dynamics of Cl+HCl(v=0,j=0) on BCMR and PK3, and Kornweitz *et al.* [31] have recently extended these calculations to j≠0. From all these studies the following picture emerges:

On BCMR, the approach of Cl′ towards HCl(v=0,j=0) is governed mainly by

Table 3. QCT integral cross sections and energy partitioning results for the PK3 surface at E_{total} = 1.183 eV (E_{avail} = 1.000 eV). The entries for $<f_j'>_{vj}$ are the same as those for $<E_j'>_{vj}$/eV.

v	j	$E_{trans}^{v,j}$/eV	$\sigma_{vj}/Å^2$	$<j'>_{vj}$	$<E_j'>_{vj}$/eV	$<v'>_{vj}$
0	0	1.000	1.13±0.08	6.8±0.4	0.99±0.08(-1)	-0.06±0.02
0	2	0.992	0.89±0.07	6.4±0.4	0.85±0.08(-1)	-0.07±0.02
0	4	0.97	0.81±0.06	8.3±0.3	0.126±0.008	-0.08±0.02
0	6	0.94	0.84±0.06	9.3±0.3	0.150±0.009	-0.09±0.02
0	8	0.91	0.82±0.06	10.2±0.3	0.172±0.009	-0.05±0.02
0	10	0.86	0.88±0.06	11.3±0.3	0.201±0.009	-0.01±0.02
0	12	0.80	1.10±0.05	12.0±0.2	0.229±0.007	0.04±0.01
0	14	0.73	1.36±0.07	12.9±0.3	0.257±0.008	0.07±0.02
0	16	0.65	1.68±0.08	14.2±0.2	0.302±0.008	0.13±0.02
0	17	0.61	1.70±0.08	14.9±0.2	0.327±0.008	0.10±0.02
0	18	0.56	2.15±0.09	15.3±0.2	0.344±0.007	0.16±0.02
0	19	0.52	2.18±0.09	15.8±0.2	0.363±0.007	0.24±0.02
0	20	0.47	1.98±0.08	16.8±0.2	0.404±0.008	0.24±0.02
0	21	0.42	1.49±0.07	18.2±0.2	0.461±0.009	0.25±0.02
0	22	0.36	1.67±0.07	19.0±0.2	0.499±0.008	0.29±0.02
0	23	0.31	0.91±0.05	19.0±0.3	0.505±0.011	0.45±0.03
0	24	0.25	0.21±0.02	17.3±0.4	0.433±0.014	0.83±0.04
0	25	0.19	<0.4(-4)	a	a	a

a No reactive trajectories were obtained for this case.

repulsive torques. Although the crossover configuration for Cl′HCl is often close to collinear, bent configurations occur while Cl′H is separating from Cl, and the repulsive energy that is released leads to highly rotationally excited Cl′H. This process is helped by rotational excitation of the reactant molecule HCl. One important mechanism for reaction for rotationally excited reactants at E_{total} = 0.70 and 1.183 eV is a "figure of eight" motion.

For PK3 on the other hand, the approach of Cl′ and HCl(v=0,j=0) is governed mainly by strongly attractive torques, which cause a reorientation of the reagents to a nearly collinear configuration. This collinear configuration persists as Cl′H and Cl start to separate, resulting in low rotationally excited products. Rotational excitation of HCl hinders this process (at least for low j) because it is more difficult for HCl to align with Cl, and the cross section decreases with j. However at high j and high E_{total} (*e.g.* Fig. 3), bent configurations for the intermediate Cl′HCl become more probable, and the cross section increases with j. In this regime for PK3, we found fewer "figure of eight" trajectories and more multiple encounter collisions than was the case for BCMR.

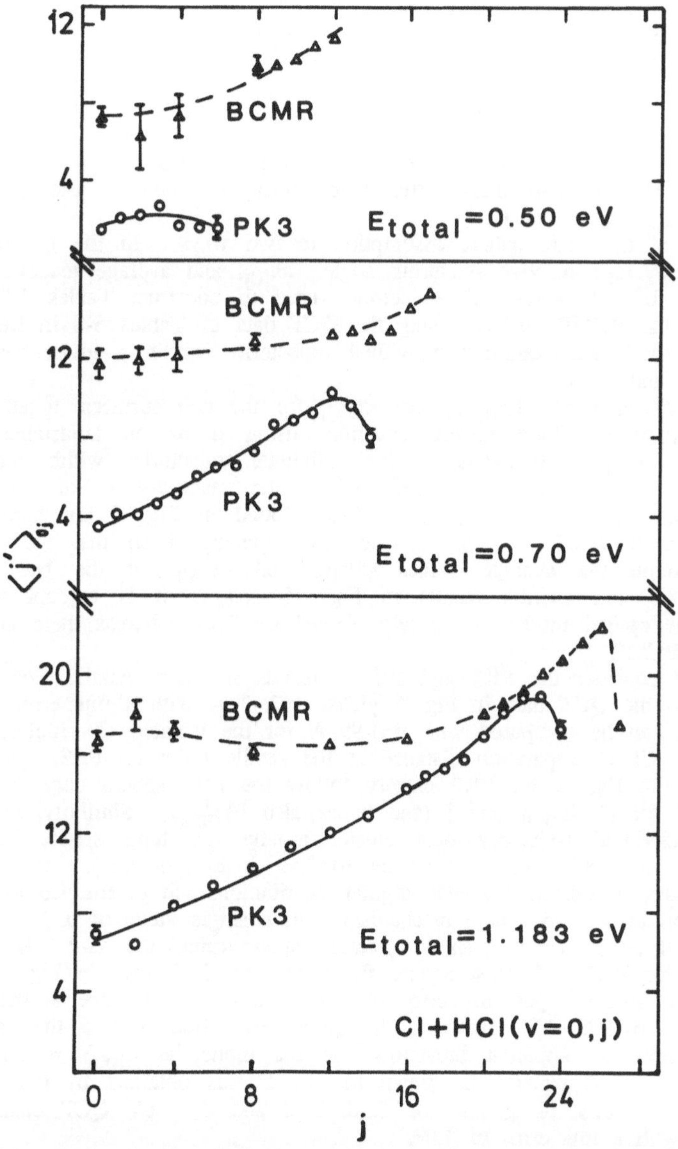

Figure 4. QCT average product rotational quantum number $\langle j' \rangle_{0j}$ versus j, at E_{total} = 0.50, 0.70 and 1.183 eV for the PK3 and BCMR surfaces. Notation and conventions are the same as Fig. 1.

Finally we report some QCT data relevant to the semiclassical optical model for differential cross sections that was discussed in Section 2. In this model, the scattering angle, θ, and impact parameter, b, are related by the hard sphere equation

$$\theta = 2 \cos^{-1}(b/d),$$

where d is the effective collision diameter. When the semiclassical optical model is used to calculate angular distributions, this hard sphere equation is assumed to be independent of E_{total}, v,v′,j and j′.

We have tested this hard sphere assumption in two ways. In the first method, we calculated the average reactive scattering angle, $\langle\theta\rangle_{0j}$, and average reactive impact parameter, $\langle b\rangle_{0j}$, for the batches of trajectories used to construct Tables 1-3. We did the same for the BCMR surface, using the QCT data of Tables 5-8 in Ref. [21]. In the second method, we examined individual trajectories within a given batch (*i.e.* for fixed E_{total}, j and v=0).

Figure 5 shows plots of $\langle\theta\rangle_{0j}$ against $\langle b\rangle_{0j}$ for the two surfaces, together with least squares fits of the hard sphere equation, using d as an adjustable fitting parameter. Figure 6 plots θ against b for individual trajectories within the batch defined by E_{total} = 0.70 eV, v=0, j=8, where the reactivity is of comparable magnitude for both surfaces (see Fig. 2). Also plotted in Fig. 6 are hard sphere deflection functions obtained for both surfaces by adjusting d so that the resulting curve passes through the average values $\langle\theta\rangle_{08}$ and $\langle b\rangle_{08}$ of the batch. On comparing the PK3 and BCMR results in Figs. 5 and 6, it is evident that the assumptions of the optical model are closely obeyed by PK3, whereas there are some discrepancies for BCMR.

Next we will consider the PK3 and BCMR results in more detail. For PK3, a least squares fit to the QCT data in Fig. 5 yields d=2.43 Å with a rms error of only 1.4°. This value can be compared with d=1.99 Å for the isolated Cl_2 molecule, and d=2.98 Å for the Cl...Cl separation distance at the saddle point on PK3. Note that all the QCT data in Fig. 5 for PK3 closely follow the hard sphere curve with this single value of d for all E_{total} and j (and hence also $E^{v,j}_{trans}$). Similarly, in Fig. 6 for PK3, the individual trajectory data cluster around the hard sphere deflection function with only a small amount of scatter (d=2.49 Å in this case). In summary, for the PK3 surface, it appears that the angular distributions are determined to a very good approximation by a hard sphere mechanism with a single value of d.

For the BCMR surface, Fig. 5 shows greater scatter around the hard sphere result than is the case for PK3. A least square fit to all the QCT data in Fig. 5 yields d=2.37 Å with the much larger rms error of 7.3° (the Cl...Cl distance at the saddle point of BCMR is d=2.93 Å). However, it can be seen from Fig. 5 that much of the error arises from an apparent breakdown of the model at E_{total} = 1.183 eV. Because of this, the least squares fit drawn in Fig. 5 was obtained by omitting the E_{total} = 1.183 eV results, using just the 0.50, 0.60 and 0.70 eV QCT data. This yields d=2.45 Å with a rms error of 4.0°.

Despite the exclusion of the 1.183 eV data from the fit, Fig. 5 shows that several data points at this energy, (those with $16 < j < 27$), follow the hard sphere curve quite closely. On the other hand, the remaining data points, which have $j < 16$, deviate strongly from the hard sphere curve. Indeed, these points cannot be fitted accurately by the hard sphere equation for any value of d. In fact this subset of the QCT data would require j dependent values of d (with d increasing rapidly with j). It has also been noted in Ref. [16] that the semiclassical optical model agrees better

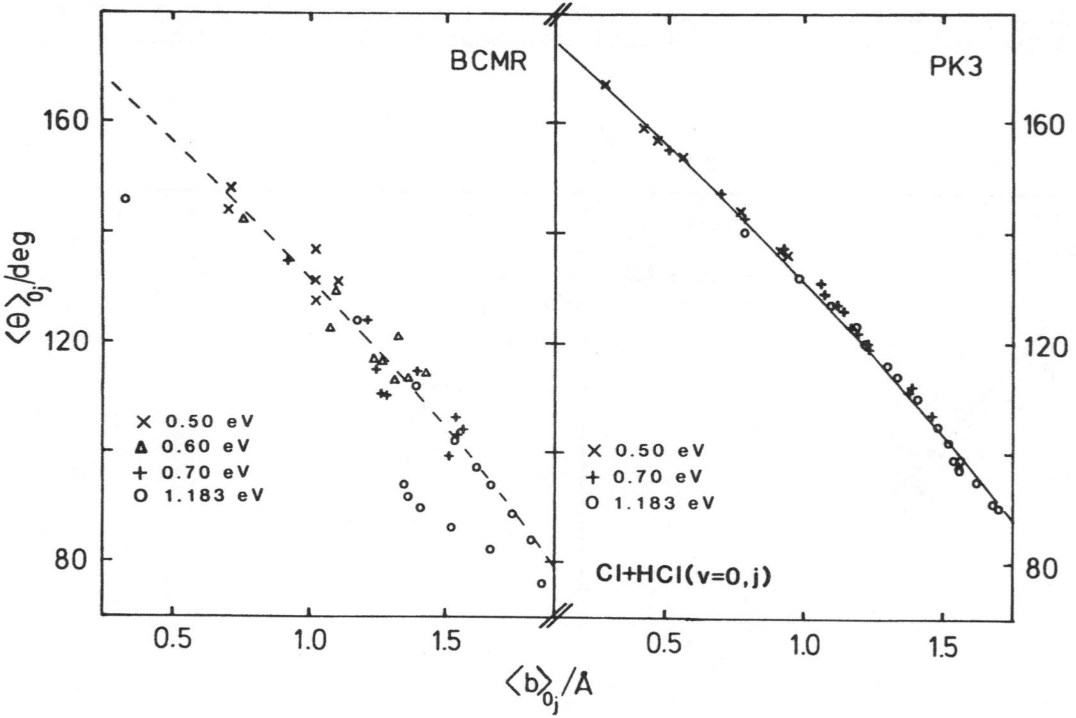

Figure 5. Average reactive scattering angle, $<\theta>_{0j}$, versus average reactive impact parameter, $<b>_{0j}$. The symbols ×, Δ, +, 0 correspond to E_{total} = 0.50, 0.60, 0.70 and 1.183 eV respectively for v=0 and various values of j. The lines are least squares fits of the hard sphere equation, with fit parameters of d=2.45 Å for BCMR and d=2.43 Å for PK3. The corresponding rms errors are 4.0^0 (BCMR) and 1.4^0 (PK3) respectively. Note that no QCT data at E_{total} = 1.183 eV has been used in the BCMR fit.

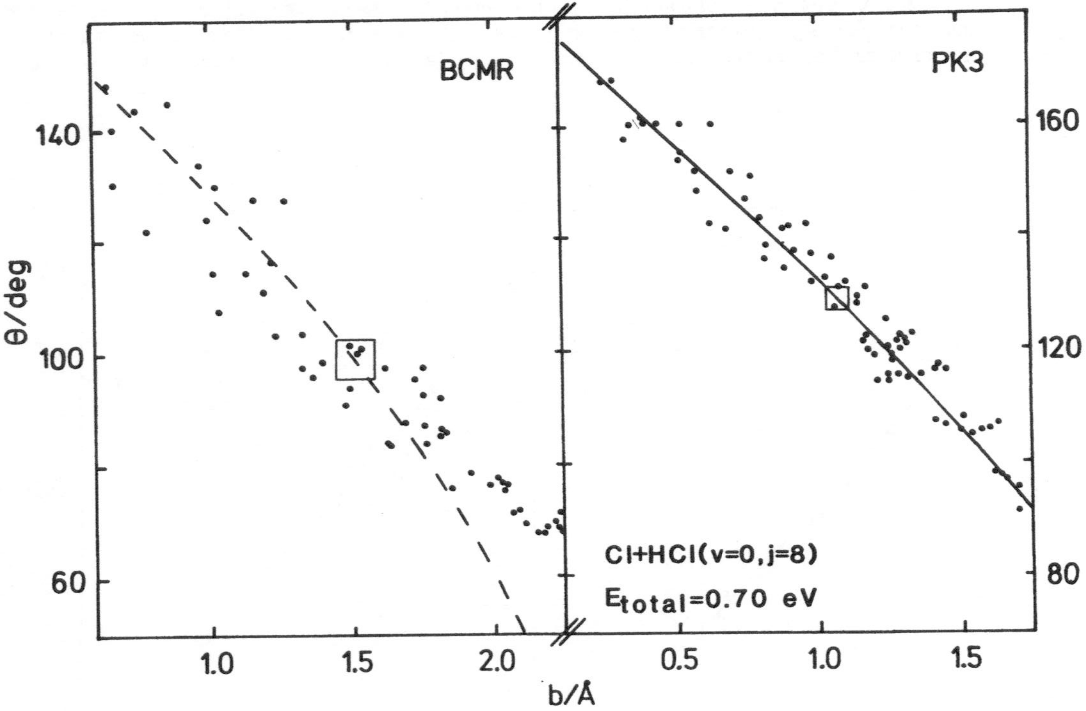

Figure 6. Reactive scattering angle, θ, versus impact parameter, b, for E_{total} = 0.70 eV, v=0, j=8. The lines are least squares fits with d chosen so that the hard sphere equation passes through the average values $\langle\theta\rangle_{08}$ and $\langle b\rangle_{08}$ for this batch of trajectories. These average values ±one standard deviation are shown as rectangles.

with the CSDW angular distributions for BCMR at E_{total} = 0.40 eV, if d is allowed to depend on j and j′.

A closer inspection of the QCT data for BCMR shows that for all four values of E_{total}, there are two distinct ranges of j, with the high j range obeying more closely the hard sphere model. This behaviour is evident in the dependence of $\langle b \rangle_{0j}$ on j: for a given E_{total}, $\langle b \rangle_{0j}$ increases with j despite the fact that E_{trans}^{vj} is decreasing. Thus the usual decrease of $\langle b \rangle_{0j}$ with E_{trans}^{vj} is more than compensated by the rotational enhancement of the reaction. At some limiting value of j (which depends only weakly on E_{total}) this trend is reversed and the usual decrease of $\langle b \rangle_{0j}$ with E_{trans}^{vj} is then observed. It is exactly for this high j range where the hard sphere model is closely followed, even at E_{total} = 1.183 eV. The maximum in the $\langle b \rangle_{0j}$ versus j plot occurs near j=10 (E_{total} = 0.50 eV), j=12 (E_{total} = 0.60 eV), j=13 (E_{total} = 0.70 eV) and j=16 (E_{total} = 1.183 eV).

The behaviour of $\langle b \rangle_{0j}$ just discussed for the BCMR surface, suggests there is an additional j dependent contribution to the angular distributions at low values of j. The QCT data shown in Fig. 6 for E_{total} = 0.70 eV, v=0, j=8 falls into the range of low j and thus can be used to investigate this point further. Three observations are of interest. (a) The value of d used in Fig. 6 to fit the hard sphere deflection function is d=2.32 Å, which is considerably smaller than the value d=2.45 Å employed in Fig. 5, or d=2.48 Å which is obtained by fitting only the high j data in Fig. 5. (b) There is greater scatter for the data points of BCMR in Fig. 6 than is the case for PK3. (c) There is a systematic deviation from the hard sphere curve at large b for BCMR in Fig. 6. Incorporation of this effect would require a b-dependent d value; however this would then invalidate the simple nature of the hard sphere model.

It is also interesting to compare the high j QCT data for BCMR with the low j=8 data in Fig. 6. A plot of θ against b for the high j range shows a considerable amount of scatter about the fitted hard sphere deflection function. However there are no systematic deviations; as a result the optical model is well obeyed on the average.

Acknowledgements

We thank our coworkers for their help with the research on Cl+HCl described in this Chapter. They are: B. Amaee, D.K. Bondi, B.C. Garrett, J. Manz, J. Römelt, G.C. Schatz, D.G. Truhlar and J.C. Whitehead. Support of this research by the Science and Engineering Research Council (UK) is gratefully acknowledged. The QCT calculations were carried out at the Computer Centre of the University of Vienna.

References

[1] J. Manz, *Comments Atom. Mol. Phys.*, 1985, **17**, 91.

[2] J. Römelt, in *The Theory of Chemical Reaction Dynamics*, Proceedings of the NATO Advanced Research Workshop, Orsay, France, June 17-30, 1985, edited by D.C. Clary (Reidel, Dordrecht, The Netherlands, 1986) pp. 77-104.

[3] J. Manz, in *Molecules in Physics, Chemistry and Biology*, edited by J. Maruani (Kluwer, Dordrecht, The Netherlands, 1988), Vol. 3, pp. 365-403.

[4] D.K. Bondi, J.N.L. Connor, J. Manz and J. Römelt, *Mol. Phys.*, 1983, **50**, 467.
[5] B.C. Garrett, D.G. Truhlar, A.F. Wagner and T.H. Dunning Jr., *J. Chem. Phys.*, 1983, **78**, 4400.
[6] D.K. Bondi, J.N.L. Connor, B.C. Garrett and D.G. Truhlar, *J. Chem. Phys.*, 1983, **78**, 5981.
[7] B.C. Garrett and D.G. Truhlar, *J. Chem. Phys.*, 1983, **79**, 4931.
[8] I. Last and M. Baer, *J. Chem. Phys.*, 1984, **80**, 3246.
[9] B.C. Garrett and D.G. Truhlar, *J. Chem. Phys.*, 1984, **81**, 309.
[10] N. Abusalbi, S-H. Kim, D.J. Kouri and M. Baer, *Chem. Phys. Lett.*, 1984, **112**, 502.
[11] M. Baer and I. Last, *Chem. Phys. Lett.*, 1985, **119**, 393.
[12] E. Pollak, M. Baer, N. Abu-Salbi and D.J. Kouri, *Chem. Phys.*, 1985, **99**, 15.
[13] I. Last and M. Baer, *Int. J. Quant. Chem.*, 1986, **29**, 1067.
[14] G.C. Schatz, B. Amaee and J.N.L. Connor, *Chem. Phys. Lett.*, 1986, **132**, 1.
[15] I. Last and M. Baer, *J. Chem. Phys.*, 1987, **86**, 5534.
[16] G.C. Schatz, B. Amaee and J.N.L. Connor, *Comput. Phys. Commun.*, 1987, **47**, 45.
[17] A. Persky and H. Kornweitz, *J. Phys. Chem.*, 1987, **91**, 5496.
[18] G.C. Schatz, *Chem. Rev.*, 1987. **87**, 81.
[19] V. Engel, R. Schinke and E. Pollak, *J. Chem. Phys.*, 1987, **87**, 1596.
[20] J.W. Tromp and W.H. Miller, *Faraday Disc. Chem. Soc.*, 1987, **84**, 441.

In Table 1, the Cl+HCl entries for κ, namely, 0.99, 0.69, 0.42, 0.27 are incorrect and should be replaced by 5.3, 2.0, 0.62, 0.33 (W.H. Miller, private communication, 1987). Also the text on p. 449, "It shows..... cf. the last column in Table 1", should be replaced by:

"It shows almost no recrossing, so that any discrepancy between conventional transition state theory and the exact quantum rate must be due to tunnelling. From Table 1(b), one sees that there is indeed significant tunnelling at 200 K.

As temperature increases, the recrossing becomes much more pronounced and the importance of tunnelling decreases. Thus the transition state theory rate - which omits the effects of recrossing - begins to overestimate the rate constant, cf. the last column in Table 1."

[21] B. Amaee, J.N.L. Connor, J.C. Whitehead, W. Jakubetz and G.C. Schatz, *Faraday Disc. Chem. Soc.*, 1987, **84**, 387.
[22] G.C. Schatz and J.N.L. Connor, *Faraday Disc. Chem. Soc.*,1987, **84**, 416.
[23] J. Wolfrum, *Faraday Disc. Chem. Soc.*, 1987, **84**, 417.
[24] B. Amaee, J.N.L. Connor, J.C. Whitehead, W. Jakubetz and G.C. Schatz, *Faraday Disc. Chem. Soc.*, 1987, **84**, 418.
[25] D.G. Truhlar, *Faraday Disc. Chem. Soc.*, 1987, **84**, 418.
[26] I. Schechter and R.D. Levine, *Faraday Disc. Chem. Soc.*, 1987, **84**, 424.
[27] R.B. Metz, T. Kilsopoulos, A. Weaver and D.M. Neumark, *J. Chem. Phys.*, 1988, **88**, 1463.
[28] G.C. Schatz, B. Amaee and J.N.L. Connor, *J. Phys. Chem.*, 1988, **92**, 3190.
[29] G.C. Schatz, *Chem. Phys. Lett.*, 1988, **150**, 92.

[30] G.C. Schatz, "Oscillating reactivity and resonances in the three dimensional Cl+HCl reaction", *Chem. Phys. Lett.*, (in press).

[31] H. Kornweitz, M. Broida and A. Persky, "Dynamics of the light atom transfer reaction Cl+HCl → ClH+Cl: Oscillating reactivity, effect of reagent rotation on reaction cross sections and rotational excitation of products", *J. Phys. Chem.*, (in press).

[32] G.C. Schatz, "Quantum effects in gas phase chemical reactions", *Ann. Rev. Phys. Chem.*, (in press).

[33] D.C. Clary and J.N.L. Connor, *J. Phys. Chem.*, 1984, **88**, 2758.

[34] G.C. Schatz, L.M. Hubbard, P.S. Dardi and W.H. Miller, *J. Chem. Phys.*, 1984, **81**, 231.

[35] J.N.L. Connor, in *The Theory of Chemical Reaction Dynamics*, Proceedings of the NATO Advanced Research Workshop, Orsay, France, June 17-30, 1985, edited by D.C. Clary (Reidel, Dordrecht, The Netherlands, 1986), pp. 247-283.

[36] D.R. Herschbach, *Appl. Opt.*, 1965, suppl. 2, 128.

[37] D.R. Herschbach, *Adv. Chem. Phys.*, 1966, **10**, 319.

[38] C. Hiller, J. Manz, W.H. Miller and J. Römelt, *J. Chem. Phys.*, 1983, **78**, 3850.

[39] D.J. Rakestraw, K.G. McKendrick and R.N. Zare, *J. Chem. Phys.*, 1987, **87**, 7341.

[40] K.G. McKendrick, D.J. Rakestraw and R.N. Zare, *Faraday Disc. Chem. Soc.*, 1987, **84**, 39.

[41] K.G. McKendrick, D.J. Rakestraw, R. Zhang and R.N. Zare, *J. Phys. Chem.*, 1988, **92**, 5530.

THE MODELLING OF COMPLEX GAS PHASE REACTIONS : FROM EXPERT SYSTEMS TO SUPERCOMPUTERS

G.M. CÔME and G. SCACCHI
CNRS (DCPR), INPL (ENSIC)
and University NANCY I
1, rue Grandville
54000 Nancy - France

ABSTRACT. The modelling of complex gas phase reactions is discussed in terms both of models and data associated to the reactants, the reactions, the reactors, and of identification techniques. The description of the reacting mixture needs to use lumping techniques and composition-properties correlations. The design of both exhaustive and simplified mechanisms is achieved by means of an expert system. Thermochemical and kinetic data are obtained from the literature and by using data models, such as BENSON's methods, analogies and lumping. The identification of the reaction model needs to solve the material, energy and momentum balances of the reactor and asks for difficult numerical problems (3D codes, stiff systems). The sensitivity analysis and the parameter estimation are computer-time consuming (large systems optimization) and calls for efficient algorithms and supercomputers.

1. Introduction

The gas phase reactions play a key role in the modern civilization, for producing CHEMICALS, THERMAL and MECHANICAL ENERGY, as well for solving ECOLOGICAL PROBLEMS. Some typical examples are shown in Table I.

All these objectives need a fundamental understanding and a capability of modelling the underlying COMPLEX REACTIONS AND REACTORS and of evaluating the associated THERMOCHEMICAL, KINETIC, TRANSPORT AND TRANSFER DATA. Of course, an indefinitly growing amount of INFORMATIONS (theoretical and experimental) has to be compiled, critically evaluated, stored and handled. These tasks cannot be achieved without having recourse to the computer and to sophisticated data bases and packages.

The general organization of a system devoted to gas phase reactions is shown in Table II. The main characteristics of the system we have been building in NANCY will be described and compared with features exhibited by other systems, essentially the SPYRO model and the CHEMKIN package.

It must be emphazised that the modelling of real gas phase reaction copes with complex mixtures, complex reactions, complex reactors, and complex data, that is to say the modelling of complexity.

The first question to be answered is related to the type of models actually used for gas phase reactions. Similar evolutions took and take place for the models used for steam-cracking, atmospheric and stratospheric chemistry, engines, etc. Very typical is the situation related to the pyrolysis reactions and more particularly the steam-cracking of hydrocarbons.

A. Laganà (ed.), Supercomputer Algorithms for Reactivity, Dynamics and Kinetics of Small Molecules, 413–432.

TABLE I. Gas Phase Reactions

PRODUCTION OF CHEMICALS

Reaction	Products
* Steam-cracking of ethane, propane, naphta	Ethylene Propylene BTX
* Pyrolysis of dichloroethane	Vinyl Chloride monomer
* Partial combustion of methane	Acetylene

PRODUCTION OF THERMAL AND MECHANICAL ENERGY

* Combustion of natural gas, acetylene, gasoline, fuel oil, kerosene
* Wedding, Engines, Airplanes, ...

ECOLOGY

Security in industrial plants.
Atmospheric and stratospheric chemistry.

TABLE II

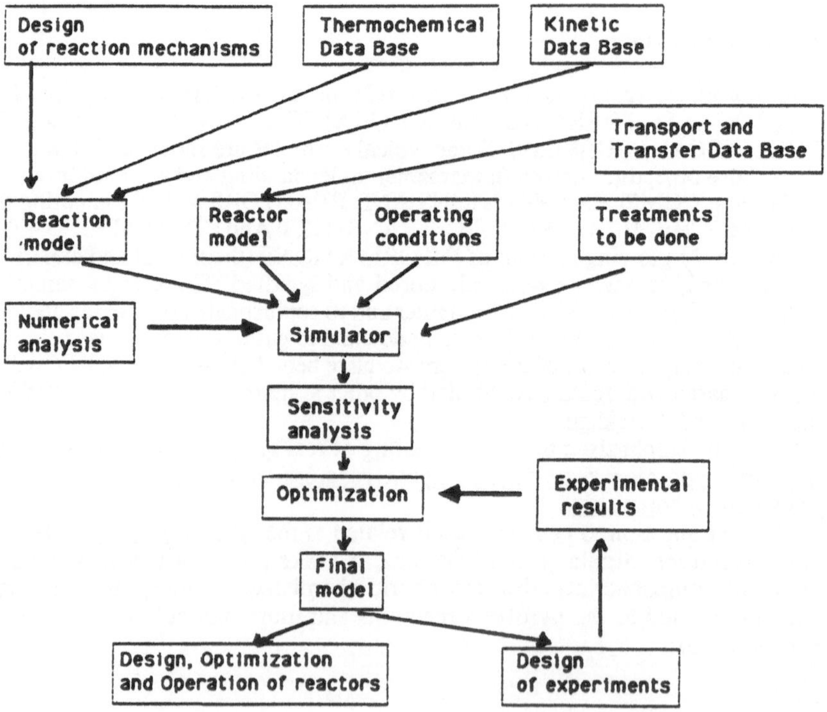

Nowadays, and probably still for a long time, the thermal cracking of hydrocarbons constitutes the main production route for olefins (see Table I), which are the basic feedstocks of the chemical industry around the world. Olefin production is achieved by pyrolysis of various feedstocks, ranging from light hydrocarbons (ethane, propane), to naphtas ; attempts are made to use gas oils and even crude oils. The variety of and change in the nature of available feedstocks due to new sources or to political problems, and the marked variation in prices and availability has emphasized the necessity of being able to predict the cracking value of a given feedstock, of incorporating a high degree of flexibility into the operation of a production unit, as well as giving the opportunity to determine optimum operating conditions.

For such goals, during the past, three kinds of models have been deviced and used in the practice.

a) Purely empirical models, i.e. mainly multilinear regression equations. These models are well adapted to optimize around an operating point.

b) A global first-order reaction

$$R \xrightarrow{k} \sum_j \nu_j P_j$$

where the reagent is treated as a single lumped constituent R, and the P_j are the reaction products. Both the rate constant k, the "stoichiometric" coefficients ν_j have to be correlated with conversion and feed specifications; these correlations are of empirical nature. Such a model allows interpolation on various experimental data.

c) The so-called "molecular" or "stoichiometric" models involve only the initial reagents and the final reaction products. The reactions are written in order to account for the "filiation" among the observed constituents; the corresponding reaction set is thus a reaction scheme, but not a real mechanism. This scheme considerably improves the preceeding approach and includes it. From a kinetic point of view, the reactions are considered as if they were elementary processes, allowing the recourse to the classical reaction orders, the ARRHENIUS law and the equilibrium constant for the reverse rate constant ($k_i = K_k k_{-i}$).

These molecular models reflect the true underlying reaction mechanism and therefore possess at least a well defined chemical sense. They are very flexible and have been proven very successful. The only restriction is due to the non-fundamental character of the rate parameters, which permits only moderate extrapolation.

d) So, it was not surprising that the further step in the modelling of reactions consisted of introducing more and more fundamental informations in the form of true elementary processes, such as $C_2H_6 \rightarrow 2CH_3.$, into the reaction model, which becomes, as a matter of fact, at least partly a real reaction mechanism. This approach has been widely used for the steam-cracking of hydrocarbons or other gas phase industrial processes, the atmospheric and stratospheric chemistry and still is in progress for the reactions in car and aircraft engines. Nonetheless, the description of complex mixtures and reactions clearly needs some lumping and therefore these models are not properly designed for the evaluation of rate parameters, but mainly for the simulation of industrial processes.

e) The purely fundamental models, which include only true elementary processes, are deviced in spite of estimating rate parameters and in some cases thermodynamic data. Usually, they are brought into operation for simple mechanisms.

Globally, the lumped fundamental models, including most information gained from truly fundamental models, possess, from a methodological point of view, the most desirable features. For these reasons, in what follows, the accent will be put onto these types of models.

2. The Design of a Reaction Mechanism

If an experimental study of the reaction of given reagents has been achieved, the purpose of most kineticists is to produce a reaction mechanism and the associated numerical data allowing to account for the experimental data. Of course, some parts of the mechanism and some numerical data can be taken from the literature.

We can classify the methods for designing an ad hoc mechanism into four types :

a) A direct design, by hand, using all the experimental and theoretical knowledges of the kineticist.

b) The cumulative use of the preceeding method, i.e. an ad hoc mechanism is considered as the sum of "block mechanisms", each block describing a well defined part of the reaction, e.g. the block 1 the reactions of CH_4, the block 2 the reactions of C_2H_6, etc. The main question with this approach is the possible "coupling" between blocks, which does not be taken into account.

***

Both the two preceeding approaches are not really systematic, because at each stage of the design, the kineticist has recourse to more or less explicit simplifications rules, which are themselves more or less valuable and/or evidenced. Thus, by these methods, one cannot be sure that a significant process has not been omitted or, conversely, that a non significant process has not been written.

c) For ensuring that a "comprehensive" mechanism (according to some criteria) has been written, the recourse to the computer is of course absolutely necessary. This approach will be shortly described here. The flow chart of the process is shown in Table III.

— As the addition of a free radical to an unsaturated molecule gives rise to a larger radical, such an elementary process could be pursued endless and a test must be carried out, to avoid too large free radicals. Another way for doing that is to lump the free radicals of a definite size.

— Generally, not all the reaction products which have been observed are generated in a first step (primary mechanism). Therefore, some products are chosen among the generated ones and reinjected as new reactants, giving rise to the following mechanisms (secondary, tertiary, etc.).

— When all the observed products have been generated by this iterative process, the computer is stopped.

Two difficulties are encountered with this type of program.

The first one concerns the rule of choice of the new reactants among the products; this question can be answered by using either experimental or theoretical knowledges on the given reaction or similar ones. Clearly here, we enter into the world of experts.

TABLE III

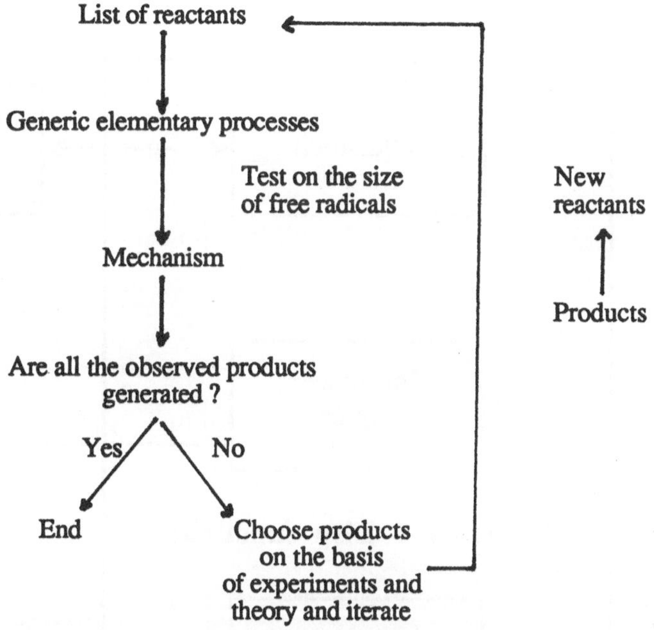

If all the products are input as new reactants, the computer program generates an exploding number of elementary steps. So, it is quite obvious that a priori simplification rules have to be used.

d) This is the main purpose of <u>expert systems</u>, which have been recently devised.
A first prototype has been built by CHINNICK et al. in LEEDS. Since then, a second prototype has been built in NANCY and will be shortly described here.
The general structure of the expert system is shown in Figure 1.
The prototype has been developed with the expert system shell EXPERKIT and the LISP language.
EXPERKIT is an expert system development environment with variables, running on PC compatibles and MACINTOSH.
The <u>fact base</u> contains chemical components, generated reactions and experimental information. The knowledge representation in the facts base is achieved by means of lists of elements, the first one being called <u>predicate</u>.

a) <u>Constituents representation</u>

(< key1 > < step > < C > < H > < CL > (< properties >))
and (< key2 > < step > < C > < H > < CL >)
with < key1 > : radical/non-radical
< key2 > : mu/beta
< step >, < C >, < H >, < CL > : integer
< properties > : saturated/unsaturated/alkyne/diene

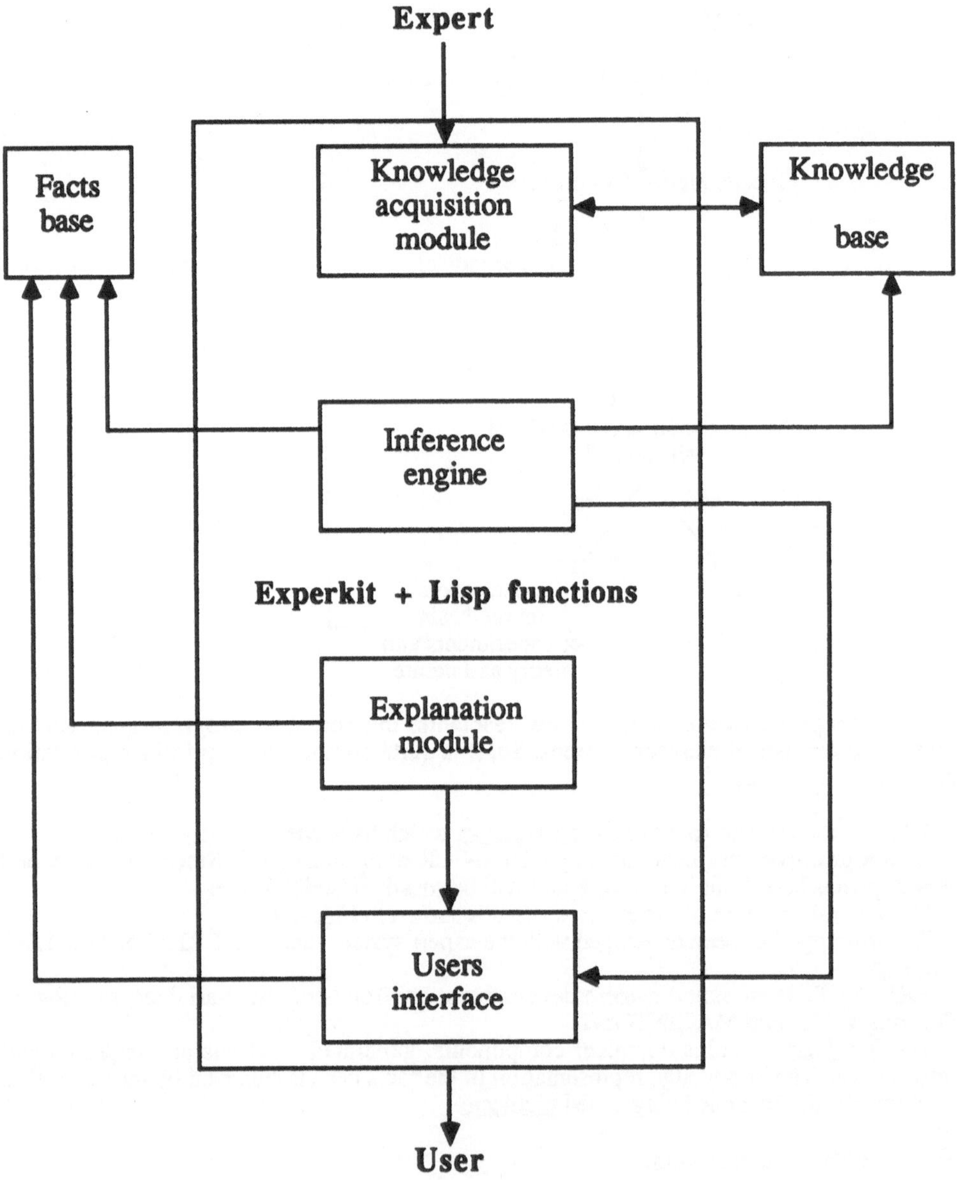

Figure 1. General structure of the expert system.

For example, the representation of CH_2Cl at step number 1 is as follows:

(< radical > < 1 > < 1 > < 2 > < 1 > (< saturated >))
(< beta > < 1 > < 1 > < 2 > < 1 >)

which means that CH_2Cl is a saturated free radical of ß type (according to the rules of GOLDFINGER, LETORT and NICLAUSE). As shown on this example, the compounds dealt with only contain C, H and Cl atoms and are described by their empirical formulae.

b) Reactions representation

```
(< reaction > < const > < const > < const > < const > (< key3 >))

with < const > : (< C > < H > < CL >)
     < key3 > : initiation/addition/decomposition/metathesis/termination
```

Example : CH_4 + Cl -> HCl + CH_3

```
(reaction (< 1 > < 4 > < 0 >) (< 0 > < 0 > < 1 >) (< 0 > < 1 > < 1 >)
        (< 1 > < 3 > < 0 > ) (metathesis))
```

The knowledge base contains the rules of mechanism generation and simplification and the rules of modification and printing.

The rules of production are represented by

```
if (condition) then actions endif
```

Example :

```
rule initiation-ht
if (step ?n and non-radical ?n-1 ?x ?y ?z ?-)
   then (call initiation ?n ?x ?y z)
endif
```

At the n step, this rule creates the initiation processes from the molecular components generated at the n-1 step.

Rules may be grouped in packages, which allows the introduction of a hierarchy in the knowledge base. The two most important packages of rules solve "high" and "low" temperature cases.

The general scheme of the expert system running is given in Figures 2 and 3.

First a level of temperature is input. Two limiting cases are considered. The case "low" temperature corresponds to long chain reactions, and the case "high" temperature corresponds to very short or rather no chain reactions. Then, the empirical formulae of the reactants are input.

Using the generic elementary processes (initiation, decomposition, metathesis, addition, termination), the system generates a primary simplified mechanism, according to the following rules :

— At low temperature, initiation steps are achieved by breaking C-C and C-Cl bonds only, and termination processes only imply chain carriers (let us notice that the system is able to recognize chain carriers among all the free radicals).

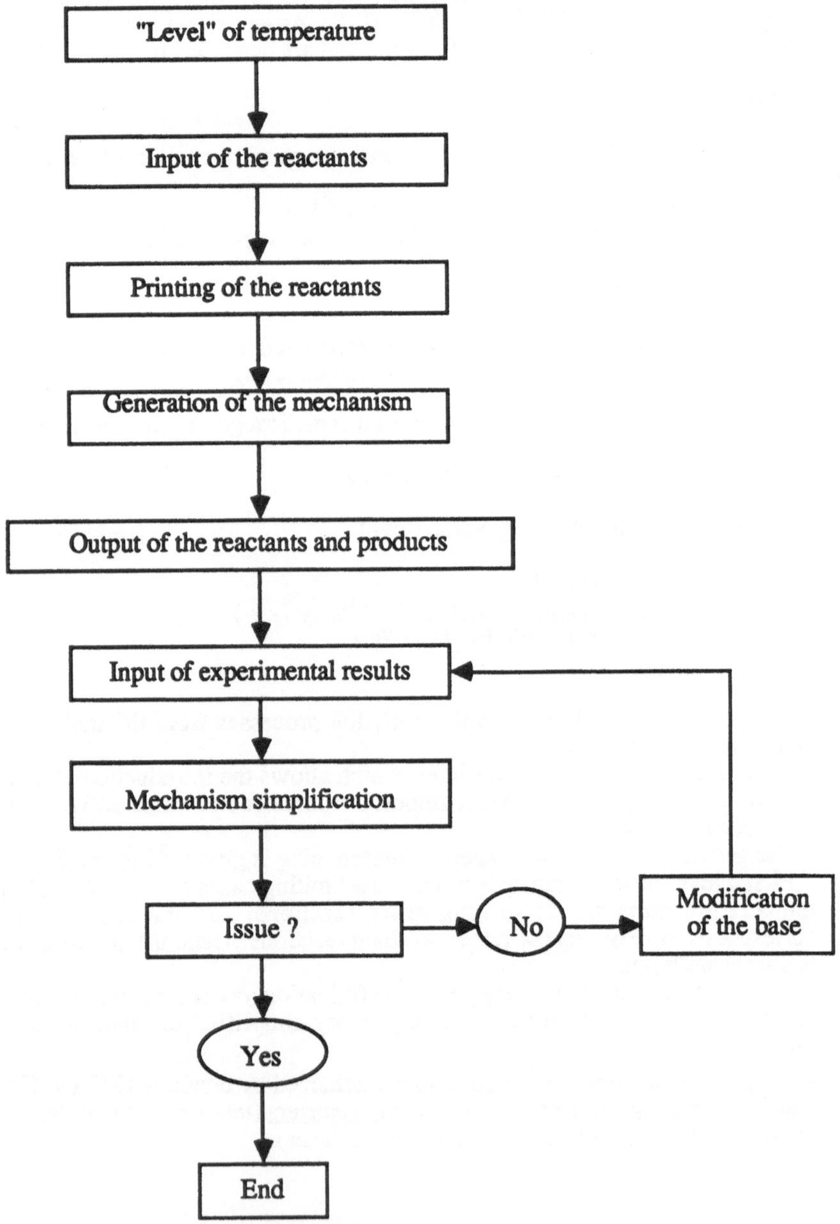

Figure 2. Running of the expert system.

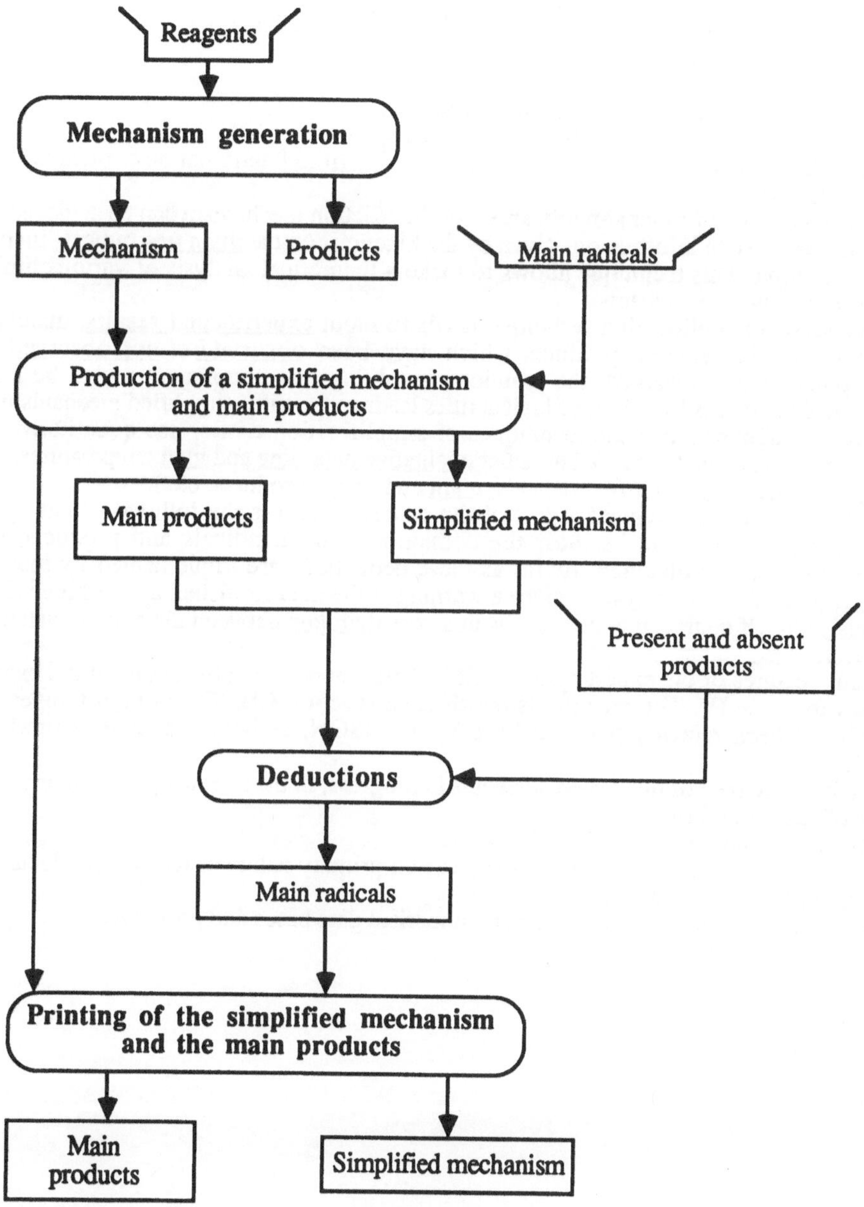

Figure 3. Generation of simplified mechanisms.

— At high temperature, initiation is done by breaking any simple bond; additions, metathesis and terminations only involve ß type free radicals, i.e. μ type free radicals decompose very fast.

The computer displays the main reaction products :

- at low temperature, they are the chain products,
- at high temperature, they are the products coming from ß termination or metathesis and μ decomposition.

At this stage, two further simplifications of the reaction mechanism can be achieved. The first has recourse to informations given by the kineticist on the main free radicals involved in the reaction. This technique allows to make a theoretical analysis of various limiting cases of reaction mechanisms.

The second simplification technique needs to input experimental results, mainly the formulae of the reaction products which have been observed or not observed. If a component is not observed, the reactions involving this component are to be killed. Iterating this process by means of logical rules leads to the most simplified mechanism.

Let us emphasize that the techniques of simplification do not use quantitative data (thermochemical and kinetics), but rather qualitative data (low and high temperatures, weak and strong bonds, ß and μ free radicals, main and minor products, etc.).

A second and very typical aspect of the expert system is the following. The system explains how it works, i.e. how the deductions (main radicals and products, main processes, etc.) are obtained. As far as these deductions are implemented by means of logical rules, the expert system gives a warning if the data furnished by the kineticist are inconsistent. Of course, this fact results in a very dialogue between the expert-system and the user.

Some features of the expert system will be better understood by means of the example shown in Table IV. The reaction is the thermal reaction CH_4/Cl_2 at high temperature, which has been recently patented by S.W. BENSON, as being capable of producing ethylene.

The improvement of this expert system is in progress, in the following directions :

- use of structural formulae,
- reactants containing C, H, O, N, X atoms,
- experimental information on the primary or non-primary nature of reaction products,
- lumping of components.

Next steps will include a connection to numerical data bases and programs.

TABLE IV

Temperature : high or low ?

high

Reactants : formulae ?

CH_4 Cl_2

Initiations :

CH_4 -> H + CH_3

Cl_2 -> Cl + Cl

Metatheses :

Cl_2 + CH_3 -> CH_3Cl + Cl

Cl_2 + H -> HCl + Cl

CH_4 + Cl -> HCl + CH_3

CH_4 + H -> H_2 + CH_3

Terminations :

H + CH_3 -> CH_4

CH_3 + CH_3 -> C_2H_6

CH_3 + Cl -> CH_3Cl

Saturated constituents :

C_2H_6 CH_4 CH_3Cl HCl H_2 Cl_2

ß free radicals :

H CH_3 Cl

WHICH ARE THE MAIN PRODUCTS ?

H_2 C_2H_6

WHICH PRODUCTS ARE NOT OBSERVED ?

CH_3Cl

I KNOW THAT

H is not preponderant

CH_3 is preponderant

H and CH_3 are preponderant

CH_3 is not preponderant or Cl is not preponderant

IS HCl IMPORTANT ?

I do not know

TABLE IV (continued)

IS Cl_2 IMPORTANT ?

No

I know that :

CH_3 is preponderant

CH_3 is not preponderant or Cl is not preponderant

Cl is not preponderant

So, the preponderant radicals are :

CH_3

C_2H_6 is produced by $CH_3 + CH_3 \rightarrow C_2H_6$

HCl is produced by $Cl_2 + H \rightarrow HCl + Cl$

H_2 is produced by $CH_4 + H \rightarrow H_2 + CH_3$

HCl is produced by $CH_4 + Cl \rightarrow HCl + CH_3$

The main products are :

H_2 HCl C_2H_6

3. Thermochemical and Kinetic Data Bases

A reaction mechanism (see Table V) is a set of p reversible elementary processes involving c components. The net rate of the ith reaction is written in the "mass action law" form. The rate constants, deduced from the activated complex theory, are characterized by an activation entropy and an activation enthalpy.

The thermodynamic data of the compounds involved in the mechanism are the molar heat capacity, the entropy and the enthalpy of formation. Very often, these data are used in computer programs in the so-called NASA form, shown in Table VI, and involving 7 coefficients.

The principles of microscopic reversibility and detailed balancing imply a fundamental relationship (see Table V) between the equilibrium constant (in terms of concentrations) and the rate constants of the reversible processes. Of course, two relations are obtained, one between entropies, and one between enthalpies.

There are mainly two types of thermochemical data : the experimental data found in the literature, and the computerized data based on a theory such as the "Thermochemical Kinetics" of S.W. BENSON. Of course, computerized data are deduced of well chosen experimental data.

These experimental data have to be compiled from the literature, critically evaluated, stored and retrieved. Furthermore, all these data have to be reconciled (see Table VI). This is a quite formidale task.

TABLE V

$$\begin{cases} \sum_{j=1}^{c} \alpha_{ij} C_j \underset{k_{-i}}{\overset{k_i}{\rightleftharpoons}} \sum_{j=1}^{c} \beta_{ij} C_j \\ i = 1, 2, \ldots, p \end{cases}$$

$$r_i = k_i \prod_j [C_j]^{\alpha_{ij}} - k_{-i} \prod_j [C_j]^{\beta_{ij}}$$

$$\begin{cases} k_i = \frac{RT}{N_A h} (R'T)^{\alpha_i - 1} \quad \exp\left(\frac{\Delta S_i^{\neq}}{R} - \frac{\Delta H_i^{\neq}}{RT}\right) \\ k_{-i} = \frac{RT}{N_A h} (R'T)^{\beta_i - 1} \quad \exp\left(\frac{\Delta S_{-i}^{\neq}}{R} - \frac{\Delta H_{-i}^{\neq}}{RT}\right) \\ \alpha_i = \sum_j \alpha_{ij} \\ \beta_i = \sum_j \beta_{ij} \end{cases}$$

TABLE VI

$$C_p^o = A + BT + CT^2 + DT^3 + ET^4$$

$$S^o = A \ln T + BT + \frac{CT^2}{2} + \frac{DT^3}{3} + \frac{ET^4}{4} + F$$

$$\Delta H_f^o = AT + \frac{BT^2}{2} + \frac{CT^3}{3} + \frac{DT^4}{4} + \frac{ET^5}{5} + G$$

$$\begin{cases} K_i = \frac{k_i}{k_{-i}} \\ \Delta S_i^o = \Delta S_i^{\neq} - \Delta S_{-i}^{\neq} \\ \Delta H_i^o = \Delta H_i^{\neq} - \Delta H_{-i}^{\neq} \end{cases}$$

The BENSON's methods for computing thermochemical data are mainly based on a group additivity scheme, but complex and often not at all negligible corrections have to be done on the preceeding "group values", in order to obtain sufficiently accurate data. Computing group values is very easy, but some corrections are very tedious and cumbersome, especially for free radicals. That is the main reason why we wrote a computer program for the automatic evaluation of thermochemical data by means of BENSON's methods. The only input is the topological formula of the molecules or of the free radicals. The results are ΔH^o_{f300}, S^o_{300}, C^o_{p300}, C^o_{p500}, C^o_{p800}, C^o_{p1000} and C^o_{p1500}. The corresponding source program is available from CNRS, upon request to the authors.

Even more complex methods have been deviced by S.W. BENSON for estimating the activation entropies and enthalpies. For $\Delta S^{\neq}$, the method works for all the processes, while it works only for a few activation enthalpies (see Table VII). Alas, no computer program has been written at the moment for doing that.

TABLE VII

COMPILATION AND CRITICAL EVALUATION (Prefered values) OF THE DATA OF THE LITERATURE

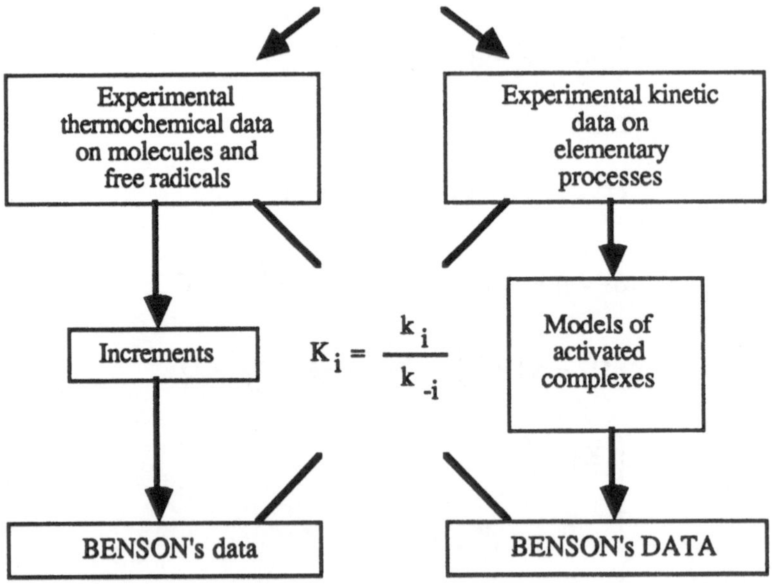

A special mention concerns the pressure-dependent rate constants, which are usually expressed in terms of the symbolism developed by TROE :

$$k = F\frac{k_0[M]}{1 + \frac{k_0[M]}{k_\infty}}$$

where

$$\log F = \frac{\log F_c}{1 + \left(\log \frac{k_o[M]}{k_\infty}\right)^2}$$

The three parameters k_o, k_∞ and F_c are recorded in the tables but, for most reactions, there are insufficient experimental data to allow the values of all of k_o, k_∞ and F_c to be derived.

Whatever, even when predictive methods, such as the thermochemical kinetics, are used, there is a lack of thermochemical and kinetic data. Therefore, we must have recourse to analogies between similar components or reactions. This can be achieved by lumping techniques, which will be discussed below in a special section.

TABLE VIII

Elementary process	$\Delta S^{\neq}$	$\Delta H^{\neq}$
Unimolecular initiation	+	+
Bimolecular initiation	+	+
Addition	+	-
Decomposition	+	-
Metathesis	+	-
Cis-trans isomerization	+	+
Intramolecular rearrangement	+	-
Combination	+	+
Disproportionation	+	+

4. Lumping

a) The lumping of components can be done in many ways and results in some cases in a drastic simplification of reaction mechanisms.

Let us consider for example the polymerization mechanism :

$$I \rightarrow 2M. \qquad (i)$$

$$M. + M \rightarrow M. \qquad (p)$$

$$2M. \rightarrow P \qquad (t)$$

where I is an initiator, M the monomer and P the polymer. The rate r of disappearance of the monomer and the chain length λ are given respectively by :

$$r = k_p \left(\frac{k_i}{k_t}\right)^{1/2} [I]^{1/2} [M]$$

$$\lambda = \frac{k_p[M]}{(k_i k_t[I])^{1/2}}$$

These simple rate laws apply quite well to some polymerization reactions.

b) By properly using Quasi or Partial Stationary State Approximation, a substantial reduction of reaction schemes can be obtained.

Let us consider for example the simple mechanism :

$$H. + O_2 \text{->} OH. + O \qquad (1)$$

$$O. + H_2 \text{->} OH. + H. \qquad (2)$$

$$OH + H_2 \text{->} OH_2 + H. \qquad (3)$$

If the PSSA is applied to O and OH free radicals, these three reactions can be lumped into one

$$H. + O_2 + 3H_2 \text{->} 2H_2O + 3H.$$

the rate of which is

$$r_1 = k_1 [H.] [O_2]$$

c) For describing the complex mixtures and reactions such as those encountered in petroleum chemistry, DENTE and RANZI use interesting lumping techniques.

- The 12 alkyl-benzenes from C_9 to C_{20} are described as a barycentre of three ones (C_9, C_{15}, C_{20}), e.g.

$$C_{16} \cong 0.8\ C_{15} + 0.2\ C_{20}$$

- The isomerization and decompositions of hexyl free radicals are considered to be very fast so that the metathesis of any free radical R. with n C_6H_{14} is described by the only equation :

$$\begin{aligned} R. + n\ C_6H_{16} \text{->}\ & RH \\ & + \alpha\ (C_2H_4 + C_4H_9.) \\ & + \beta\ (C_3H_6 + C_3H_7.) \\ & + \gamma\ (C_4H_8 + C_2H_5.) \\ & + \delta\ (C_5H_{10} + CH_3.) \end{aligned}$$

The parameters α, β, γ, δ are estimated from data of the literature and from thermochemical kinetics, e.g. at 850 °C :

$$\alpha = 0.114,\ \beta = 0.419,\ \gamma = 0.276,\ \delta = 0.141$$

For describing the properties of complex mixtures, and the complex reactions thereof, we can expect that lumping techniques will know a growing development.

5. Simulator

The simulator has to achieve the mass and energy balances and the momentum balance.

The corresponding equations of course strongly depend on the nature of the reactor. Here again, as for chemical kinetics, two questions arise : - which model ? which data ? These questions are relevant to chemical reaction engineering and are outside the scope of this paper. Nevertheless, it must be emphasized that the reaction and reactor models and their characteristic parameters are undissociable, which means that an error in the design of one of the models (reaction or reactor) induces a systematic error in the parameters of the two models, as does a bias in the experimental results.

The mathematical models of reactors are very different :

- the simplest case (often called zero-dimensional) is the steady state isothermal continuous flow stirred tank reactor. The material balance leads to a system of c implicit non-linear algebraic equations, if c components are involved in the mechanism.

This ideal reactor does really exist, mainly in the laboratory.

- the worst case is a non isothermal, non isobar, non isochor, unsteady state three-dimensional reactor.

The mathematical model is made of partial differential equations with boundary conditions, the partial derivatives being taken with respect to space coordinates and time.

In fact, most real gas phase reactors are relevant of this category : engines, meteorology, etc.

Once again, it is outside the scope of this paper to enter in this fascinating world, from both chemical and numerical points of view. We will focus on a simple ideal case, the adiabatic plug flow reactor to put in the light the most prominent features of such models.

The mathematical model is shown in Table IX. It consists of a system of c+1 ordinary differential equations with initial conditions. these ODE's are stiff, even in the isothermal case, and of course stiffer in the adiabatic case, because of the exponential term due to ARRHENIUS law.

For solving these equations, the most popular code is the one designed by GEAR and implemented under various versions called LSODE (for Linear Solver of ODE's).

This linear solver is used in particular in the computer program for chemical kinetics called CHEMKIN and written by KEE, MILLER and JEFFERSON in the SANDIA National Laboratories (LIVERMORE, Ca.).

So, we can conclude that this probelm is solved, as far as we are concerned with numerical analysis. But, there is another question, if we are interested in the modelling of reactions, since we have to cope with an easy change of the model and of the parameters. In other words, we need a chemical language very easy to use by the chemist, and a compiler, which automatically encodes the mathematical model from the chemical model.

These requirements were first recognized by EDELSON and, surprisingly, are not met in the NASA and SANDIA programs.

We have devised in NANCY a chemical language which possesses three characteristics: it is linear, non-ambiguous, non canonical. From a practical point of view, it is very close to the classical notation, e.g. $CH_2//C(CH_3)_2$ for isobutene. With the associated compiler, the chemist has only to write the chemical reactions, such as :

< 1 > CH_3/CH_3 -> 2 CH_3(.) , AD = 16.3 , ED = 87.5 ;

TABLE IX

$$\frac{dF_j}{dV} = \sum_{i=1}^{p} (\beta_{ij} - \alpha_{ij})\, r_i$$

$$\frac{dT}{dV} = \frac{-\sum_{j=1}^{p} r_i \Delta H_i^o}{\sum_{j=1}^{c} F_j C_{pj}^o}$$

$V = 0 : F_j = F_{jo},\ T = T_o,\ [C_j] = [C_j]_o$

$j = 1, 2, \ldots, c$

F_j : molar flow rate of component C_j
V : current reactor volume.

6. Model Identification

Let us assume that the models of the reaction and of the reactor are well defined, and, moreover, that the parameters of the reactor model are known. Thus, if the mechanism includes p elementary processes, 2p parameters (p preexponential factors and p activation energies) are to be estimated from the experimental results.

Indeed, if the reaction model has been written a priori, it includes negligible processes, and also non-negligible processes, whose rate parameters cannot be determined from the experimental results (we call them non-determining processes), and a few determining processes, whose rate constants could be estimated from the experimental results.

The discrimination between these three types of processes is achieved by means of a sensitivity analysis. In the direct method, we compute the sensitivity matrix, defined by the general element :

$$\sigma_{ij} = \frac{\partial \ln [C_j]}{\partial \ln k_i}$$

for given experimental conditions. Therefore, if we want to compute the matrix line after line, p simulations are needed for only one operating point.

A careful analysis of sensitivity allows us to detect the determining parameters which could be estimated from the results. But we are not sure that all these determining parameters are independent, for given experimental results. Unfortunatly, it seems that there is no general answer to the question of finding the independent determining parameters.

The last point deals with the numerical process of estimation of the parameters, which usually involves to find the minimum of a least-squares type criterion. Whatever the optimization technique used for this task, calling or not the Jacobian matrix of the model, repeated simulations are needed, to compute the theoretical values corresponding to each experimental point.

For complex reaction and reactor models, the sensitivity analysis and the parameter estimation by optimization are computer-time consuming and call for more efficient algorithms and computers. Here clearly, any improvement in the speed of such computations is desirable, and even necessary, for the practical use of fundamental models. The requirements of speedness would be rather increased if a fundamental model, instead of a black box, were used for optimal control purposes. So, we think that supercomputers will be more and more useful for solving the numerical problems involved in the mechanistic modelling of complex gas phase reactions.

7. Conclusion

Gas phase reactions play a key role in the modern way of life. Therefore the modelling of these complex phenomenons is useful for improving old processes or designing new ones for the production of energy and chemicals, and for taking decisions in ecological problems.

The design of reaction and reactor models and the obtaining of the associated numerical data is, at the moment, the most limitating step. Progresses in this way are expected from expert systems, lumping techniques, data models as well as experimental and bibliographic data evaluation, and composition-properties correlations.

The identification of the models involves the repeated solving of large systems of non-linear differential or partial derivatives equations, for the necessity of sensitivity analysis and optimization processes. These goals could be better reached by using more efficient algorithms (progresses in this way are expected) or supercomputers (some works are in progress in this line).

8. Bibliography

ALBRIGHT L.F., CRYNES B.L., CORCORAN W.H., Ed. (1983), *Pyrolysis. Theory and Industrial Practice* (Acad. Press).

ALRAN D., CÔME G.M., CUNIN P.Y., GRIFFITHS M. (1979), 'Mechanistic modelling of homogeneous reactors : a chemical compiler', *Comp. and Chem.*, **3**, 87.

BAMFORD C.H., TIPPER C.F.H., Ed. (1983), *Comprehensive Chemical Kinetics*, **24**, 'Modern Methods in Kinetics' (Elsevier).

BENSON S.W. (1976), 'Thermochemical Kinetics (Wiley).

BITTKER D.A., SCULLIN V.J., 'General chemical kinetics computer program for static and flow reactions, with application to combustion and shock-tube kinetics', Lewis Research Center, NASA, Cleveland, Ohio (USA).

CÔME G.M., in BAMFORD and TIPPER (1983), Chap. 3, p. 249-332, 'The use of computers in the analysis and simulation of complex reactions.

DENTE M.E., RANZI E.M., in ALBRIGHT et al. (1983), Chap. 7, p. 133-175, 'Mathematical modelling of hydrocarbon pyrolysis reactions.

EBERT K.H., DEUFLHARD P., JAGER W., Ed. (1981), 'Modelling of Chemical Reactions Systems, Springer Ser. Chem. Phys., **18**.

GARDINER W.C., Ed. (1984), 'Combustion Chemistry' (Springer).

HAUX L., CUNIN P.Y., GRIFFITHS M., CÔME G.M. (1985), 'Construction automatique d'un mécanisme de réaction radicalaire - I. Principe', *J. Chim. Phys.*, **82**, 1027.

KEE R.J., MILLER J.A., JEFFERSON T.H., 'CHEMKIN : a general purpose, problem independent, transportable, FORTRAN chemical kinetics code package', SANDIA National Laboratories, Livermore, Ca. (USA).

LYKOS P., SHAVITT I., Ed. (1981), 'Supercomputers in Chemistry', ACS Symp., Ser. 173.

MULLER C., SCACCHI G., CÔME G.M. (1985), 'A computerized thermochemical data base of molecules and free radicals in the gas phase', in *The role of data in scientific progress* (Elsevier).

PIERCE T.H., HOHNE B.A., Ed. (1986), 'Artificial intelligence applications in chemistry', ACS Symp. Ser. 306.

SALMI T., LINDFORS L.E. (1986), 'A program package for simulation of coupled chemical reactions in flow reactors', *Comp. and Ind. Eng.*, **10**, 45.

STEWART W.E., RAY W.H., CONLEY C.C., Ed. (1980), 'Dynamics and modelling of reactive systems', (Acad. Press).

WESTERBERG A.W., CHIEN H.H., Ed. (1984), Proceedings of the second international conference on "Foundations of computer aided process design" (CACHE).

Index